Estates Gazette Law Reports

1980

Volume 1

Series edited by
Barry Denyer-Green LLM PhD FRICS
of the Middle Temple, barrister

Cases edited by
J Muir Watt OBE MA
of the Inner Temple, barrister

Audrey Boyle BA
Publisher's editor

Estates Gazette

Routledge
Taylor & Francis Group
LONDON AND NEW YORK

EG Books is an imprint of Routledge
2 Park Square, Milton Park, Abingdon, Oxon OX14 4RN
711 Third Avenue, New York, NY 10017, USA
Routledge is an imprint of the Taylor & Francis Group, an informa business

ISBN for complete set of 2 volumes: 0 7282 0230 1
ISBN for this volume 0 7282 0231 X
ISSN 0951-9289

For information on all EG Books publications
visit our website www.routledge.com

CONTENTS

TABLE OF CASES

Titles of cases shown in bold type
Names of parties reversed shown in ordinary type

INDEX OF SUBJECT-MATTER

TOWN AND COUNTRY PLANNING — *continued*
Development — *continued*

Material change of use

Bakery, first-floor residential use — Planning authority's refusal to grant permission to use first floor as solicitors offices — Whether inspector applied proper test — Whether perverse and disclosed errors of law on its face — Held, no ground to hold that decision unreasonable — No ground for saying that wrong questions asked — No error on face of decision.
***Sears Blok* v *Secretary of State for the Environment (QB)*, 142**

Whether use of house as holiday accommodation for mentally-handicapped persons within Class XIV of Town and Country Planning (Use Classes) Order 1972 — Whether material change of use from previous permitted use as hotel/guest-house — Held, minister wrong to decide case within Class XIV — Purpose of order not to define what is development, but what is not.
***Rann* v *Secretary of State for the Environment (QB)*, 139**

Worked-out quarry — Condition in 1950 planning permission that only approved materials could be tipped into quarry — From 1950 materials tipped including rubble in breach of condition — Whether other inert material could be tipped — Held, to be material change of use — Breach in 1950 — Further tipping did not constitute breach or development — Enforcement action time barred because breach before 1964.
***Bilboe* v *Secretary of State for the Environment (CA)*, 130**

Planning agreement

Construction of agreement

S52 agreement that development should not be commenced until such time as necessary for works of improvement to sewage disposal system "to be placed in contract" — Planning permission that development to be begun not later than three years from date of permission — Sewage contract not let within the three years — Literal result that there was never any time when, in accordance with agreement and planning permission, development could be commenced — Held, planning permission expired.
***LAH Ames Ltd* v *North Bedfordshire Borough Council (CA)*, 119**

Planning permission

Change of use

Erection by limestone quarrying company of "ready-mixed concrete batching plant" — Whether permitted development — Whether plant required in connection with treatment or disposal of limestone — Held, operation neither "treatment" nor "disposal" of limestone but "utilisation" — Not permitted development — Class XIX, para 2 of Town and Country Planning Order 1977.
***South Glamorgan County Council* v *Hobbs (Quarries) Ltd (Ch)*, 121**

Inspector's decision

Inspector "appointed person" under Sched 9 to Town and Country Planning Act 1971 — Application to quash minister's decision dismissing planning appeal relating to development of property — Whether inspector erred — Held, inspector erred in making comparison which is ill-founded on material point.
***Millard* v *Secretary of State for the Environment (QB)*, 133**

Tree preservation order

Validity

Coppice — Farmer unsuccessful in obtaining planning permission for housing development — Order challenged on ground that it is not possible to make order in respect of coppice — Whether order valid — Held, trees in coppice can be subject to preservation orders.
***Bullock* v *Secretary of State for the Environment (QB)*, 140**

TREE PRESERVATION ORDER — See **Town and country planning** (Tree preservation order)

TREES — See **Negligence** (Duty of care and liability)

TRESPASS

Injunction

Works on neighbouring land

Premises in dangerous condition — Permission sought to go on to neighbouring land to carry out remedial works — Refusal ignored by bank and trespass committed — Whether prohibitory injunction could be granted to remove bank off land — Held, injunctions both prohibitory and mandatory granted.
***John Trenberth Ltd* v *National Westminster Bank Ltd (Ch)*, 102**

TRIBUNALS

Jurisdiction — See **Rates and rating** (Assessment — Appeals)

VALUATION

Compensation — See **Compensation for acquisition of land** (Valuation)
Rates and rating — See **Rates and rating** (Valuation)

WATER

Trunk main

Contribution for cost of construction

Whether water authority entitled to contributions in respect of main — Whether "necessary main" within s37 of Water Act 1945 — Held, to be within Act — Developers' liable to pay contributions.
***Royco Homes Ltd* v *Southern Water Authority (HL)*, 108**

AGRICULTURE

A

HOUSE OF LORDS

December 19 1979

(Before Viscount DILHORNE, Lord EDMUND-DAVIES, Lord FRASER OF TULLYBELTON, Lord RUSSELL OF KILLOWEN and Lord LANE)

JACKSON v HALL AND ANOTHER
WILLIAMSON v THOMPSON AND ANOTHER

Estates Gazette January 12 1980

(1979) 253 EG 145

Agricultural holdings—Succession provisions of Agriculture (Miscellaneous Provisions) Act 1976—House of Lords settle controversy as to date for determining eligibility to apply for
B
new tenancy after death of tenant—Two appeals (one under "leap-frogging" procedure) decided together—Unanimous decision that eligibility is to be determined at the date of deceased tenant's death—Applicant must, however, also be eligible at the date of the application to the agricultural land tribunal and (per Viscount Dilhorne) the date of the hearing—Eligibility cannot be achieved by procedure of divesting applicant's interest in a "commercial unit" between date of death and application—House of Lords by a majority (Lord Russell of Killowen dissenting) decide that a person is rendered ineligible if he is a joint occupier of a commercial unit at the date of death—Ineligibility not confined to sole occupiers—
C
Jackson appeal from Court of Appeal allowed (thereby restoring decision of Sir Douglas Frank QC and the tribunal)—Williamson appeal dismissed (thereby affirming under "leap-frogging" procedure the decision of Michael Kempster QC, sitting as a deputy High Court Judge)

In the first of these appeals William Horner Hall and Geoffrey Alan Marr, the trustees of Geoffrey Edwards Marr deceased, owners of Grange Farm, Keyingham, Humberside, appealed against the decision of the Court of Appeal (1978) 249 EG 955 in favour of Graham Christopher Jackson. The Court of Appeal (Brandon LJ dissenting) had held that Mr Jackson was eligible to apply for a tenancy of the farm,
D
notwithstanding the fact that, at the date of the deceased tenant's death, he was a joint occupier of another farm, White House Farm, which was a "commercial unit." In the second appeal Lewis Williamson appealed against a decision of Michael Kempster QC (1979) 251 EG 955 in favour of the landlords, Peter Howard Thompson and James Montagu Carpenter, of certain riverside land forming part of the Eardington estate. Mr Kempster, sitting as a deputy judge of the Queen's Bench Division, held that the applicant was ineligible to apply for a new tenancy on the ground that at the date of his father's death the applicant was a joint occupier with his brother of a commercial unit.
E
Leolin Price QC and G O A Sebestyen (instructed by Collyer-Bristow, agents for Chambers, Thomas & Williamson, of Hull) appeared on behalf of the appellant landlords in the first appeal; E C Evans-Lombe QC and C Joseph (instructed by Warren, Murton & Co, agents for Stamp, Jackson & Procter, of Hull) represented the respondent, Graham Christopher Jackson. In the second appeal Derek Wood QC and Miss J Moss (instructed by Ellis & Fairbairn, agents for Burges, Salmon & Co, of Bristol) appeared on behalf of the appellant, Lewis Williamson; Gavin Lightman and Mitchell Caller (instructed by Wedlake Bell, agents for Ivens & Morton, of Kidderminster) represented the respondent landlords. F

In his speech, VISCOUNT DILHORNE said: Section 24(2)(g) of the Agricultural Holdings Act 1948 enabled a landlord of an agricultural holding to give notice to quit that holding if the tenant had died within three months before the date of the giving of the notice, and it was stated that the notice was given for that reason. This provision was replaced by a provision that such a notice could only be given where a person was the sole or sole surviving tenant under a contract of tenancy immediately before his death (Agriculture (Miscellaneous Provisions) Act 1976, section 16(1)).

Part II of that Act made provision for succession on the G
death of the tenant of such a holding. Section 18(1) provides that where the sole or sole surviving tenant dies and is survived by any of the following persons:

(*a*) the wife or husband of the deceased.
(*b*) a brother or sister of the deceased.
(*c*) a child of the deceased.
(*d*) any person (not within (*b*) or (*c*) above) who, in the case of any marriage to which the deceased was at any time a party, was treated by the deceased as a child of the family in relation to that marriage,

the following provisions of that Part of the Act are to apply.

Section 20(1) reads as follows: H

> Any eligible person may within the relevant period apply to the Tribunal for a direction entitling him to a tenancy of the holding.

"Eligible person" is defined by section 18(2) as meaning, subject to subsection (3) of that section (which does not apply in these cases), and without prejudice to section 21:

> A survivor of the deceased in whose case the following conditions are satisfied—
> (*a*) he falls within paragraphs (*a*) to (*d*) of subsection (1) above;
> (*b*) in the seven years ending with the date of death his only or principal source of livelihood throughout a continuous period of not less than five years, or two or more discontinuous periods together amounting to not less than five years, derived from his J
> agricultural work on the holding or on an agricultural unit of which the holding forms part; and
> (*c*) he is not the occupier of a commercial unit of agricultural land within the meaning of Part II of the Agriculture Act 1967 or, if he is, occupies it as a licensee only.

Section 21 gave the tribunal discretion to treat a survivor as an eligible person even though he was unable to show that condition (*b*) above was fully satisfied. If he was able to show that it was satisfied to a material extent, and in all the circumstances it appeared fair and reasonable for the applicant to be able to apply for a direction entitling him to a tenancy of the holding, the tribunal could allow him to apply.
No such discretion is given to the tribunal in relation to condition (*c*). If the applicant is the occupier of a commercial unit as defined, he is not an eligible person and so is not qualified to apply to the tribunal for such a direction. K

On March 28 1977 Mr James William Jackson died. He was tenant of Grange Farm, Keyingham, which he farmed with his two sons. Following upon his death notice to quit was given and then his two sons, Mr Graham Christopher Jackson and Mr Martin James Jackson, lodged an application for a direction entitling them to a tenancy of that holding. At the opening

A of the hearing before the tribunal, Mr Martin James Jackson withdrew his application.

Adjoining Grange Farm lies White House Farm which had been farmed by the deceased and his two sons in partnership. On September 19 1967 he had entered into a tenancy agreement letting it to himself and his two sons on a yearly tenancy. White House Farm was certified on August 19 1977 to be a commercial unit within the meaning of Part II of the Agriculture Act 1967. "Commercial unit" was defined by section 40 of that Act as meaning:

> an agricultural unit which in the opinion of the appropriate Minister is capable, when farmed under reasonably skilled manage-
> B ment, of providing full-time employment for an individual occupying it and for at least one other man (or full-time employment for an individual occupying it and employment for members of his family or other persons equivalent to full-time employment for one man).

This Part of the 1967 Act made provision for, *inter alia*, grants towards expenditure incurred in securing amalgamations of farm land so as to form "an intermediate unit" (also defined in section 40) or a commercial unit. Such a unit may of course provide full-time employment for a large number of people. The definition states the minimum number for
C whom employment is to be provided for that unit to be treated as commercial.

At the time of Mr Jackson's death Graham Christopher Jackson and his brother were in joint occupation of White House Farm and it appears then to have been thought that he was consequently ineligible, if he remained a joint occupier, for applying for a direction entitling him to a tenancy of Grange Farm. On June 22 1977 he assigned all his interests in White House Farm to his brother, and it was his contention that, as he had ceased to be in occupation at the date of the application, condition (*c*) did not render him ineligible to make the application. The tribunal thought that the meaning of
D sections 18 and 20 was plain and that the date that mattered in deciding whether an applicant was the occupier of a commercial unit was the date of the death of the tenant of the holding in respect of which the application was made. Mr Jackson appealed to the High Court and Sir Douglas Frank, sitting as a deputy High Court judge, dismissed his appeal on the ground that he had to be eligible at the date of his father's death. The Court of Appeal, however, held by a majority (Lord Denning MR and Lawton LJ: Brandon LJ dissenting) that it sufficed that the applicant should be eligible at the date of the application and at that of the hearing. Throughout this litigation until this case came to this House
E it appears to have been assumed that joint occupation of a commercial unit was a bar to eligibility.

I now turn to the facts of the other appeal. Mr George Williamson died on January 17 1968. He was the tenant of land known as riverside land which formed part of the Eardington estate. This land was part of an agricultural unit farmed by the two sons of Mr Williamson. That agricultural unit excluding the riverside land was a commercial unit within the meaning of Part II of the 1967 Act. Mr Lewis Williamson, one of the sons, applied for a direction entitling him to a tenancy of the riverside land. The agricultural land tribunal which heard his application held that he was eligible to do so. On appeal Mr Kempster, sitting as a deputy High Court judge,
F allowed the appeal, holding that a joint occupier of a commercial unit is the occupier for the purposes of condition (*c*). He was given leave to appeal directly to this House in the belief, that now appears was erroneous, that in *Jackson v Hall*[1] it had been decided, when it was merely assumed, that joint occupation of a commercial unit was a bar to eligibility. At the commencement of the hearing of these appeals, application was made on behalf of Mr Jackson to take the point that he was not debarred by joint occupation of the commercial unit from applying to the tribunal for a direction. Leave to do so was granted.

So there are two questions to be decided in these appeals. The first, common to both, may be expressed thus: G

> Does condition (*c*) render only a sole occupier of a commercial unit ineligible.

and the second thus:

> If a joint occupier of such a unit at the time of the death of the tenant of the holding in respect of which a direction is sought is ineligible, can he make himself eligible by divesting himself of that occupation before he applies for a direction.

I propose to consider the second question first. Lord Denning MR based his conclusion that a person was eligible provided conditions (*a*) and (*b*) were satisfied, if he was not the occupier of a commercial unit at the date of his application H
and at the date of the hearing, primarily on the use of the present tense in section 20. Lawton LJ based his conclusion to the like effect on the ground that being a survivor was a continuing state, and on the absence in condition (*c*) of any reference to the date of death. Brandon LJ, as I have said, dissented. In my opinion Brandon LJ came to the right conclusion. Section 20, which provides that any eligible person may apply within the relevant period (in the majority of cases within three months beginning with the day after the date of the death), to my mind gives no indication of when eligibility must arise, though I agree with Lord Denning that the applicant must be eligible both at the date of the application and J
at the date of the hearing.

Section 18(1) defines the persons who may satisfy condition (*a*). It provides that "Where . . . the . . . tenant of an agricultural holding dies and is "survived by any of the following persons . . ." This to my mind signifies that one has to determine who were the survivors at the date of the death. True it is that those who fell within paragraphs (*a*) to (*d*) of section 18(1) then, will remain within that class as survivors of the deceased for the rest of their lives, but the conjunction of the words "dies and is survived by" seems to me a clear indication that the point of time that has to be considered is the time of death. Condition (*b*) states that the period to be considered in relation to that condition ends at the date of the death, and K
it would I think be very odd if condition (*c*) could be satisfied at a different time.

I entirely agree with the conclusion reached by Brandon LJ and with his reasons. In my view eligibility has to arise at the date of the death. The persons who are then eligible can apply, though eligibility must continue to exist at the date of the hearing and at the date of the application. I do not think that Parliament can ever have intended that a person who was at the date of death ineligible as the sole occupier of a commercial unit could bring himself later within the class of eligible persons by divesting himself of that occupation. So my answer to the second question I have posed is in the negative. L

I now turn to the first question, which is one of more difficulty. The answer to it depends on the meaning to be given to the words "he is not the occupier." Do they mean the sole occupier? If that had been the intention of Parliament it could easily have been expressed. On this construction great weight has to be placed on the article "the." Or should they be read as if instead of "the" the word "an" was there? So read a joint occupier would be ineligible. If it was Parliament's intention that he should be, why was "an" not the word used?

The argument was advanced that the rules made by the Agricultural Land Tribunals (Succession to Agricultural Tenancies) Order 1976 might be relied on as an aid to the M
construction of the statute. Those rules provide that the applicant for a direction must complete a form which requires him to give particulars of the holding in respect of which the application is made, and paragraph 11 of the form reads as follows:

> (*a*) I occupy as owner-occupier/tenant/licensee the following agricultural land [*give particulars of land occupied, including area*]:—
>
> (*b*) I do not occupy any other agricultural land.

The applicant is told to strike out whichever of these two statements is inapplicable. The "any other agricultural land" in (*b*) must I think mean land other than the holding in respect of which the application is made.

My Lords, I have no hesitation in rejecting the contention that rules made in the exercise of a statutory power can be relied on as an aid to the construction of a statute. As Lawton LJ said in the Court of Appeal in the *Jackson v Hall* appeal "The form cannot, of course, be a guide to the construction of section 18(2). Such value as it has lies only in showing what a draftsman, used to looking closely at words, thought the ones used in section 18(2) meant."

Our task is to give effect to the intention of Parliament as revealed by the words of the enactment. The difficulty is that the Act gives no clear indication as to whether the intention was to exclude only sole occupiers or to exclude both sole and joint occupiers, to exclude anyone in occupation of a commercial unit. The Act shows that Parliament recognised that there might be a number of persons who fell within paragraphs (*a*) to (*d*) of section 18(1) and the Act makes provision for dealing with claims by individuals who are eligible applicants, but no provision was made for a direction entitling two applicants to a joint tenancy of a holding. It may be, one does not know, that Parliament just overlooked the possibility that there might be joint ownership of a commercial unit and so made no provision for that.

It is clear from the terms of the Act that it made, and was intended to make, a further encroachment on the rights of landlords of agricultural holdings. It gave a person who came within section 18(*a*) to (*d*) and who could show that he had derived his livelihood either entirely or principally from work on the holding for the requisite period the right to apply for a direction provided that he was not the occupier of a commercial unit. Whether the land was a commercial unit does not depend on the number of persons who can gain their livelihood from it. However many or few they may be, the test of whether it is a commercial unit depends on the number of persons employed thereon. What in my view Parliament has in effect enacted is that a survivor may apply if he comes within section 18(2) and can satisfy condition (*b*), but that he cannot do so if he already occupies a commercial unit which must I think be taken to be a viable farm. If he does occupy such a unit, then Parliament has not thought it right that a landlord should be deprived by the grant of a direction of the advantages which may enure to him from vacant possession.

That is a short and I hope not inaccurate summary of the Act, and in the light of it I cannot think that Parliament intended to give each of two joint occupiers of a commercial unit, which it may be is of such size and character as to provide them with a good living, the right to apply for a direction entitling them to a tenancy of the holding. To hold that only a sole occupier is barred, is I think putting weight on the word "the" which it does not bear. A person in joint occupation of a farm if asked "do you occupy that 'farm' could not truthfully answer 'no.'" A truthful answer would be "yes, I occupy it and I occupy it with someone else."

My Lords, it is to be regretted that this lengthy and no doubt expensive litigation has been brought about by the inadequacy of the drafting of this Act. I do not find the answer to this first question easy. My conclusion is that the words "he is not the occupier" is just another way of saying "he does 'not occupy'," and so read the result is that the applicant in each of these cases was ineligible to apply for a direction.

For the reasons I have stated, in my opinion the appeal of the appellants, Mr Hall and Mr Marr in the *Jackson v Hall* appeal should be allowed with costs here and below and in the case of *Williamson v Thompson* the appeal should be dismissed with costs.

In the course of a concurring speech, LORD FRASER, referring to the *Jackson* appeal said: I should say that I regard the respondent's act in assigning his interest in White House Farm to his brother as perfectly genuine and legitimate. Sir Douglas Frank regarded it as subterfuge, but with respect I do not agree with that view. Nevertheless, I am of opinion that it cannot avail the respondent for the purpose of this application. If a person in the position of the respondent is not eligible at the date of his parent's death, he cannot in my opinion acquire eligibility thereafter. The whole of section 18 seems to be defining and referring to a state of affairs existing at or before the parent's date of death. The scene is set by the opening words of section 18(1) which are "where . . . the sole . . . tenant . . . dies and is survived by any of the following persons." Plainly that is looking at the date of death. It then defines "the following persons" by their relationship to the deceased, which obviously cannot change after his death. Then in subsection (2) the definition of "eligible person," after referring back to the family relationship, sets out the two economic qualifications that I have already mentioned in relation to the *Williamson* question. The qualification set out in paragraph (*b*) is necessarily dependent on events that have occurred "in the seven years ending with the date of death." Nothing occurring after the date of death can affect that qualification. The qualification in paragraph (*c*) is that the person "is not" the occupier of a commercial unit, and the present tense must, I think, mean "is not at the date of death." Any other meaning seems to me hardly possible. Similarly the use of the present tense in section 20(2) in the phrase "the tribunal, if satisfied that the applicant *is* an eligible person, shall determine whether he *is* in their opinion a suitable person" must mean that the applicant must be eligible and suitable also at the date of the tribunal's decision. Accordingly he may lose eligibility before that date. But it is accepted that the applicant must have been eligible for at least some period before the date of the decision. The respondent says that the period begins with the date of his application to the tribunal, but for the reason I have stated, I consider that it begins with the date of death.

A separate and powerful reason why eligibility cannot be acquired after the date of death is that otherwise it would be easy for a child of a deceased tenant to divest himself of an interest in a commercial unit immediately after the parent's death and the condition in paragraph (*c*) would for practical purposes be almost ineffective. He would have three months in which to acquire eligibility and he might do so by divesting himself of the disqualifying interest. I do not consider that Parliament can have intended to leave open such an obvious and comparatively easy method of circumventing the requirement of paragraph (*c*).

In a speech agreeing that eligibility must be present at the date of death, but dissenting from the view that joint occupation of a commercial unit was sufficient to exclude eligibility, LORD RUSSELL OF KILLOWEN said: The policy of this part of the 1976 statute is to meet (for the first time) in a particular way the event of the death of a tenant farmer, which death would otherwise give to the landowner an unrestricted right to determine the tenancy. The particular way adopted was in that event to give to a member of the deceased's family a chance of obtaining a tenancy of the holding against the will of the landowner, when that member had derived his principal source of livelihood from his agricultural work on the holding or on any agricultural unit including the holding. In general this was a recognition of a family's interest in a family-worked farm, and no doubt it was thought to encourage good farming by the prospect of succession on the death of the tenant. Section 18(2)(c) states a situation in which it was thought proper to negative a right to apply for a new tenancy: and that situation was related to an applicant's ability to continue with security of tenure to farm elsewhere, notwithstanding the loss of his principal livelihood in farming the holding. To be eligible to apply he must *(inter alia)* show that he satisfies the condition that "he is not the occupier of a commercial unit of agricultural land within the meaning of Part II of the Agriculture Act 1967, or, if he is, occupies it as a licensee only." It would obviously have been too stringent an exclusion to relate the occupation to too small or otherwise unviable an area of farming land. So the device was chosen of relating it

A to a commercial unit, which was defined in the 1967 Act for a quite different purpose, in connection with rationalisation of the farming industry by encouraging the formation of sufficiently viable farm holdings. In the 1967 Act "'commercial unit' means an agricultural unit which in the opinion of the appropriate minister is capable, when farmed under reasonably skilled management, of providing full-time employment for an individual occupying it and for at least one other man (or full-time employment for an individual occupying it and employment for members of his family or other persons equivalent to full-time employment for one man)." Now this definition was, as I have said, originally for a wholly different purpose, and
B constitutes a minimum objective yardstick for that purpose. But its form is to be expected to have been in the mind of the draftsman of section 18(2)(c), and in construing that latter provision and determining whether "the occupier" is to be taken to indicate "an occupier" or "the sole occupier" it is not I think without some significance that the definition of commercial unit speaks of one individual occupier of the agricultural unit.

In the present case no doubt the appellant in joint occupation of the particular land with his brother has ample opportunity for earning a good living in farming it without any necessity to extend (or continue) his farming activities to
C or on the holding of the deceased tenant. But this particularity of the case should not lead your Lordships to lean in favour of exclusion from eligibility of a joint occupier. In another case the applicant may be in joint occupation of an agricultural unit only just within the definition of commercial unit, and with more than one other.

Of course there are many contexts in which the definite article is not used as an indication of uniqueness: to say for example of X that he is the son of Y is not to be taken as indicating that Y had no other son. Many examples may be given in which the definite article means no more than would the indefinite article: some of them in cases cited in this appeal.
D The arguments in this appeal really rest on what is *not* said. On the one hand it is pointed out that if the draftsman meant sole occupier he could have said it: and other provisions of this part of the statute show that (in a different context) the draftsman was aware of and used the word "sole." On the other hand if joint occupation were in mind in section 18(2)(c) the question is why the draftsman selected the definite article, which I take to point *prima facie* to singularity? He might have said "he is not an occupier": or "He is not in occupation of": or "He does not occupy . . . or if he does, occupies it as a licensee." Three choices of language were readily available to make it plain that a joint occupation was
E intended, and none were chosen.

This, as is sometimes said, is a matter of impression: I conclude that the joint occupation of the appellant does not deprive him of eligibility to apply for a tenancy of his deceased father's holding and I would allow this appeal.

LORD EDMUND-DAVIES and LORD LANE agreed with the speeches of Viscount Dilhorne and Lord Fraser.

The appeal in the case of *Jackson v Hall and Another* was accordingly allowed and the appeal in the case of *Williamson v Thompson and Another* dismissed.

F

COURT OF APPEAL

March 28 1980

(Before Lord Justice BUCKLEY, Lord Justice SHAW and Lord Justice BRIGHTMAN)

ELSDEN AND ANOTHER v PICK

Estates Gazette May 17 1980

(1980) 254 EG 503

**Agricultural holdings—Section 23 of Agricultural Holdings Act 1948 (now section 1 of Agricultural Holdings (Notices to Quit) Act 1977)—Effect of statutory provision requiring a minimum of 12 months' notice to quit in ordinary cases— G
Whether this provision overrode and rendered ineffective an agreement between the parties that a notice which had been served of slightly less than the statutory length should be accepted as valid—Held that there was nothing to prevent the parties to an agricultural tenancy from waiving any defect in a notice to quit, including a failure to comply with the statutory provision as to length of notice—Parties cannot, however, agree in advance, either in the tenancy agreement or otherwise, that a tenancy can be terminated by less than the statutory length of notice—Such advance agreement would conflict with the words in the section "notwithstanding any provision to the contrary"—Declaration in favour of landlords**

H

This was an appeal by the landlords, Richard William Hilary Elsden and Charles Michael Watson-Smyth, from a decision of Robert Wright QC, sitting as a deputy High Court judge in the Chancery Division, refusing the landlords' claim for a declaration that they were entitled to possession of Woodside Farm, Newton, Lincolnshire. The tenant was John Shelbourn Pick, the defendant in the action before Mr Wright and the respondent to the present appeal. The notice to quit which was the subject of these proceedings was dated April 4 1977 but was served on April 7 1977 to determine the tenancy on April 5 1978. (In fact the correct term date was April 6, but this point was not a matter which formed an issue either at J
the trial or the appeal.)

Maurice Price QC and H M Harrod (instructed by Dawson & Co) appeared on behalf of the appellants; Jonathan Parker QC (instructed by Roythorne & Co, of Spalding, Lincolnshire) represented the respondent.

Giving the first judgment at the invitation of Buckley LJ, SHAW LJ said: This is an appeal by the plaintiffs in the action from a judgment of Mr Robert Wright QC, sitting as a deputy High Court judge, given on February 26 1979 whereby he refused their claim for a declaration that they were entitled to possession of Woodside Farm, Newton, Lincolnshire, as on April 6 1978. K

The plaintiffs were at all material times the trustees of a settlement which included the Welby Estate. The principal beneficiary under the settlement is Sir Bruno Welby, Bart. He was concerned with the management of the estate, of which a Mr R M Cawthra [BA FRICS] was the land agent.

Included in the estate were a number of farms some of which were let to tenant farmers. Two, which were known respectively as Woodside Farm and Welby Warren Farm, were so let to the defendant, Mr Pick, who by 1977 had occupied them for many years. For reasons which are not relevant to the present appeal, the Welby Estate had at some time towards the end of the 1960s adopted the policy of not L
reletting farms on the estate when they fell vacant. Sir Bruno had formed a family company called D & S Farms Ltd; it took over the tenancies of such farms and farmed them in partnership with a Mr Giles Halfhead under the partnership title Sapperton Farming Co.

At the beginning of 1977 Mr Pick, who had until then farmed Welby Warren Farm and Woodside Farm with reasonable success, found himself surrounded by misfortune, both matrimonial and financial. His wife was in the process of divorcing him. He was confronted by burdensome obligations and liabilities. He was in a state of great mental stress and consulted his solicitor about his various troubles. On April 1 M
1977 the solicitor wrote advising him that from what he had been told there was a real possibility that Mr Pick might have to cease business. Mr Pick himself foresaw that he might find himself so destitute of resources that he would not be able to pay the rent of his farm, or at any rate the rents of both of them. In that situation he did what an honourable man would do. On April 4, which was a Monday, he went to see Mr Cawthra, who acted for the landlord trustees, and told him what his position was. Their respective versions of what took place as recounted in their evidence before the learned deputy judge did not altogether tally. The judge's assessment

A of them as witnesses was that each was honestly seeking to recall what was arranged between them at that meeting but that Mr Cawthra's recollection was the more reliable. This assessment is not challenged. According to Mr Cawthra he was shown the letter from Mr Pick's solicitor. A half year's rent in respect of each of the two farms held by Mr Pick was to fall due on April 14. He was concerned as to whether he should give notice to terminate one or both tenancies. The requisite notice was one year's notice in writing, expiring on April 6 in any year. If, therefore, Mr Pick wished to bring either tenancy to an end by notice he would have to give that notice before April 6 1977 so as to expire on April 6 1978.

B Mr Cawthra was not then anxious to recover possession of the farms held by Mr Pick, who had been over many years a good tenant as well as a good farmer. He suggested that Mr Pick might bide his time at least for a few days before coming to a final decision as to relinquishing either or both of his tenancies. In pursuance of this suggestion he proposed that Mr Pick should make out notices in respect of each farm, dated April 4 1977 to expire on April 5 1978. This latter date was erroneous as it should have been April 6, but this slip appears to have passed unnoticed even at the trial and has not been the foundation of any of the many submissions addressed to this court. Mr Cawthra's suggestion was that if
C by the end of the week (that is by Saturday April 9 1977) Mr Pick had resolved to give notice to determine one or both of his tenancies he might then give the notice antedated as proposed and it would be accepted. Mr Cawthra, when cross-examined at the trial, said he was not absolutely certain whether or not late service would invalidate the notice. He said, "I thought if we agreed delayed service it would be valid." According to the learned deputy judge's findings, Mr Pick accepted Mr Cawthra's proposals and went away to consider them in relation to his own financial circumstances and prospects. It seems also that Mr Cawthra promised that if later on Mr Pick's position improved and he wished after all
D to continue as before, then Mr Cawthra would recommend to the landlords that Mr Pick be allowed to retract his notice.

On Thursday April 7, Mr Pick returned to Mr Cawthra's office. He took with him two letters. Each bore the date stamp "April 4 1977."

The first letter read: "Dear Mr Cawthra, Following our conversation this morning, after giving more thought to it, and talking to father about it, I enclose a form of words which I hope you will accept as notice on the Newton Farm. The Welby Farm, I think, I should not give up unless I have to. I need a little more time to work out the implications on that one. Many thanks for your sympathy and understanding.
E Yours sincerely, John Pick."

The second document was a notice in respect of Woodside Farm; the body of it was in these terms:

> To the Trustees of the Newton Settlement, the Estate Office, Denton. I, John Shelbourn Pick, tenant of the farm known as Woodside Farm, situate at Newton, near Sleaford in the County of Lincolnshire hereby give you notice of my intention to terminate my tenancy of the above holding on April 5 1978.

Underneath is written: "Dated April 4 1977. Signed J S Pick, Woodside Farm, Newton, Sleaford."

F Mr Cawthra enquired whether that notice represented Mr Pick's final decision. Mr Pick said it did. He asked for some grace in regard to the payment of rent due for Woodside Farm on April 12, and was accorded it. In the event the rent for each of the farms was duly paid by Mr Pick. According to his version of the arrangement between him and Mr Cawthra the notice to quit should then have been torn up. However, this assertion was rejected by the deputy judge and it is unnecessary to consider it further. However, when early in the following August he got a letter from Mr Cawthra the contents of which indicated that his tenancy was regarded as coming to an end, he wrote a reproachful letter. It was dated August 12 1977 and read thus:

> Dear Mr Cawthra, I am very distressed by your letter received this morning. I cannot believe that you would do this, after 40 years here as loyal tenants, because of a hasty, unnecessary act done without any advice. I was very close to a nervous breakdown at the time, a time of extreme stress the reasons for which I discussed with you, and I must appeal to you not to try to hold me to the consequences of such an irrational action. Perhaps we could discuss the matter. Incidentally, I trust that this will now be treated reasonably confidentially G

—and he goes on to say why.

It is evident from the tone and tenor of this missive that Mr Pick believed that he had given a notice which would be effective to determine his tenancy of Woodside Farm if it was to be taken at its face value. Mr Cawthra's response was that H matters had progressed too far to allow the notice to quit to be retracted.

Thereafter Mr Pick, having procured a copy of that notice, consulted his solicitors. He has since maintained that the purported notice was ineffective and did not determine his tenancy. Accordingly he has asserted a right to remain in possession. Hence the claim for a declaration. Mr Pick's basic contention has been that the purported notice to quit Woodside Farm was not an effective notice because it was not in conformity with the requirements of the tenancy agreement and contravened section 23(1) of the Agricultural Holdings Act 1948, which I now read: J

> A notice to quit an agricultural holding or part of an agricultural holding shall (notwithstanding any provision to the contrary in the contract of tenancy of the holding) be invalid if it purports to terminate the tenancy before the expiration of twelve months from the end of the then current year of tenancy.

This is the root of the controversy which was ventilated in the court below. The learned deputy judge came to the conclusion that the effect of that statutory provision was to vitiate a notice which did not conform to the requirement of being a year's notice expiring on April 5 (*sic*). The judge regarded the provision as mandatory and as permitting of no contractual relaxation in any circumstances. He dismissed the K plaintiffs' contention that the arrangement or agreement made between Mr Cawthra and Mr Pick served to validate the notice as and when served in these terms: "Finally, as a matter of legal analysis, whereas in this case the tenancy agreement provides for twelve months' notice, I find difficulty in seeing how an arrangement to accept short notice can be treated otherwise than as a variation of the tenancy. If so, it falls within the prohibition in section 23 against contracting out."

As I have said, this was the primary issue; but the plaintiffs were not without other resources. Apart from any question L of due statutory notice, Mr Price QC contended on their behalf that a tenant could at any time surrender his interest and if that surrender was accepted the interest would thereupon cease. This contention was rejected on the ground that the parties had not got a surrender in contemplation. The learned judge said:

> I am afraid I cannot accept it because as a matter of fact Mr Pick and Mr Cawthra did not intend to enter into a conditional contract of surrender. Mr Cawthra thought he was accepting late service of a notice under the tenancy agreement. This is made clear by the passage from his evidence which I have set out above: "I thought if we agreed delayed service it would be valid." His M requirement that the notice should be dated before April 5 points in that direction. So also does his evidence generally. He told me that he considered that this was going to be a termination of the tenancy in the ordinary way. This is also confirmed by Mr Cawthra's attitude when he had written to Mr Pick the letter dated August 10 1977 and it had become apparent that Mr Pick did not wish to leave. Mr Cawthra wrote in a letter dated August 15 1977: ". . . it would appear that you are now asking for the notice to quit Woodside Farm next Lady Day to be withdrawn." In a letter of September 7 1977 he wrote that Mr Pick on a certain occasion had not even suggested that he might "be regretting giving his notice." The form of the notice to quit which I have already read indicates a notice to quit under the tenancy agree-

ment. That was also Mr Pick's understanding of the position. I find this to be so on the evidence of what took place on this point at the meeting of April 4. There is really no dispute about this part of what took place.

On this basis the notice falls fairly and squarely within the prohibition in section 23.

The learned judge goes on to say:

I realise, however, that what the parties may have said is not conclusive. They are not lawyers. It is the duty of the court to ascertain the legal consequences of what has been done. The name parties give to a transaction may not determine its legal nature. Nevertheless, a contract is a matter of the intention of the parties. In this case I am satisfied that the intention was to give a notice under the tenancy agreement but to accept service of it late. I do not think the court can hold the arrangement to be a contract of surrender. That would be to distort what I find to be the agreement which the parties reached.

Moreover, I do not think the court ought to be astute to analyse a transaction so as to bring it outside a statutory provision when the consequence would be (a) to distort the true intention of the parties and (b) to drive a coach and four through the statutory provision.

Having reached that conclusion it is unnecessary for me to deal with the questions of law which arose on the basis that the contract was one of surrender.

In this connection there had arisen also problems as to consideration and whether or not there was a sufficient memorandum in writing to satisfy section 40 of the Law of Property Act 1925, since a surrender involved, if only incidentally, the transfer of an interest in land. As to consideration, Mr Price submitted that mutual promises to forgo existing contractual rights constituted consideration passing from each of the parties related as landlord and tenant. The learned judge accepted this, but was of the view it did not resolve the problem. Mr Price conceded that there was no sufficient memorandum in writing. This appeared to me, as to the other members of this court, a somewhat surprising concession, since the notice dated April 4, and the letter which accompanied it, contained all the essential elements of a memorandum for the purposes of section 40. Mr Price asked leave to retract his concession and as it was made in relation to the construction of documents, which is a matter of law, that leave was granted. If, therefore, surrender may provide a solution there is no inhibiting evidential factor. This being so, it is unnecessary to pursue the somewhat desperate proposition, as it seems to me, that there was part performance on the part of the plaintiffs which was sufficient to support their claim to specific performance of the agreement between them and Mr Pick that he would surrender his tenancy on April 5 1978. The part performance relied upon consisted of acts done by the plaintiffs in relation to collateral matters such as employing the services of a valuer to arrive at figures for the tenant's interest so as to determine compensation and so on. It was contended that these acts done, not in pursuance of the contract of surrender itself, but in reliance on its having been made and fulfilled, came within the concept of acts of part performance as delineated in *Steadman v Steadman* [1976] AC 536. I do not myself read their Lordships' opinions in that case as enlarging the fundamental concept of acts of part performance. Such acts must still be in furtherance of the contract and not merely a recognition of its existence or its contemplation. However, it is unnecessary, in the light of the view I have reached as to the primary issue, to dwell further on this aspect. I content myself by saying that I do not regard the speeches in *Steadman v Steadman* as introducing a radical change in the basic concept of what may be regarded as an act of part performance. It must still necessarily be an act in furtherance of the contract alleged to have been made, although it may not go so far as to amount to the discharge of any primary obligation imposed by it.

Another argument for the plaintiffs was founded on estoppel. In this court it was enlarged so as to take in the subtle ramifications of promissory estoppel. Mr Parker QC, who appeared for the defendant before the deputy judge as well as on this appeal, met this with the retort that estoppel cannot aid a party to overcome what was a statutory prohibition. This was indeed the essential theme of the defendant's case in refutation of every contrary argument. It was the simple statement that section 23(1) precluded the efficacy of any notice to quit which was of less than 12 months' duration and that was the end of any argument.

I therefore turn back to the section itself in order to examine its language and to deduce its true tenor. It is true that it is designed principally to protect the tenant farmer from peremptory or unduly prejudicial ejectment on the part of the landlord. Nonetheless it serves also to protect a landlord from the abandonment of a tenancy in circumstances which may cause a discontinuity in cultivation or a lapse from proper standards of husbandry. Thus the time for the ending of a tenancy is a matter of common interest both to a landlord and to his tenant. It may suit them both to determine a tenancy without waiting for what may be as long as nearly two years to bring it to an end. No statute could have so absurd an intention as to constrain a landlord and a tenant of an agricultural holding to remain bound in that relationship at a time when neither desires that it should endure. If they are in accord, can it matter whether they demonstrate that accord by an agreement to surrender or an agreement to accept short notice?

I have read section 23(1); the first matter to observe is that the subsection uses the word "invalid" and not "unlawful." Thus there is no penal prohibition; it is simply that provision in the tenancy agreement for a shorter notice than 12 months is nugatory. So also any variation of a tenancy agreement in relation to an agricultural holding which purports to make a shorter period of notice than 12 months effective will fail of its purpose. This produces the situation that there can be no operative provision whereby notice can in prospect be made effective if it is less than 12 months; but a notice is of course a unilateral act available to one party or the other without the ad hoc consent of the party to whom the notice is given. There seems to me to be no impediment created by section 23(1) to the party in receipt of a notice to quit to waive his strict right that the notice should expire on a particular day or that it should be of a particular duration. A contractual provision which enures for the benefit of a party can be waived by that party albeit that his right to that benefit is reinforced by statute. There may be circumstances which might qualify this situation, as where an element of public interest is involved, but in such a case one would expect the statutory provision to speak in terms of illegality (thus: "it shall be unlawful") rather than of mere invalidity.

I would wish to pay respectful tribute to the careful analysis of the learned deputy judge; but in my judgment he erred in concluding that as between the landlord and the tenant of an agricultural holding there could not be a waiver of the requisite term of notice. The statutory provisions do not extend to a situation such as that which developed between Mr Cawthra, acting on behalf of the plaintiffs, and Mr Pick. The outcome of their discussions on April 4 and the acceptance by Mr Cawthra of the notice and the accompanying letter handed to him on April 7 was that the landlords effectively waived, as I am of the view that they were entitled to do, the requirements of the tenancy agreement as to the term and the expiry of the notice. The plaintiffs were not precluded from such waiver by section 23(1) of the Agricultural Holdings Act 1948 or in any other way. It follows that in my judgment Mr Pick's tenancy was duly determined as at April 5 1978, and that the plaintiffs were entitled to the declaration claimed in their writ.

I would allow the appeal and make that declaration in their favour.

Agreeing, BRIGHTMAN LJ said: On April 7 1977, in the circumstances outlined by Shaw LJ, the tenant of Woodside Farm, Newton, handed to the landlords' agent his notice of intention to quit, dated April 4. Accordingly the length of the

A notice fell short of one year by a matter of three days. The notice was accompanied by a letter from the tenant to the agent, "I enclose a form of words which I hope you will accept as notice on the Newton Farm. The Welby Farm, I think, I should not give up unless I have to." Both the landlords' agent and the tenant fully understood that less than one year's notice was being given. The agent agreed with the tenant that he would accept the notice to quit as a valid notice notwithstanding that the length of the notice was short. The landlords' agent also accepted the notice despite another defect, namely, that the notice did not strictly accord with the tenancy agreement in that the date of termination was
B expressed to be April 5 and not April 6 (a point which was probably not appreciated by either side, and upon which nothing turns). It was the intention of the landlords' agent, and also the intention of the tenant, that the tenancy should end on April 5 1978 by virtue of the defective notice so given.

The tenant now seeks to resile from the mutually intended consequence of the defective notice to quit. To do so the tenant relies on section 23 of the Agricultural Holdings Act 1948. The question is whether he can lawfully do so. The trial judge held that he could.

Section 23(1) provides that a notice to quit an agricultural holding shall be invalid if it purports to terminate the tenancy
C before the expiration of 12 months from the end of the then current year of the tenancy. If, for example, an agricultural holding is held on a yearly tenancy which is silent as to the length of notice to quit required to terminate the tenancy, the notice will be invalid if it purports to be less than a 12 months' notice, calculated in the specified manner, although only a six months' notice is required at common law. Even if the tenancy agreement specifies that a shorter length of notice may be given, a notice of less than 12 months would still be invalid, because the subsection has effect "notwithstanding any provision to the contrary in the contract of tenancy of the holding."

D In the instant case the tenant's counsel argued that a notice which is defective because it is short is defective in all circumstances and for all purposes, and is incapable of valid acceptance by the recipient. If a notice served by the tenant is short by a day, but the landlord is content to accept it and so agrees with the tenant, nevertheless the notice is waste paper and the agreement of the parties is valueless. It is no good the tenant saying "Please accept this notice as a valid notice to quit although it is short by a day" and the landlord saying "I am willing to accept it." They have to go off to their solicitors and ask for a written agreement to be drawn up to end the tenancy. Such an agreement, says counsel, will
E be perfectly valid. But not an agreement by the recipient of the notice to accept the notice though defective in length.

Take this extreme case. The landlord serves on his tenant notice to quit. The notice is one day short of the statutory 12 months. The tenant spots the defect and informs the landlord of it, but states that he will accept the notice despite the defect. The tenant intends to emigrate and makes all his plans to leave the country. I will assume that the landlord does not know about the tenant's plans, so as to avoid any complications about estoppel. A day before the notice expires the landlord tells the tenant that the notice to quit was indeed invalid, but that the invalidity was not cured by the landlord's express
F acceptance of the notice and that the tenant is legally liable for another year's rent and performance of the covenants, or more likely two years having regard to the timing which I have assumed in this extreme example. On the argument put forward by the respondent's counsel, the landlord will have a cast-iron case against his tenant.

I ask myself whether this can possibly be the law. I do not think it can be. What section 23 means is that a short notice to quit is invalid as against the recipient. A tenant is not bound to accept less than the statutory 12 months' notice to quit served by his landlord (nor vice versa) even if the tenancy agreement so provides. If the tenant chooses to do so, he can simply ignore a short notice served on him and resist any G
attempt by the landlord to recover possession on the strength of it. But if the tenant wishes to do so, he can bind himself to accept it. The parties are entitled to agree that the notice shall be treated in all respects as if it were a notice of the statutory length. If the parties so agree, the tenancy will come to an end on the agreed date by virtue of the defective notice to quit which it is agreed shall be treated as valid. Such an agreement could not effectively be made before a notice to quit is served, because the parties cannot agree that the tenancy shall be capable of being terminated by a short notice. Neither the landlord nor the tenant can bind himself in advance to accept a short notice from the other of them. That would H
be a "provision to the contrary" in, or supplemental to, the contract of tenancy and would not be effective. But once an invalid notice has been served, which the recipient is entitled to ignore, I see nothing in section 23 to prohibit an agreement between landlord and tenant that the notice shall be followed by the same consequences as if it were valid.

There is an alternative approach, which in my view is equally tenable. Although in the instant case the notice to quit is invalid qua notice to quit, it contains all the ingredients which are needed to constitute a valid written acceptance on April 7 of the oral offer made on April 4 by the landlords' agent, or (if this is the correct interpretation of the events) a J
valid written offer on April 7 which the landlords' agent accepted orally on the same day. The difference between the two interpretations is of no moment because the legal consequences are the same. On the assumption that the latter interpretation is correct, the defective notice was an expression of the intention, and therefore of the willingness, of the tenant that the tenancy agreement should terminate on April 5 1978, which intention the tenant communicated to the landlords' agent in the hope or expectation that the agent would accede thereto. The agent did accede thereto, as a result of which there was *consensus ad idem* between tenant and agent that the tenancy should end on April 5 1978. Such a course of K
dealing seems to me on analysis to be indistinguishable in any relevant respect from an offer by the tenant to terminate the tenancy on April 5 1978, which offer is accepted by the landlords. Such an agreement is clearly outside section 23.

I find support for this approach in the judgment of Bayley J in *Johnstone v Hudlestone* (1825) 4 B & C 922 at p 935 to which my Lord, Buckley LJ, invited counsel's attention during the course of the argument.

For myself, I do not mind which way the case is put and I doubt whether there is any substantial difference between the two interpretations of the events.

I would allow the appeal for the reasons which I have L
endeavoured to express.

Also agreeing, BUCKLEY LJ said: The Agricultural Holdings Act 1948, section 23, is clearly designed to protect an agricultural tenant against eviction from his holding against his will by shorter notice than the section prescribes. It may also confer advantages upon the landlord by ensuring that he will receive at least 12 months' notice from the tenant of termination of the tenancy. It does not, like sections 3 and 24 of the Landlord and Tenant Act 1954, provide that a tenancy shall not come to an end except by being terminated in accordance with the provisions of the Act. It does not, in my judgment, preclude determination of an agricultural tenancy on shorter M
notice than the Act requires if both parties agree to this. Mr Parker concedes, rightly in my view, that the parties to an agricultural tenancy can bring the tenancy to an end by an agreement that the tenant shall surrender the tenancy to the landlord forthwith or by an agreement that the tenancy shall come to an end at some future date which is less remote than the earliest date at which it could be brought to an end by a notice to quit in accordance with the section.

In a case which is unregulated by any statutory provision, if a tenant gives notice to quit in circumstances which make the notice ineffective—perhaps because it is too short or

A because it is given for the wrong day—the landlord can, if he chooses, agree to accept the notice as valid notwithstanding the defect in it: he can waive his right to rely on the defect. If he does so in a manner which is legally binding, the notice will take effect as if it were not defective. There is, in my opinion, nothing in section 23 of the 1948 Act to prevent the parties to an agricultural tenancy from waiving any defect in a notice to quit, including a failure to comply with the requirements of the section itself.

It is, in my opinion, clear upon the evidence that both Mr Cawthra and Mr Pick thought that the notice to quit handed by Mr Pick to Mr Cawthra on April 7 1977 was an
B effective notice, notwithstanding that it was a day late, because Mr Cawthra on the landlords' behalf agreed to accept it as effective. Mr Cawthra in evidence said that he had had some doubts about this, but the learned judge found that Mr Cawthra believed that the notice would terminate the tenancy. What then occurred cannot, in my judgment, have constituted a surrender of the tenancy, for it is well settled that a tenancy cannot be surrendered *in futuro*. I agree with the learned judge that in the circumstances of this case one could not find that there was an agreement for a future surrender of the tenancy. An agreement to surrender a term or to terminate the relation of landlord and tenant on a particular date and
C an effective notice to terminate the tenancy on that date may have the same legal effect, but they are different transactions; one operates bilaterally by way of contract, the other unilaterally by setting a term to a contract (viz the tenancy) in exercise of a power under that contract.

A contract to render an existing defective notice to quit effective by waiving the defect is not, in my view, a contract to vary the tenancy agreement. An agricultural tenancy could not, consistently with section 23, be modified by agreement between the parties in such a way that either party to it could terminate it by a notice to quit which did not accord with the terms of the section. Such an agreement would conflict
D with the words of the section which are within brackets. But it does not follow from this that, when a notice to quit has been given which fails to satisfy the requirements of the section, the parties cannot effectually agree that, notwithstanding the defect in the notice, it shall take effect as though it were a valid notice. Such an agreement would not, in my opinion, conflict with any part of the section. It would not alter any of the terms of the tenancy agreement. Its effect is that the parties agree to the tenancy coming to an end on the date specified in the notice, subject to which the tenancy remains in full force and effect. So viewed—and in my opinion rightly so viewed—the agreement comes within Mr Parker's con-
E cession.

I therefore agree with the learned judge that in the present case the intention of the parties was that Mr Pick should give notice under the tenancy agreement and that consequently the acceptance by Mr Cawthra of late service of the notice did not give rise to a contract for the surrender of the tenancy. With deference, however, I do not agree with him that the agreement to accept short notice involved a variation of the tenancy falling within the prohibition in section 23 against contracting out of the section. In my judgment, Mr Cawthra's agreement on the landlords' behalf on April 7 1977 to accept the notice to quit as a valid notice effectually determining the
F tenancy on April 5 1978 had the effect of a binding waiver of any defect in the notice and of a binding agreement that the tenancy should accordingly come to an end on April 5 1978. Such waiver and such agreement did not, in my judgment, conflict with section 23. It is not contended that, if that transaction can legally take effect, it was not contractually binding and supported by mutual consideration. It is consequently, in my judgment, binding on the defendant.

For these reasons and for those contained in the judgments already delivered, with which I agree, I would consequently allow this appeal.

The appeal was allowed and a declaration made that the appellants became entitled to possession of Woodside Farm G
on April 6 1978. The respondent was ordered to pay the costs below and three-quarters of the appellants' costs in the Court of Appeal.

QUEEN'S BENCH DIVISION

March 12 1980

(Before Sir Douglas FRANK QC, sitting as a deputy judge of the division)

DAGG v LOVETT

Estates Gazette June 14 1980
(1980) 254 EG 993 H

**Agriculture (Miscellaneous Provisions) Act 1976—Widow's application for a tenancy under succession provisions—Appeal from decision of agricultural land tribunal—Tribunal decided that widow should be treated as eligible by virtue of section 21 of Act, but held that she was not suitable to become the tenant of the holding—Tribunal's finding based on the view that widow's experience was too limited to enable her to farm to the standard which would produce the anticipated future level of market rent—Tribunal's assessment of a probable five-fold increase in rent not put to parties for comment during hearing and not based on evidence—Tribunal guilty of breach of "audi alteram partem" rule of natural justice and in error J
in excluding evidence of market rent—Question also as to adequacy of tribunal's reasons—Decision quashed and case sent back for matter to be redetermined—Parties to be allowed to lead evidence on economic rent and potential income from farm—Submission rejected that tribunal's decision could not be quashed unless court was satisfied that no reasonable tribunal could have come to such a conclusion**

This was an appeal by case stated by the Agricultural Land Tribunal for the Northern Area under section 6 of the Agriculture (Miscellaneous Provisions) Act 1954. The tribunal's
decision was challenged by the applicant to the tribunal, K
Elizabeth Ann Dagg, widow of Matthew Smith Dagg who had been the tenant of Wark Farm, Cornhill-on-Tweed, Northumberland. The applicant had been unsuccessful in obtaining the tribunal's direction entitling her to a tenancy of the farm, following the death of her husband. The landlord, the respondent, was Rowland Haddon Lovett. A report of the tribunal's decision was published in (1979) 251 EG 75.

Leolin Price QC and Grant Crawford (instructed by Heald & Nickinson, agents for Sanderson, McCreath & Edney, of Berwick upon Tweed) appeared on behalf of the applicant; Peter Langdon-Davies (instructed by Travers Smith, Braith-
waite & Co) represented the respondent. L

Giving judgment, SIR DOUGLAS FRANK QC said: This is an application under section 6 of the Agriculture (Miscellaneous Provisions) Act 1954 by case stated by the chairman of the Agricultural Land Tribunal (Northern Area). The material facts are as follows:

The farm concerned is known as Wark Farm. It is situated largely adjoining the River Tweed and is mainly arable land. Matthew Smith Dagg occupied the farm from 1951, first as a joint and subsequently as sole tenant. In 1972 Mr Dagg began to suffer ill-health and his state deteriorated over the year. In 1974 he underwent a serious operation and thereafter his
involvement with the farm became less and his wife virtually M
took over the management of the farm. In June 1976 he and his wife entered into a deed of partnership. On September 14 1977 Mr Dagg died and his widow made applications under sections 20 and 21 of the Agriculture (Miscellaneous Provisions) Act 1976 for a direction entitling her to a tenancy of the holding. The landlord thereupon served notice to quit on the personal representatives and there followed the various statutory notices. The hearing before the tribunal took place for three days in September 1978 and for three days in January 1979. The tribunal gave their decision in writing on February 14 1979. The decision is in three parts. The first deals

with the application under section 21, in which the tribunal held that Mrs Dagg was not an eligible person under section 18(2)(b). However, they then went on to determine under section 21(2) of the Act that she should be treated as a suitable person. The third part of the decision, and this is the material part, was directed to the question to be determined under subsection (2) of section 20 whether Mrs Dagg in the opinion of the tribunal is a suitable person to become the tenant of the holding.

In paragraph 33 of their decision the tribunal stated that "the serious challenge to her suitability, in summary, was that she had no formal training in agriculture and her experience was too limited to enable her to farm Wark to a satisfactory level and enable her to find the increased rent which the landlord would require and market forces would determine." It is apparent from that statement and from other parts of the decision that the crucial question was whether the applicant would, and again I quote, "have sufficient ability to farm the land to a level where the landlord would receive a fair rent for the farm and the tenant a reasonable standard of living." That was the test applied by the tribunal and they found "Mrs Dagg produced no evidence to rebut the landlord's suggestion that her output would barely meet the envisaged rent level", and accordingly they dismissed the application.

In the case stated the tribunal set out four grounds for dismissing the application, and they can be summarised as follows:

(a) There was no evidence that the applicant would be able to farm the land profitably if an economic rent were charged.

(b) The applicant had taken no steps to improve or expand the existing records, which the tribunal considered were inadequate by any standards.

(c) The applicant had only a shallow knowledge of farming and no practical experience and when tested, albeit not to any great length, her knowledge was found to be lacking.

(d) The amount to which the applicant would be forced to rely on her steward was considered undue in a farm of this size and importance.

I deal first with what was undoubtedly the crucial question, namely, whether the applicant would be able to meet the market rent. It was common ground that the present rent is too low and that upon the grant of a tenancy to the applicant the landlord would be entitled to charge a substantially higher rent which, if not agreed, would fall to be determined by arbitration. However, there was no argument as to what the market rent is, and the tribunal declined to allow evidence to be directed to that question. It is true that their decision was prompted by an objection voiced, albeit faintly, by Mr Price who appeared for the applicant. It may be that if the matter had rested there, as Mr Price had raised the objection, it would not lie with the applicant to complain. However, the tribunal, relying on the knowledge of what it called its lay members, came to the conclusion that it was likely that the rent would increase in the region of five times the present level. That assessment was not put to or communicated to the parties during the hearing, and so they had no opportunity of dealing with it, although it was essential to the tribunal's conclusion that the farm's output would barely meet the envisaged rent level. In my judgment the tribunal's actions in assessing the possible rent in the way they did was a clear breach of the rules of natural justice, and in particular the rule known as *audi alteram partem*. Indeed, Mr Langdon-Davies, who appeared for the respondent, did not seriously challenge that proposition. On that ground alone I would quash the decision. Furthermore, however, it appears from the case as stated that, taking into account the last farm accounts and the rent as estimated by the tribunal, the applicant would be able to pay the rent and would be left with what was described as "a comfortable net profit of £9,000." In that respect, therefore, there seems to have been a misdirection or miscalculation by the tribunal, or at least there is an ambiguity.

I must now consider whether the tribunal were right in excluding evidence of market rent. On the one hand the tribunal said they would not allow the proceedings to become a trial run for rent arbitration, and on the other hand that the question of an economic rent did affect the applicant's ability to make a success of the farming enterprise. This was an element in considering her suitability. It would appear necessarily to follow from the second limb of the tribunal's reasoning, and indeed from the main contention of the landlord, that the amount of the market rent was an essential question for decision. I appreciate that that might be a question to be determined later by arbitration, but that does not seem to render it inadmissible in the tribunal's proceedings, and there was certainly no ground for the tribunal making their own assessment without hearing evidence. In my judgment, evidence on market rental was admissible and should not have been excluded.

Another ground of criticism advanced by Mr Price was that the tribunal had imposed a burden of proof not imposed by Parliament. Without rehearsing the arguments, in my judgment the position is this: The tribunal had to form an opinion whether the applicant is a suitable person and by necessary inference whether, on the other hand, she is an unsuitable person. That is an opinion to be derived from the facts and no question of burden of proof there arises. That burden is upon he who asserts any fact which he wishes the tribunal to take into account in arriving at their opinion. However, I cannot say with confidence that in holding that "the onus of satisfying us that she is a suitable person (under section 20 of the Act) to be a tenant of the farm in question falls on the applicant" necessarily establishes that the tribunal misdirected themselves on the burden of proof. On the other hand, I am left in real doubt as to whether the tribunal expected more than is envisaged by the scheme of the Act. It cannot have been the intention that only those having graduated at Wye or Cirencester who had, say, five years' experience of management would qualify. The very fact that widows are included in the scheme negates any such test. Yet on the decision as it stands, apart from the question of rent, it is not clear what facts the tribunal relied upon for their conclusion that the applicant's knowledge of agriculture was found to be lacking. Moreover, I find it difficult to understand how the tribunal, having found that the farm had been reasonably managed since the death of the applicant's husband, were not convinced that the applicant was competent to farm the holding adequately on a long-term basis. It is true that in the case stated the tribunal say: "Our decision was made in the light of all the evidence we heard and the facts we found: It should not be viewed in isolation in respect of certain items of the evidence but taken as a whole."

A "blanket" reason of this kind will not do. It is incumbent on the tribunal, having regard to the Tribunals and Inquiries Act, to give reasons for their decision. It is well settled that those reasons must be adequate and not only intelligible but also can reasonably be said to deal with the substantial points that had been raised: see *Re Poyser and Mills' Arbitration* [1964] 2 QB 467. Further, the reasons must not leave in the mind of an informed reader such real and substantial doubt as to the reasons for the decision and to the matters which were or were not taken into account: see *Givaudan & Co Ltd v Minister of Housing and Local Government* [1967] 1 WLR 250. If, in fact, the tribunal, as is said in the case, relied only on the applicant's failure to recognise the meaning of the term pH and correctly to name the three nutrients which make up a normal compound fertiliser, then having regard to the applicant's past performance, I would have difficulty in concluding that the decision was reasonable. Moreover, I do not think that the applicant could reasonably have been expected, without notice that it was to be an issue at the hearing, to have improved the farming records without knowing whether she was to be allowed to retain the farm.

A I am left in no doubt that the decision must be quashed and that the correct course is to send the case back to the tribunal to redetermine the matter. It will be necessary for them to allow the parties to lead evidence on such matters as the economic rent and the potential income from the farm. I do not accept Mr Langdon-Davies' submission that I am not entitled to quash the decision unless I am satisfied that the only decision a reasonable tribunal could have come to was to allow the application. It is for the tribunal, having heard the evidence and applying their minds to the relevant questions, to reach an opinion: that jurisdiction is B
vested in them, not in the court, and in my opinion rightly so.

It is with some regret that I reach this conclusion, for the tribunal are to be congratulated on the form of their decision and the great trouble they have taken with this matter. So far as the first two parts of the decision are concerned, not only is it unassailable but, if I may say so, should serve as a model in dealing with this kind of application.

The applicant was awarded costs. The respondent was refused leave to appeal.

COMPULSORY PURCHASE AND COMPENSATION

A **HOUSE OF LORDS**

November 8 1979

(Before Lord HAILSHAM OF ST MARYLEBONE LC, Lord WILBERFORCE, Lord FRASER OF TULLYBELTON, Lord RUSSELL OF KILLOWEN and Lord KEITH OF KINKEL)

LONDON & CLYDESIDE ESTATES LTD v ABERDEEN CITY COUNCIL AND ANOTHER

Estates Gazette March 8 1980

(1979) 253 EG 1011

**Compensation for compulsory acquisition—Certificate of alternative development—Certificate issued by the local plan-
B ning authority failed to include the statement of the applicants' rights of appeal required by the relevant statutory instrument —Effect of omission—Certificate "vitiated" by omission, as requirement was mandatory and not merely directory—Certificate not, however, a nullity, so that it could not be alleged that the position was as if no certificate had ever been issued—Authority's argument that applicants were out of time for appealing, because the time limit applicable in the failure to issue cases had passed, accordingly rejected by House—Other arguments by authority rejected—Lord Chancellor's comment on "the rapidly developing jurisprudence of administrative law"—Authority ordered to issue fresh certificate complying
C with the statutory provisions**

This was an appeal by London & Clydeside Estates Ltd and a cross-appeal by Aberdeen City Council against a decision of the Second Division of the Inner House of the Court of Session in Scotland. The Second Division on an appeal from the Lord Ordinary (Lord Dunpark) decided that a purported certificate of alternative development issued by the council as local planning authority was invalid and should be reduced. The Second Division, however, refused a decree ordering the council to issue a fresh certificate, on the ground that such an issue was contrary to the statutory provisions. London & Clydeside Estates Ltd appealed against this refusal and the
D council cross-appealed against the decision to reduce the certificate. The land which was the subject of the litigation consisted of three sites at Scotstoun, Bridge of Don, Grampian. The statutory provisions mentioned by Lord Hailsham are in substance the same as those applicable to England and Wales.

A M Morison QC and A G C McGregor (instructed by Stephenson Harwood, agents for Biggart, Baillie & Gifford WS, of Edinburgh) appeared for the appellants; M S R Bruce QC and J A Cameron (instructed by Martin & Co, agents for Shepherd & Wedderburn WS, Edinburgh) for the respondents.

In his speech, LORD HAILSHAM OF ST MARYLEBONE LC said: My task in this case is rendered considerably lighter by reason
E of the fact that I have had the advantage of reading in draft the opinions prepared by my noble and learned friends Lord Fraser of Tullybelton and Lord Keith of Kinkel. With them I agree, and accordingly I am of the opinion that this appeal should succeed, the cross-appeal be dismissed, and the appellants be allowed their expenses throughout these proceedings, including those of the proceedings before your Lordships' House, other than the expenses relating to the joinder of the second defender as to which it is not now sought to disturb the order of the Second Division of the Court of Session. Nevertheless, I wish to frame my own reasons for F
coming to this conclusion.

It will be logical to deal first with the cross-appeal, which seeks to reverse the interlocutors to the extent to which the pursuers succeeded below. These interlocutors were in the terms of the first conclusion of the pursuers' summons in these proceedings, which sought the reduction of a purported certificate by the respondents. The appeal itself is confined to the contention that the Second Division of the Court of Session were wrong to refuse the second conclusion in the appellants' summons after sustaining their contention that they were entitled to succeed on the first (which is the subject of the cross-appeal). G

The proceedings relate to three areas of ground at Scotstoun, Bridge of Don, which the respondents or their predecessors desire to acquire for educational purposes.

On September 9 1974, the appellants applied through their architects to the respondents' predecessor authority for a certificate of alternative development pursuant to section 25 of the Land Compensation (Scotland) Act 1963. The form of this certificate necessarily affects the amount of compensation payable for the acquisition, and, although this does not appear directly from the record, we were told that, independently of these proceedings, a reference to the Lands Tribunal for Scotland following an agreement for sale has duly taken place H
and has resulted in an award in an alternative form on each of two alternative bases [LTS/APP/3/59 dated September 24 1975; W A Elliott QC and Wm Hall FRICS]. On October 22 1974, and in response to the appellants' application, the respondents' predecessors issued what purported to be the appropriate certificate. But this purported certificate was admittedly defective (to use a neutral word) because, contrary to the terms of article 3(3) of the Town and Country Planning (General Development) (Scotland) Order 1959 (SI 1959 No 1361), which is admitted to apply to the case, it did not "include a statement in writing . . . of the rights of appeal to the Secretary of State. . . ." These rights, under the relevant terms J
of article 4(1) of the order, provided for notice of appeal to be given within one month from the date of receipt of the certificate.

In the events which happened, the appellants purported after the expiry of the time limit of one month to intimate an appeal to the Secretary of State. This they did by letter dated January 9 1975. But by letter dated January 15 1975, the Secretary of State declined to accept this letter as a valid appeal on the grounds that it was out of time, and adhered to this decision despite a further letter on behalf of the appellants complaining in effect of the defective character of the respondents' certificate. K

The outcome was the raising, on April 11 1975, of the present proceedings in which the appellants concluded (1) for a reduction of the purported certificate of October 22 1974, and (2) for a declarator that the respondents were bound to issue an amended, or more properly a fresh, certificate complying with article 3(1) of the order, and a decree ordaining the respondents to issue such a certificate within two months of the decree. These are the only two conclusions still alive in the proceedings before your Lordships. There was a third and

A alternative conclusion, now no longer effective, which resulted from a provisional view framed by the Lord Ordinary in the course of the proceedings before him. The Lord Ordinary had at first been disposed to consider that the Secretary of State was wrong to decline jurisdiction to hear the attempted appeal, but changed his mind on hearing argument for the Secretary of State, who had been joined by amendment for the purposes. No point on this abortive solution remains to be decided on this appeal, the appellants expressly refraining from pursuing the argument as to expenses raised in their case to your Lordships' House.

In the event, the appellants succeeded in their first conclusion B (for the reduction of the purported certificate) both before the Lord Ordinary and the Second Division, and this result forms the subject of the respondents' cross-appeal. But before the Second Division the appellants failed in their claim to the decree concluded for in their second claim for relief (the subject of the appeal itself) on the ground, as the Second Division held, that to ordain in accordance with the second conclusion would be "flying in the face" of article 3(2) of the development order. This provided that the time within which the relevant certificate was to be issued by the respondents was to be "the period of two months from the date of receipt" of the relevant application, and from this the Second C Division were of opinion that the respondents had no remaining power to issue a certificate in the form required by the second conclusion of the appellants' summons. In passing, I should remark that the point was a novel one before the Second Division, the respondents having conceded before the Lord Ordinary that the two conclusions stood or fell together and having withdrawn this concession on the reclaiming motion in the Second Division.

It will be convenient to deal with the points raised in what I conceive to be their logical order rather than the order in which they were argued by the respective counsel.

On this basis, the first question for consideration is the D consequence of what was admitted to be a defect in the purported certificate of October 22 1974, namely the failure by the predecessors of the respondents to include in the certificate information in writing as to the appellants' rights of appeal to the Secretary of State. Was this requirement, which has the authority of Parliament behind it, mandatory or was it in some sense directory only? I have no doubt that it was mandatory and that the failure to include this information was fatal to the certificate. In the course of argument counsel for the respondents candidly conceded that the only purpose of the requirement was to inform the applicant of his rights of appeal, including the time limit within which they should E be exercised. The present appellants aver that they were misled by this defect and that it was as a result of this that their appeal was out of time. The averment has never been put to the proof, and one of the respondents' alternative arguments was that, in the event of otherwise total failure, the appellants should be put to the proof of this. But in my view this argument is without foundation. The validity of the certificate itself is in question, and if, as I believe, the requirement is mandatory, the certificate falls independently of whether the appellants were in fact misled. I find it impossible to accept that a requirement by an instrument of statutory force designed for the very purpose of compelling a public F authority to inform the subject of his legal rights can be treated as simply regulatory if the requirement is not complied with. If I required authority for this proposition I would refer to *Agricultural, Horticultural and Forestry Industry Training Board v Kent* [1970] 2 QB 19 (CA), *Rayner v Stepney Corporation* [1911] 2 Ch 312, and *Brayhead (Ascot) Ltd v Berkshire CC* [1964] 2 QB 303 (DC) notwithstanding that it relied on *Edwick v Sunbury UDC* [1962] 1 QB 229 which was disapproved in *James v Secretary of State for Wales* [1968] AC 409, which was decided on an argument irrelevant to the present appeal. However, I am content to assert a general principle to the effect that where Parliament prescribes that an authority with compulsory powers should inform the subject of his right G to question those powers, *prima facie* the requirement must be treated as mandatory. For the reasons which follow, however, this does not dispose the matter in the appellants' favour.

If the requirement that the subject should be informed of his legal rights was mandatory, what follows? The respondents attempted, as I thought at one time, to argue that it thereupon became a nullity, and that therefore a decree of reduction was inappropriate because there was nothing upon which it could operate. But I do not accept this argument. The certificate was effective until it was struck down by a competent authority (cf *Brayhead (Ascot) Ltd v Berkshire CC supra; James v Secretary of State for Wales, supra*). In the course of argu- H ment I ventured to draw attention to the passage at p 445 of the opinion of the Judicial Committee in *Calvin v Carr* [1979] 2 All ER 440 in which Lord Wilberforce says of a contention that a decision of the stewards of the Australian Jockey Club was void for breach of natural justice:

> This argument led necessarily into the difficult area of what is void and what is voidable, as to which some confusion exists in the authorities. Their Lordships' opinion would be, if it became necessary to fix on one or other of these expressions, that a decision made contrary to natural justice is void but that, until it is so declared by a competent body or court, it may have some effect, or existence, in law. J This condition might be better expressed by saying that the decision is invalid or vitiated. In the present context, where the question is whether an appeal lies, the impugned decision cannot be considered as totally void, in the sense of being legally non-existent. So to hold would be wholly unreal.

The subject-matter of that case was wholly different from the present, but my opinion is that the thinking behind it is applicable. The certificate was vitiated in the sense that it failed to comply with a mandatory requirement. But the subject could not safely disregard it as not having been issued. Had he done so, he might well have fallen into the very trap of losing his right to complain of the vitiating factor, which K has caught other subjects in the reported decisions, and, in my view, he was not only wise but bound to seek a decree of reduction or some other appropriate remedy striking down the offending certificate.

A similar line of reasoning disposes of the next contention of the respondents, also rejected in the Second Division, to the effect that, if the certificate is vitiated, the position is the same as if no certificate had been issued and that section 26(4) of the Land Compensation (Scotland) Act 1963 then operates in such a way that, no certificate having been issued under section 25, the preceding provisions of the section as to appeals should apply at the expiry of the prescribed period "as if" the local L planning authority had issued a certificate "containing such a statement as is mentioned in" section 25(4)(b) of the Act. The effect of this read with articles 3 and 4 of the order would have put the appellants out of time for appeal on the expiry of one month after the expiry of the prescribed (two months) for the due issue of the certificate by the respondents. The fallacy in this argument lies in the assumption (for it is no more) that the issue by an authority of a certificate vitiated by failure to comply with a mandatory requirement is the same thing as the failure by that authority to issue any purported certificate at all.

The respondents were at pains to argue that the issue by M the authority of a certificate vitiated for want of compliance with a mandatory requirement was a *casus omissus* from the Act and that, in this context, the law of Scotland (unlike the law of England) afforded no remedy at all unless it be by the invocation of the jurisdiction peculiar to Scottish law, which goes by the imposing name of "nobile officium." I was utterly unpersuaded by this argument or that there was any difference between Scottish and English law in this respect, and my want of belief is reinforced by what my two learned and noble friends, Lord Fraser of Tullybelton and Lord Keith of Kinkel, have to say about the more arcane aspects of "nobile

A officium." In my opinion, in both jurisdictions the law is the same. The first task is to construe the statute, and ask the question whether the duty in question is mandatory or directory. If it be mandatory, the second task is to ask what remedy is available for non-compliance. If the statute specifies the remedy, well and good. If it is silent, the ordinary remedies available in each jurisdiction, eg proceedings for declaration or prerogative order in England, summons for declarator or reduction in Scotland, should be pursued as appropriate. There is no room for a *casus omissus* in either case. Counsel for the appellants called in aid of this part of the case the authority of *Maitland* 1961 161 SC 291, but I do not think B authority is required for a proposition to my mind so evident on general principle.

More persuasive, in some ways, was the argument for the respondents that it was an odd sort of statute which first provided that an applicant should look at the Act in order to ascertain his right under section 25 (now amended and printed as a Keeling schedule in Schedule 9 to the Community Land Act 1975) to make an application and then, if no certificate were forthcoming, look at section 26(4) of the Act and the order in order to note and exercise his right of appeal but, as regards a purported certificate failing to apprise him of his rights of appeal (which by that time one would have supposed C him to know), that he should be in the position to rely on the invalidity of the certificate in the way he now seeks to do. That there is a certain paradox in this I do not deny. But I do not think we are entitled to play fast and loose with statutory requirements designed to inform the subject as to his legal rights against an authority possessed of compulsory powers. There would be an even greater paradox in allowing an acquiring or planning authority first to flout such a requirement and then to be heard to say that its non-compliance had no effect on the validity of its legal documents. I do not think that prescriptions for the benefit of the subject are so to be disregarded.

D At this stage I should notice a contention on the part of the respondents, which, though, as will be seen, I partly agree with it, does not seem to me to be relevant to the disposal of the cross-appeal. The contention was that in the categorisation of statutory requirements into "mandatory" and "directory," there was a subdivision of the category "directory" into two classes composed (i) of those directory requirements "substantial compliance" with which satisfied the requirement to the point at which a minor defect of trivial irregularity could be ignored by the court and (ii) of those requirements so purely regulatory in character that failure to comply could in no circumstances affect the validity of what was done. E The contention of the respondents was that, even on the assumption against themselves that the requirement of the order was that the certificate should include a notification of the appellants' rights to appeal to the Secretary of State, the rest of the certificate was so exactly in accordance with the provisions of the order that the remaining defect could be safely ignored.

I do not consider that this argument assists the respondents in the present appeal. I have already held that the requirement relating to notification of the appellants' rights of appeal was mandatory and not directory in either sense contended for by the respondents. But on the assumption that I F am wrong about this, a total failure to comply with a significant part of a requirement cannot in any circumstances be regarded as "substantial compliance" with the total requirement in such a way as to bring the respondents' contention into effect.

Nevertheless I wish to examine the contention itself. In this appeal we are in the field of the rapidly developing jurisprudence of administrative law, and we are considering the effect of non-compliance by a statutory authority with the statutory requirements affecting the discharge of one of its functions. In the reported decisions there is much language presupposing the existence of stark categories such as "mandatory" and "directory," "void" and "voidable," a "nullity," and "purely regulatory." Such language is useful; indeed, in G the course of this opinion I have used some of it myself. But I wish to say that I am not at all clear that the language itself may not be misleading in so far as it may be supposed to present a court with the necessity of fitting a particular case into one or other of mutually exclusive and starkly contrasted compartments, compartments which in some cases (eg "void" and "voidable") are borrowed from the language of contract or status, and are not easily fitted to the requirements of administrative law.

When Parliament lays down a statutory requirement for the exercise of legal authority it expects its authority to be obeyed down to the minutest detail. But what the courts have H to decide in a particular case is the legal consequence of non-compliance on the rights of the subject viewed in the light of a concrete state of facts and a continuing chain of events. It may be that what the courts are faced with is not so much a stark choice of alternatives but a spectrum of possibilities in which one compartment or description fades gradually into another. At one end of this spectrum there may be cases in which a fundamental obligation may have been so outrageously and flagrantly ignored or defied that the subject may safely ignore what has been done and treat it as having no legal consequences upon himself. In such a case if the defaulting authority seeks to rely on its action it may be that the subject J is entitled to use the defect in procedure simply as a shield or defence without having taken any positive action of his own. At the other end of the spectrum the defect in procedure may be so nugatory or trivial that the authority can safely proceed without remedial action, confident that, if the subject is so misguided as to rely on the fault, the courts will decline to listen to his complaint. But in a very great number of cases, it may be in a majority of them, it may be necessary for a subject, in order to safeguard himself, to go to the court for declaration of his rights, the grant of which may well be discretionary, and by the like token it may be wise for an authority (as it certainly would have been here) to do everything in its power to remedy K the fault in its procedure so as not to deprive the subject of his due or themselves of their power to act. In such cases, though language like "mandatory," "directory," "void," "voidable," "nullity" and so forth may be helpful in argument, it may be misleading in effect if relied on to show that the courts, in deciding the consequences of a defect in the exercise of power, are necessarily bound to fit the facts of a particular case and a developing chain of events into rigid legal categories or to stretch or cramp them on a bed of Procrustes invented by lawyers for the purpose of convenient exposition. As I have said, the case does not really arise here, since we are in the presence of total non-compliance with a requirement which I L have held to be mandatory. Nevertheless I do not wish to be understood, in the field of administrative law and in the domain where the courts apply a supervisory jurisdiction over the acts of subordinate authority purporting to exercise statutory powers, to encourage the use of rigid legal classifications. The jurisdiction is inherently discretionary and the court is frequently in the presence of differences of degree which merge almost imperceptibly into differences of kind.

There was only one other argument for the respondents on their cross-appeal that I need notice. This was that the requirement not complied with was separable from the rest M of the requirements as to the certificate. I do not read it as such. It was an integral part of the requirement that the certificate should "include" a written notification of the rights of appeal.

Once the cross-appeal is disposed of, I do not find much difficulty in stating my reasons for allowing the appeal. In my view the Second Division only refused the second conclusion of the summons because in their view of article 3(2) of the order the respondents had no power to issue the new certificate demanded. Again, I do not so read the order. The duty under section 25 is a continuing duty. The fact that article 3(2)

A of the order is not complied with in time does not put an end to the obligation of the authority to comply. That this is so is apparent from a construction of section 25 (as amended) in the light of section 26, which expressly allows the parties to agree an extension of time, which would not be possible if an extension of time was *ultra vires* the authority.

In my view, therefore, the appeal succeeds, and the cross-appeal fails with the results indicated in the first paragraphs of this opinion. B

LORD FRASER OF TULLYBELTON and LORD KEITH OF KINKEL delivered concurring speeches and LORD WILBERFORCE and LORD RUSSELL OF KILLOWEN agreed.

DAMAGES

A **COURT OF APPEAL**

December 21 1979

(Before Lord Justice MEGAW, Lord Justice BROWNE and Lord Justice DONALDSON)

DODD PROPERTIES (KENT) LTD AND ANOTHER v CANTERBURY CITY COUNCIL AND OTHERS

Estates Gazette March 29 1980

(1979) 253 EG 1335

Assessment of damages—Damage caused to garage premises by works carried out in construction of adjoining multi-storey car park—Claims by owners and lessees of garage premises
B **against city council, main building contractors and subcontractors—Main issue date as at which damages should be assessed—Whether date at which cause of action arose or date when repairs could reasonably have begun—If later date correct in principle, should it be taken to be a date in 1970, as Cantley J held, or the date in 1978 when the plaintiffs' action was heard?—Held by Court of Appeal that the later date was correct—It was reasonable to defer repairs until the defendants' liability had been established—Judge wrong in considering that the Liesbosch decision compelled him to select the earlier date—Criticism of Philips v Ward—Appeal allowed—Large difference in amount of damages**

C This was an appeal and a cross appeal from a decision of Cantley J, reported at (1978) 248 EG 229, in an action brought by Dodd Properties (Kent) Ltd, the owners, and Marlowe Garage (Canterbury) Ltd, the lessees, of a building in Rose Lane, Canterbury, known as Marlowe Garage. The first defendants were Canterbury City Council; the second defendants, Truscon Ltd, were the main contractors; and the third defendants, Frankipile Ltd, were the subcontractors. The respondents to the appeal were the third defendants, who had undertaken to indemnify the other two defendants.

Roger Titheridge QC and Michael McMullan (instructed by Lewis & Dick) appeared on behalf of the appellants;
D Oliver Popplewell QC, Stephen Desch and Antony Edwards-Stuart (instructed by Ponsford & Devenish, Tivendale & Munday) represented the respondents (third defendants).

Giving judgment, MEGAW LJ said: This is an appeal from a judgment of Cantley J.

The first plaintiffs, Dodd Properties (Kent) Ltd, are the owners of a building in Rose Lane, Canterbury, known as Marlowe Garage. The second plaintiffs, Marlowe Garage (Canterbury) Ltd, have been the occupiers of Marlowe Garage as lessees of the first plaintiffs. They carry on their business there as motor-car dealers and they sell petrol, oil and car accessories.

E In 1968 the first defendants, the mayor, aldermen and citizens of the City of Canterbury, erected a large multi-storey car park close to Marlowe Garage. The second defendants, Truscon Ltd, were the main contractors, the third defendants, Frankipile Ltd, were their subcontractors for the foundations of the car park. As a result of their operations, damage was caused to the plaintiffs' building. Liability was for long denied, but shortly before the action came on for hearing before Cantley J in 1978 liability was admitted in nuisance by the second and third defendants, though the extent of the damage was in issue, and also the basis of F
assessment of the amount of the damages to which the plaintiffs were entitled. The first defendants did not formally admit liability. But they took no part in the proceedings, having received an undertaking of indemnity from the other defendants.

The judge held that the first defendants also were liable. They are not parties to the appeal. There is no dispute as to liability. The issues are as to damages. No question of fact is now in dispute, the judge's findings of fact are accepted as to the extent of the physical damage and as to other matters.

On the question of the extent of the damage, the judge G
to a large degree accepted the evidence of the defendants' experts. On their evidence, the necessary repairs would, at the prices prevailing at the time of the hearing in 1978, cost about £30,000. On the evidence of the plaintiffs' expert, the repairs required were much greater and the cost much higher. The question which remained, and which is the primary issue before us, is this: by reference to which of two dates is the cost of the repairs to be ascertained, for purposes of arriving at the amount of the defendants' liability for their tort? The plaintiffs say that the relevant date for this purpose is the date of the hearing, or of the judgment: that is, that the 1978 prices are relevant and decisive. The defendants say H
that the relevant date is 1970 and the relevant prices are the 1970 prices. As a result of inflation, the difference between the computations at those respective dates is very large. The 1978 figure, for the repairs which the judge held to be required, is £30,327. The 1970 figure, for the same work, is approximately £11,375.

The second plaintiffs also have a claim. It gives rise to the same issue as to the proper date of assessment. The second plaintiffs' claim arises out of prospective interruption of their business during the time that would be required for the carrying out of the appropriate repairs, if and when that work is done. The figure, if the repairs were to be J
carried out in 1978, would be £11,951. In 1970 the corresponding amount would have been £4,108.

Taking the first and second plaintiffs' potential entitlements together, the sums payable by the defendants as damages (apart from any question of interest) would be: on the 1970 assessment, £15,483; on the 1978 assessment, £42,278.

Cantley J held that in law, in the circumstances, judgment had to be given on the 1970 basis. He also awarded interest, making the total payable by the defendants to the first and second plaintiffs £22,974.20. Against that judgment the plaintiffs appeal and the defendants cross-appeal. The plaintiffs K
say that the judge was wrong in law to make his assessment of damages on the basis of the cost of the repairs in 1970. They say that he should have taken the 1978 computation. They say, in the alternative, that, if they should be wrong on this, which is their first and main contention, then the judge ought to have awarded interest from an earlier date and at a higher rate. They accept that, if they are right on their first contention—that is, the acceptance of 1978 as the date by reference to which the cost of the repairs is to be assessed—then they could not claim interest.

A The defendants' cross-appeal raises an issue affecting the damages of the second plaintiffs only. The defendants say that, since the judge held that it was only "just about established" that it was probable that the repairs would in fact be carried out after his judgment, he ought not to have awarded to the second plaintiffs the full amount of the prospective loss to them arising from the interruption of their business which would be caused by those potential repairs. The judge, say the defendants, should have awarded the second plaintiffs only, say, 60 per cent of the total prospective loss by interruption, because the chance that the loss would in fact occur was no greater than a chance of that
B order.

On the first, and main, issue raised by the plaintiffs, it is necessary to see what the learned judge found were the reasons why the repairs for this damage to the Marlowe Garage, caused in 1968, had still not been carried out when the action was heard in 1978. Because I think it is important to see precisely what the judge held in this respect, I shall quote the judge's own words from the transcript of his judgment:

> I find that the first plaintiffs could probably have raised the money for repairs but this would have increased their annual losses and their financial stringency. As a commercial decision, judged
> C exclusively from the point of view of the immediate and short-term welfare of the companies, it was reasonable to postpone incurring the very considerable expense of these repairs while no harm was being done to the building by the delay in repairing it and while these three rich defendants with apparent if not genuine belief in the validity of their defences were firmly denying liability to make even a contribution.

The learned judge then referred to the well-known, much-discussed, case, *Liesbosch (Dredger) v SS Edison* [1933] AC 449. He said, at page 11-B: "In the case of destruction of a chattel, the normal measure of the damage is the market value at the time of the loss. That was the measure of
D damage applied in *Liesbosch*." The learned judge then cited from the judgment of Denning LJ in *Philips v Ward* [1956] 1 WLR 471 at p 474: "The general principle of English law is that damages must be assessed as at the date when the damage occurs, which is usually the same day as the cause of action arises. . . . A fall thereafter in the value of money does not in law affect the figure, for the simple reason that sterling is taken to be constant in value". Although this may not affect the statement of "the general principle," I think that the reasoning as to sterling having to be taken to be constant in value is unfortunately no longer good law, having regard to the facts of life and the
E recent authoritative decisions, including *Miliangos v George Frank (Textiles) Ltd* [1976] AC 443.

Cantley J then said:

> No authority has been cited to me and in my very limited opportunity lately I have discovered none for myself where a court has considered the time at which damages are to be assessed in the cases of buildings damaged and put in need of repair by a tortious act. If there is no authority on that precise point, it may be because no-one has ever before thought to contend that the general principle did not apply to it. The general principle is that damages must be assessed as at the date when the damage occurs. In my view, that general principle applies here. It is not,
> F of course, to be rigidly applied as a rule of thumb, fixing the time rigidly by the calendar and the clock. The damage may be concealed by some fault of the wrongdoer or not reasonably discoverable by the victim until some time after it has first appeared, see eg *East Ham Corporation v Bernard Sunley & Sons Ltd* [1966] AC 406 and *Applegate v Moss; Archer v Moss* [1971] 1 QB 406.
>
> Moreover, repairs cannot usually be put in hand at once and at prices ruling at the very date of damage. There may have to be inspections and specifications and tenders and an available contractor may have to be found before the work can be started.
>
> Furthermore, the nature and circumstances of the damage may be such that it would be imprudent and possibly wasteful to begin the work before waiting longer to ensure that no further damage
> is going to develop from the same cause. This is particularly true G
> when the foundations of a building have been disturbed by vibrations.
>
> I would put it in this way. The appropriate damages are the cost of repairs at the time when it was reasonable to begin repairs. Whether the time is reasonable must be judged objectively and not taking into account such matters as impecuniosity or financial stringency which, in the words of Lord Wright in the *Liesbosch* case, are extrinsic.

The learned judge then held that it was reasonable for the plaintiffs not to begin repairs until 1970 even though the damage had all occurred, and was known, in 1968. On that basis he adopted "as the measure of damage . . . the cost H
of repairs on the prices ruling in 1970"; that is, £11,375. There is no dispute in this case but that the appropriate measure of damages on this claim of the first plaintiffs is by reference to the cost of the repairs required. The defendants do not challenge the judge's acceptance of the 1970 figures. That means that they do not now contend that the judge should have taken the lower prices for the repair work prevailing in 1968 when the tort was committed.

It is important to bear in mind that we are not concerned with any suggestion that the plaintiffs were under a duty towards the defendants to repair the premises damaged by the defendants' wrongdoing. The plaintiffs did not lose their J
right to recover damages from the defendants because they did not effect the repairs. True, in certain circumstances with which we are not concerned here, such as the building being destroyed by fire before the repairs were carried out, the amount of the plaintiffs' entitlement to damages might have become nil. But what we are concerned with here is: by reference to what date is the amount of the recoverable loss to be calculated, during a period when the cost of the necessary work is rising as time goes on? Since the defendants do not suggest that the judge was wrong in taking the 1970 prices instead of the 1968 prices, it is accepted, and I think necessarily and rightly accepted, by the defendants K
that there are circumstances in which the proper amount of damages, where, as here, the damages are to be computed by reference to the cost of repairs, have to be computed by reference to that cost at a date later than the date of the wrongdoing which caused the damage.

The general principle, referred to in many authorities, has recently been recognised by Lord Wilberforce in *Miliangos v George Frank (Textiles) Ltd* [1976] AC 443 at p 468E: namely, that "as a general rule in English law damages for tort or for breach of contract are assessed as at the date of the breach". But in the very passage in which this "general rule" is there stated, it is stressed that it is not L
a universal rule. That it is subject to many exceptions and qualifications is clear. Cantley J in the present case rightly recognised that that was so, in the passage from his judgment which I have recently read. Indeed, where, as in the present case, there is serious structural damage to a building, it would be patently absurd, and contrary to the general principle on which damages fall to be assessed, that a plaintiff, in a time of rising prices, should be limited to recovery on the basis of the prices of repair at the time of the wrongdoing, on the facts here, being two years, at least, before the time when, acting with all reasonable speed, he could first have been able to put the repairs in hand. Once that M
is accepted, as it must be, little of practical reality remains in postulating that, in a tort such as this, the "general rule" is applicable. The damages are not required by English law to be assessed as at the date of breach.

The true rule is that, where there is a material difference between the cost of repair at the date of the wrongful act and the cost of repair when the repairs can, having regard to all the relevant circumstances, first reasonably be undertaken, it is the latter time by reference to which the cost of repairs is to be taken in assessing the damages. That rule conforms with the broad and fundamental principle as to

A damages, as stated in Lord Blackburn's speech in *Livingstone v Rawyards Coal Co* (1880) 5 App Cas 25, at p 39, where he said that the measure of damages is "that sum of money which will put the party who has been injured, or who has suffered, in the same position as he would have been in if he had not sustained the wrong for which he is now getting his compensation or reparation."

In any case of doubt, it is desirable that the judge, having decided provisionally as to the amount of damages, should, before finally deciding, consider whether the amount conforms with the requirement of Lord Blackburn's fundamental principle. If it appears not to conform, the judge should B examine the question again to see whether the particular case falls within one of the exceptions of which Lord Blackburn gave examples, or whether he is obliged by some binding authority to arrive at a result which is inconsistent with the fundamental principle. I propose to carry out that exercise later in this judgment.

The judge has held, in a passage which I have already read, that as a commercial decision, judged exclusively from the plaintiffs' point of view, it was reasonable to postpone incurring expense of the repairs up to—for so I understand what the judge says—the time when the action had been heard and liability decided, resulting in a judgment which, C when complied with, would have put the plaintiffs in funds. The reasons why that deferment of repairs was reasonable from the plaintiffs' point of view included the fact, not that they were "impecunious," meaning poverty-stricken or unable to raise the necessary money, but that the provision of the money for repairs would have involved for them a measure of "financial stringency." Other reasons, consistent with commercial good sense, why the repairs should be deferred include those mentioned in evidence by a director of the plaintiff companies, whose evidence was accepted by the judge as truthful and reliable. If there had been no money problem, he said, he would still not have spent money on D the building before he was sure of recovering the cost from the defendants. It would not have made commercial sense to spend this money on a property which would not produce corresponding additional income. So long as there was a dispute, either as to liability or amount of compensation, he would have done no more than to keep the building weatherproof and "in working order."

If that was, as the judge held, reasonable from the point of view of the plaintiffs as being grounds for deferring the carrying out of repairs, and if the time at which the cost of the repairs falls to be completed in order to ascertain E the amount of damages is the time when it has become reasonable to do the repairs, why did the judge reject 1978, for which the plaintiffs contended, and accept 1970 for which the defendants contended?

There are, as I see it, two possible answers to that question. The first answer is that what is reasonable has to be looked at from the point of view of both parties and a balance struck. The judge's findings of reasonableness of the deferment from the point of view of the plaintiffs does not, therefore, conclude the matter. But I do not think that that was the answer intended to be given by Cantley J. He nowhere refers to the question in any such form and there F is no indication of any attempt by him to strike a balance. If a balance had to be struck, surely it would be right, even in a climate of indulgence to contract-breakers or tortfeasors, that the scales should move heavily in the favour of the innocent party as against the wrongdoer in any comparison of respective disadvantages or unfairnesses? It has to be borne in mind that these were defendants who were wrongly maintaining a denial of any liability and thereby leaving the plaintiffs faced with all the potentially heavy expenditure of money required for the mere purpose of establishing by litigation what we now know to have been their rights. Moreover, as the plaintiffs concede, they could not claim interest on the amount of their compensation starting to G run before the date when the money was expended on repairs. So the defendants, being liable, as we now know, to recompense the plaintiffs for the tort which the defendants committed in 1968, will have enjoyed the free use for their own account of the money which would have been the appropriate compensation at the date, with the opportunity of earning compound interest thereon, from 1968 until the date of judgment. If that were the ground on which the judge held in favour of the defendants on this issue, I would respectfully hold that it was a wrong ground. But I do not think that the judge did so hold.

The second possible answer is that which I believe to have H influenced the learned judge. He thought that the decision in *Liesbosch* [1933] AC 449 precluded him from taking into account, in considering the reasonableness of the deferment of repairs, any part of the deferment which was caused by "financial stringency."

Liesbosch has been the subject of much debate and much speculation, and a considerable measure of disagreement, as to its *ratio decidendi* and the scope of its application, particularly in the light of later House of Lords decisions. See, for example, the discussion of the case by the learned author of the article on "Damages" in *Halsbury's Laws of England* (4th ed) vol 12, para 1144, footnote 4. (I J agree with the analysis of *Liesbosch* and the comments thereon in the judgment which Donaldson LJ will deliver hereafter.) I do not think that, on any fair view of the *ratio decidendi* of *Liesbosch*, it applies to the issue with which we are concerned. Amongst other reasons, there are these two. First, it was not "financial stringency," let alone "impecuniousness" as in *Liesbosch*, which on any fair view, on the judge's findings, was the cause, or even, I think, an effective cause, of the decision to postpone repairs. The "financial stringency" which would have been created by carrying out the repairs was merely one factor among a number of factors which together produced the result that K commercial good sense pointed towards deferment of the repairs. The second reason which I would mention is that, once it is accepted that the plaintiff was not in any breach of any duty owed by him to the defendants in failing to carry out repairs earlier than the time when it was reasonable for the repairs to be put in hand, this becomes, for all practical purposes, if not in theory, equated with a plaintiff's ordinary duty to mitigate his damages. Lord Wright in his speech in *Liesbosch* [1933] AC 449 at p 461 accepted Lord Collins's dictum in *Clippens Oil Co Ltd v Edinburgh & District Water Trustees* [1907] AC 291 at p 303: ". . . in L my opinion the wrongdoer must take his victim *talem qualem*, and if the position of the latter is aggravated because he is without the means of mitigating it, so much the worse for the wrongdoer. . . ." I agree with the observations of Oliver J in *Radford v de Froberville* [1977] 1 WLR 1262, at p 1268, as to the relationship between the duty to mitigate and the measure, or amount, of damages in relation to a question such as the question with which we are here concerned. A plaintiff who is under a duty to mitigate is not obliged, in order to reduce the damages, to do that which he cannot afford to do: particularly where, as here, the plaintiffs' "financial stringency," so far as it was relevant at M all, arose, as a matter of common sense, if not as a matter of law, solely as a consequence of the defendants' wrongdoing.

My provisional answer to the question raised in the first issue would, thus, be that the damages in this case are to be assessed by reference to the 1978 cost of repairs. I now carry out that exercise which I mentioned earlier. Once it is accepted, as it is accepted by the parties, that the damages fall to be computed by the cost of repairs to the building, and once *Liesbosch* and *Philips v Ward* are out of the reckoning, there is no exception of which I am aware which is relevant here to exclude the application of Lord Blackburn's fundamental principle. On the relevant facts as found by the judge, the 1978 cost of the repairs gives the answer which accords with that principle. The calculation of damages by reference to the 1970 cost of repairs would not so accord.

On that issue, I would allow the appeal.

The result is that the plaintiffs' alternative ground of appeal, as to the appropriate calculation of interest, does not arise. For it is a necessary part of their submission on the first issue that, damages being referable to the deferment of repairs, interest is not payable up to the date of the hearing. In the circumstances, I think it better to say nothing, on that point, on which the argument on either side was commendably brief.

If I am right on my conclusion on the main issue in the appeal, I think that the cross-appeal, whatever its merits in law might otherwise have been, ceases to have validity. Early in this judgment, I summarised the issue raised by the cross-appeal. I need not repeat it. I find difficulty in reconciling two passages in the judgment as to the probability of repairs being carried out. The learned judge, referring to the evidence of Mr Smith, a director of the plaintiff companies, said: "However, Mr Smith, who I think was being careful to say no more than the truth and no more than he knew, said that in all probability the repairs will be done." The acceptance of the truthfulness and the reliability of that evidence by the man who was in the best position to give the best evidence on that question would appear to be conclusive. How then can the judge's acceptance of it be reconciled with what he says a few lines later: "Having regard to what Mr Smith said, I think the probability . . . is just about established"? If Mr Smith was truthful and reliable in saying "in all probability," that results in much more than probability being "just about established."

However, fortunately, it is not necessary to resolve that difficulty, since the learned judge went on to say: "although I would be more confident of the extent of the second plaintiffs' loss if the first plaintiffs were recovering the present-day costs of the repairs, so that they could, with a light heart, carry out the full repairs." As in my judgment the plaintiffs are entitled to recover the 1978 cost of the repairs, this court should not remit this Methuselah of an action for a further hearing by the judge on that issue. We should make our own assessment what the discount, if any, should be. In my opinion, the discount, if the law requires any discount—I do not find it necessary to decide this—would be *de minimis*. As the law requires us to disregard the trivial, I propose to disregard it, as I think Cantley J would have disregarded it, if he had reached the same conclusion as I have reached on the main issue.

So I would allow the appeal, and direct that judgment be entered for the first plaintiffs for £30,327, without interest up to the date of Cantley J's judgment, and for the second plaintiffs for £11,951, also without interest up to that date. I would dismiss the cross-appeal.

Agreeing, BROWNE LJ said: I agree that this appeal should be allowed and the cross-appeal dismissed, for the reasons given by Megaw LJ and the reasons which will be given by Donaldson LJ in the judgment he will deliver very soon. I can summarise my own reasons fairly shortly, because they are in substance the same as theirs.

The first principle for the assessment of damages is that the injured person should, so far as money can do it, be put in the same position as if the wrong—in this case the tort—had not been committed against him: (see *Halsbury's Laws* (4th ed) vol 12, title "Damages" para 1129, and, for example, the authority cited by Megaw LJ, *Livingstone v Rawyards Coal Co* (1880) 5 App Cas 25, Lord Blackburn at p 39). This the damages of £11,375 awarded to the first plaintiffs, for the cost of repairs in 1970, glaringly fail to do. By the time of the hearing in 1978 the cost had risen to £30,327. In fact, the repairs had not been done by that time, and the cost will probably have risen still further by the time they are done, but the plaintiffs do not make any further claim beyond the cost at the date of the hearing.

It is not disputed that in this case the measure of the first plaintiffs' damages is the cost of repair, as opposed to the other possible measure in a case of this sort, ie the diminution in the value of the building. The only question is the time as at which that cost shall be taken.

The general rule, both in contract and tort, is that damages should be assessed as at the date when the cause of action arises, but they may be assessed as at some later date. In my view, Cantley J was plainly right in saying "The appropriate damages are the cost of repairs at the time when it was reasonable to begin repairs." In *Johnson v Agnew* [1979] 2 WLR 487 Lord Wilberforce said (at p 499 E-H):

> The general principle for the assessment of damages is compensatory, ie that the innocent party is to be placed, so far as money can do so, in the same position as if the contract had been performed. Where the contract is one of sale, this principle normally leads to assessment of damages as at the date of the breach—a principle recognised and embodied in section 51 of the Sale of Goods Act 1893. But this is not an absolute rule: if to follow it would give rise to injustice, the court has power to fix such other date as may be appropriate in the circumstances. In cases where a breach of a contract for sale has occurred, and the innocent party reasonably continues to try to have the contract completed, it would to me appear more logical and just rather than tie him to the date of the original breach, to assess damages as at the date when (otherwise than by his default) the contract is lost. Support for this approach is to be found in the cases. In *Ogle v Earl Vane* the date was fixed by reference to the time when the innocent party, acting reasonably, went into the market; in *Hickman v Haynes* at a reasonable time after the last request of the defendants (buyers) to withhold delivery. In *Radford v de Froberville*, where the defendant had covenanted to build a wall, damages were held measurable as at the date of the hearing rather than at the date of the defendant's breach, unless the plaintiff ought reasonably to have mitigated the breach at an earlier date.

Lord Wilberforce, of course, was there speaking of damages for breach of contract, but I have no doubt that the same principle applies to this case, where it is common ground that the measure of damages is the cost of repairs. I think this view is supported by analogy by the decision of the House of Lords in *Birmingham Corporation v West Midland Baptist (Trust) Association* [1970] AC 874.

In this case, it was common ground, and the judge accepted, that it was reasonable to postpone the doing of the repairs from 1968, when damage was first discovered, until 1970, and that 1970 was the earliest date as at which the cost of repairs should be assessed. The defendants/respondents contended that the assessment should not be any further postponed; Cantley J accepted this contention, and assessed the damages on the cost of repairs in 1970.

In the course of the passage of the judgment which Megaw LJ has already read, and I will not read it in full again, the judge held that "As a commercial decision, judged exclusively from the point of view of the immediate and short-term welfare of the companies, it was reasonable to postpone incurring the very considerable expense of these repairs . . . " Like Megaw LJ, I understand this to mean that it was in this sense reasonable to postpone doing the repairs until after the hearing. This finding was based on the evidence of Mr Smith, a director of both the plaintiff companies and a chartered accountant, which is set out in the judgment and has been summarised already by Megaw LJ. He gave a number of reasons for the decision. Only one of what he said were the relevant factors was financial, and I think that the judge's finding on this point falls far short of "impecuniosity" or "financial embarrassment" in the *Liesbosch* sense.

The judge said: "Whether the time is reasonable must be

A judged objectively and not taking into account such matters as impecuniosity or financial stringency which, in the words of Lord Wright in the *Liesbosch* case, are extrinsic." I am afraid I do not clearly understand what the judge meant by "objectively" in that sentence. If he meant that the decision to postpone, although reasonable from the point of view of the plaintiff companies, was not reasonable from the point of view of a hypothetical reasonable commercial man, I cannot agree; it seems to me that any commercial man in the circumstances with which Mr Smith was faced could reasonably, and probably would, have come to the same decision.

The judge relied on *Philips v Ward* [1956] 1 WLR 471 B and *Clark v Wood* [1965] 2 All ER 353, in which Lawton J (as he then was) simply followed and applied *Philips v Ward*. I agree with Megaw LJ that the reasoning of the Master of the Rolls (Lord Denning) in *Philips* at p 474 can no longer be regarded as good law.

That leaves only *Liesbosch* [1933] AC 449. I do not propose to analyse that difficult case, because I entirely agree with Megaw LJ and Donaldson LJ that, for the reasons they give, it did not compel Cantley J to take the 1970 cost of repairs. I will only say that, like Megaw LJ, I agree with the observations of Oliver J in *Radford v de Froberville* [1977] 1 WLR 1262 at p 1272 as to the relationship between C the duty to mitigate and the measure of damages in a case such as this.

I would, therefore, allow the first plaintiffs' appeal and vary the judgment by substituting £30,327 for £11,375. I think it necessarily follows that the appeal of the second plaintiffs should also be allowed, and their damages increased from £4,108 to £11,951. The plaintiffs' alternative ground of appeal as to interest therefore does not arise, and, like Megaw LJ, I think it better to say nothing on that point.

I agree that the cross-appeal should be dismissed, for the reasons given by Megaw LJ.

D Also agreeing, DONALDSON LJ said: The general object underlying the rules for the assessment of damages is, so far as is possible by means of a monetary award, to place the plaintiff in the position which he would have occupied if he had not suffered the wrong complained of, be that wrong a tort or a breach of contract. In the case of a tort causing damage to real property, this object is achieved by the application of one or other of two quite different measures of damage, or, occasionally, a combination of the two. The first is to take the capital value of the property in an undamaged state and to compare it with its value in a damaged state. The second is to take the cost of repair or E reinstatement. Which is appropriate will depend upon a number of factors, such as the plaintiffs' future intentions as to the use of the property and the reasonableness of those intentions. If he reasonably intends to sell the property in its damaged state, clearly the diminution in capital value is the true measure of damage. If he reasonably intends to continue to occupy it and to repair the damage, clearly the cost of repairs is the true measure. And there may be in-between situations. Happily there is no issue in the present case as to which measure of damage falls to be applied. It is the cost of reinstatement. The primary issue is as to how and, more particularly, on what date those costs are to be F assessed. This is a very significant issue in the light of the increase in costs over the period between the occurrence of the damage in 1968 and the trial in 1978.

Mr Popplewell, for the respondents, submits, and I for my part would readily accept, that the general rule is that damages fall to be assessed as at the date when the cause of action arose. The rule is so stated by Lord Wilberforce in *Miliangos v George Frank (Textiles) Ltd* [1976] AC 443 at p 468. And I am inclined to think that in normal circumstances this would be applicable where the relevant measure of damage was diminution in the capital value of the property. But it is only a general or basic rule and is subject G to many exceptions. Thus damages for personal injury, excluding consequential loss to which other principles apply, are assessed in the light of the value of money at the date of the hearing. The issue here is whether the assessment of damages based upon the cost of repair or reinstatement is another exception as Mr Titheridge for the appellants contends. I think that it is.

In *Birmingham Corporation v West Midland Baptist (Trust) Association* [1970] AC 874 the House of Lords was faced with the problem of whether the reasonable cost of equivalent reinstatement, which was the basis of compensation under the Land Compensation Act 1961, involved taking costs which H prevailed at the date of the notice to treat or those which prevailed at the earliest date when the claimants might reasonably have begun rebuilding. The decision was in favour of the latter. Lord Reid, at p 894, cites with approval various statements on the measure of compensation which assume that the assessed cost of reinstatement would be sufficient to enable the owner to undertake the work if he acted reasonably. In an era of rising costs, this could only happen if compensation was assessed on the basis of costs applicable at the time at which reinstatement would in fact occur, on the assumption that the owner acted reasonably. Again Lord Morris of Borth-y-Gest at p 903 said that "The reasonable J cost, depending upon the facts of particular cases, will be the actual reasonable cost which a claimant has incurred or can be expected to incur; it will be such cost at the time when equivalent reinstatement reasonably does or should take place."

While this is not a decision on the measure of damages in tort, I think that the reasoning is directly applicable to the present problem. It is also only commonsense, for, as Mr Titheridge pointed out, it would be wholly unfair to the defendant to charge him with the costs applicable to reinstatement in 1968 when the damage occurred, if the actual reinstatement took place at a later date when improved K technology had reduced the cost. That this happy situation has not in fact arisen does not affect the point of principle.

In the absence of special and extraneous factors, there is no divergence between the interest of a plaintiff and a defendant on the choice of the most propitious moment at which to effect reinstatement. Both wish to achieve the maximum economy, at least so long as the plaintiff is in doubt whether he will be entitled to a full indemnity from the defendant. It follows that in a case in which a plaintiff has reinstated his property before the hearing, the costs prevailing at the date of that operation which were reasonably incurred by him are *prima facie* those which are relevant. Equally in a L case in which a plaintiff has *not* effected reinstatement by the time of the hearing, there is a *prima facie* presumption that the costs then prevailing are those which should be adopted in ascertaining the cost of reinstatement. There may indeed be cases in which the court has to estimate costs at some future time as being the reasonable time at which to reinstate, but that is not this case. This is, however, only a *prima facie* approach. It may appear on the evidence that the plaintiff, acting reasonably, should have undertaken the reinstatement at some date earlier than that in fact adopted or, as the case may be, earlier than the hearing. If so, the relevant costs are those ruling at that earlier date. Whether M this is regarded as arising out of the primary measure of damage, ie that the relevant time is when the property should have been reinstated or whether it is regarded as being a reflection of a plaintiff's duty to mitigate his loss, may not matter.

In the present case Cantley J accepted that the relevant date was when it was reasonable to begin repairs. However, in deciding what was reasonable, he considered himself bound by the decision of the House of Lords in the *Liesbosch* case to disregard such factors as impecuniosity or financial

A stringency experienced by the plaintiffs. Accordingly, although he considered that the plaintiffs had acted reasonably in deferring reinstatement until after the hearing, he felt constrained to adopt September 1970 costs, the delay until then being justified exclusively on other grounds, namely, the need to make sure that no further damage would occur before repairs were started. Dealing with the latter delay he said:

> I find that the first plaintiffs could probably have raised the money for repairs but this would have increased their annual losses and their financial stringency. As a commercial decision, judged exclusively from the point of view of the immediate and B short-term welfare of the companies, it was reasonable to postpone incurring the very considerable expense of these repairs while no harm was being done to the building by the delay in repairing it and while these three rich defendants with apparent if not genuine belief in the validity of their defences were firmly denying liability to make even a contribution.

Whatever the difficulties inherent in the *Liesbosch* decision —and it is not at once apparent why a tortfeasor must take his victim as he finds him in terms of exceptionally high or low profit-earning capacity, but not in terms of pecuniosity or impecuniosity which may be their manifestations—it binds this court as much as it bound Cantley J unless and until it is C reviewed by the House of Lords. However, it is important to see precisely what it did decide.

The *Edison* fouled the *Liesbosch's* moorings, carried her out to sea and sank her. The ordinary measure of damage was the cost of buying another similar vessel, the cost of getting her to the *Liesbosch's* old moorings and any loss of profit consequent upon the disruption of commercial operations while the substitute vessel was being obtained and delivered. However, the plaintiffs contended for a different and special measure of damage. Substitute dredgers were available for purchase but the plaintiffs could not afford to buy them. Instead they hired another dredger, the *Adria*, which D was larger than the *Liesbosch*, more expensive to operate and for which they had to pay a very high rate of hire. Eventually the port authority, for whom the plaintiffs were working, bought the *Adria* and resold her to the plaintiffs under a credit sale contract. The plaintiffs claimed the cost of hiring the *Adria* until the port authority bought and resold her to them, the cost of purchasing the *Adria* and the excess cost of working her as compared with the *Liesbosch* together with unrecovered overhead charges and lost profit while they were without any dredger. The ordinary measure of damage is E based upon market rates. The measure of damage claimed by the plaintiffs was quite different, namely, one based upon their actual loss and expenditure.

As I understand Lord Wright's speech, he took the view that in so far as the plaintiffs in fact suffered more than the loss assessed on a market basis, the excess loss flowed directly from their lack of means and not from the tortious act, or alternatively it was too remote in law. In modern terms, I think that he would have said that it was not foreseeable. The position of the plaintiffs in the present case seems to me to be quite different. They were not impecunious in the *Liesbosch* sense of one who could not go out into the F market. On the contrary, they were financially able to carry out the work of reinstatement in 1970. However, on the learned judge's findings, they were commercially prudent in not incurring the cash flow deficiency which would have resulted from their undertaking the work in the autumn of 1970 and waiting for reimbursement until after the hearing, particularly when the defendants were denying liability and there was a dispute as to what works could and should be done by way of reinstatement. In my judgment, the decision in the *Liesbosch* has no application to such a situation which is distinguishable.

If the decision whether to adopt 1970 or 1978 costs turns G upon whether, bearing in mind the likelihood that prices would rise, the plaintiffs should have undertaken the work in 1970 in pursuance of their duty to mitigate their damage, there is another ground for distinguishing the *Liesbosch* and for taking full account of the plaintiffs' financial position. This is that Lord Wright's explanation of the decision in *Clippens Oil Co Ltd v Edinburgh and District Water Trustees* [1907] AC 291, where Lord Collins at p 303 said that the tortfeasor must take his victim as he found him, including any lack of means, was that that decision represented the rule in relation to the duty to minimise damage. H

I would therefore allow the appeal by the plaintiffs and, for the reasons stated by my Lord, Lord Justice Megaw, would dismiss the defendants' cross-appeal.

The order was made as stated at the end of Megaw LJ's judgment. The appellants (plaintiffs) were awarded the costs below and the costs of the appeal and cross-appeal. Leave was refused to the respondents to appeal to the House of Lords.

FEES AND COMMISSIONS

QUEEN'S BENCH DIVISION

(Before Mr Justice MICHAEL DAVIES)

July 17 1979

MILTON MARLOWE & CO v SOUTHCUT LTD (TRADING AS STAR JEANS)

Estates Gazette April 26 1980

(1979) 254 EG 293

Estate agents' action for commission for introducing leasehold shop property to defendant company—Claim by plaintiffs that, the property not then being on the market and the plaintiffs not being retained by the owners, the defendants agreed to pay plaintiffs RICS scale fees in the event of the property introduced being acquired by the defendants—Property subsequently acquired by defendants—Sharp conflict of evidence as to whether there was a specific agreement to pay scale fees if plaintiffs' introduction resulted in the acquisition of the interest—Plaintiffs' account accepted by judge and judgment given in their favour for amount claimed

In this action Milton Marlowe & Co, a firm of estate agents of New Bond Street, London W1, sued Southcut Ltd, a company which traded as "Star Jeans," for a sum of £2,943 which the plaintiffs claimed was due to them as commission for their introduction to the defendants of a prime leasehold shop property at 20 New George Street, Plymouth, which the defendants later acquired.

Michael J Burton (instructed by Laurence Brass & Co) appeared on behalf of the plaintiffs; Barry Payton (instructed by Austin Ryder & Co) represented the defendants.

Giving judgment, MICHAEL DAVIES J said: This is a very unfortunate dispute indeed. It is a dispute between two young men, one an estate agent, one a businessman, and both of whom, I have not the slightest doubt, exhibit in their work and business all the best qualities of energy and integrity which is characteristic of people like them.

There is not a great deal of money in dispute—under £3,000, which, of course, is not to anyone a trifle, but, in relation to the cost of litigation and the time-consuming nature of it, not a great sum. They have had to be here for the best part, or parts, of three days and, quite obviously, from having been on christian name terms at the time of the matters with which this case is concerned, they are now at daggers drawn. However, although I express regret at that situation, it is nothing to do with the court. The dispute having arisen, my task is to resolve it on the facts and with the assistance of counsel, and by applying the law as I understand it.

Mr Milton Marlowe is the principal in effect, although in law I suppose the dominant partner—the other partner being his wife—in the estate agents' firm of Milton Marlowe & Co. He is not qualified professionally, but he had a lengthy experience before he set up his own firm a little time in advance of the matters involved in this case. His experience had included a good deal of work with shop properties, and I have no doubt that he had developed by 1974-75 a considerable expertise in that field.

The defendants are, in fact, Southcut Ltd, but I do not think that name has been mentioned at all in the course of the case for this reason, that they trade as "Star Jeans," and they have been referred to throughout this case, and will be in this judgment, as "Star Jeans." Ian Michael Anthony Lyons (as he is truly named) is and was at all material times the general manager of Star Jeans. For entirely understandable and creditable reasons he was and is always known as Michael Star (that being a family name) and is referred to in the correspondence and evidence, and will be in this judgment, as "Mr Star."

Star Jeans dealt in casual clothing and, in particular, in the retailing of that ubiquitous uniform of the modern day, "jeans." Star Jeans, as Mr Star explained to me—and of course everybody knows that this is right—looked in the various places where they traded or wished to trade, and that was in the south west of England generally, for the best places consistent with the rent they were prepared to pay, and able to pay, having regard to the size of their company, that is to say, the best sites in the trading centres of the towns and cities concerned. I think they are known as "prime sites." As in every property deal, I suppose, in relation to retail shops, very often those prime sites if they come on the market are disposed of before they are advertised, publicised, circularised, or whatever. Of course there is nothing wrong with that.

Mr Marlowe, among his other activities, and having built up over the previous years a connection and, no doubt, a friendship with the property managers of various large concerns, kept his ear very close to the ground. As well as, no doubt, acting in the ordinary way as a retained agent for the buying and selling of properties (in particular, shops), if he heard of shop premises which had not come on the market but were likely to be available for disposal he would endeavour to place those by interesting potential purchasers or tenants. It is for that reason specifically that Mr Marlowe began to develop a relationship with Star Jeans and, in particular, Mr Star. He was anxious to foster that relationship because, as he said, he hoped that it might in the end result in his becoming, in effect, the sole retained agent for Star Jeans in the various buying and selling of shops and leases in which they might be concerned as they expanded or entered different towns or cities, or resited their shops more favourably by disposing of one and acquiring another.

A good deal of time has been spent in this case exploring the evidence and documents relating to transactions, or potential transactions, of that nature in which Mr Marlowe and Mr Star were involved. There are a number of documents relating both to the opportunity given to Mr Marlowe by Mr Star to sell properties—three or four of them—belonging to the defendants, and also in relation to the offer by Mr Marlowe to Star Jeans, through Mr Star, of possible properties. There are also a good many documents, particularly letters, where Mr Marlowe was suggesting properties to other companies, and the court spent some time investigating them because counsel appeared to desire it. I do not propose to dwell on those matters save in so far as they are material to the point in this case—which I have thought for a long time during the course of this case, and have said so, and I think counsel now agree, to be the only point of major importance and certainly the point on which the decision in this case turns—and that is as follows.

A The plaintiffs' claim is for a total sum of £2,943, including VAT, which tax they have already paid, I am told. The plaintiffs claim, as I say, the sum of £2,943 as commission to which they claim to be entitled from the defendants because they introduced to the defendants in November 1976 a prime shop property in Plymouth at 20 New George Street which, in fact, the defendants did buy. The actual contract was not signed until May 27 1977.

The issue is this, and it is quite simple and stark, although there is a hot contest as to the answer at which the court should arrive. Did the plaintiffs on November 22 1976, or thereabouts, introduce the defendants to that property and
B did the defendants agree to pay scale fees? The sum to which I have referred is calculated on the scale fees of the Royal Institution of Chartered Surveyors, which the pleader of the statement of claim, Mr Michael Burton, who appeared on behalf of the plaintiffs, like so many people, wrongly describes as the "Royal Institute" in the statement of claim. I do not hold that against the plaintiffs.

I put the issue in that way because Mr Barry Payton, on behalf of the defendants, says with force that certainly at some stage in the events which followed on November 22, when the dispute arose, probably both parties—certainly his client Mr Star, and probably the plaintiffs as well—approached
C the case on the basis of (a) whether there was an introduction and (b) whether that introduction led to the eventual purchase. They overlooked in their minds the necessity of there being an agreement to pay the scale fees. Mr Burton now acknowledges that there must be an agreement—either a general agreement in relation to any property introduced or a specific agreement in relation to this particular property—before the plaintiffs are entitled to recover their fees.

In addition to that which is pursued now, which is (a) "We, the defendants, already knew of this property before November 22 and the plaintiffs did not introduce it to us," and (b) "We never agreed to pay fees," various other defences
D are pleaded or foreshadowed, in what I am bound to describe as the very verbose and unconventional defence which was settled in this case by counsel other than Mr Barry Payton, and which has, in one or two particulars to which I shall have to refer very briefly later, somewhat recoiled on the defendants' head, as Mr Burton pointed out. In fact, and I say this rather putting the cart before the horse, those other defences, such as that of causation, now have disappeared from the case. Without making any formal abandonment, Mr Payton realistically concedes that if the plaintiffs introduced this property, and if the defendants agreed to pay fees, then the fact that subsequently the defendants did not use the
E plaintiffs to negotiate for the property and went on their own, as it were, to deal with the vendors is neither here nor there. The issues are those which I have stated.

I have said there was a hot contest about this, and there is. It is necessary to summarise what the two protagonists say. First Mr Marlowe, who says that he knew that Mr Star was very keen on getting a shop on a prime site in Plymouth. Star Jeans were already trading in Plymouth—I think in Royal Parade—but, says Mr Marlowe, they wanted to get, if possible, into New George Street, which is the best shopping street. He gathered that information from what he says were the frequent telephone conversations he was having at this
F time with Mr Star in connection with the various other matters and properties which I have mentioned, and of which I do not intend to go into the details.

As a result (as they say) of information received, Mr Marlowe found out that a shop belonging to the well-known company of clothes retailers, Dorothy Perkins Ltd, at 20 New George Street in Plymouth, was likely to, but had not yet, come on the market. He telephoned, did Mr Marlowe, a Mr Burton, who was the property manager of Dorothy Perkins Ltd and whom he knew as a result of his work with other surveyors in earlier years. After the conversation with Mr Burton he was referred to a Mr Fenton, who was a representative of the firm of estate agents who were retained by Dorothy G
Perkins Ltd, possibly generally, but certainly in connection with this property, 20 New George Street. So Mr Marlowe telephoned Mr Fenton.

Mr Fenton was not called as a witness. I think he is, in fact, at the moment in Canada, but certainly abroad on holiday. But in the agreed bundle of documents there is what is accepted to be a note made by Mr Fenton on November 22, which is the day Mr Marlowe telephoned him. There was, according to Mr Marlowe, a second telephone call to Mr Fenton—which I shall mention in a moment—on that day, and there was some suggestion that the note on a later page may have been a note made by Mr Fenton of the second H
conversation. I think it is more probable that it was the first, but it does not matter very much. Be that as it may, the note in so far as it is relevant reads as follows: "M Marlowe Star Jeans Plymouth." Then Mr Marlowe's telephone number appears on Mr Fenton's note. "November 22. Around £10,000 quoted." That £10,000 is in fact the figure—which is supported by the documents—which Dorothy Perkins and Mr Fenton had discussed as being the figure which should be asked for the premium on the disposal of the lease.

The evidence of what Mr Fenton did is, of course, restricted to some extent by the rules against hearsay, and in so far as inadmissible evidence from both Mr Marlowe and Mr Star J
was given at various times (sometimes, I am bound to say, in effect at my invitation, because the head of steam in the witness was rising high from the layman's understandable refusal to forgive the lawyer's insistence on the exclusion of hearsay evidence) I ignore it in so far as it contains matters which in the case of any interview or telephone conversation might be in dispute. But I do not think that there is any dispute that Mr Fenton did give that information and he gave it to Mr Marlowe on the basis that Mr Marlowe was, in effect, going to pass it on to one client only and that was Star Jeans, whose name Mr Marlowe gave to Mr Fenton.

I repeat that the property had not yet come on the mar- K
ket and it is accepted to have been a very good site for the defendants to move into if they could acquire it. "And," says Mr Marlowe, "having done that I telephoned Mr Star, and I said I had come across an extremely good property in Plymouth, a prime position in New George Street. I said, 'It is not on the market'." Then he said, "Following my usual practice," (whether he used these words or not does not matter, they are my words but that was the effect of Mr Marlowe's evidence) "having dangled the bait of a very attractive property in front of the person to whom I was telephoning," (and in using those words I do not suggest that there was anything improper in this) "before giving him any L
details I said to him, 'It is not on the market. I have to look to you for payment of scale fees if there is a sale.' He said 'OK'." Now that is the specific agreement, or the basis of it, which the plaintiffs rely on. "And so," says Mr Marlowe, "I gave him the particulars I got from Mr Fenton. I told him the rent was £26,000 per annum and that they were looking for a premium of £10,000. That was correct information and Mr Star was very enthusiastic about the property. He wanted me to tell Mr Fenton he was very interested. He did not want them to sell to anyone else and he would come back to me. As a result of that I telephoned Fenton and he was pleased to hear that Star Jeans were interested." That M
was the second conversation I have mentioned between Mr Marlowe and Mr Fenton which took place on that day.

On the following day Mr Marlowe undoubtedly wrote a letter dated November 23 1976 to Mr Star. But before I refer to that letter let me, by comparison, put the version of Mr Star alongside that conversation according to Mr Marlowe. Mr Star says this, "Yes, we did have a telephone conversation on that day November 22." He said at one point in his evidence, "I agree, we did quite often speak on the telephone, although not as frequently as Mr Marlowe says, about various matters." Mr Star says this, "He" (Mr Marlowe) "said he

A had a property in Plymouth. I said I was already dealing in one which I thought was coming to fruition." Indeed, there was another property, at 19 New George Street, which, although next door in number to no 20, in fact was on the opposite and (in a metaphorical sense) less sunny side of the street from the retailing point of view.

Continuing with Mr Star's account of the conversation, he told the court this: "Mr Marlowe, although I told him I was not interested, said, 'You must listen to this, it is very good.' I was on the other line and very busy, but Mr Marlowe was very insistent on giving me the information. I was not really taking it in in detail because I was not really interested and B told him so. I just did not want to listen." Then he told him that it was no 19 which he was interested in, of which the rent was about £15,000 a year. "But," said Mr Star, "nevertheless, Mr Marlowe persisted. He said, 'It is a tremendous property, absolutely prime. The rent is £26,000 per annum.' I said, when he mentioned the rent, 'I know about that,' because this suggested to me the Dorothy Perkins unit, and I knew that there might be something coming on the market at a rent of £26,000." The source was a Mr Webster, who I think was another estate agent. Mr Webster was not called, but that is just a fact, and I do not draw any adverse inference from the absence of Mr Webster or Mr Fenton for or against C either side. There are often very good reasons why witnesses cannot be called. I have not had any assistance from Mr Webster and, again, anything Mr Star says Mr Webster told him is hearsay, but he says he had some sort of knowledge about this in an informal way.

Then Mr Star said to me, "I told Mr Marlowe that £26,000 per annum was way out of our league. It was ridiculous over-rentalisation." Those are Mr Star's exact words. "He went on talking. I think he may have mentioned the premium, but I do not recollect it clearly." That was Mr Star's account in summary. He added, "I didn't ask him to convey my interest to Fenton. I did not authorise him to name us to Mr Fenton D at any time, and I wasn't interested in Dorothy Perkins. I certainly didn't agree to pay any fee." There are the two sides of the case in regard to that conversation.

I have already mentioned the defence, and Mr Burton said —and in an ordinary case of this sort, whatever the subject matter, I agree with his attitude—that he would not normally take technical pleading points. Provided no party is prejudiced by some error in a pleading, technicalities are certainly not welcomed in this division of the High Court in regard to pleadings. "But," said Mr Burton, "in this case the pleading is more like a solicitor's letter than the conventional pleading." He points to paragraphs 11 and 12—particularly paragraph 12— E which has been amended, as have several paragraphs in the defence. The circumstances of this were that Mr Star did not see the original defence before it was served, he told me. When he did see it he pointed out various errors, and amendments took place.

But there still appear in paragraph 12 these words relating to the conversation on the telephone, which is admitted to have taken place:

> Accordingly the Defendants by Star expressed interest in the said property during the conversation with Marlowe on November 22 1976.

F That, of course, is completely contrary to Mr Star's sworn evidence which I have just recited. Then it goes on in its amended form:

> On or shortly after November 24 1976 the Defendants by Star informed the Plaintiffs by Marlowe orally that the premium of £10,000 and the rent of £26,000 per annum quoted by the Plaintiffs was too high and at that price the Defendants were not interested.

So there is the case being pleaded as though Mr Star had shown interest in the first telephone conversation, and had then, on or about the 24th—two days later—indicated that the rent and the premium were too high. Of course, what he now says is that he was not interested right from the very beginning G and rebuffed or attempted to rebuff Mr Marlowe.

Those are the two accounts and, of course, notwithstanding the various matters on the periphery of the case upon which they were cross-examined, to choose between those two accounts would be very difficult if there were no contemporaneous documents, or no contemporaneous admitted or proven behaviour which assisted the court. Fortunately there is such material to bear in mind. There is, first of all, the letter written by Mr Marlowe to Mr Star on November 23. It says this:

> Dear Michael,
> *20 New George Street, Plymouth* H
> I am just dropping you a line to confirm the details of the Dorothy Perkins unit that we discussed.

Then he describes in detail the premises and the nature of the lease and the rent, and so on. He goes on to say:

> Dorothy Perkins have not yet placed these premises "on the market," but they would consider doing a deal at a premium of £10,000.
>
> Finally, I would just confirm that this is an instance where we are not retained by the Assignors and in the event of a transaction taking place, we would be looking to you for payment of our scale fees. The appropriate scales would be. . .

and he sets them out. J

> . . . I look forward to hearing from you after you have been to Plymouth.

In the bottom right-hand corner appear some words which are of no importance whatever, but Mr Marlowe seems to put them on pretty well every letter he writes. I am very doubtful whether in that way they have the slightest legal effect—any more than putting "Merry Christmas" at the bottom of a business letter would—but he puts the words "Subject to Contract." I only mention them because they have not been mentioned in this case before. In case anybody should be interested in this judgment for any purpose other than the present case, they are nothing to do with the K case and are not relied on. But the letter is, of course.

There are several points about that letter which seem to me to be important. It is true that even if Mr Marlowe had been rebuffed on November 22, he is sufficiently zealous—to use one of his own words—and enthusiastic, and sufficiently the salesman, if he will allow me to say that, that he might still have written a follow-up letter giving particulars of the property, even if Mr Star had rebuffed him; so that the mere writing of the letter is not very significant. But, in my view, the terms of the letter are somewhat significant in some respects.

First of all, there are a number of other letters in the bundle L —I am not going to read any of them—in which both to Mr Star of Star Jeans and to other individuals and other companies, in relation to other properties (and indeed it will be seen very shortly in relation to this property), Mr Marlowe wrote similar letters giving particulars of a property, but he usually ended up his letters with some such words as this: "Finally we would mention that this is an instance where we are not retained by the Assignors and in the event of a transaction taking place we would be looking to you for payment of our scale fee." In other words, I do not think in any other letter are the words, "We" or "I confirm that in this instance. . . ." M

It seems to me that (although, of course, in the correct sense of the word no one can corroborate himself by writing a letter) the fact that he said that he *confirmed* that situation in relation to the fees does in my judgment lend some consistency and support to his account of the reference to the scale fees in the interview on November 22, which reference Mr Star denies ever took place. Then, as Mr Burton puts before me and submits, there are the words at the end, "I look forward to hearing from you after you have been to Plymouth." It seems to me that that seems reasonably consistent with Mr Marlowe thinking that Mr Star was interested.

Also there is the pleading point, saying at that stage that the defendants were interested.

There was no reply to that letter. That is not of vast significance because Mr Star (a) was a very busy man and (b) did have letters not only from Mr Marlowe but from other agents about properties. Probably he did not reply to many of them. But it is a fact—a small fact, maybe, but a fact—that there was no letter sent saying, "Deal Milton," or "Dear Mr Marlowe, Your letter of November 23—what on earth are you wasting time writing to me about this for? I told you on the telephone that I was not interested. We have got another property in view, and the rent was too high in any case."

Well, what happened after that? According to Mr Marlowe, "About a week later" (in Mr Marlowe's words in the witness box) "in the course of one of my conversations with Mr Star on the telephone, it became clear that the defendants' interest had evaporated. They weren't interested any more. He did not say why. He certainly didn't say it was because the terms were unacceptable."

Mr Star said, about that, that it was mentioned again about a week later, or some time in a conversation and he simply said "No" again. He said "I am not interested. No, I repeat 'no'." So he maintained his attitude of no interest.

That being so, said Mr Marlowe—and this part of the case I can deal with quite shortly—he wrote to other possible purchasers about this property. The first letter, addressed to a Mr Holt of K Shoe Shops Ltd, is dated December 1 1976. As it turns out, and as I think should have been obvious to Mr Marlowe all along—as he now agrees—he ought never to have offered the property to anybody else. It was not on the market at that time, and the particulars had been given to Mr Marlowe by Mr Fenton simply for passing on to one specified possible purchaser.

As a result of those further letters, starting on December 1, Mr Marlowe got a well-deserved rebuke from Mr Burton of Dorothy Perkins. The letter is in the bundle. He accepted the rebuke and acknowledged it in a letter to Mr Burton, which again I think is consistent with his case, but I need not read it, on December 29. That is by the way; I am not concerned with the propriety or otherwise of his doing that. But the fact is that it seems to me that the writing of letters to other people offering the property on December 1 is again consistent with his evidence that Mr Star withdrew his interest in 20 New George Street about a week after November 22; because if Mr Marlowe was going to offer it to other companies, being, as I am satisfied he is, a very go-ahead sort of person in this way, I think he would have started writing straightaway on November 22, 23 or 24. But he did not, and I think that that is entirely consistent with Mr Star withdrawing his interest after about a week.

That was that, until—picking up again the story at what seems to me a relevant matter—in May 1977, according to Mr Marlowe, there was a telephone conversation about something else and Mr Star mentioned that he was looking at a prime site in New George Street, Plymouth. "I said"—this is Mr Marlowe's evidence—"'It's not the Dorothy Perkins shop, is it?' He said it was, and I said, 'This is the shop we introduced to you.' He said, 'I can't remember if that is the case, but not to worry as we probably wouldn't go ahead with it anyway,' and so I did not pursue it."

Mr Star says that evidence is a fabrication. There was no conversation of that sort whatsoever in May 1977. Of course, if there were such a conversation it would show that, at the very least, Mr Star was not being entirely frank with Mr Marlowe, because Mr Star, having expressed horror at a rent of £26,000, had in fact—as he frankly admitted in the witness box—had a change of heart. He says, "I agree; I changed my mind about 20 New George Street. Another 'Jeans' retailer had moved into New George Street and I was fearful that we would be unsuccessful in our business in Plymouth." So he negotiated for the purchase of this shop, or for the lease of it, and, as I have already mentioned, a contract was in fact signed at the end of May, but of course the decision to buy had been made considerably earlier.

Here I can get rid of one point which is not of importance as the case has turned out. In the end the defendants paid the rent, which was not negotiable, being the rent under the lease which was already in existence, and a premium which in effect included the fixtures and fittings, or most of them, of £5,000 and not £10,000. It is not suggested that that is material in this case, because it is plain from the documents—and I am not going to take up time, as it is not a live issue, in reading them out—that Dorothy Perkins would always have been happy to accept £5,000, as they in fact did. I have no doubt that if the matter had remained in the hands of Mr Marlowe the same agreement would very quickly have been reached. But, of course, Milton Marlowe & Co had no part in those negotiations. Nevertheless, it is not suggested that if there was an agreement between the plaintiffs and the defendants in the first place, any reduction in the premium or other variation of the terms upon which the lease was sold would deprive the plaintiffs of the benefit of it.

What happened next is that (I think at the Jubilee weekend) Mr Marlowe is quietly walking down the main shopping centre of Plymouth, when suddenly he gets a shock. The shock is this, that he looks into no 20, and what does he see? Mr Star. He goes in, and, according to Mr Marlowe, he said—following the May conversation—"I was pleased he had taken the shop and that it was one I knew we had introduced to him, and I told him so. Mr Star said he couldn't remember if that was right, but he didn't carry all that knowledge in his head, and he was too busy with the shopfitters to discuss it, and he said, 'Well, write to me when you get back to London.' He said he would try to find our original letter of November 1976."

Mr Star agreed that Mr Marlowe did come in and see him, but there was, he said, an unpleasant discussion. "I was surprised to see him. He insisted that he had introduced the property to us. I insisted that he had not done so. I was busy and I told him to send us a letter later." That is what Mr Marlowe did, and on June 8 by recorded delivery he sent the letter. He says:

> I would refer to our meeting in Plymouth . . .

(that is the one I have just mentioned)

> . . . and as requested by you . . .

Mr Star denies that he so requested.

> . . . attach a copy of my original letter dated November 23 1976, when I first offered you the premises.
>
> As you will see from my letter, the premises had not been placed "on the market" at the time and this point has been confirmed by Dorothy Perkins' agents, and so there can be no question in my mind of your having been offered the premises previously.

Then he goes on to deal with the history of the matter in some detail, reciting information that he had obtained from Mr Fenton. Then he says this, in the second paragraph:

> Since January we have spoken about the above premises and it is only a few weeks ago that I brought to your attention once again the fact that the property had been introduced by us, although at that time you informed me that you were not interested in taking this property. At no time during that conversation or any other time did you indicate that you had made an offer and were actively negotiating for the premises.

So there is Mr Marlowe setting out, in that letter of June 8, the May conversation—or the effect of it—which I have recited and which Mr Star says is a complete invention. Then he goes on to set out his complaint and sets out his claim.

On June 22, Mr Star wrote a succinct reply, and, of course, there is nothing wrong in keeping a letter short if appropriate. He said:

> In reply to your letter of the 8th instant and notwithstanding your letter of November 23 1976, at no time was there an agreement for you to negotiate on our behalf.
>
> You were not instructed as an acquiring agent for Southcut Ltd, in this or any other case, and your sole connection with us at that time was as a selling agent.

What is said on behalf of the plaintiffs, by Mr Burton, is: "Well, it is rather strange, although this was a succinct letter, and although, of course, it was a denial of any liability, there was no comment on the allegation about the May conversation —nothing to the effect of 'You even fabricate alleged conversations,' or anything like that. This is really the letter of a man who knew that the plaintiffs had introduced the property to him and his company but thought that he was not liable for a fee because his company had not used Mr Marlowe to negotiate." I think that particularly in the second of those two points there is substance. I think that it does read as the letter more of a man who is taking that attitude.

This judgment has already taken a long time, and if I went into every peripheral issue then it would take a great deal longer, but I do not think it is necessary to do so.

The plaintiffs' case is first put on the basis that there was a general agreement for the payment of fees for introductions. I think the evidence is insufficient to establish that. I think that the position was that there was an understanding of a general nature between the plaintiffs (through Mr Marlowe) and the defendants (through Mr Star) that he would expect to get paid for any introductions which resulted in a purchase. But I should be very reluctant to base any judgment in favour of the plaintiffs on a general agreement legally enforceable.

The much more powerful submission, on much stronger grounds, which depends, as I have said more than once in this judgment, on my assessment of who is accurate about the interview on the telephone on November 22, is that it is said that there was a specific agreement in this case to pay scale fees of the Royal Institution if the introduction resulted in a purchase. That is how the case is pleaded and clearly pleaded in paragraph 3 of the statement of claim.

Having, I hope, fairly summarised the evidence of each party and pointed out the main documents, and of course reminding myself constantly that the burden of proof of the case is on the plaintiffs, it seems to me that the history of the matter and the documents which I have taken time to recite and the contemporaneous behaviour of the parties strongly points to the conclusion that the court should prefer, as I prefer, the account given by the plaintiffs. That being so, in my judgment, an agreement is proved as pleaded in the paragraph to which I have just referred.

Very interesting points might have arisen on, for example, causation, on whether the plaintiffs were entitled to scale fees or only a *quantum meruit*, and Mr Burton, in spite of gentle protests from me, insisted on opening them to me. When I chided him with that again in his final submission he said, "Oh well, it was a good thing I did in the end because, even if it did not impress your Lordship, it must have impressed my learned friend because he in the end did not pursue those defences." That is by the way. There is no dispute, as I have said, about that matter. Once the agreement is proved, that is it. I was a little troubled, particularly after I had been referred overnight to the judgment of O'Connor J in *Sinclair Goldsmith v Minero Peru Comercial* (1978) 248 EG 1015, given on July 17 1978 as to whether this was a case in which it would be appropriate, in view of the fact that what the plaintiffs did was "introduce" and no more, that they should receive their full fees, or whether they should receive less on a *quantum meruit* basis. But Mr Burton has pointed out to me the various regulations. Mr Payton realistically does not argue to the contrary, and I am satisfied that in this case, once the agreement is found, as I find it, that the introduction should result in payment at the scale fees which are applicable, then the plaintiffs are entitled to the fees as claimed.

Various frightful possibilities were dangled in front of me by Mr Payton as to the effect on the offer of properties if I found for the plaintiffs. I do not think those fears are justified; it depends entirely on the facts of a particular case whether an agent can prove that in the circumstances he is entitled to commission.

I should add one further point, that it is not suggested that any other agents were involved or that the defendants, in fact, were put in the position of having to pay, or of being liable to pay, commission to more than one agent in respect of the introduction.

In those circumstances, as I have said, the plaintiffs are entitled to judgment. They have paid, as I have said, the VAT, and so, subject to anything counsel may want to say, I think the plaintiffs are entitled to judgment in the sum of £2,943.

COURT OF APPEAL

December 12 1979

(Before Lord DENNING MR, Lord Justice BRIDGE and Sir David CAIRNS)

A A DICKSON & CO v O'LEARY

Estates Gazette May 31 1980

(1979) 254 EG 731

Estate agents' claim for commission on sale of house—Commission payable "should we be successful in either directly or indirectly introducing a person ready, able and willing to purchase on terms authorised by you"—Two sets of negotiations taking place at same time, one through plaintiffs and the other through competing agents—Vendor's solicitors exchanged contracts with purchaser introduced by plaintiffs' competitors before the solicitors acting for plaintiffs' prospective purchaser had received her signed contract—Plaintiffs' claim fails—Property sold before person put in touch by the plaintiffs was "ready, able and willing" to purchase—Observations in E P Nelson & Co v Rolfe cited—Implied term that commission is not payable if, before introduction, the principal has entered into a binding contract to sell to another—Appeal from deputy county court judge dismissed

This was an appeal by A A Dickson & Co, a firm of estate agents, of 392 Lillie Road, London, SW6, from a decision of Deputy Judge Gerald Owen QC, at West London County Court. They were plaintiffs in an action for payment of commission alleged to be due to them from the defendant, M B O'Leary, the respondent to the present appeal, in respect of the sale of a house at 25 Edgarley Terrace, London SW6. The deputy judge dismissed the action.

C Beaumont (instructed by Cree, Godfrey & Wood) appeared on behalf of the appellants; D P Friedman (instructed by Donald Nelson & Co) represented the respondent. Mr Friedman was not called on.

Giving judgment, LORD DENNING MR said: Mr O'Leary was a bricklayer. He bought a house at 25 Edgarley Terrace in London. He did it up himself. He renovated it and he decorated it. He then determined to sell the first floor as a flat. He put it into the hands of house agents—more than one. One was a firm called A A Dickson & Co. He put it into their hands in June 1976 at £14,000 as an asking price. The agents confirmed his instructions by a letter of June 22 1976 in which they said:

> Whilst writing, we would also mention that should we be successful in either directly or indirectly introducing a person ready, able and willing to purchase on terms authorised by you, we shall look to you for payment of our commission on the basis of 2½ per cent of the purchase price

Reading between the lines, it is pretty clear that some other agents got ahead of A A Dickson & Co. It appears that, by the end of August, arrangements had got so far through those other agents that a contract had already been sent to a proposed purchaser. A A Dickson & Co did not find a person likely to come in until September 6 1976. Then they said that they had arranged a sale to a Miss Kyd at the sum of £13,250 subject to contract. Mr O'Leary had not actually agreed that sum. He would probably have accepted it in due course, but he had not agreed it. It was sufficiently promising, however, that he instructed his solicitors, and negotiations went on with Miss Kyd's solicitors.

So, in effect, there were two negotiations taking place at the same time. One through the other agents with other purchasers—which was hanging fire a bit, but the contract had been sent off to them—and the other through Messrs Dickson with Miss Kyd and her solicitors.

I need not go through all the negotiations until I come to the date of October 28. There had been negotiations "subject to contract," but on October 27 Miss Kyd's solicitors sent her —through the post—a contract to sign. She signed it on the 28th, and returned it to her solicitors, who received it on the 29th. It was unconditional this time—showing that she was ready, able and willing to purchase for the sum of £13,250. That was the first time on the evidence that it was shown unequivocally that she was ready, able and willing to purchase for the sum of £13,250.

But, as it turned out, Mr O'Leary's solicitors had still been carrying on negotiations with the other purchaser and the other purchaser's solicitors. So much so that they had already signed and exchanged contracts on October 28 with the other solicitors. No doubt commission will be payable on that deal.

On the next day, October 29, Miss Kyd's solicitors said to Mr O'Leary's solicitors, "We have obtained Miss Kyd's signature, and we are ready to exchange contracts." But the answer was, "We are sorry; you are too late. We have already sold to another purchaser, and contracts have already been exchanged."

There it is. Mr O'Leary said he could not pay Messrs Dickson any commission. Messrs Dickson were aggrieved. They said they had done all their work. They were entitled to commission. They immediately sent in a debit note for £357.75 for commission. Mr O'Leary's answer to that was, "The property had been sold before your client was herself ready, able and willing to enter into an unconditional contract."

The deputy county court judge decided against the agents and in favour of Mr O'Leary. He had the case argued twice. He wanted to consider particularly the observations made in the case of *E P Nelson & Co v Rolfe* [1950] 1 KB 139. In that case there was a contract in precisely the same terms as this one. Again "a person able, ready and willing to purchase." Again there were two prospective purchasers, but with this difference: there was no binding contract with one of them as there was in this present case. Geoffrey Howard at p 142 put forward three implied terms:

> Commission is not payable if, before the introduction, (1) the authority to the agent is withdrawn, (2) if the property has already been sold

—I gather that means "conveyed"—

> or (3) if the principal has entered into a binding contract to sell to another.

As I read the judgments of the Court of Appeal, those three terms were accepted by the court as being correct. Bucknill LJ said at p 145: "Another implied term which Mr Howard admitted must be read into this contract was that commission would not be payable if the property had been already sold before the person able, ready and willing to purchase the bungalow had been introduced." Cohen LJ at p 146 repeated the very words of Mr Howard's argument and threw no doubt upon them. It seems to me that the third implied term is correct. It is good sense and is applicable to this case: because, when a house is in the hands of more than one agent, it cannot be supposed that the vendor is to be made liable for double commission. In the ordinary way commission is payable to the agent who is the first to find a purchaser who enters into a binding contract which both parties accept. If a binding contract is made before another agent has produced "a person able, ready and willing to purchase"—or before any contract is made with that person—the first agent gets commission, the second does not. It is a race as to which agent wins. He wins who first gets the binding contract. I am afraid that in this particular case the other agent won and Messrs Dickson did not.

I think that the deputy county court judge was right, and the appeal should be dismissed.

Bridge LJ agreed.

Also agreeing, Sir David Cairns said: I am satisfied that the learned deputy circuit judge in his very careful judgment rightly held, mainly on the authority of the case of *Nelson v Rolfe*, that this particular term should be implied. It is right to say that although it was not a matter of decision in that case, it was clearly the view of both Bucknill LJ and Cohen LJ that a term to that effect should be implied; and I feel quite sure that, if either of those two judges had thought that the concession which was made with regard to it was not rightly made, they would have said so. Certainly it does not appear that since 1950 there has been any dissent from the view that such a condition could be implied.

I accept, for my part, that there may be cases where there are two estate agents both entitled to commission. In that respect, I accept as accurate what was said by Orr LJ in his judgment in the case of *Christie Owen & Davies Ltd v Rapacioli* [1974] 1 QB 781 at p 790. But it is a situation which reasonable people would be likely to seek to avoid, and therefore it seems to me that the implication of such a term as is relied upon here which would contribute to that end is permissible. I therefore agree that the appeal should be dismissed.

The appeal was dismissed.

HIGHWAYS

QUEEN'S BENCH DIVISIONAL COURT
May 17 1979

(Before Lord WIDGERY CJ, Lord Justice GEOFFREY LANE and Mr Justice ACKNER)

R v SURREY COUNTY COUNCIL, EX PARTE SEND PARISH COUNCIL

Estates Gazette February 9 1980
(1979) 253 EG 579

Highways Act 1959—Dispute between parish council and county council—Application by parish council for order of mandamus to compel the county council to take action under section 116 to prevent the obstruction of a footpath—Various obstructions erected by "frontagers" to impede passage on what was probably a public footpath—Although proceedings were begun by county council under section 124 to require removal of obstructions, they were not pressed and subsequently the council acceded to a suggestion from the frontagers that the path should be extinguished and an alternative way provided—County council's "pusillanimous attitude" promoted the interests of the frontagers who were obstructing the path instead of protecting the rights of the public to the use and enjoyment of it—Order of mandamus granted.

This was an application by Send Parish Council for a judicial review, the remedy sought being an order of mandamus directed to Surrey County Council to compel them to carry out their duty as a highway authority under section 116 of the Highways Act 1959 to remove obstructions from a highway. The path in question ran from Highcotts Lane to Dedswell Drive in West Clandon, Surrey.

F M Ferris (instructed by Pettit & Westlake) appeared on behalf of the applicant parish council; J Sullivan (instructed by the Clerk, Surrey County Council) represented the respondent county council.

Giving the first judgment at the invitation of Lord Widgery CJ, GEOFFREY LANE LJ said: This is an application by the Send Parish Council for a judicial review in the shape of an order of mandamus directed to the Surrey County Council to carry out their obligations under the provisions of the Highways Act 1959 and particularly of section 116 of that Act.

The material portions of the section read as follows:

(1) It shall be the duty of the highway authority to assert and protect the rights of the public to the use and enjoyment of any highway for which they are the highway authority, including any roadside waste which forms part of it.

Subsection (3) provides:

Without prejudice to subsections (1) and (2) of this section, it shall be the duty of the council which is a highway authority to prevent, as far as possible, the stopping-up or obstruction of highways for which they are the highway authority. . . .

Subsection (5) provides:

Without prejudice to their powers under section 276 of the Local Government Act 1933 a council may, in the performance of their functions under the foregoing provisions of this section, institute legal proceedings in their own name, defend any legal proceedings and generally take such steps as they deem expedient.

Subsection (6), which is the most important part of this section, reads as follows:

If the council of a parish, or community or, in the case of a parish or community which does not have a separate parish or community council, the parish meeting or a community meeting, represent to a local highway authority—(a) that a highway, being one as to which the local highway authority have the duty imposed by subsection (3) of this section, has been unlawfully stopped up or obstructed . . . it shall be the duty of the council of the local highway authority, unless satisfied that the allegations are incorrect, to take proper proceedings accordingly.

This case is about a long-standing altercation. It concerns a "path" (to use a neutral term) at the village of West Clandon near Guildford in the county of Surrey. The path, in so far as it has a name at all, has at its northern end the name of Highcotts Lane. It runs to the west of and roughly parallel to the A247 Clandon Road. Between the path and Clandon Road there have been built over the years a number of houses with large gardens. What has happened is this. The owners of some of those houses have bought part of the land over which the path runs, and have extended or purported to have extended their gardens to include those portions of the path which abut upon the western boundary of the gardens. They have erected wattle fences and various obstructions which have been demonstrated in photographs before us so as to make it impossible in effect for anyone to walk along the path unimpeded. We have been told that on the western side of one of the gardens there is a summer house, which has been planted, if not actually put into the soil, on the line of the path; and at another point it is said that part of a swimming pool or a wall abutting on a swimming pool also impinges upon the line of the pathway.

The altercation has involved those house-owners (to whom I will refer for a matter of convenience as the "frontagers"), and they have been in bitter opposition to the parish council (and the predecessors in title of the parish council), a society which seeks to preserve the village of West Clandon called the Clandon Society, and indeed with most of the villagers, it seems, other than those who own these particular houses.

The Surrey County Council, who are admittedly the relevant highway authority, occupy a position, or should occupy a position, somewhere midway between those various protagonists, a position which has over the years been somewhat ambivalent as events will show.

Since 1969 various efforts have been made to persuade the Surrey County Council to act under their powers under the Highways Act which I have described and to persuade them to remove the obstructions. At one stage it seemed certain that they were going to do just that. In autumn 1977, however, as the result of pressure from the frontagers, the Surrey County Council's previous decisions, which were in favour of removing the obstructions, were revoked, and they determined to accede to the frontagers' suggestion that they should exercise powers to extinguish the path under section 110 of the Act in return for the frontagers providing an alternative way to the west of the existing path. That involved a passage over land which did not belong to the frontagers, but they were prepared to, and in effect eventually did, make the necessary purchase from the adjoining farmer in order that they would

A have the land over which that suggested alternative path could run.

The parish council had—and this is admitted—on August 16 1977 made the representation which they are entitled to make under section 116 (6)(a) and they contend that it is the duty, in the terms of the Act, of the Surrey County Council to take proper proceedings accordingly; that is to enforce the removal of the obstructions on the pathway.

The parish council contend, first of all, that the Surrey County Council cannot be, and are not, satisfied that the allegations are incorrect, which is the one loophole the Act provides for their not acting to remove the obstructions. B Indeed, it is said (and it is not seriously contended to the contrary) that both the Surrey County Council and the frontagers have had opinions from counsel whom they have instructed, indicating that if the matter is litigated the chances are heavily in favour of it being held that this path is in fact a way over which the public have the right to walk. Secondly, the parish council contend that the Surrey County Council have not taken proceedings, and, even if they have, have not taken proper proceedings to put matters right, as it is contended they are obliged by the Act to do. They could, it is said, have taken proceedings under section 124 or for a declaration and injunction in the Chancery Division, had they been so minded.

C I do not find it necessary to go through the detailed history of events since 1969. We have been taken, so to speak, with a fine-tooth comb through almost every incident which happened between these parties over those 10 years. It is sufficient perhaps for the purposes of this judgment to make one or two points and to describe one or two incidents which will fill out the bare bones of that summary which I have just given.

In 1958 a Mr Claydon bought from the Onslow Estate (which up to that date had been the owners) land which included the land over which this pathway ran. He sold the material section of the path to a Mr Ekins, who at that stage owned one of the frontagers' houses. It matters not which. D In 1968 or 1969 Mr Ekins sold to the other seven frontagers with whom we are concerned that portion of the path which adjoins their respective gardens on the west side.

There is a body of evidence, so it seems, to the effect that the path was used by the public as a footpath or bridleway for very many years before these obstructions were placed upon it. In 1962 Mr Ekins erected notices on the pathway saying "No public right of way," but representations were made to him by the parish council, among other people, and in, I think, May of 1962 Mr Ekins agreed that the path was to be maintained as a public right of way thereafter.

It was in 1969 that the frontagers started to erect the E obstructions which I have described. Representations were made to the Guildford Rural District Council, who at that time were the appropriate authority, who accepted that this was a public footpath, and indeed they succeeded in persuading some, but not all, of the frontagers who had placed obstructions upon it to remove those obstructions.

For reasons which it is not possible, at any rate at this stage, to discover with any accuracy (it may have been a selfish attempt to restrict the use of the path to villagers only) the path was not included in the Surrey County Council map prepared under the National Parks and Access to the Countryside Act 1949. But the evidence of the use of this path for F 20 years and upwards prior to the 1959 sale of Onslow Estate was furnished to the Surrey County Council.

Eventually, in May 1972, the Clandon Society referred this matter to the county council, and on June 28 1972 the relevant committee of the county council recommended action under section 124 of the Act.

Section 124 reads as follows:

> (1) Where a structure has been erected or set up on a highway otherwise than under a provision of this Act or some other enactment, the highway authority for the highway may by notice require the person having control or possession of the structure to remove it within such time as may be specified in the notice.

There are three other subsections which go on to describe G the procedure and remedies which are in furtherance of that particular power.

The Surrey County Council, however, did not adopt the suggested course. They favoured at that stage an informal inquiry, but when this proposal was put to the interested parties they did not share the county council's enthusiasm for the idea, and indeed they rejected it.

Once again, in November 1976, the committee resolved that the path should be recognised as a public footpath and should be included in the next definitive map. Secondly, they agreed that notices should be served under section 124 requiring removal of the obstruction; and, thirdly, the committee H suggested or recommended that such further actions should be taken as might be necessary to protect the public use of the footpath.

In January 1977 notices under section 124 were served, and in a letter sent a day or two later to the various frontagers this was said to them, among other things: "The county council have received representations that such a right of way has been established and are satisfied on the evidence supplied that the public have acquired rights. The Area Highways Subcommittee have considered this matter and have authorised this action in an attempt to resolve this dispute." It was at this stage that the frontagers started to make their counter- J proposals for the diversion, as I have indicated.

On June 20 1977, something like five or six weeks after the original notices, further notices were served under section 124. It is not necessary to enquire why. The Surrey County Council at that stage were certainly not impressed, or did not seem to be impressed, by the counter-offer put forward by the frontagers. In the meantime the diversion which was being proposed was hotly contested and strongly objected to by the parish council and all the other interested parties on the side opposite the frontagers. It was opposed as being much less commodious than the existing path, opposed as being of indifferent foundations whereas the original path was on solid K foundations and so on. They also objected to the way in which it was apparently being suggested that the frontagers should be allowed to obstruct what was a public footpath in their submission, and then, so to speak, buy their way out of the trouble by purchasing this strip of land from the adjoining farmer and then suggesting that a footpath should be run over there and so extinguish the rights of those who were entitled to use the footpath.

To cut a long story short, the letter under section 116(6) was sent by the parish council on August 16 1977, as I say, requiring the county council to act under section 116. Yet again the matter came before the subcommittee of the county L council, this time in early October 1977. They concluded that the clerk and chief executive officer be authorised to enforce the section 124 notice. It was resolved: "That the clerk and chief executive be authorised to enforce the notices served under section 124 of the Highways Act 1959 upon the frontagers of land affected by the footpath from Highcotts Lane to Dedswell Drive, West Clandon."

There then ensued a series of letters passing chiefly between the frontagers' solicitor and the local authority. They start with a letter of October 5 1977 and end with a letter of October 17. Those letters are letters discussing what action it should be proper for the county council on the one hand to M take and the frontagers on the other. The result was that the frontagers through their solicitor issued proceedings against the county council claiming a declaration that the path is not a public footpath, and claiming injunctions to prevent the county council from exercising their rights or their duties under section 124.

It is fair comment, and one that puts it objectively, that that writ claiming the injunction was to a great extent stimulated by the letters which the county council had written. They seem, at least on a cursory reading of the correspondence, to be inviting the frontagers to take that action. Be that as it

A may, the action was taken, and since then the Surrey County Council have maintained rigid inactivity.

Nothing has been done with regard to the writ. No one has suggested that the hearing might be expedited. No one has sought to dismiss the proceedings for want of prosecution, and not unnaturally those who are anxious to have the benefit of the footpath are getting a little anxious. It is the result of that anxiety that we find ourselves here, deciding whether or not an order of mandamus should issue in order to stimulate the county council into activity. The writ by the frontagers was issued on October 19, after, as I say, the Surrey County Council had persistently invited the frontagers to take
B that action. It is interesting to note that a letter from the frontagers' chief protagonist was dated the very same date as the writ. That is to be found as an exhibit to an affidavit. I will not read it, but it is an interesting comment on the situation. Since then, as I say, nothing has happened.

If one turns to correspondence one finds the present situation. It is a letter dated December 13 1977 from the clerk and chief executive to Sturges Trotter & Co, solicitors:

> Footpath from Highcotts Lane to Dedswell Drive, West Clandon. I thank you for your letters of November 28 and December 12 and I am obliged to you for arranging this meeting at such short notice. As you are aware the representatives of the sub-
> C committee recently met members of the Parish Councils and the Clandon Society and it was clear that they were totally opposed to the suggested compromise. The main concern of the Parish Council was that the alternative route was not sufficiently wide and the public had established rights over the wider area of land that has now been enclosed in your clients' gardens. They also considered that the surface of the alternative route was unsatisfactory. Generally the representations were that the county council should see that the existing route is opened before any diversion is considered.

The evidence, as I say, that this is a public footpath is very strong. What has now happened is simply this. The Surrey County Council are proposing to try to stop up what is alleged
D to be the existing footpath and to use the suggested new footpath as an alternative. They have been persuaded by the frontagers that this is a proper exercise of their powers; and that the villagers and the Clandon Society hotly contest. The parish council, the Clandon Society and the villagers maintain that the proposed new path is much less commodious. It is a good deal less wide, and they point out that the county council have been persuaded by the frontagers to cut down even the original narrow width of the new path from 10 ft to 8 ft, and they have been persuaded to give way on the question of fence or no fence between the farmer's field and the path.

E We do not have to decide in this court whether this is a public footpath or not. The basis upon which we can approach the problem is that the evidence that it is a public footpath is very strong indeed, and if the matter were to be litigated it is agreed on all hands (and we have seen the various opinions of counsel) that the chances of the frontagers succeeding in showing there was no such footpath are pretty remote. The only two points in issue are whether the Surrey County Council have taken proper proceedings accordingly under section 116, and, if they have not, whether the court should exercise its undoubted discretion to order mandamus to issue.

The submissions of the applicants are these. First of all, they
F suggest that section 116(6)(b) is clear. It obliges the county council in the circumstances of this case to take proper proceedings and to take proper proceedings accordingly. They submit that the word "accordingly" must refer to the earlier subsections and mean that the county council must take steps to assert and protect the rights of the public to enjoy the use of the highway and to prevent or have removed any obstructions of that highway. The county council have, so it is submitted, no discretion. It is their duty to take proceedings, and it is their duty not to compromise those proceedings or to settle them.

I do not find that an acceptable argument. It seems to me that the use of the word "proper" indicates that the local G authority must have not only a discretion as to the form of the proceedings which they choose to take, for example, section 124 for an action in the Chancery Division—that is not in dispute—but also a further discretion as to the way in which and the extent to which those proceedings ought to be prosecuted. To that extent I would agree with the submissions of Mr Sullivan on this point, Mr Sullivan appearing as he does for the county council. But then one asks oneself the next question: what are the limits of that discretion? Those limits must be culled from the words of the Act themselves. The local authority must at all times act with the object of protecting the highway and of preventing or removing any H obstruction, and, more broadly speaking, of promoting the interests of those who enjoy the highway or should be enjoying the right of way; and the county council must likewise operate against the interests of those who seek to interrupt such enjoyment of the highway.

It is trite law now, and needs hardly to be expressed, that this court must not, and will not, interfere merely because it thinks, if it had been a local authority, it would have acted in a different way. Before we can properly interfere by way of mandamus, or indeed by any other method, we must be satisfied that the local authority have acted under a misapprehension as to their duties or else in a way no reasonable J local authority could have acted had they had their duties as set out in the 1959 Act properly in mind. This seems to me to be in accordance with the speech of Lord Reid in *Padfield v Minister of Agriculture, Fisheries and Food* [1968] AC 997. The passage to which I should like to refer is at p 1029 as follows:

> There are a number of reasons which would justify the minister in refusing to refer a complaint. For example, he might consider it more suitable for arbitration, or he might consider that in an earlier case the committee of investigation had already rejected a substantially similar complaint, or he might think the complaint to be frivolous or vexatious. So he must have at least some K measure of discretion. But is it unfettered?
>
> It is implicit in the argument for the minister that there are only two possible interpretations of this provision—either he must refer every complaint or he has an unfettered discretion to refuse to refer in any case. I do not think that is right. Parliament must have conferred the discretion with the intention that it should be used to promote the policy and objects of the Act; the policy and objects of the Act must be determined by construing the Act as a whole and construction is always a matter of law for the court. In a matter of this kind it is not possible to draw a hard and fast line, but if the minister, by reason of his having misconstrued the Act or for any other reason, so uses his discretion as to thwart or run counter to the policy and objects of the Act, then our L law would be very defective if persons aggrieved were not entitled to the protection of the court. So it is necessary first to construe the Act.

Perhaps for the sake of completeness I should also refer to the decision of the Court of Appeal on March 13 1979 in *Meade v Haringey London Borough Council*, which was the "schools" decision, as yet unreported. We have a transcript, and at page 11 of the transcript, in the judgment of the Master of the Rolls, appears this passage: "But if the public authority flies in the face of the statute, by doing something which the statute expressly prohibits, or by failing to do something which the statute expressly enjoins, or otherwise so conducts itself—by omission or commission—as to frustrate or hinder M the policy and objects of the Act, then . . . it is acting *ultra vires*."

It is at this point that, in my judgment, the Surrey County Council have fallen down. One can see how it happened. They were influenced by the so-called hardship to the frontagers in having to remove their summer houses and wattle fences and so on, though the extent of those hardships seems from the photograph to have lost nothing in the telling. They were alarmed by the length of time and the amount of money which legal proceedings would involve. But as to the length of time, this argument started at the latest in 1969, and

A another month or two would not make very much difference there. Whatever form of proceeding is taken, and some form of proceeding must be taken, it is all going to cost money. They were impressed by the simplicity of the solution which was apparently provided by the new and alternative path.

What it seems to me they did not realise, or if they realised it they disregarded it, was the fact that, looked at objectively, they were acting in the interest of the frontagers, in the interest of the people who had in fact obstructed this public footpath, and not in the interest of those who should have been enjoying the use of the right of way over the footpath. It seems to me no reasonable local authority could have so
B acted if they truly had in mind the ambit of their duty under section 116. What they have in fact attempted to do is to adopt a course of action, that is to say proceedings under section 110, a proceeding which is most likely to result in the extinguishment of the footpath without any hope of successful appeal on the part of the parish council or the villagers. This method of procedure avoids the necessity of having to prove that the new path is shorter or more commodious, which they would have had to show were they to proceed under section 111 rather than section 110. In other words, if the old path were open, free for access, as *ex hypothesi* it should be, and if one started from that basis,
C it seems to me the Surrey County Council could not succeed. As it is, the parish council, the Clandon Society and the villagers are left with the barren possibility of persuading the minister, in the teeth of opposition from the county council at that, that it is not expedient to confirm the proposed extinguishment order. The frontagers have, therefore, by their skilful manoeuvring, or the skilful manoeuvring of their solicitor, engineered a situation, or brought about a situation, which might be a kinder way of putting it, where the Surrey County Council are in effect playing their game for them.

That does not seem to me to be the duty of the local authority as set out in section 116. Mr Sullivan, in a meticulous
D speech putting forward the suggestions of the Surrey County Council, submitted a number of points to which I will make a brief reference, and I hope he will not think I am discourteous if it is only brief.

He suggests, first of all, that, if there is the slightest chance that the existence of the public right of way will be disputed, then the whole of the county council's duty under section 116 disappears. The court subjected him to some fairly close questioning on that point, and it seems to me that he was driven to the position of having to rewrite subsection (6) of section 116 in order to make good his point on that. That argument did not impress me.

E Then he makes the point, or endeavours to make the point, that the county council were acting reasonably in endeavouring to compromise the matter in the way that they did. He suggests the outcome of the section 124 proceedings could not be predicted with absolute certainty. Very few things can be predicted with absolute certainty, but so far as certainty does exist in litigation, in so far as one is entitled to rely on the opinions of counsel instructed by the frontagers and by the Surrey County Council, it seems to me that certainty is as nearly attainable in this case as it can be.

He points to the expense and protracted nature of the litigation. I have dealt with that. He suggests that, on all these
F matters, coupled with the debated width of the right of way, the county council were entitled to take the somewhat pusillanimous attitude which they undoubtedly did. That adds to the consequences to the landowners who have to remove their summer houses and so on.

Those points do not seem to me to outweigh the other considerations in the contrary sense which I have already detailed.

Then, finally, he urges the court in its discretion not to grant an order of mandamus because, he said, it is contrary to principle to compel anyone to institute legal proceedings by way of an order of mandamus. He drew to our attention the case
of *R v Southampton Port Commissioners* (1870) LR 4 HL G
449 and he cited in particular a passage at p 485, which is in the speech of Lord Chelmsford, as follows:

> As to the form of the mandamus, it appears to me that it may be supported upon the ground that it is not to be construed so as necessarily to import a command to institute legal proceedings. The words are not to take the necessary and legal measures and proceedings for *enforcing* and recovering payment, but for obtaining and recovering. The distinction is rather fine, but I think it admits of the construction that the defendants are merely ordered to take proper measures for obtaining payment, but are not commanded absolutely to bring an action.

In the present case there is no command to institute pro- H
ceedings. It may be necessary to defend the proceedings brought by way of injunction and declaration by the frontagers. It may be necessary possibly to counter-claim in those proceedings. But what the order of this court is aimed at securing is the performance by the county council of their duties under section 116, namely, the removal of these obstructions. That duty may carry in its train all sorts of subsidiary necessary actions. What they are it is impossible to foretell, but for those various reasons, and I am afraid rather rudely brief, my view is that the submissions made by Mr Sullivan do not avail against the duties plainly set out in section 116.

I would grant this application. J

Agreeing, ACKNER J said: Send Parish Council (the applicants) pursuant to section 116(6) of the Highways Act 1959 represented to the respondents, the Surrey County Council that the footpath from Highcotts Lane to Dedswell Drive, West Clandon, Surrey, was a highway being one as to which the local authority have the duty imposed by subsection (3) of section 116 and that it had been unlawfully stopped up or obstructed.

Subsection (3) reads:

> Without prejudice to subsections (1) and (2) of this section, it shall be the duty of the council which is a highway authority
> to prevent, as far as possible, the stopping-up or obstruction of K
> highways for which they are the highway authority. . . .

Such representations having been made, what is the result? The answer is to be found in the self-same subsection (6): ". . . it shall be the duty of the local authority, unless satisfied that the allegations are incorrect, to take proper proceedings accordingly."

It has never been suggested in this case that the Surrey County Council were satisfied that the representations to which I have referred were incorrect.

The allegations having been made and the proviso not
availing, the county council then have to accept and act upon L
them. Mr Sullivan's first submission was that, as it could be said that there was a dispute as to the footpath being such a highway and, therefore he urged, a doubt as to the validity of the assertion, then there is a discretion in the county council as to whether or not to take any action at all.

He sought to reply for that proposition upon one case in this court, *R v Lancashire County Council, ex parte Guyer* (1977) 76 LGR 290. I need not go in any depth into that case. Suffice it to be said that it is a decision purely on the obligation of the county council in relation to section 116(3). It does not relate to this case where the representations were
made not by an individual but by a parish council. M

What Mr Sullivan in his careful submission seeks is the redrafting of subsection (6) so that it reads in this way: "Where a highway, being one as to which the local authority have the duty imposed by subsection (3), then, if the council of a parish represent that the highway has been unlawfully stopped up or obstructed. . . ." If Parliament had intended that to be the effect of subsection (6), then it could have, with the simplicity which I have suggested, so expressed its intention. It failed to do so.

Mr Sullivan's next main submission was that there was discretion in the council and it involved a discretion to com-

A promise any proceedings or actions it had initiated. I accept that there must be some discretion arising out of the words "proper proceedings" taken together with the duty of the council under subsection (3) to prevent "as far as possible" the stopping-up or obstruction of highways. So I accept that in a proper case a proper compromise of proceedings may be made. But such a compromise must be intended to promote the policy and the object of the Act. The compromise which was here suggested, on the contrary, would have resulted in the objects of the Act being frustrated. It would have *extinguished* the highway and provided something else in its place—a pathway which is clearly open to the various criticisms
B that have been referred to in the affidavits.

On the available material, including in particular the C
photographs exhibited to the third affidavit of the clerk to the appellants, it appears to me that the respondents have been overimpressed by the frontagers' plea of hardship. In a sentence, the county council has, in my judgment, in effect shut its ears to the parish council's application made pursuant to the provisions of section 116(6). It has thus refused to exercise its discretion and, in my judgment, mandamus accordingly lies.

LORD WIDGERY CJ also agreed.

The court ordered mandamus to issue as prayed, with costs against the county council. A stay of proceedings pending a possible appeal was refused. D

HOUSING ACTS

A

EDWARDS v UPTON

(Before His Honour Judge MARK SMITH at Kingston County Court on July 26 1979)

Estates Gazette February 23 1980

(1979) 253 EG 818

Tenant's claim for repayment of rent paid after making of closing order under Housing Act 1957—Money paid under illegal contract—Tenant continued to reside in house until order for possession was made more than five years after closing order—Tenant's failure to complete appropriate forms for rehousing by council—No intention of leaving unless B compelled—Claim dismissed—Claim for damages for breach of implied covenants to repair also dismissed

This was a case arising from a closing order made in respect of premises at 72 Kings Road, Teddington, Middlesex. The tenant was William G Edwards and the landlord was Mrs Doris Upton, whose interest was represented by Messrs A H Sneller, who became managing agents in 1971.

The plaintiff claimed the return of all moneys paid by him as rent since the date of the closing order, on the basis that it was rent paid under an illegal contract. He further claimed that the landlord was in breach of implied covenants to repair by allowing the house to fall into disrepair so that the closing C order was made whereby he lost the protection of the Rent Acts and possession and that therefore he was entitled to damages for that breach. The defendant counterclaimed for arrears of rent amounting to £109.60.

K Raftery (instructed by Myers, Ebner & Deaner) appeared for the former tenant; T R G F Ryland (instructed by Sherwood Cobbing & Williams) represented the landlord.

Giving judgment His Honour JUDGE MARK SMITH said: This is perhaps an unusual case in that the plaintiff Mr Edwards was a tenant of a property upon which, unfortunately, a closing order was made by Richmond upon Thames Council under section 27(1) of the Housing Act 1957 as a result of D which he had to leave. He was found other accommodation by the council. He was given the appropriate money by way of disturbance allowance. Nevertheless, he says that is not enough, that he is entitled to damages, indeed £2,000, and one can but sympathise with him.

In 1939 he became the tenant of the ground floor at 72 Kings Road, Teddington, Surrey, a house owned by the defendant's father. In 1952 the plaintiff remarried, and his second wife went to live there with him. In 1957 Mrs Upton's father transferred the house to her and she became the landlady. In 1969 the plaintiff took over the whole house at a rental which was then £5 11s, inclusive of rates. On April 14 1972 the closing order was made. By virtue of that, the E defendants should have left the premises at once, because section 27(1) of the Housing Act 1957 says that a closing order should be an order prohibiting the use of premises in respect of which the order is made for any purpose other than one approved by the local authority and any person who, knowing that a closing order had become operative and applies to any premises, uses those premises in contravention of the order or permits them to do so shall on summary conviction be liable to a fine and so on.

The plaintiffs knew of the closing order very soon after it had been made, probably because Mrs Edwards heard about F it from a neighbour. They did not leave, Mrs Upton did not tell them to leave and the council turned a blind eye, and it is abundantly clear from the evidence of Mr Sneller that it was the council's policy, and is the policy of councils throughout the country, to turn a blind eye and give a period of grace so that either they can rehouse the tenant or he can find somewhere else to live. I am in no doubt that Mr Edwards knew the council would rehouse them when the time came when they would have to leave their home.

The defendant sympathised with the plaintiff's predicament and for a long time allowed him and his wife to stay. Mr Sneller made it clear to the Edwards that he would help them G by approaching the council and assisting their claim for rehousing and he telephoned the council on their behalf at least once every six months. What they really wanted was to stay permanently at 72 Kings Road, but they could not, as they knew sufficient work would have to be done to satisfy the council that the closing order should be revoked. They knew from Mr Sneller that Mrs Upton would not evict until the council did rehouse them. They would not be homeless, and I have no doubt this partly motivated their thinking, and for some years they did not complete the appropriate forms, which Mrs Edwards says was her fault, as the form contained questions about private matters which as a matter of principle H she could not answer.

Eventually Mr Sneller found out that the forms were not being completed and told Mr and Mrs Edwards that under the circumstances, five years or more having elapsed since the closing order, proceedings for possession would be taken, which they were, and as a result the Edwards were rehoused. They do not like the new house and Mr Edwards in his claim says that but for Mrs Upton's failure to keep the house in sufficient repair they would not be in this situation, and that has caused them great distress and inconvenience. As a matter of pleading by his advisers I would say one might think that credit would be given for the payment made by the council J for disturbance.

Mr Edwards also says in his claim that the contract under which he paid weekly what I shall refer to as rent was an illegal contract immediately the closing order was made, ie after April 14 1972, and therefore he says he is entitled to the return of moneys paid as rent. Regarding that illegality, I say in passing that two applications were made to the rent officer after April 14 1972 for him to fix a fair rent and I am satisfied he knew on both occasions that there was a closing order in existence, but neither he nor anyone else thought that what was happening was in any way illegal. Mr Sneller said the policy of leaving tenants in possession K until further accommodation was found was a humane gesture by the council which benefited the tenant, as he kept a roof over his head, and enabled the landlord to draw some money during a period when he did not have the use of the house.

Mrs Upton in answer says she did nothing wrong, that she and her father did practically everything the tenants asked them and that is not doubted apart from some sash cords, which old Mr Upton used to see to, not being repaired when

Mr Upton was too old to deal with them. In 1966 the plaintiff had a bathroom added which the defendant paid for. She paid a bill of £844 for that and added £1.50 to the weekly rent, which seems to be within the 12½ per cent then permitted by law to be added to rent in respect of improvements. I should mention that at the date of the closing order the rent had risen to £4.53 per week. I have no doubt, on the evidence that I have heard, that the defendant, her father, and Mr Sneller did everything the tenants asked them to do and I am also quite sure the Edwards asked for work to be done only if it was really necessary.

Both counsel agree that after April 14 1972 the contract under which the plaintiff remained in the premises was an illegal contract and I shall approach the matter on that basis. It may help if I say what happens to money paid under illegal contracts. I need not go into the cases read to me, as in my view they fall within, and indeed are the basis of, the principles as I understand them. Money which has actually been paid under an illegal contract can be recovered only in certain specific circumstances: (i) if the money has been paid in ignorance of the facts which created the illegality; (ii) if statute specifically provides that the money should be returned; (iii) if the parties are not *in pari delicto* and it is not against public policy for the money to be returned; and (iv) if there is a course of action which does not depend upon the illegal contract. It is the third ground which is applicable here.

The situation, as I see it, is this. That if anything the defendant is in a stronger position than the plaintiff for two reasons. Firstly, she genuinely believed it would only be a short time before the plaintiff and his wife would be rehoused and she was acting out of kindness in allowing them to stay for what she thought was a short period, whereas the Edwards were not intending to leave at all unless compelled. Secondly, the effect of a closing order is to impose upon a tenant the duty of leaving—they have to go; they did not, but the defendant did not encourage them to stay, nor did she make them believe they were entitled to stay. It was the plaintiff who was responsible for the fact that he stayed so long. He should have filled in the forms properly; that he left it to his wife and she did not is no excuse. The blame, if that is the right word, is on him; but even if no fault be attached to him, equally none can be attached to the defendant and that is sufficient to dispose of that course of action and there is no basis for the recovery by the plaintiff of any sums which he paid to the defendant.

Turning to the implied covenant, section 32 of the Housing Act reads: "In any lease of a dwelling-house being a lease to which this section applies there should be implied a covenant by the lessor (a) to keep in repair the structure and exterior of the dwelling-house. . . ." Subsection (3) says "in determining the standard of repairs . . . regard is to be had to the age, character and prospective life of the dwelling-house and the locality. . . ." As already stated, the defendant and her father did all repairs necessary as a result of matters of which they had knowledge. In my view she had no notice of sash cords needing repair, and it would be contrary to justice to saddle her with the consequences of not putting right something she did not know about. On the basis of what I have heard about the house I have the gravest doubts whether in order to comply with section 32 she would have had to do anything more than she did even if other defects had been brought to her notice. It follows that she is not in breach of that implied covenant and even if she were, the fault would lie at the plaintiff's door for not telling her what work needed doing. The plaintiff's claim alleges further that there should be another implied term to the effect that the defendant would not cause or permit the condition, design or other characteristics of the premises to be or become such that a closing order under section 27 of the Housing Act 1957 would be made. When statute specifically sets out what covenant is to be implied in relation to a particular subject-matter, there is no room, save possibly in special and exceptional circumstances, for there to be an implied term dealing with that same subject-matter. The implied term alleged here is so wide as to be almost breathtaking in its implications. What control would a landlady have over natural disasters or acts of war? That implied term is so wide it could not possibly be implied. There comes a time when any house reaches the end of its useful existence, and no landlord/lady can be expected to pour money into it practically to rebuild it to ensure the accommodation remains suitable for a tenant, and that is precisely, as I understand it, what section 32(3) is there for in determining the standard of repair required by the lessors' repairing covenant.

The only matter I have not dealt with is the situation as it was after the council's letter of January 20 1972 which they sent to the defendant's agents, Messrs. Sneller, enclosing a schedule of defects saying the house was unfit for human habitation having regard to the standards which are set out in section 4 of the Housing Act 1957 and saying it could not be rendered fit at a reasonable expense. The plaintiff argues that the defendant therefore knew about the repairs from that date and ought to have done them. My view is that those repairs go far beyond what is envisaged in section 32 and she was under no duty to do them. In all those circumstances the claim must be dismissed.

There is a counterclaim under which £109.60 is claimed allegedly as arrears of rent from December 5 1977 to April 5 1978 at £6.34 per week. I doubt if the moneys claimed can properly be described as rent, and while a court may or may not help a party to reclaim money paid under an illegal contract in certain circumstances I cannot believe that it will assist anybody to obtain the repayment of money due or allegedly due under the illegal contract and our rather simplistic approach to the counterclaim. I dismiss the counterclaim as well.

Claim dismissed with defendant's costs Scale 4. Counterclaim dismissed with plaintiff's costs Scale 1. Order that both costs be set off against each other.

LANDLORD AND TENANT

GENERAL

CHANCERY DIVISION

November 22 1979

(Before Mr Justice WALTON)

FROBISHER (SECOND INVESTMENTS) LTD v KILORAN TRUST CO LTD AND ANOTHER

Estates Gazette March 22 1980

(1979) 253 EG 1231

Landlord and tenant—Long lease of flat—Service charge—Effect of section 91A of Housing Finance Act 1972—Covenant by lessees to pay interim sums on account of prospective liability for annual contributions towards lessors' expenditure in carrying out their obligations under lease—As a result of section 91A a service charge can be recovered only in respect of expenditure already incurred, not prospective expenditure—Interim sums held to be part of service charge and hence, being payments in advance, not recoverable by lessors—Alternative suggestion that lessors might recover from lessees cost of borrowing money to meet expenditure required by lease—Held that there was nothing in lease to authorise such a claim—No help to be derived from the doctrine of "The Moorcock" or from the implied term principle in Finchbourne v Rodrigues—Lessors not entitled to require lessees to make interim payments provided for in lease on account of estimated or proposed expenditure—An "unfortunate conclusion" for these and other lessors

This was an originating summons taken out by the lessors, Frobisher (Second Investments) Ltd, to which the lessees, Kiloran Trust Co Ltd and Vernon Aidan Ravenscroft were defendants. The summons, which related to flat 34, Campden House, Sheffield Terrace, Kensington, London W8, sought four declarations, namely, (1) that interim payments provided for in the lease dated November 25 1965 between the parties' predecessors in title were charges to which section 91A of the Housing Finance Act 1972 applied; (2) that the lessors were entitled to require such payments to be made in advance of expenditure; (3) that the lessors were entitled to recover from the lessees the appropriate proportion of the cost of borrowing moneys to meet expenditure under the lease until the contributions to the lessors fell due under the lease; and (4) that the lessors were entitled to recover from the lessees the amounts required to reimburse the lessors' managing agents for the cost of borrowing in order to effect expenditure for the purpose of the lease.

B K Levy (instructed by Lieberman, Leigh & Co) appeared on behalf of the plaintiff lessors; Paul de la Piquerie (instructed by Thwaytes) represented the defendants.

Giving judgment, WALTON J said: The matter before me arises out of the impact of section 91A of the Housing Finance Act 1972, which was added by the Housing Act 1974, upon a lease. in a fairly standard form. The lease was made on November 25 1965 between Campden House Chambers Ltd, as lessors, and George Maylam and Deborah Smith, as lessees. The reversion is now vested in the plaintiffs, Frobisher (Second Investments) Ltd, and the lease is now vested in the defendants, Kiloran Trust Co Ltd and Vernon Aidan Ravenscroft.

The lease was a demise in consideration of the payment of a premium of a flat in a block of property described in the first schedule to the lease as Campden House, 29 Sheffield Terrace in the Royal Borough of Kensington. The lease is a long lease at a low rent, so that the provisions of the Rent Acts do not apply to it. By the fourth clause of the lease the lessors covenanted with the lessee "from time to time and at all times during the said term to observe and perform each and every of the obligations on the lessors' part set out in the Seventh Schedule." That is in substance the services to be provided by the lessors, which I will have to look at a little more closely in a moment. The lessees undertook to pay in respect of that "the contribution," and we will see in a moment how "the contribution" is in fact defined.

The schedule of matters which the landlord undertook to do were the kind of things that one frequently finds in a block of flats; for example, keeping the reserved property in good order, repair and condition, keeping the property insured, paying all premiums necessary for effecting and maintaining insurance, to apply all moneys received under insurance in respect of loss or damage in rebuilding and reinstating the property, to pay the charges incurred in the water rates on the whole of the property, electricity charges in connection with part of the property, the cost of maintaining and renting a public GPO telephone, and a clause, to which I shall have to refer in more detail later, but for present purposes may be taken as the reasonable and proper fees payable by the lessors to its managing agents for the time being for carrying out the general management and administration of the property. Then there was another clause concerning the cleaning and maintaining of the part of the reserved property, the dustbins, paying the rent and performing the covenants in the head lease. Then there are matters with which we are not concerned, employing and maintaining a porter, providing a supply of hot water, and repairing and keeping in runing order the lifts.

Now, one of the covenants on the part of the lessee contained in the sixth schedule is that the lessee will pay the rent and "the contribution" and the interim sums on the days and in the manner stipulated in this lease. That takes us on to the eighth schedule, which deals with "the contribution," and "the contribution" is a yearly sum equal to a percentage, which would vary from flat to flat, but in the case of the flat with which we are concerned is 0.3 per cent

of the amount which the Lessors shall from time to time have expended during the year immediately preceding the date next hereunder mentioned in (a) meeting the outgoings costs expenses and liabilities incurred by them in carrying out their obligations under the provisions of the Seventh Schedule hereto (except paragraphs 6 and 7 thereof) and (b) in paying from time to time the costs and expenses of and incidental to making repairing maintaining amending and cleansing all or any ways roads pavements gutters sewers channels drains pipes watercourses walls party walls party structures fences and other conveniences which shall belong to or be used for the Premises in common with any other part or parts of the Property or which shall form part of the Reserved Property such contribution to be paid on the 25th day of March in every year. The amount of such Contribution shall be ascertained and certified by the Lessors' Managing Agents acting as experts and not as arbitrators once each year throughout the term on the 25th day of March in each year (or if such ascertainment shall not take place on the said 25th day of March then

A the said amount shall be ascertained as soon thereafter as may be possible as though such amount had been ascertained on the aforesaid 25th day of March) commencing on the 25th day of March 1964 and such certificate shall contain full details and figures relating to all the component elements comprised therein and a copy thereof shall be supplied to the Lessee at the request of the Lessee and without charge to him (but not more frequently than once in every yearly period. . . .)

Then it is provided first of all:

> That the Lessee shall (if required by the Lessors) Pay such a sum on account of the Contribution payable by the Lessee under this Schedule as the Lessors' Managing Agents shall certify as being a reasonable interim sum (in this Lease referred to as
> B "interim sum" or in the plural as "interim sums") to be paid on account of the Contribution. Such interim sums shall be paid (if required as aforesaid) half yearly in advance on the 25th day of March and the 29th day of September in every year commencing on the quarter day next following the date of this Lease in respect of the period from the date hereof to the next following half yearly day.

Then (2):

> That the Contribution payable by the Lessee hereunder (or such balance as shall remain after giving credit for any interim sum) shall be paid by the Lessee or any proper balance found to be repayable to the Lessee shall be so repaid to him on the 24th day of
> C June next following the year ending on the 25th day of March to which such contributions shall relate or as soon thereafter as may be possible.

Then there is a proviso limiting the contribution for a comparatively short period from the commencement of the lease; and it is finally provided:

> That (without prejudice to any other remedy or right) the Lessors shall not be entitled to re-enter under the provisions in that behalf herein contained in respect of non-payment only of any interim sum.

So the general scheme is quite clear: twice a year, on March 25 and September 29, the lessees are required to pay interim
D sums certified by the managing agents as being reasonable. At the end of the year at the next March 25 or shortly thereafter there is worked out what the contribution properly comes to. The sums which have been paid on account are set against that, and then either the lessee has to pay the balance, if it is a balance against him, or if it is a balance in his favour, he receives the balance back.

There can, I think, be no doubt whatsoever, and it is not challenged in any way, that the contribution proper is in fact a service charge. For present purposes—that is to say, for the purposes of section 91A of the Housing Finance Act 1972—the expression "service charge" means

> E any charge for services, repairs, maintenance or insurance, being a charge which is payable as part of, or in addition to, the rent, and which varies or may vary according to any costs (including charges for overheads) incurred from time to time by or on behalf of the landlord or any superior landlord.

That is a definition to be found in section 90(12) of the Housing Finance Act 1972, and it applies to section 91A, because in subsection (7) of section 91A there is to be found this definition:

> In this section "chargeable items" means any items for which a service charge may be payable and other expressions used in this section have the meanings assigned to them by section 90
> F (12) above.

The difficulty which is created by section 91A lies in this, that if one reads the opening words of section 91A, it is I think abundantly clear that a service charge *strictu sensu* can be recovered only in respect of expenditure which has already been incurred or defrayed by the landlord and cannot be recovered in respect of prospective matters, however clear it is that those matters, when actually carried out, will be properly chargeable items and costs proper to be brought into account as part of the service charge.

That is because section 91A in its opening words—and I intend to read only the opening words because there are very grave difficulties thereafter, with which, fortunately, I am in G
no wise concerned in this case—begins as follows:

> A service charge shall only be recoverable from the tenant of a flat—(a) in respect of the provision of chargeable items to a reasonable standard; and (b) to the extent that the liability incurred or amount defrayed by the landlord in respect of the provision of such items is reasonable. . . .

Mr de la Piquerie, on behalf of the tenants, has argued that that result is brought about not only by the wording of section 91A(1)(b) but also by the wording of 91A(1)(a), "A service charge shall only be recoverable from the tenant of a flat—(a) in respect of the provision of chargeable items to a reasonable standard." But, in my judgment, he is putting too H
great a burden on that wording, and if that wording stood alone, there would be no reason to restrict the service charge that may be recovered to past items. But, however, when one gets to (b), "to the extent that the liability incurred or amount defrayed by the landlord in respect of the provision of such items is reasonable . . .", it appears to me perfectly clear that the landlord must have defrayed the cost, or at any rate incurred liability to pay the cost, before it can be recovered from the tenant.

Therefore, I am faced in this case with the simple point—simple, that is, to state, but difficult to solve—as to whether the interim sums are or are not properly to be designated as J
a service charge. In my judgment, they are to be so designated. It seems to me that the wording of the lease is too strong to enable me to do anything else, because the first proviso to the eighth schedule, which I have already read, provides that the interim sums are to be paid on account of the contribution or, spelling that out, the interim sums are sums paid on account of the service charge, and it appears to me that sums paid on account of the service charge are part of the service charge.

Mr Levy, for the landlords, has sought to argue that the interim sums are a kind of trust fund or moneys held by the lessors' managing agents as stakeholders, and that therefore K
they really have nothing to do with the service charge, save that when the contribution is ascertained and becomes payable, then the lessors' managing agents, on behalf of the tenants, are entitled to put those moneys towards the discharge of the contribution, and that is the first time that those moneys become part of or are used to discharge the service charge.

But that seems to me to ignore the realities of the matter. Clearly the interim sums are paid so as to enable the managing agents to have money in hand to discharge the costs of the lessors, which are properly recoverable by means of the contribution, as and when they arise without waiting for the contribution to be ascertained; and because that is so, it L
appears to me that the suggested analogies of the lessors' managing agents being stakeholders or alternatively trustees of the money just will not fit the circumstances of the case in any way at all. They are not stakeholders of the money, because stakeholders hold the money of third parties, to be applied in a particular way when a particular event happens. But I cannot see that there is anything in the lease to prevent the lessors' managing agents from applying the moneys as they receive them to whatever appropriate purpose they think fit. Indeed, we know that that is precisely what happens, because in fact the managing agents maintain a separate service charge bank account, and they maintain that to meet M
the day-to-day costs during the year. Similarly it seems to me that there is no possibility here of treating the moneys as trust moneys in any way. First of all, one does not import the question of trust into a matter of this nature, which is a purely commercial matter, unless one has to do so, but, even more importantly, if those moneys were trust moneys to be applied in the way that has been suggested, only at the end of the year in meeting the contribution, it seems to me that it would be the duty of the managing agents to keep all those moneys together all during the year until the specified event happened, and that that is not what is intended is perfectly apparent.

A So it seems to me that the provision for repayment if by any chance the interim sums do exceed the contribution payable by the lessee is a matter which very sensibly rests purely in contract as between the landlord and the tenant and does not in any way rest in trust or on some concept of the managing agents being stakeholders.

I realise, of course, that that is a most unfortunate conclusion so far as the landlords are concerned, but I have reached that conclusion on the argument which has been placed before me. Having reached that conclusion, I did not require counsel to address me upon it, but I am by no means certain that that is the only way of putting the matter
B against the landlords. At the very lowest, the interim sums are some form of security for the fact that the contribution will ultimately be paid; and the opening words of section 91A are that a service charge shall only be recoverable from the tenant of a flat to the extent that the liability incurred etc; and it seems to me that even if one could, by some attribution of a label, call the interim sums something other than "service charge," the exemption of the interim sums would still be a matter of recovering the service charge and thus invalidated by the opening words of section 91A(1).

This now leads me on to the second point, which is this: Mr Levy says: "Well, if now it is impossible to recover the
C costs in advance in this way, this is going to create very difficult problems for the landlords," and I should not like the landlords to think that I do not fully sympathise with the fact that that is precisely and exactly what my decision will do. I can see their very great difficulties which will result. What Mr Levy says is that it will mean that the landlords, or the landlords' managing agents, will have to borrow moneys in order to carry out their obligations under the lease, and he naturally wishes on behalf of the landlords to include the costs of the borrowed money—that is to say, the interest which will be payable upon it—in the matters which can be recovered from the tenants.

D The only clause in the lease, I think it is fair to say, to which he can really point as providing any ground for such a submission is to be found in clause 3(iv) of the seventh schedule. Of course, if it is in clause 3(iv) of the seventh schedule, then it will be one of the matters which the tenant will have to reimburse pursuant to the provisions of the eighth schedule. But I must now read the clause in detail:

> The reasonable and proper fees payable by the Lessors to its Managing Agents for the time being for carrying out the general management and administration of the Property including (but without prejudice to the generality of the foregoing) all fees payable to such Agents in connection with the collection of the rents the
> E Contribution and the interim sums payable by the respective owners of the Flats the payment of the rent under the Head Lease and other outgoings payable by the Lessors in respect of the property or any part thereof and the preparation of all accounts in connection with the calculation and assessment of the Contribution and interim sums and arrangements for the supervision of any works which may be carried out pursuant to paragraph 1 of this Schedule and (if undertaken by such Agents) preparation of Specifications in connection therewith.

I am afraid that at the end of the day I cannot find in there any real reference of any description to "interest." It seems to me that what that clause is dealing with are the fees payable to the managing agents. They are fees payable for carrying
F out the general management and administration of the property. Then the clause goes on to specify, without prejudice to the generality of that phrase, various matters: fees payable to the agents in connection with the collection of the rents, fees payable to the agents in connection with the collection of the contribution, fees payable to the agents in connection with the collection of the interim sums, fees payable to the agents in connection with the payment of the rent under the head lease, fees payable to the agents in connection with the payment of other bills payable by the lessors in respect of the property or any part thereof, fees payable to the agents in respect of the preparation of accounts and in connection with the calculation and assessment of the contribution and interim sums G
and fees payable to the agents in respect of the arrangements for the supervision of any works and preparation of specifications. It does not seem to me that one can fairly say that in that wording there is any general phrase which one could fairly read as including interest payments of the general nature with which we are dealing.

Moreover, it would be very surprising indeed if one could, and for this reason: it may very well be that the present landlord is an impecunious landlord—I know not—but the landlord might, far from being impecunious, have very large sums of money under his control, in which case he would not need to borrow any other money at all; and if that were the H
situation and interest was going to be charged, one would have expected to find a clause specifically dealing with it and mentioning the interest on sums expended, or words to that effect, somewhere in the lease, and one does not, unfortunately, find any such words.

Mr Levy then says: "Well, be that so, it must be implied as a matter of necessary implication, following the doctrine of *The Moorcock* (1889) 14 PD 64, in this lease that if the statute were to render the payment of the service charge in advance unlawful, the tenant would immediately start paying interest upon moneys expended in the manner in which it will now become necessary for the landlord to incur that interest." J

While the doctrine of *The Moorcock* is a very useful doctrine indeed, and I would be the last person to refuse to apply it if I thought it could be properly applied, Mr de la Piquerie for the tenant says that so far from the tenant saying "Of course that must be the case," if he was asked at the commencement of the lease "Supposing the statute intervened to make your payments in advance unlawful, I assume that you will be quite willing to pay interest upon the moneys that the landlord has to borrow?" so far from saying "Of course," the tenant might very well reply "Well, I don't know about that. I shall want to know a very great deal more about the whole matter. I am not sure I am going to agree to that.' K

Whether that be so or not, I do not think that in fact one can apply the doctrine of *The Moorcock*, because I know of no case—and Mr Levy has been unable to cite any case to me—in which the doctrine of *The Moorcock* has been applied when there has been a disturbance to contractual arrangements as the result of a statute. It seems to me that if there is a disturbance of contractual relationships because a statute intervenes, then it must be left to the statute to say what is to happen consequentially upon its intervention, and that one cannot foist upon the parties what some outside body thinks would have been what they would have agreed to in circumstances which neither of them can possibly have contemplated L
under any circumstances. Indeed, if I am right on the construction that I have given to section 91A—and I say "if I am right," because, unfortunately, this appears to be the first case on that section, and it is likely, I imagine, to go to a higher court—the legislature has set its face against the tenant paying in advance for anything except liabilities actually incurred by the landlord; and in order to imply a clause taking the fullest possible advantage of that fact, which could undoubtedly be done in future leases, one would have to reconstruct the whole system of liability completely; and it seems to me that it is not a proper use of *The Moorcock* to imply a term reconstructing the position as between the land- M
lord and tenant only partially.

Therefore, for those reasons, I feel unable to accept Mr Levy's submissions on *The Moorcock* on that point of principle. Moreover, although I do not, having done that, need to decide it, I feel at any rate a lurking suspicion that if the tenant had been asked, "What are the parties going to do if statute intervenes to frustrate any part of your bargain?" he would have said, "Well, that must be left to statute. I am not going to agree to any particular term in advance," because it must be recalled that implied terms do not, as it were, get written into the contract as one goes along. One goes back to the inception of the contract and asks the question then and says "Is that the term to which both of you would agree if asked by the officious bystander?" It seems to me that it must at the very least be dubious whether the tenant would not have said, "Well, if statute intervenes, I shall have to read the statute and see what it says. I am not going to agree in advance what is going to happen."

Mr Levy pressed me on this aspect of the case very much with the decision in *Finchbourne Ltd v Rodrigues* [1976] 3 All ER 581, where the Court of Appeal had no hesitation in implying a term that various matters should be carried out fairly and reasonably in a lease and that maintenance of the property should not be left to the landlords' discretion to adopt the highest conceivable standard of maintenance for the block of flats and to charge the tenant with that cost. It seems to me that that case, as it were, speaks for itself. Obviously, in order to give business efficacy to a block of flats leased in the Mile End Road, one assumes a Mile End Road standard of maintenance, just as if the flats were in Park Lane one would assume a Park Lane standard of maintenance. It seems to me that the readiness of the Court of Appeal to adopt, as it were, the current standard for the block of flats in that particular case does not in any way assist me in this matter at all.

Therefore, at the end of the day, for those reasons, I have come to the conclusion that the interim payments for which provision is made in the proviso to the eighth schedule to the lease are charges to which section 91A of the Housing Finance Act 1972, as amended by the Housing Act 1974, applies, and I have come to the conclusion that the plaintiff is not entitled to require the said, or any, interim payments to be made in advance on account of estimated or proposed expenditure to be incurred for the purposes referred to in the seventh schedule to the said lease, or for any of such purposes, before such expenditure has been incurred; and also I have come to the conclusion that the plaintiffs are not entitled to recover from the defendants as lessees the appropriate proportion of the cost of borrowing in the financial market the moneys required to make the expenditure referred to in the said seventh and eighth schedules until such time as the payments referred to in the said eighth schedule are made to the plaintiffs, nor would that in the slightest be changed if instead of the plaintiffs as lessors incurring the interest charges, those charges were incurred by the managing agents themselves.

Declarations were made accordingly. There was no order as to costs.

CHANCERY DIVISION

November 1 1979

(Before Mr Justice DILLON)

MOORCROFT ESTATES LTD v DOXFORD AND ANOTHER

Estates Gazette June 7 1980

(1979) 254 EG 871

Landlord and tenant—Lease of part of a building to a firm of surveyors, valuers and estate agents—Provision in lease for payment by lessees of a contribution towards the costs of various services and works of maintenance and repair carried out by lessors—Contribution to be a "due proportion" of the costs, based on the ratio of the rateable value of the demised premises to the total rateable value of the building —Question raised as to whether "rateable value" in this context meant the rateable values at the date of the lease, so that the proportion would then be fixed once and for all, or the rateable value from time to time of the premises demised and the total rateable value from time to time of all parts of the building—Held that the latter was the correct construction and that the exact point of time for calculation was the time or times when each item of expense was paid

This was a construction summons to determine the true construction of a clause in a lease dated May 11 1971 relating to part of St James' House, 13 Kensington Square, London W8. The plaintiffs were the present lessors, Moorcroft Estates Ltd, and the defendants were R Doxford and A Colvin, who carried on business as Robert Bruce & Partners at St James' House.

Miss H E Williamson (instructed by Lawrance Messer & Co) appeared on behalf of the plaintiffs; P J Talbot (instructed by Stones, Porter & Co) represented the defendants.

Giving judgment DILLON J said: This originating summons raises a short question of construction of a lease which is dated May 11 1971 whereby certain rooms and offices on the ground and lower-ground floors of St James' House, 13 Kensington Square, London W8, were demised to the present defendants, Mr Doxford and Mr Colvin, who carry on business there as surveyors and estate agents under the name of Robert Bruce & Partners. The plaintiffs, Moorcroft Estates Ltd, are currently the lessors entitled to the reversion in the demised premises.

The lease was for a term of 14 years from March 25 1971. By it, after the reservation of the main rent, there was a reservation by way of further or additional rent of such sum as may from time to time be determined by the surveyor for the time being of the lessors as a due proportion of the annual premium or premiums paid by the lessors for the insurance of the building against certain risks and also there was reserved by way of further or additional rent during each year of the said term a due proportion of the cost to the lessors of providing the services mentioned or referred to in the second schedule. These were a range of services covering recurrent matters such as costs of maintenance and repair of the structure of the building, cost of lighting the common parts and keeping them clean and tidy, window cleaning of the common parts and such like. Then in clause 4 (vi) of the lease the lessees covenanted to contribute and pay on demand a due proportion of the costs and expenses of repairing and cleansing sewers, drains, pipes, water courses, wires and other easements used or to be used in common. The phrase "due proportion" used in the passages to which I have referred is defined in clause 1 of the lease as follows:

> The expression a due proportion wherever used and employed in this lease shall mean a proportion of the total expenses and/or expenditure in question payable, paid or incurred by the lessors in respect of the building hereinafter mentioned and defined as is in the ratio which the rateable value of the premises hereinafter demised bears to the total rateable value or values of all parts of the building inclusive of common parts and areas.

The rateable values of the constituent parts of the building were increased in consequence of a rating revaluation which took effect from April 1 1973 and they were not increased in the same proportion, one part to another, because parts of the building, including the parts demised by this lease, are business premises but other parts are residential. The question I have to consider is whether the reference to rateable value of the premises in the definition of due proportion means the rateable values as at the date of the lease, so that the proportion was then fixed once and for all, or means the rateable value from time to time of the premises demised and the total rateable value from time to time of all parts of the building so as to fluctuate with any fluctuations in the relative rateable values.

In my view the commonsense of the clause requires the expression "rateable value" to be construed as meaning the rateable value from time to time so that the proportions reflect any changes that there may be in rateable values. The due proportion is to be paid in respect of the current expenditure. They have chosen rateable value as a means for calculating the proportion, and it seems to me to be sensible that they should be referring to rateable value from time to time. If they had intended to refer to a proportion fixed once and for all at the date of the lease, it would have been sensible to set out what that proportion was, namely, as it has been calculated

A and agreed, 38.82 per cent, without using a formula at all. Furthermore, there is the possibility that there might be additions, or subdivisions in the building, such as the building of a penthouse flat, and if that were to happen and the proportions had been fixed once and for all at the date of the lease, the result would be either that the landlords would make a profit or that no part of the burden would be attributable to the assumed penthouse flat. Mr Talbot, for the lessees, has pointed out that in various other parts of the lease the parties have been careful to use the words "from time to time" or "for the time being" when they meant those words. He has given me quite a number of examples. For instance,
B on p 5 of the lease, there is reference to additional insurance premiums paid by the lessors "by reason of or consequent upon the occupation and user from time to time carried on upon the demised premises." On p 4 in the reservation of the insurance rent the reference is to "such sum as may from time to time be determined by the surveyor." And on p 6 where there is a rent review clause the question of the revised rent in default of agreement between the surveyors and the parties is to be referred for determination to a third independent surveyor to be appointed by the president for the time being of the Royal Institution of Chartered Surveyors. There are other instances. Now, it is natural enough for the
C words "from time to time" or "for the time being" to be used in those other clauses, but I do not think the fact that they have been used in such clauses is sufficient to override what I regard as the natural meaning of the expression "a due proportion."

Mr Talbot has also invited me to look at the circumstances surrounding the execution of the lease and he has referred me to evidence that prior to the execution of the lease the demised premises were in very poor order. The evidence is that it was a condition of the defendants entering into the lease that substantial refurbishment to the basement would be carried out by the landlords, but that the defendants
D would be responsible for the refurbishment of the ground floor.

Looking at that information, however, as part of the matrix of facts does not afford me any assistance in concluding one way or the other whether the rateable values referred to in the definition of "a due proportion" are the rateable values at the date of the lease or the rateable values "from time to time." Mr Talbot has also urged as one of the surrounding circumstances that the defendants knew the comparative rateable values of the constituent parts of the premises at the date of the lease and were inferentially content to bear that proportionate burden of the service
E charge and insurance rent, but to my mind if one is to draw any inferences from the assumption that the parties knew the comparative rateable values, the inference I would have drawn would be that if they intended the proportion to be fixed once and for all, then they would have referred to the fixed proportion and not used a formula.

Mr Talbot has also invoked the *contra proferentem* rule of construction saying that this is a lease put forward by the landlords, but he has frankly said that that is a last-ditch argument and I do not regard this as a case to be determined only in the last ditch. I take the view that the due proportion means the due proportion of the rateable value from time
F to time of the premises relative to the total rateable value or values from time to time of all parts of the building.

The originating summons raises a second question which is as to the date at which the due proportion of any expense is to be calculated. Miss Williamson has argued that it is to be calculated at the date on which demand for payment is made, but that must be a wholly arbitrary date, and I prefer Mr Talbot's argument on this point, that as what is to be calculated is a due proportion of various expenses paid or incurred by the lessors, the rateable values for calculating the due proportion are to be taken at the times when each item was paid or incurred.

His Lordship made declarations accordingly. It was agreed by counsel that the calculation of the due proportion should G
be made at the dates when the relevant expenses were paid, not incurred. The defendants (lessees) were ordered to pay the costs of the summons.

CHANCERY DIVISION

February 1 1980

(Before His Honour Judge RUBIN, sitting as a judge of the High Court)

ALAN ESTATES LTD v W G STORES LTD AND ANOTHER H

Estates Gazette June 14 1980
(1980) 254 EG 989

Landlord and tenant—Question as to the date of commencement of lease, being the date when rent began to be payable—Difficulties caused by the fact that both the lease and the counterpart were delivered in escrow—Subsequently both became unconditional when an outstanding term under which the counterpart had been delivered was satisfied—Whether escrow takes effect from the date when it becomes unconditional or, by relation back, from the date of its original execution—Authorities reviewed and a number of ancient authorities in favour of the latter view rejected—An instrument J
delivered as an escrow takes effect from the satisfaction of the relevant condition and not from original date of delivery as an escrow—Plaintiffs' claim based on earlier date dismissed.

In this action the plaintiffs, Alan Estates Ltd, sought a declaration that a lease of premises at 32 Goldhawk Road, London W12, granted by the plaintiffs to the first defendants, W G Stores Ltd, commenced on November 1 1976, when a counterpart was sent by the defendants' solicitors as an escrow to the plaintiffs' solicitors. The second defendant, Derek Desmond Solomons, who was named as the lessee in the original draft
lease, was subsquently described as a surety, although the lease K
contained no covenant by him of guarantee or otherwise. The date when the delivery of the lease and the counterpart became unconditional was November 18 1977. The plaintiffs in these proceedings also asked for an order for payment of arrears of rent and the apportionment of an insurance premium, based on the commencement date of the lease being November 1 1976.

J B W McDonnell (instructed by Martin Boston & Co) appeared for the plaintiffs; L F R Cohen (instructed by Paul Woolf & Co) for the defendants.

Giving judgment, JUDGE RUBIN said: By this originating summons the plaintiffs are seeking a declaration that the lease mentioned in its title commenced on November 1 1976 and an L
order for payment of arrears of rent and an apportioned part of the insurance premium which would be due upon the basis of that commencement date.

The history of this ill-fated matter starts in the early autumn of 1976 when Mr Solomons, the second defendant, was negotiating with the plaintiff company for the grant of a lease of 32 Goldhawk Road, London W12. No contract was entered into, but a time arrived when a draft lease was prepared. Mr Solomons then decided that he did not want to take the lease in his own name but in the name of the first defendant. Accordingly, the drafts were amended and the first defendant became tenant under the lease and Mr Solomons was joined as the M
party of the third part. Although he is there described as the surety, nowhere in the lease did he enter into any covenant and accordingly there seems to have been no purpose in joining him as a party.

The lease and counterpart were duly engrossed and were sealed respectively by the plaintiff company and the first defendant. There is no evidence before me which establishes the dates of either of these sealings, other than each must have occurred on or before November 1 1976.

In September 1976 the defendants had been allowed to have a key of the premises on payment of a quarter's rent by way

A of deposit. On October 28 the plaintiffs' solicitors sent a completion statement to the defendants' solicitors. This statement is difficult to understand. It credits the defendants with the deposit of £937.50. It then debits rent £3,250 per annum, being 59 days to September 29 amounting to £525. Counsel for the plaintiffs says that such description of the debit must be wrong and suggests it should read " from October 27 " instead of " to September 29." He bases that submission on the calculation that October 27 is 59 days before the quarter. Even if that is so, I fail to understand the relevance of October 27, as no one now suggests that the rent runs from that date.

On November 1 1976 Paul Woolf & Co, the defendants'
B solicitors, sent the counterpart of the lease sealed by both the defendants to the plaintiffs' solicitors together with a cheque for the balance of £4.54 shown on the completion statement I have just mentioned. In the letter which accompanied the counterpart the plaintiffs' solicitors were asked to hold the counterpart to the order of the defendants' solicitors pending clarification of certain points. I need not deal with those points in any detail because in an action between the same parties Whitford J, on November 15 1977, declared that the counterpart had been delivered as an escrow upon terms, one of which remained unsatisfied at the date of that judgment. It is common ground between the parties that the final term was
C satisfied on November 18 1977, when the delivery of the counterpart became unconditional.

Also on November 1 1976, but after receipt of the counterpart, the plaintiffs' solicitors sent the lease sealed by the plaintiffs to Paul Woolf & Co. From the letter which accompanied the lease, the plaintiffs' solicitors show that their view was that the lease and counterpart had become unconditional, since they asked the question whether the lease could be dated November 1 1976.

It is not established on the evidence before me precisely when the defendant company went into possession of the premises, but in my judgment I do not have to determine that date.
D The argument before me has been whether the rent should be treated as running from November 1 1976, the date of the physical exchange of the counterpart and the lease, or November 18 1977, the date upon which the final condition of the escrow was satisfied.

On November 9 1976 Paul Woolf & Co informed the plaintiffs' solicitors that the defendants had decided to withdraw from the matter and required repayment of the deposit. This act by the defendants was described in the first action as a repudiation. On November 11 1976 the plaintiffs issued the writ in the first action, claiming a declaration that the lease and counterpart had taken effect according to their terms, with
E an alternative claim for an inquiry whether the counterpart had been delivered as an escrow and, if so, upon what conditions it was delivered, and whether and upon what date or dates the conditions had been satisfied. Whitford J delivered his judgment in that action on November 15 1977, holding that the counterpart had been delivered as an escrow and that one of the conditions remained outstanding. The order entered as a result of that judgment contained a declaration that upon the plaintiffs on or before November 18 1977 producing to the defendants a certified copy of the transfer of mortgage from Northern Commercial Trust Ltd to Knowsley & Co Ltd the conditions attached to the delivery in the escrow of the counter-
F part lease and statement of claim mentioned are complete.

At the conclusion of Whitford J's judgment Mr Picarda, who was appearing for the plaintiffs, attempted to raise the question of arrears of rent. The judge had been critical of the pleadings in the first action and summarily disposed of the question saying, " There is nothing about this in the pleadings in this connection at all." The question is now before me on the present originating summons.

The only parts of the lease and counterpart, which fortunately are in identical terms, to which I need refer are the habendum, which reads: " To hold the demised premises unto the tenant for a term of 24 years from the 29th day of September 1976, determinable as hereinafter provided, paying G
therefor unto the landlords *from the date thereof* and throughout the said term hereby granted yearly and proportionately for any fraction of the year the several rents hereinafter referred to." Also I must refer to the covenant contained in clause 2(1), which reads: " To pay the respective yearly rents hereinbefore reserved and made payable at the times and in the manner at and in which the same are hereinbefore reserved and made payable without any deduction."

Mr McDonnell, for the plaintiffs, argued that as a matter of construction of the lease I should reach a conclusion that the rent was reserved from a date not later than November 1 1976. I accept that rent does not have to run from the date of the H
lease and can, if its terms so provide, run from a date prior to its execution. Authority for this proposition is to be found in the recent judgment of the Vice-Chancellor in *Bradshaw v Pawley,* which is reported in [1980] 1 WLR 10. However, the terms of the present lease seem to be perfectly clear, that is to say, " from the date hereof "; and the question to be decided is not so much a question of construction but of ascertainment of the date of the lease.

Mr McDonnell argued that the completion statement is a document from which it can be deduced that the parties intended the rent to run from October 27. Even if that were right, I fail to see how it assists the plaintiffs. There is no J
claim for rectification in the present action, and as no such claim was raised either in this action or in the former action, in my judgment it is now too late, quite apart from the added difficulty that the plaintiffs have now parted with the reversion and the present reversioner is not a party to the present proceedings.

The plaintiffs argued that notwithstanding the prior delivery of the counterpart as an escrow, the subsequent physical delivery of the lease to the defendants' solicitors was unconditional and took effect at once. Mr McDonnell attempted to avoid the impossible situation which would arise if the condition upon which the counterpart had been delivered as an K
escrow was never satisfied by arguing either that the tenant would be bound by the covenants in the lease because he had entered into possession, or the term which had come into being would become divested. I am unable to accept that argument.

Beesly v Hallwood Estates Ltd [1961] Ch 105 is authority for the proposition that when a company affixes its seal to a deed that is execution of the deed by the company and no further delivery is required. It is not, however, any authority for the proposition that the execution is necessarily absolute and can well be as an escrow. Indeed, it clearly appeared from the judgment of Harman LJ at p 116 that the grantor company executed its deed conditionally upon the due execution and L
delivery by the grantee of the counterpart. At p 118 Harman LJ summarised the effect of the delivery of the deed as an escrow in these words:

> ... if you do deliver a document as an escrow it is your act and deed and is not recallable by you. If, of course, the condition be never performed, it never becomes binding, and I suppose there must come a time, if there be unreasonable delay in the performance of the condition, when, in these days at any rate where equitable principles govern the actions of the court, the person or firm that has executed the escrow would be released from its obligation.... In my opinion, however, where it is the common intention of the parties as, indeed, is almost universal in these days to proceed by means of M
> lease and counterpart, it is readily to be inferred that delivery of the counterpart is a condition of the escrow.

One can see a like inference being drawn by Sir John Pennycuick in delivering the judgment of the Court of Appeal in *Glessing v Green* [1975] 1 WLR 863 where at p 867 he said:

> The manifest intention of the vendor in such circumstances is that conveyance shall not be operative in advance of completion, and, apart from special circumstances, the inference is inescapable that the conveyance is executed as an escrow conditional upon completion, ie against payment of the purchase price and where appropriate execution by the purchaser.

So here it seems to me that the lease must be regarded as having been executed by the plaintiff in escrow, the condition being the unconditional delivery by the tenant and the surety of a counterpart. As the counterpart delivered by the defendant was itself delivered as an escrow, both lease and counterpart remain conditional until the outstanding term under which the counterpart had been delivered was satisfied on November 18 1977.

On the basis that the delivery of the lease became unconditional on November 18 1977, in order that I can answer the question "What is the date thereof?" I must determine whether an escrow takes effect when it becomes unconditional, or ought to be treated as having taken effect on the date of its original execution.

In support of his argument that an escrow should be treated as having taken effect on its first delivery, Mr McDonnell relied upon a decision of Gorell Barnes J in *Edmunds v Edmunds* [1904] p 362. Undoubtedly the learned judge decided that the proper date was the date of first execution and not of the satisfaction of the term. Unfortunately, he deals with the point very shortly at p 334 by reference to a passage in *Williams on the Law of Real Property*, 1901 ed at p 151, which shortly states that an escrow operates from the date of its execution. In support of the proposition there are a number of authorities contained in a footnote. Mr McDonnell has referred me to some of those cases, but for my part I have considerable doubt whether they provide much support. The first case was *Coare v Giblett* (1803) 4 East 85. The relevant passage from the judgment of Lord Ellenborough appears on p 95, where he says:

The principle is that an inchoate act, which is to be consummated upon the performance of a conditional act required to be first done by the party who was the object of such inchoate act, and where the performance rests only with such party, becomes, when consummate by the performance of his part of such conditional act, an effectual act for the benefit of the inchoate actor by relation from the time of such inchoate act done.

The problem with that case, as Lord Ellenborough pointed out a little earlier in his judgment, is that he was not dealing with an escrow at all, as there had been an actual delivery to a party, and, as the law was then understood, such delivery was unconditional.

The next case was *Graham v Graham* (1791) 1 Ves Jun 272. At the end of p 274 Eyre LCB, after saying that the bonds must be considered as escrows, said: ". . . and then that they take effect from their original sealing and delivery." The word "from" in that context can, I think, have two meanings: that is, "from the date of" or "by reason of," and for my part I think the latter is to be preferred.

Finally, Mr McDonnell referred to *Perryman's Case* (1599) 5 Co Rep 84a. In that case the words used were "takes effect by force of the first delivery without any new delivery."

Mr Cohen, for the defendants, preferred to rely upon more modern authority and in particular upon the decision of Walton J in *Terrapin International Ltd v Inland Revenue Commissioners*, which is reported in [1976] 1 WLR 665. It is right that I should bear in mind that none of the cases I have already mentioned appear to have been cited, and also the case involved a question of stamp duty. However, it seems to me that the decision is based upon principles generally referable to escrows. At p 669 in his judgment Walton J said this:

In my judgment, it is essential to go back to first principles. There are, unfortunately, a large number of dicta to be found in the books which are clearly not consistent with first principles, including one notable one by Lord Reid in *Wm Cory & Son Ltd v Inland Revenue Commissioners* [1965] AC 1088 at p 1108 . . .

—and then the reference reads in this form:

. . . "Then Lord Denning instanced an executed conveyance delivered in escrow. In my understanding there would then be nothing binding: both parties would have a *locus poenitentiae*. So long as that position continued I do not see that there would be any obligation to stamp the instrument." That is, quite clearly, a total misunderstanding of the doctrine of an escrow. The true position, in my judgment, is accurately stated in *Prestons Abstracts*, 2nd ed vol III (1824) p 65, where that work of authority states the law as follows: "The rules respecting escrows are, first, the writing will not operate as a deed till the second delivery. Secondly, the party deputed to make the second delivery cannot give effect to the writing by delivering the same before the conditions are performed. Thirdly, on the second delivery of the writing, it will have relation, for the purpose of title, and not for the purpose of giving a right to immediate rents, etc from the delivery. Fourthly, so as the conditions be performed, and the deed delivered a second time, the deed will be good, notwithstanding the death of both or either of the parties before the second delivery."

Walton J then made a reference to *Halsbury's Laws of England*, which I need not repeat, and he went on to say:

I think that there can really be no dispute as to the law. A document which is intended to take effect as a deed when certain conditions have been fulfilled may be executed as an escrow: that is to say, with all the formalities of a deed save that the vital unconditional delivery, which is essential for the proper execution of a true deed, is missing; it is replaced by a conditional delivery, usually express, but capable of being assumed. At this stage, the document is not a deed; and although of course it contains within itself the possibility of becoming an effective deed, a deed rising phoenix-like from the ashes of the escrow, at the stage before the condition is fulfilled it is of no effect whatsoever. Thus, even in a case where a condition is subsequently fulfilled, the grantee of land under an escrow would never be entitled to the intermediate rents: see, for example, *Thompson v McCullough* [1947] KB 447. When the condition is fulfilled, under the old strict practice there would undoubtedly have been, for the first time, an unconditional delivery of the instrument, which would then become, again for the first time, a deed effecting the purpose therein stated. The theory of the old law was that an escrow had to be delivered to a third party—not to the person intended to take thereunder because, if so, he took immediately under the deed—and on the fulfilment of the condition, the third party would deliver it to the person intended to take thereunder. Indeed, so obvious was this that mere possession of the escrow by such party was taken to be prima facie evidence of the fulfilment of the condition.

This seems to me to be the clearest possible authority, which I ought to follow, that a deed, delivered as an escrow, takes effect from the satisfaction of the condition and not from the date of its delivery as an escrow. Applied to the present case, in my judgment both lease and counterpart took effect together on the satisfaction of the condition of November 18 1977 and, accordingly, such date is "the date hereof," and the rent ought to be calculated from and become payable from that date.

It is common ground that rent has in fact been paid from that date, and therefore the claim in the present originating summons fails and the summons should be dismissed.

At the end of the argument, Mr Cohen realised for the first time that the plaintiffs had parted with the reversion, and sought at that stage to raise a counterclaim to recover the deposit which had been paid before exchange and could no longer be set off against a future rent. When I indicated that such a claim was not open to him in the present proceedings as they stood, Mr Cohen no doubt very wisely did not persist.

For these reasons I propose to dismiss the originating summons with costs.

COURT OF APPEAL

April 1 1980

(Before Lord DENNING MR, Lord Justice BRIDGE and Lord Justice OLIVER)

REGALIAN SECURITIES LTD v RAMSDEN

Estates Gazette June 28 1980

(1980) 254 EG 1191

Landlord and Tenant Act 1954, Part I—Long tenancy at low rent—Maisonette and flat included in one lease—Flat originally

A **used for lessee's domestic staff—Subsequently maisonette sublet for term of head lease less one day and occupied by sublessee —Position at end of lease for purpose of application of the security of tenure provisions of Part I—Maisonette and flat two separate dwellings used and occupied as such—Not let as one dwelling—Lessee accordingly entitled to the protection of Part I only in respect of the flat of which he was in occupation at the end of the lease—Appeal from decision of county court judge allowed**

This was an appeal from a decision of Her Honour Judge Rowland at Westminster County Court who held that the
B respondent, Jack R Ramsden, was entitled under Part I of the Landlord and Tenant Act 1954 to protection in respect of both a maisonette and a flat comprised in a 42-year lease expiring on September 29 1978. The maisonette and flat were located on the top floors of a large block of flats in Petty France in the City of Westminster. The appellants were the lessors, Regalian Securities Ltd.

R Pryor (instructed by Thornton, Lynne & Lawson) appeared on behalf of the appellants; M Blackett-Ord (instructed by Watkins, Pulleyn & Ellison) represented the respondent.

Giving judgment, LORD DENNING MR said: This case con-
C cerns a "penthouse" at the top of a big block of flats in Petty France in the City of Westminster. The block of flats was erected in 1936. An enterprising man, Victor Ramsden—who was about 46 years of age in 1936—saw the block of flats being put up. He inspected the incomplete building and suggested that instead of the two top floors being divided into three flats, as was originally intended, a penthouse should be built for himself and his family, with a domestic flat attached. Nos 83 and 84 were made into a maisonette for his own family and personal use. No 85 was made into a separate and self-contained flat. He used that flat for his domestic staff—the nannies who looked after the children, and so forth.

D There were no Rent Acts operating in 1936 on new buildings like this. Mr Ramsden took a 42-year lease. He thought it would see him out. It was to expire on September 29 1978.

In 1973 Mr Ramsden was about 81 years of age. He thought he would live permanently in the country. So he decided to get rid of the maisonette. His son was living in the domestic flat. He negotiated with his son, and eventually he assigned to him the whole of the lease which was to expire on September 29 1978. I ought to say that Victor Ramsden had paid a very low rent over those 42 years—£300 a year. His tenancy was clearly outside the Rent Act altogether at any time. Mr Ramsden's son bought the lease from him for £10,000—
E presumably with the contents of both maisonette and flat.

The son did not want to use the maisonette himself. He sublet it to a Mr Debye—a friend from Holland—for £8,600. The sublease was for the whole of the head lease less one day. Mr Debye stayed there only a year or so. About the beginning of 1975 Mr Debye assigned the sublease to a Mr Carter. So from 1975 onwards Mr Carter was the undertenant of the maisonette—that is, less one day of the lease. That remained the position always thereafter. Mr Carter was in the maisonette, occupying it. (The position with regard to the furniture is a little uncertain, but I need not bother about it.) Mr Carter's underlease of the maisonette came to an end on September 28
F 1978: and Jack Ramsden's lease of the whole came to an end on September 29 1978—just one day between the two.

Then the question arose as to whether any of the premises were protected by the Landlord and Tenant Act 1954. Clearly none of them was protected by the old Rent Acts, because the rents were less than two-thirds of the rateable value in each case. So the whole question arises on the interpretation of a section of the Landlord and Tenant Act 1954. Section 2(1) of that Act gives protection to a person who holds a long lease at a low rent, as Mr Ramsden did here, if the tenancy had been such that, if it had been one at a low rent, he would have been entitled to retain possession under the old Rent Acts. Under the old Rent Acts there was protection to a person if G
premises were let to him as a separate dwelling, and he was in residential occupation of them himself.

The important section for consideration here is section 22(3) of the Landlord and Tenant Act 1954, which provides:

> In determining, for the purposes of any provision of this Part of this Act, whether the property comprised in a tenancy, or any part of that property, was let as a separate dwelling, the nature of the property or part at the time of the creation of the tenancy shall be deemed to have been the same as its nature at the time in relation to which the question arises, and the purpose for which it was let under the tenancy shall be deemed to have been the same as the purpose for which it is or was used at H
> the last-mentioned time.

That is a very complicated section. I need not go into it all again now, because we have had two or three cases upon it. We considered it in *Herbert v Byrne* [1964] 1 WLR 519, and later on in *Crown Lodge (Surbiton) Investments Ltd v Nalecz* [1967] 1 WLR 647. Taking it quite shortly, we have to look at the position on the term date at the end of the lease. That is, you have to look at it on September 29 1978. You have to look at the nature of the premises then, and you have to look at the purpose for which they were used then.

The nature of the premises on September 29 1978 is clear to my mind. The maisonette, nos 83 and 84, was a separate and J
self-contained dwelling in the occupation of Mr Carter. Although his underlease had come to an end, he was claiming to be there in possession under the Rent Acts. As to the domestic flat, no 85, that was separate and self-contained. It was a separate dwelling in the occupation of Jack Ramsden himself. Looking at the nature of the premises and the use of the premises, it seems to me plain that they were not used as one separate dwelling. They were two separate dwellings—and used as two separate dwellings. So, if one applies the old Rent Act test here, the premises were not let as one dwelling: they were let as two separate dwellings. Such a tenancy under the old Act was not protected at all: because the premises K
were not let as a separate dwelling.

The judge thought that the matter depended on the intention of Jack Ramsden. She was influenced by the intention he had of using the whole penthouse (maisonette and domestic flat) as one dwelling again in the future. But that was not correct. The facts of occupation are more important than the intent. It seems to me plain that there were two separate dwellings. Jack Ramsden is not protected in respect of the whole of the premises: because the maisonette and the domestic flat were used as separate dwellings.

It is different with regard to the domestic flat which Jack Ramsden occupies himself. It is quite plain from section 22(3) L
that that has to be considered separately from the rest. Section 3(2)(b) of the Act expressly says:

> . . . if the premises qualifying for protection are only part of the property comprised in the tenancy, the tenancy while continuing after the term date shall have effect as a tenancy of those premises to the exclusion of the remainder of the property. . . .

The statute there contemplates that there may be protection in respect of part of the premises. It seems to me that Jack Ramsden was in occupation at the end of the tenancy of the domestic flat. If it was a Rent Act case in the old days, it would be regarded as let as a separate dwelling to him in which M
he was in residential occupation. He would certainly be entitled to protection of it. The result is that Mr Ramsden is entitled to protection in respect of the domestic flat. The rent and all other consideration should be determined in accordance with the provisions of the statute.

Mr Ramsden is not entitled to the continuance of the lease of the whole of the premises for the simple reason that he was not in any way in occupation of the maisonette. That was in the separate and self-contained occupation of Mr Carter.

I should say a word about the furniture. The judge was influenced by that. She said: "The family furniture remained and still remains in the maisonette." As a result of the further evidence which we have admitted, it is not altogether clear as to whom the furniture belonged. It seems to me that it would be wrong to place any reliance upon the title to the family furniture. It looks as though quite a good deal of it may have passed first to Mr Debye and then to Mr Carter. Some may belong to Jack Ramsden or to the family. At all events, no help can be gained from the furniture. The plain fact is that at the end of the lease Mr Carter was in occupation of this separate dwelling.

It seems to me that Jack Ramsden cannot claim the continuance of the lease of the whole premises, but only in respect of the domestic flat.

I would allow the appeal accordingly.

Agreeing, Bridge LJ said: If premises are constructed as a single dwelling-house and let as such, the tenant may be protected under the Rent Act in his possession of the whole notwithstanding that parts of the dwelling-house are sublet as separate dwellings: see *Berkeley v Papadoyannis* [1954] 2 QB 149 and *Herbert v Byrne* [1964] 1 WLR 519. Conversely, if two structurally separate and distinct units are let to a single tenant for his single occupation, he may be protected in his possession of both units so long as he is in occupation of both: see *Haines v Herbert* [1963] 1 WLR 1401. But when two structurally separate and distinct units are let, albeit under a single lease, for the purpose of being occupied as two separate and distinct dwellings, the tenant cannot claim to retain possession of the whole, because the whole is not let to him as a single separate dwelling. That, I think, must follow from the decision of this court in the case of *Horford Investments Ltd v Lambert* [1976] Ch 39, where it was held that in the phrase "let as a separate dwelling" as found in section 1 of the Rent Act 1968 the word "dwelling" must be construed as not including the plural.

That is exactly the situation we have to contemplate here, since by section 22(3) of the Landlord and Tenant Act 1954 the premises, which have always been structurally divided into the separate and distinct maisonette and flat, are deemed to have been let for the purpose for which they were used at the date of expiry of Mr Ramsden's contractual lease. At that date Mr Ramsden used the flat as his dwelling, but Mr Carter continued to use the maisonette as his dwelling. It is true that Mr Carter's contractual tenancy expired on the previous day and he enjoyed no statutory protection save that he could not be evicted without an order of the court. But these considerations do not prevent the conclusion that his use of the maisonette at the material date gave rise to the deemed letting of the maisonette for the purpose of that use. It follows in my judgment that Mr Ramsden can claim no Rent Act protection in respect of the maisonette.

On the other hand, I have no difficulty in reaching the conclusion that Mr Ramsden is entitled to protection in his possession of the flat. Section 2(1) and section 3(2) of the 1954 Act both contemplate that a tenant may be entitled to protection in respect of part only of the premises comprised in a long tenancy at a low rent, and the language of section 22(3) clearly envisages that this partial protection will be attracted when part only is, at the termination of the lease, let as a separate dwelling which the tenant occupies.

Accordingly I, too, would allow the appeal so far as it concerns the maisonette.

Oliver LJ also agreed.

A declaration was made the effect of which was that the tenancy to which Part I of the Landlord and Tenant Act 1954 applied was the tenancy of the flat only. The appeal was allowed with costs in the Court of Appeal, no order to be made for costs below. Leave to appeal to the House of Lords was refused.

QUEEN'S BENCH DIVISION

December 21 1978

(Before Mr Justice NEILL)

MELVILLE AND ANOTHER v GRAPELODGE DEVELOPMENTS LTD

Estates Gazette June 28 1980

(1978) 254 EG 1193

Landlord and tenant—Whether obligation under lease to pay rent and obligation under an undertaking by landlords to carry out repairs were dependent or separate—Whether a claim for unliquidated damages for breach of the undertaking was capable of being set off against a claim for rent—Held that the obligation of the tenants under the lease to pay the rent and the obligation of the landlords to carry out repairs were separate, albeit closely linked, obligations—Held also that the claim for unliquidated damages was capable of being set off against the claim for rent in view of the close connection between the two claims—Authorities on equitable set-off considered—Further development of the principle

These proceedings were concerned with the issue of liability only, not with the issue of damages, in an action by the intermediate landlords of premises at 23 and 24 Easton Street, London WC1, against their tenants for arrears of rent and payment of a proportion of an insurance premium. The tenants, the defendants, counterclaimed for alleged breach of an undertaking to carry out certain repairs. The matter came before Neill J following a master's order that there should be a trial on the issue of liability on the claim and counterclaim before the issue of damages was tried.

K Topley (instructed by Routh, Stacey) appeared on behalf of the plaintiffs; D Naish (instructed by Belmont & Co) represented the defendants.

Giving judgment, Neill J said: The plaintiffs in this action are the mesne landlords of premises at 23 and 24 Easton Street, London WC1. The defendants are in occupation of the third floor of those premises under a lease dated March 21 1977 whereby the plaintiffs demised the third floor to the defendants for the term of 25 years from September 26 1976. The plaintiffs' claim is for arrears of rent alleged to be due under the lease for the four quarters of 1977, that is the rent due in March, June, September and December 1977. In addition the plaintiffs claim the sum of £267.68 in respect of what is called insurance rent, being the sum alleged to be due as the proportion payable by the defendants in accordance with clause 1(5) of the lease toward insurance premiums on the premises. The total claim set out in the writ is therefore for the sum of £1,967.68. The defendants deny that this sum or any sum is due to the plaintiffs; they rely on a written undertaking given by the plaintiffs in a letter dated March 21 1977; they also counterclaim in respect of an alleged breach of that undertaking.

The matter comes before me for trial on the issue of liability only. Thus by an order of Master Elton for Master Waldman dated June 7 1978 it was ordered that the issue of liability on the claim and counterclaim should be tried before the trial of the issue of damages. Before me no oral evidence has been given and the evidence has been limited to (a) the lease dated March 21 1977; (b) an agreed bundle of correspondence containing a number of letters passing between the parties and their solicitors between September 8 1976 and August 11 1978; (c) an agreed schedule showing the weather conditions in central London between March 21 and 25 1977.

The defendants trade under the name of Clements Press; they are printers and platemakers. On September 6 1976 they went into occupation of the third-floor premises at 23 and 24 Easton Street. At the outset the defendants were allowed to occupy the premises rent-free; this period of rent-free occupation was originally from September 6 1976, but the period was subsequently extended by a month until January 6 1977.

Beginning in December 1976 there was correspondence between the solicitors for the parties concerning the condition

A of the premises. I shall refer only to the material paragraphs. On December 14 1976 the solicitors for the defendants wrote as follows:

> Our clients have consulted a surveyor who has suggested that a schedule of conditions is inserted in the lease. Our clients further inform us that the roof leaks in places and that your clients' agents have promised to decorate and repair the entrance hall and staircases but that they have not done this as yet.

On December 15 1976 the plaintiffs' solicitors replied:

> We have your letter dated 14th inst and note the various points you raise with regard to the condition of the above premises and as to the common parts of the building etc. We are inquiring from
> B Messrs Walker Son & Packman with regard to the question of decorations of the common parts to which you refer but your letter is the first time that this question has ever been raised although your clients have been in the premises since September 6 last. Our clients only allowed your clients to go into possession on the date mentioned in order that they might be able to make them ready for their business purposes and we have little doubt that they have done this and are carrying on business on the premises notwithstanding the defects to which you refer and apart from the fact that more than three months have elapsed and rent should now be payable on the terms which had been agreed and which are set out in the lease.

On December 22 the plaintiffs' solicitors wrote again (in
C reply to a letter which I have not seen, dated December 20):

> Your client has stated that the roof of the property is leaking. We are informed by the agents that repairs have taken place to the roof and that any leaks have now been stopped. If further leaks are occurring we suggest that your clients contact the agents once again, and they will no doubt do what they can to rectify the matter. As far as any other repairs are concerned our information is that no undertaking has been given by the agents at any time that repairs or redecorations would take place within a specific time scale. The ability or otherwise to undertake repairs will of course depend on financial considerations, which will naturally improve as soon as your clients are paying rent.

D On January 4 1977 the defendants' solicitors replied:

> We have referred to our clients on the question of the condition of the premises. They are adamant that they went into the premises on a clear undertaking by a representative of your clients' agent and of Mr Sheffield of your client company that these defects would be remedied and that no rent would be payable until the defects were properly remedied. It is for this reason we feel that a schedule of condition is appropriate.

On January 5 the plaintiffs' solicitors wrote:

> Concerning the condition of the premises we have today spoken to the agents who are equally adamant that no undertaking of any kind was given to your clients. Indeed, if any such undertaking had been given, the agreement for a three-month rent-free period would
> E surely have been meaningless. Your letter to us of December 14 contains the first reference to a schedule of conditions, which at that time was "suggested" by your client's surveyors some three months after they had had possession of the premises, and we feel that this hardly supports the existence of an undertaking. As we have stated in previous correspondence, your client was aware of the terms on which the lease would be granted at the time possession was obtained for the purposes of carrying out decorations, and the three-month rent-free period was not given for any other reason and to make up for any sums which needed to be spent. Thus, we remain of the opinion that a schedule of conditions should not be incorporated.

Finally on January 6 the defendants' solicitors replied:

> F However our clients' instructions are clear to the effect that they only moved into the premises on the basis of representations made by your clients and/or their agents and it is unfortunate that there appears to be this conflict between us. As we are so near each other, would it not be possible to arrange a meeting, attended by solicitors and clients to resolve this outstanding matter?

The suggestion of a meeting was adopted and on Monday, January 17 1977 a meeting duly took place. I have no record of the meeting, though there is a reference to it in a letter from the plaintiffs' solicitors. It is clear that at that meeting there were discussions to resolve the outstanding differences between the parties and it was agreed that rent should be payable from January 6 1977 to March 25 1977. On January 19 G
1977 the plaintiffs' solicitors wrote to say that on completion of the lease rent payable for that period would be £363.29. There was then further correspondence between January and March in which reference was made to an undertaking to be signed by the plaintiffs to carry out certain repairs to the roof and windows. On February 22 1977 the plaintiffs' solicitors sent a draft undertaking to the defendants' solicitors for their consideration.

I now come to the undertaking which was finally signed by the plaintiffs' solicitors on behalf of their clients. It will be remembered that the lease itself between the plaintiffs and the defendants was dated March 21 1977. On the same day the H
plaintiffs' solicitors signed a letter in these terms:

> Lease of Third Floor: 23/24 Easton St, London WC1. In consideration of your completing this matter today, we hereby undertake that our clients will at their own expense (a) have repairs to the present roof defects carried out as soon as possible, and (b) have repairs to the present defects in the window surrounds carried out as soon as the weather permits but in any event such works to commence not later than three months from today and such work to be completed to the reasonable satisfaction of the Lessees. It is agreed that the matter is completed on the basis that the Tenants' contribution to all future repairs under Clause 2(5) of the Lease shall not be in excess of 20 per cent of the total cost
> thereof. J

The subsequent events can be stated shortly. The lease was duly signed and completed by the defendants. On March 25 1977 a quarter's rent became due under the lease. This rent was not paid nor were any subsequent payments made in respect of subsequent quarters. Furthermore it is conceded for the purposes of the present case that no repairs, or at any rate no sufficient repairs, were carried out by the plaintiffs in fulfilment of the undertaking contained in the letter of March 21 1977.

It is in these circumstances that the plaintiffs bring this action for arrears of rent and for a further sum in respect of insurance rent. The claim, as I have already stated, amounts K
to £1,967.68 and includes four quarters of rent amounting to £1,700 for the year beginning March 25 1977. The contentions on behalf of the parties can be summarised as follows. The plaintiffs say (1) the sums claimed are due under the lease; (2) they are not in breach of the undertaking, because their obligation under that undertaking was suspended as soon as the defendants defaulted in payment of the rent; (3) even if the obligation to carry out repairs continued, the defendants are not entitled as a matter of law to set off any claim for damages for breach of the undertaking against the claim for the rent. The defendants on the other hand contend (1) their
obligation to pay rent was suspended as soon as the plaintiffs L
became in breach of the undertaking; (2) the plaintiffs were in breach of the undertaking of March 21 on or before March 25 because (a) the obligation to carry out repairs to the roof defects was to carry them out as soon as possible and that that obligation was not subject to the words "in any event such works to commence not later than three months from today"; (b) the work could have started in the four days between March 21 and 25; (c) the weather conditions as set out in the schedule were not such as to prevent work beginning; (3) if, as is conceded in the defence and counterclaim, the rent is due, the defendants are entitled to set off against
the claim the amount of their damages. M

It is to be noted that the first argument on behalf of the defendants, that the obligation to pay rent was suspended by the plaintiffs' breach of the undertaking, was not pleaded. Nevertheless it was argued on behalf of the defendants without any objection on behalf of the plaintiffs, and I therefore propose to deal with it.

I can deal very shortly with the plaintiffs' contention that their obligations under the undertaking were put into suspension by the failure of the defendants to pay the rent due on March 25 1977. I am satisfied that this argument is wrong. The undertaking was given in consideration of the defendants' "completing this matter today." The defendants did duly complete the matter by signing the lease. The obligation of the defendants under the lease to pay the rent and the obligations of the plaintiffs under the undertaking to carry out the repairs are to my mind separate, albeit closely linked, obligations. I am reinforced in this view by the decision of du Parcq J in *Taylor v Webb* [1937] 2 KB 283. The decision of du Parcq J was reversed by the Court of Appeal, but there is nothing in the judgments of the Court of Appeal to throw any doubt on the views of du Parcq J on this point. It is further to be noted that the decision in *Taylor v Webb* in the Court of Appeal was itself overruled by the House of Lords in *Regis Property Co Ltd v Dudley* [1959] AC 370.

I am also satisfied that there is no sound basis for the defendants' argument that their obligation to pay rent was suspended because of the plaintiffs' breach of the undertaking. By the lease dated March 21 the defendants undertook an obligation to pay rent in accordance with the terms of the lease, and in my view the obligation was not removed or altered by any breach by the plaintiffs of the undertaking. Furthermore I would find it impossible on the present evidence to come to the conclusion that the plaintiffs were in breach of that undertaking by March 25 1977. Discussions about repairs to the roof had been continuing since December 1976, and it seems to me to be quite unrealistic to treat the plaintiffs as in breach of the undertaking by reason of the fact that they had not started work by March 25. In any event, however, I consider that the obligation to carry out the repairs was subject to the words "but in any event such works to commence not later than three months from today."

I turn therefore to the second main argument put forward by Mr Topley on behalf of the plaintiffs. He argued that even if the obligation to carry out the repairs continued, the defendants are not entitled as a matter of law to set off any claim for damages for breach of the undertaking against the claim for rent. In support of this submission Mr Topley referred me to the decision of Goff J (as he then was) in *Lee-Parker v Izzet* [1971] 1 WLR 1688. In that case Goff J held that where a landlord has failed to comply with his obligations to carry out repairs under the terms of a tenancy agreement and the tenant has carried out the work himself, the tenant is entitled to treat the expenditure as a payment of rent. The learned judge referred to the old case of *Taylor v Beal* (1591) Cro Eliz 222 as authority for the proposition that the right to recoup the expenditure out of future rents was an old common law right. Mr Topley submitted that it was implicit in the decision of Goff J that unless the tenant had actually expended money on repairs in fulfilment of the landlord's obligation, there was no right to any set-off against his obligation to pay the rent. With respect to Mr Topley, I do not consider that one can make any such deduction from Goff J's judgment. Indeed he said in terms at p 1693 that the ancient common law right of recoupment had nothing to do with technical rules of set-off. Counsel were unable to refer me to any cases which threw light on the right to claim a set-off against a claim for rent and therefore, as the question of set-off was a matter of some general application, I thought it right in the circumstances to reserve my judgment.

It might be argued that the obligation to pay rent is an obligation of a special nature and that the ordinary rules of equitable set-off as explained by Morris LJ in *Hanak v Green* [1958] 2 QB 9 do not apply. It is certainly a matter for surprise that, certainly as far as the searches of counsel can find, there does not seem to be any reported authority since the Judicature Acts in which a claim for damages for breach of a repairing covenant has been allowed by way of set-off for an action for rent. Indeed in *Hart v Rogers* [1916] 1 KB 646 Scrutton J held that a cross-claim by a tenant for breach of an implied covenant to repair did not constitute a defence to a claim by the landlord for unpaid rent of the premises. It is to be noted, however, that in that case Scrutton J was not referred to the principles underlying the doctrine of equitable set-off; in particular he was not referred to the decision in *Beasley v D'Arcy* ((1800) noted in 2 Sch & Lef 403) in which the court in Ireland granted relief in a case of ejectment by way of injunction where the tenant was able to set up a claim for unliquidated damages against the landlord for breach of covenant. The report of the decision in *Beasley v D'Arcy* appears as a note to the report of *O'Mahoney v Dickson* (1805) 2 Sch & Lef 400, a case decided before Lord Redesdale.

It can be further argued in this case that the claim for damages arises not under the lease itself but under a separate contract constituted by the undertaking contained in the letter of March 21. Accordingly, it could be said, the decision in *Beasley v D'Arcy* is distinguishable.

I have come to the conclusion that at the present day there are no adequate grounds for treating an obligation to pay rent as an obligation which is different in kind from that imposed by other contracts. It follows therefore that in my view a claim for unliquidated damages is *capable* of being set off in equity against a claim for rent provided there is a sufficiently close connexion between the two claims.

In *Hanak v Green* [1958] 2 QB 9 Morris LJ referred at p 24 to the decision in *Bankes v Jarvis* [1903] 1 KB 549. In that case the plaintiff's son had bought a veterinary surgeon's practice from the defendant. For the purpose of carrying on the practice the plaintiff's son made use of a house of which the defendant was the lessee and the son covenanted to pay the rent and to perform the covenants and to keep the defendant indemnified. After carrying on the practice for a time the plaintiff's son left the country after giving the plaintiff authority to sell the practice. The plaintiff sold the practice to the defendant. The sum of £50, being part of the agreed price for the practice, was not paid by the defendant, but he had a claim against the plaintiff's son for £51 consisting of £21 by way of rent which the plaintiff's son had failed to pay in respect of the premises and a further £30 which the defendant had had to pay because the son had failed to perform the covenants in the lease. The defendant was held to be entitled to set up the £51 by way of a set-off against the claim for the balance of the purchase price. Morris LJ said that he regarded the decision as clearly correct and added: "There was a close relationship between the dealings and transactions which gave rise to the respective claims."

I should also bear in mind the words of Lord Denning MR in *Federal Commerce & Navigation Co Ltd v Molena Alpha Inc* [1978] 3 WLR 309 where at p 338 he said:

> The courts of equity . . . allowed deductions—by way of equitable set-off—whenever there were good equitable grounds for directly impeaching the demand which the creditor was seeking to enforce. . . . These grounds were never precisely formulated before the Judicature Act 1873. It is now far too late to search through the old books and dig them out. Over 100 years have passed since the Judicature Act 1873. During that time the streams of common law and equity have flowed together and combined so as to be indistinguishable the one from the other. We have no longer to ask ourselves: what would the courts of common law or the courts of equity have done before the Judicature Act? We have to ask ourselves: what should we do now so as to ensure fair dealing between the parties? See *United Scientific Holdings Ltd v Burnley Borough Council* [1977] 2 WLR 806, p 812 per Lord Diplock. This question must be asked in each case as it arises for decision: and then, from case to case, we shall build up a series of precedents to guide those who come after us. But one thing is quite clear: it is not every cross-claim which can be deducted. It is only cross-claims that arise out of the same transaction or are clearly connected with it. And it is only cross-claims which go directly to impeach the plaintiff's demands, that is, so closely connected with his demands that it would be manifestly unjust to allow him to enforce payment without taking into account the cross-claim.

Lord Denning MR used similar language in *The Brede* [1974] QB 233, where he said at p 248: "It (equitable set-off) is available whenever the cross-claim arises out of the same transaction as the claim; or out of a transaction that is closely related to the claim." In the present case I am satisfied that

A the claim to rent was closely connected with the rights given to the tenant by the undertaking contained in the letter of March 21.

It should be appreciated that I have heard no evidence about the quantum of the tenants' claim and that I am deciding this matter simply as one of principle. On that basis I have reached the conclusion that the defendants' counterclaim for damages for breach of the undertaking is capable in law of B constituting a valid equitable set-off against the plaintiffs' claim for rent.

The defendants were awarded the costs of the issue decided by the judge. It was directed that the action should be tried by a master if one was available; if not the case should go into the ordinary list.

LANDLORD AND TENANT

BUSINESS TENANCIES

A **COURT OF APPEAL**

October 17 1979

(Before Lord Justice MEGAW, Lord Justice SHAW and Lord Justice WALLER)

FAMILY MANAGEMENT v GRAY

Estates Gazette January 26 1980

(1979) 253 EG 369

Landlord and Tenant Act 1927, section 18(1)—Action by reversioners on expiry of two long leases for damages for failure to comply with repairing covenants—Subtenancies had been granted to business subtenants of shops forming part of the premises who had rights under Part II of the Landlord and B Tenant Act 1954—Subtenants negotiated full repairing leases for 20 years with the reversioners—Difficulties in the assessment of damages under section 18(1) of the 1927 Act discussed—Whether real loss in respect of reversion so far as it related to the shops was substantial or was minimal when regard was had to the new leases granted to the tenants of the shops and the impact of section 34 of the 1954 Act—Official referee in error in calculating the damages on the basis of so many years' purchase of assumed loss of rental value without taking account of the new full repairing leases at rents in accordance with section 34—Appeal allowed

C This was an appeal by the defendant, Mr Alfred William Gray, from the decision of the late Judge Laughton-Scott, an Official Referee, in favour of Family Management (a company incorporated with unlimited liability), the plaintiffs, in an action for damages for breach of repairing covenants relating to premises at 137 and 139 South Lambeth Road, London. The plaintiffs had acquired the reversion in respect of the two properties on two leases granted in January 1887 for terms ending in December 1974. The residue of the terms under the leases was vested in Mr Gray. The premises consisted of shops on the street level with living accommodation above, the shops being the subject of subleases to a dry cleaner and a D person carrying on a delicatessen business respectively.

M Stimpson (instructed by Chapman & Wilson, of Brighton) appeared on behalf of the appellant; R R Russell (instructed by Gale & Phelps) represented the respondents.

Giving the first judgment at the invitation of Megaw LJ, SHAW LJ said: This is an appeal by the defendant in the action from a judgment of His Honour the late Judge Laughton-Scott, given in London on April 7 1978. The action was brought by the reversioners of two leases which had been granted as building leases in January 1887 for a term of 90 years, and they related to premises in South Lambeth Road in London, which was then a neighbourhood of higher standing E and amenity than it is at the present time. The premises in question were nos 137 and 139 South Lambeth Road and they consisted of shops on the street level with living accommodation above. The leases, when originally granted, contained full repairing covenants on the part of the lessees.

At some time in the course of the history the plaintiffs acquired the reversion in respect of both properties. In about 1947, or perhaps earlier, the defendant, Mr Gray, acquired the residue of the term of the two leases, which expired in December 1974. The time came when he, in turn, sublet the shops which were part of those premises to sublessees. No 137 was let in 1963 to a gentleman who carried on the business F of dry cleaner, and no 139 was let to a lady who carried on the business of a delicatessen shop. In the subleases to them there were covenants also by the sublessees to keep the premises in proper repair.

When the defendant's leases came to an end and the reversion fell in the covenants to repair contained in his head lease and in the subleases had not been complied with. It is true that in the course of the history the defendant had spent a good deal of money on improvements to the property, but it was quite evident that in December 1974 the premises were not in the state called for by the covenants in his leases.

In June 1974, that is some six months or so before their G respective leases were due to expire, the business subtenants of the shops on the ground floor were served with notices by the landlords, the plaintiffs in this action, who indicated that they would not object to a renewal of the tenancies under the Landlord and Tenant Act of 1954. Those are matters either to be negotiated or, if any dispute arose, to be determined by the court under the provisions of section 34 of that Act.

What happened thereafter was that the subtenants of the respective shops negotiated leases for 20 years upon terms which were arranged between them and the plaintiffs, and their respective leases also contained covenants to put and keep the H premises in repair. When the defendant's leases came to an end, despite their state of disrepair, the plaintiffs did not serve any notice or schedule of dilapidations. No action was then taken in respect of the failure by the defendant to comply with his covenants to keep the premises in repair. It was not until February 1976 that the plaintiffs issued their writ. By that time, the possession of the sublessees under their leases had become established. By the amended statement of claim the plaintiffs asked for damages, first relating to the cost of painting and repairing the premises amounting to £6,110.6 and there was added to it various incidental costs relating to the fees of surveyors and so on. So that the total amount that J was claimed was something over £6,500.

Actions in respect of the failure to comply with repairing covenants resulting in a loss to the reversion present peculiar complications, and the provisions of section 18 of the Landlord and Tenant Act 1927 have given rise to expressions of judicial opinion which are not always easy to reconcile. Subsection (1) of that section provides that:

> Damages for a breach of covenant or agreement to keep or put premises in repair during the currency of a lease, or to leave or put premises in repair at the termination of a lease, whether such covenant or agreement is expressed or implied, and whether general or specific, shall in no case exceed the amount (if any) K by which the value of the reversion (whether immediate or not) in the premises is diminished owing to the breach of such covenant or agreement as aforesaid; and in particular no damage shall be recovered for a breach of any such covenant or agreement to leave or put premises in repair at the termination of a lease, if it is shown that the premises, in whatever state of repair they might be, would at or shortly after the termination of the tenancy have been or be pulled down, or such structural alterations made therein as would render valueless the repairs covered by the covenant or agreement.

A That section presents its own difficulties. When the matter is further complicated by the impact of the provisions of section 34 of the Landlord and Tenant Act of 1954, which provides that:

> The rent payable under a tenancy granted by order of the court under this Part of this Act shall be such as may be agreed between the landlord and the tenant or as, in default of such agreement, may be determined by the court to be that at which, having regard to the terms of the tenancy (other than those relating to rent), the holding might reasonably be expected to be let in the open market by a willing lessor,

B and then there follow certain factors which are to be disregarded in determining what that figure is—the problem may almost defy resolution. In the present case the related impact of those two provisions in those respective statutes gave rise to the problems which the learned judge at the trial had to consider.

On the one hand, evidence was called in support of the plaintiffs' claim to show that the want of repair, if it were redressed, would require the expenditure of sums amounting to some £6,000. If that were the true basis for an assessment of the damage to the reversion there would be no difficulty, but, as section 18 of the 1927 Act expressly says, that is not
C necessarily the answer to the question.

Evidence was called by an expert on behalf of the defendant who said that, having regard to the local conditions and to other contemporary factors, the actual loss to the value of the reversion, as opposed to any loss in the intrinsic value of the premises, was something in the region of a few hundred pounds. It might be £500 or £800, or something of that kind.

That was the conflict that presented itself to the learned judge. He came to the conclusion, after a great deal of evidence had been given on both sides. After reviewing that evidence he said:

> But the matter does not end there because I am required, as
> D is conceded by both sides, to ensure that no damages should be awarded to any extent beyond that by which, in view of the dilapidations at the end of the lease, the value of the reversions has been diminished, that is the value of the premises to the landlords. On that a great deal of evidence has been called. This lease happened to come to an end at the end of 1974 when there was (and only then was at its beginning) a considerable business slump, and I have to have regard for the consequences of that business position. I have been shown by surveyors, and so forth, for both parties—and they have done all they could to put before the court comparable figures—details of a number of houses and shops, including one comparable house and shop at the very end of this terrace, no 127 of the relevant road. There it so happens that we see a lease of that property was
> E granted in 1976 at the annual sum of £1,600 a year. That compares with rentals for these two properties, of no 139 with a yearly rent of £1,200 rising to £1,400 from December 25 1975, more or less one year after its inception; and in respect of the other at a commencing rent of £1,100. A great deal of evidence has been given to show how the position might have been affected had these repairs been carried out. Much of it is speculation, much of it is supposition; it has to be, in the very nature of things. But I have to keep in the forefront of my mind the actual characteristics of the business situation which confronted all concerned at Christmas 1974.
>
> At the same time I have to envisage a situation as to what better the landlords might have done for themselves if at
> F that time they had, or the tenant had, expended upon two of these properties a sum in excess of over £2,000.

He then dealt with the upper part of the premises. I omit that because it is not the subject-matter of any part of this appeal. He went on:

> I now turn to the business premises. The tenants there, who were former tenants of Mr Gray's, have undertaken 20-year leases, copies of which are before me, in which they have undertaken a full repairing obligation. That, however, is nothing new for them because they have recently, at the end of 1974, held tenancies which then expired, which themselves had full repairing covenants in them. It so happens that Mr Gray, out of terms of friendship or affection, has not seen proper to bring in any of the tenants of the properties as third parties in these proceedings; G
> that is a matter entirely for his judgment. What I have to consider is simply, had an appropriate amount of repair been done to these premises, what extra rents in terms of net emolument would these premises have afforded? Again, I think, in view of the uncertainties in the matter, and it being for the plaintiffs to prove their case, in respect of each of these premises I should ascribe a sum of £100 a year more for each, had the premises at the expiry of the tenancy (as they should have done) been handed over in proper condition.

Leaving out the domestic part of the premises, that meant that he had attributed to each of the shop premises a loss of
£100 a year to the landlords in respect of the value of the H
reversion. He continued to the effect that taking what might normally be in the case of new premises a nine years' purchase, an eight years' purchase was appropriate here. He therefore multiplied the £100 for each of the premises by eight. That resulted in a total of £1,600 as the figure attributable to the damage to the reversion of the shop premises. It is that figure which is the subject-matter of this appeal, Mr Stimpson not having sought, either by his notice of appeal or by any argument in this court, to attack in any way the figures relating to the residential premises.

The argument put forward on behalf of the appellant is that J
in arriving at that figure the learned judge had erroneously founded himself upon an artificial situation, and that he had probably been led into so doing by a passage in the cross-examination of the expert called on behalf of the defendant. That witness, Mr N J Winbourne [FRICS], was asked, assuming the ability to convey vacant possession of the premises at the end of December 1974, what difference there might be in the rental value of the premises. Mr Winbourne, who was the defendant's witness, agreed that in such a hypothetical situation he might attribute £100 per annum loss in respect of each of the shop premises at 137 and 139 South Lambeth Road. But
that answer was clearly given diffidently and with much K
expostulation that the question related to a situation which was not the real situation which had to be considered. It was not the real situation because of the operation of section 34 of the Landlord and Tenant Act 1954, and because there were business tenants in occupation who obviously had an interest themselves to protect and of whom it might be said that the almost certainty was that they would claim new leases under the provisions of section 34 when the occasion arose. As Mr Stimpson has urged in this court, when the negotiation of the terms of those new leases came to be dealt with between the reversioners and the prospective lessees they could not, in
diminution of what was the proper rent to be paid, urge their L
own default in having failed to comply with the repairing covenants under the lease as a justification reason for a lower rent, whether arrived at by negotiation or determined by the court.

That has been the crux and gravamen of Mr Stimpson's submissions in this court, and he has said that it is clear from a reading of the judgment, in which section 34 of the 1954 Act is nowhere specifically referred to, that the learned judge had taken Mr Winbourne's evidence as if it were evidence relating to the actual facts of the case, instead of being evidence which related to a hypothetical situation which had been put to him
by counsel for the landlord. The figures which the learned M
judge determined indicate that he was in this respect looking at the evidence of Mr Winbourne rather than at the evidence of Stuart Glyn [FRICS ACIArb], who was the expert called on behalf of the plaintiffs. Mr Glyn's evidence was so far removed from anything like £100 a year in respect of each of the shop premises as to make it quite impossible that the learned judge accepted his figures as a tenable foundation for assessing what was the damage to the reversion.

In the course of the argument complicated and difficult problems were suggested. I need refer only to one of the many authorities cited, namely *Smiley v Townshend* [1950] 2 KB 311.

A The headnote reads:

> The assignee of a long lease of premises, containing full covenants to repair, which was assigned to him two years before the date of the termination of the lease, when the premises were already requisitioned, was sued by the landlord on the termination of the lease for breaches of the covenants to repair. The premises at the date of termination of the lease were still requisitioned, and the date of the termination of the requisition was undetermined.

It was contended that one had to have regard to the fact that the premises being requisitioned at the date when the reversion fell in, the loss to the landlord must be assessed by reference
B to the time when they actually physically reverted to his possession.

The relevant holdings set out in the headnote are first, "The damages must be assessed not by reference to the time when the land reverted into the possession of the landlord at the termination of the requisition, but as at the date of the termination of the lease." Next is holding number (5) which reads: "Future events, after the termination of the lease, could not in themselves reduce or extinguish the damages, but they might be taken into account in so far as they threw light on the value of the reversion as at the date of the end of the lease." At first reading that appears not only cryptic but possibly self-
C contradictory. However, when one comes to look at the relevant passage in the judgment of Denning LJ at p 320, the principle becomes clear.

I read from the top of that page:

> What has to be considered is the diminution in value of the reversion at the end of the lease; and I take the word "reversion" to mean the landlord's then interest in the premises. If he is the owner and the premises are requisitioned, then the reversion is his freehold, subject to the existing requisition, because that is the landlord's then interest in the premises. The question is, therefore: how much was the market value of the landlord's interest diminished at the end of the lease by reason of the disrepair for which the then assignee was responsible?
> D Any other view would give rise to great difficulty. For instance, if the future cost of repair to the landlord were the measure, the court would be asked to deduct the estimated amount of repairs which will be done by the corporation after the lease comes to an end.

I turn from there to the top of p 322, where Denning LJ continues:

> Mr Shelley took another instance which is not within the last half of section 18 (1). He took the example of a house let on a long lease which was, by reason of its large size and great number of rooms, not suitable for occupation as a dwelling-house at the present day, but only for use as a warehouse, for which
> E decorative repairs would be valueless. If the decorations were not in fact done, he said, the court would take that fact into account, even though it did not fall within section 18. I agree that the court would do so, but the reason would be because in such a case the damage to the reversion would be nil. At the end of the lease the premises would be of as much value as a warehouse as a highly decorated dwelling-house. These illustrations show that, although future events do not in themselves reduce or extinguish the damages, nevertheless they may properly be regarded in so far as they throw light on the value of the reversion at the end of the lease.

So events which follow upon the determination of the lease
F but which are independent of any fact or consideration which was either operative or potential at the date when the lease expired cannot affect the determination of the loss to the value of the reversion. As I read the judgment cited it decides that factors which existed, either as operative or potential factors but which had substance and reality at the date when the reversion fell in, were to be considered as having a bearing on the valuation of the loss or damage to the reversion by the failure on the part of a lessee to comply with his covenants under a lease.

Upon that basis it was requisite in this case for the learned judge to address his mind to a factor which had become a reality by the date of the hearing, but which existed potentially at the date when the reversion fell in in December 1974, G
namely the right of the respective tenants of the two shop premises to look for a fresh lease under section 34 of the Landlord and Tenant Act of 1954. Such leases were, indeed, negotiated between the plaintiffs and the respective sublessees (as they had hitherto been) and those leases themselves incorporated full repairing leases putting the respective lessees under an obligation to put and keep the premises in good and substantial repair. That in my view is a factor which must be taken into account at the time the court is asked to determine the proper figure for compensation for the damage to the reversion. It may produce the result that the reversioner has sustained no damage at all. H

Mr Russell, in his able argument to this court, has contended that this begs the question. He relied on the necessity to look at the position as if, in a free market, there were in prospect a sale with vacant possession to a willing tenant. Thus although there may be some force in the suggestion that the prospect of the court's granting a lease to a prospective purchaser if there was no lease negotiated might have its impact, in the circumstances of a case such as this that ought to be disregarded. What should be looked at was the actuality of the case in relation to the condition of the premises at the end of 1974. It was thus relevant to look at comparable premises in that area to find what were comparable rents in order to J
determine whether or not the landlords might have done better with the bargain that they had made if, in fact, the premises had been kept in the state of repair which was required by the leases and which had been held by the defendant.

That as I see it does not fall within the principle enunciated by Denning LJ in *Smiley v Townshend*. Moreover, it denies recognition of what were the realities of the situation. The fact that 20-year leases were granted perhaps does not in itself matter, and to that extent I would accept the argument put forward by Mr Russell, but that there were leases in prospect and that there was a right on the part of the sublessees to look for new leases, and that the chances were that they K
would arise by negotiation or be granted by the court was a reality which ought and had to be recognised. From my reading of the learned judge's judgment, that was not given its due weight.

He accepted the proposition put forward that one had to look at the situation as though this was a case where vacant possession could be given in an open market to a willing prospective tenant, and upon that basis he said that the decrease in the annual rental value was £100 in respect of each shop, and so on. If he had taken into account, as was called for, the full impact of section 34 of the Landlord and Tenant Act 1954 in the history of this matter, he would have been led to L
the conclusion that when the reversion fell in at the end of 1974 the damage to the reversion had been reduced either to nothing or something so close to it as to be *de minimis* and not to require or call for judicial assessment.

In my judgment, therefore, the learned judge was in error adopting, as he apparently did, what he thought was the effect of the evidence of Mr Winbourne as applied to the situation. He accordingly assessed the annual loss at £100 a year in respect of each of the shops and so arrived at a total of £1,600 for both of them over an eight-year period. I think that that figure was not merely in excess of the right figure, but was £800 more in respect of each shop than it ought to have M
been. In other words, the proper figure was really zero or minimal. There had been no damage to the reversion which was capable of real assessment save as a minute figure. Accordingly, I would allow this appeal.

Agreeing, **Waller** LJ said: It appears from the transcript of the evidence of Mr Winbourne, the witness called by the defendant before the learned judge, that the substantial difference between the plaintiffs and the defendant was the difference between valuation on a vacant possession basis and valuation on the basis provided by section 34 of the Landlord and Tenant Act 1954, that is to say a basis which takes into

A account the fact that there are in the commercial premises (because there are two shops concerned here) tenants who have a right to stay and who, in the circumstances of this case, as at Christmas 1974, were highly likely to remain, having negotiated a rent under the terms of that section.

Mr Stimpson has submitted that the judge, in taking a figure of £100 as being the addition which should be made to the rent which was actually paid as a partial measure of the damage to which the plaintiffs are entitled, had misunderstood the evidence. It appears that in some ways the judge broadly accepted Mr Winbourne, or at any rate very nearly accepted. For example, while Mr Winbourne had suggested seven years' B purchase for the way of calculation of the value of reversion, the judge took eight years. One does not know what the valuer on the other side said.

But the misunderstanding is perhaps clearly shown by one question and answer which was given by Mr Winbourne in cross-examination. He had been asked by counsel for the plaintiffs a long question, to which Mr Winbourne said, "So it follows from your question, does it not, in order that I give a correct answer to this, it follows from your question that the repair liabilities would have been carried over to Mrs Selsdon?" I quote that to show that there was really a very substantial difference between them as to the basis. Then followed this: C "What I am asking you is this. On Christmas Day 1974 there were two possibilities to be considered: (1) that Mrs Selsdon would be granted a new tenancy, the landlord carrying out all the work set out in the schedule; the second possibility being that Mrs Selsdon would be granted a new tenancy without such work being done and without there being any requirement on her part to put it in repair." Pausing there, neither of those possibilities were the possibilities that, in fact, took place, because the landlords did not carry out all the work set out in the schedule, and Mrs Selsdon was not granted a new tenancy without there being any requirement on her part to put it in repair.

D Just going on with the question, "Taking those two cases, do you say that the same rent, namely £1,100 a year, would have been the market rent? (A) Of course it would not. Taking it that you are basing it on a hypothesis, I would agree with you. I do not believe, in the terms of the lease and the Landlord and Tenant Act, that the landlord would have done the repairs or been liable to do the repairs."

Then he goes on to deal with the hypothesis and in the next question: "What increased rent?" asked counsel, and Mr Winbourne said: "Say another £100 a year." I quote that question in full because it shows the absolutely fundamental difference between the basis on which counsel was asking the E question and that which is the proper basis in this case.

The matter is important because although Mr Winbourne had mentioned in evidence section 34 of the 1954 Act, there is no mention whatever in the judgment of the learned judge of the existence of that section. In my view, Mr Stimpson has made out a powerful case for the judge having misunderstood those answers of Mr Winbourne because it does seem that there was nowhere else from which he could have obtained the figure of £100 additional rent.

For the reasons which my Lord has already mentioned, citing the case of *Smiley v Townshend*, the proper basis for the valuation of this reversion is based on the rent which it F would be likely to produce immediately following the termination of the lease, there being disregarded—and I quote section 34(1)(a)—"any effect on rent of the fact that the tenant has or his predecessors in title have been in occupation of the holding."

Mr Stimpson has submitted that if the tenants, who had in their lease a repairing covenant, were to get the benefit of the fact that no repairs had been done, they would be getting a double benefit. He submits that the effect of Mr Winbourne's evidence and the effect of that section really drive one to the conclusion that if the judge had applied the proper test, he would have come to the conclusion that the rent—and again I quote, "... the holding might reasonably be G expected to be let in the open market by a willing lessor, there being disregarded. . ." that factor and, indeed, the factor of goodwill—would be the rent at which these premises were, in fact, let. Therefore, the value of the reversion as at Christmas 1974 would be that at which the premises were, in fact, let, it being almost certain that the tenants in occupation with a right to renew the lease if the rent could be properly negotiated would remain in occupation at that rent.

I do not see any conflict between that and the passage in *Woodfall's Law of Landlord and Tenant* (27th ed, p 1382) to which Mr Russell drew our attention, namely, dealing with paragraph (a) of section 34 of the Act: H

> This seems to dispose equally of any accretion to the rent attributable to such occupation (the premises being envisaged as empty premises on the market) and of the "sitting tenant" concession which was sometimes allowed for in assessing a "reasonable" rent under the Leasehold Property (Temporary Provisions) Act 1951.

As it seems to me, it disposes of those, but it does not dispose of the situation where, under the Landlord and Tenant Act 1954, the rent has been agreed by somebody who is remaining there and who is taking a full repairing lease.

I agree with my Lord that this appeal should be allowed.

MEGAW LJ also agreed. J

The appeal was allowed with costs in the Court of Appeal and with a variation in the appellant's favour of the order for costs in the court below.

CHANCERY DIVISION

April 9 1979

(Before Sir Robert MEGARRY, Vice-Chancellor)

BRADSHAW AND ANOTHER v PAWLEY K

Estates Gazette February 16 1980

(1979) 253 EG 693

Landlord and tenant—Point of law of general interest following grant of new lease of business premises to existing lessee in pursuance of consent order in proceedings under Part II of Landlord and Tenant Act 1954—Covenant in new lease to pay rent at a higher rate from a date anterior to date of execution —Whether lessee liable to pay higher rent from that anterior date or only from date of execution—Period in dispute covered continuation of old tenancy under 1954 Act and interval between termination of old tenancy and execution of new lease —Nearly four years involved—Natural construction that higher rent was payable from anterior date correct—Lease cannot retrospectively vest an estate in a lessee or make him retrospectively liable for a breach of covenant, but there is nothing to stultify agreement to make payments for periods anterior to grant—Vice-Chancellor's important summary of the law where a term is granted to run from a date anterior to the execution of the lease L

In this action the landlords, John Richard Bradshaw and Norman Frederick Bradshaw, claimed from the tenant, Alexander Christopher Pawley, payments of rent at the rate mentioned in a lease executed on March 10 1978 for a period commencing almost four years earlier, namely, from March 24 M 1974, when the previous lease, which was the subject of proceedings under Part II of the Landlord and Tenant Act 1954, expired.

D R G Hands (instructed by Kingsley, Napley & Co) appeared for the landlords; R W Moxon Browne (instructed by C A Maddin & Co, of Surbiton) for the tenant.

Giving judgment, MEGARRY V-C said: This case raises a short point of principle on the law of landlord and tenant. The basic question is whether on the grant of a new lease to an existing lessee a covenant to pay rent at a certain rate from a date anterior to the date when the lease was executed can make the lessee liable for rent at that rate from that anterior date or only from the date when the lease was executed. The question arises in an action brought by the landlords against the tenant claiming rent at the agreed rate from the anterior date; and a striking feature is that the two dates are nearly four years apart.

The relevant facts may be briefly stated. The premises in question are business premises which the tenant formerly held under a lease for 21 years less a day from March 25 1953. The natural expiration of the term was thus on March 24 1974, but of course the Landlord and Tenant Act 1954, Part II, applied to the tenancy. The landlords accordingly gave a statutory notice on February 13 1974 to determine the tenancy on August 14 1974; and on June 12 1974 the tenant applied to the county court for the grant of a new tenancy under the Act.

Prolonged negotiations then took place between those advising the parties. From early on these were on the basis of a new term for 10 years, and after a while this emerged as being 10 years from March 25 1974. Finally, on January 11 1977, a consent order was made in the county court for the grant of a new lease of most of the premises for a term of "Ten Years from the 25th March 1974 at a rent of £1,750" a year, with provision for "upward reviews" during the term. The rent payable under the previous lease, I may say, was £312 a year, with no provision for rent revision. In each case there was in addition an insurance rent which does not affect what I have to decide. It is regrettable that the order does not state on the face of it that it was a consent order, as every consent order ought to do; but it is common ground that in fact it was a consent order. The order also provided that the parties should jointly apply to the court under section 38(4) of the Act for an order authorising an agreement to exclude sections 24 to 28 of the Act in respect of the new lease; and it was agreed that this agreement should be included in the lease. This, of course, would prevent the new lease from conferring any right to a new tenancy under the Act when it expired. On May 16 1977 the court made the requisite order to this effect.

A week later, on May 23 1977, the tenancy flowing from the combined effect of the old lease and the Act came to an end. Section 24(1), when read with section 64, had prolonged the tenancy until the expiration of three months after the end of the period of six weeks allowed for an appeal from the order of January 11 1977. The Judicature Act 1925, section 31(1)(h), does not prohibit appeals from consent orders, though it does require leave for such an appeal. The new lease, however, still had not been executed, for dissension arose on the point now before me for decision. Finally, on March 10 1978, the lease was duly executed; and on June 29 1978 the writ was issued.

It will be seen that there are thus four main periods with which I am concerned.

(1) Until March 24 1974 the old lease was running at a rent of £312 a year; and this period is not in issue.

(2) From March 24 1974 until May 23 1977 the tenancy created by the old lease was being continued by force of the Act. This second period is very much in issue. The tenant contends that his liability for this period is at the rate of £312 a year, whereas the landlords contend that it is at the rate of £1,750.

(3) From May 24 1977 until March 10 1978 the old lease had come to an end and the new lease had yet to be granted. This third period is also in issue. There is the same dispute about the amount of the rent, and Mr Moxon Browne, who appeared for the tenant, was forced to contend that either the tenant was holding over during that period, without any lease or tenancy, or else he was there by virtue of the agreement for a lease evidenced by the consent order, on the terms of that order.

(4) From March 10 1978 onwards the new lease was in force, and there is no dispute that thereafter the tenant was liable for rent at the rate of £1,750 a year. Accordingly, what I have to decide is the liability of the tenant during the second and third periods.

With that, I turn to the provisions of the new lease, which is expressed in terms of "the Lessor" and "the Lessee." By the habendum, the premises are demised to the lessee to hold the same unto the lessee "from the 25th day of March 1974 for the term of ten years but determinable as hereinafter provided YIELDING AND PAYING THEREFOR during the said term yearly and proportionately for any fraction of a year the rents hereunder set out." There are then specified "for the first four-and-three-quarter years of the said term" the rent of £1,750 a year, and "for the next three years of the said term" a rent determined in accordance with clause 3, which contains provisions for rent review. There is then a similar provision "for the remainder of the said term"; and all the rents are to be paid quarterly in advance on the usual quarter days. The lessee's covenants are made "to the intent that the obligations may continue throughout the said term"; and the first of these is "to pay during the said term the said reserved rents at the times and in manner aforesaid," and so on.

In considering the rival contentions, I think that the starting point is to construe the lease in the ordinary way, before considering whether there is anything in the law to modify that construction or to prevent its being effectuated. It seems perfectly clear that the obligation to pay a rent of £1,750 a year is an obligation to pay that rent for the period of 10 years from March 25 1974. The reddendum is in terms of paying rent at that rate "during the said term," and so is the covenant to pay rent: and there is no term in the document that can be called the "said" term except the "term of 10 years" running "from the 25th day of March, 1974."

That construction seems to me to be supported (if support be needed) by the consequences of an answer which Mr Moxon Browne gave to a question from the Bench. He was constrained to accept that the time-table for increases of rent for the specified periods "of the said term" was to operate during the 10 years from March 25 1974 and not the six years and a few days running from the execution of the lease on March 10 1978. Accordingly, the revised rent was to run from the expiration of "the first four-and-threequarter years of the said term," and so would commence at Christmas 1978, and so on. Mr Hands, who appeared for the landlords, said that on this footing it was impossible to see how it could be said that "the said term" meant one thing for the purpose of laying down a rent of £1,750 a year and another for the purpose of the time-table of rent increases, especially when the first four-and-threequarter years of that time-table provides for the rent to be £1,750 a year. The "said term" must mean the same term, whether the phrase is "during the said term" or "of the said term."

That contention seems to me to be unanswerable. Mr Moxon Browne's contentions reduced him to contending that the obligation to pay £1,750 a year "during the said term" really meant not "during the said term of 10 years from March 25 1974," but "during so much of the said term of 10 years from March 25 1974 as subsists while the parties are in the relationship of landlord and tenant under this demise," and so when the lease was executed meant during the next six years and a few days. Similarly, the provision in the consent order for the grant of a new lease "for a term of ten years from the 25th March 1974, at a rent of £1,750" was really an agreement to pay a rent of £1,750 not for 10 years but for so much of the then unexpired period of approximately seven years and two months out of that 10 years as still remained unexpired when the lease was finally executed. As the difference between the two rents is nearly £4 a day, each day's delay in executing the lease would save the tenant some £4. As a matter of construction I cannot accept these contentions. I feel no doubt that on ordinary principles of construction the true meaning of the consent order and the lease is that a rent of £1,750 a year was to be paid throughout the period of 10 years from March 25 1974 irrespective of the date on which the lease was actually executed.

The question, then, is whether there is anything in the law to

A require the lease to be construed differently, or to prevent effect being given to the natural construction of its language. When asked, Mr Moxon Browne agreed that on this he relied not on any provisions of the Act but on the general principles of the law. The only qualification to this was that he relied to some extent on the provisions of the Act which prevented the old tenancy from determining when the old lease expired; and he also contended that section 24A, when read with section 64, would be pointless if the landlords were right. This latter point, however, did not withstand investigation, and I need not pursue it further. In the main, Mr Moxon Browne relied on authorities which for the most part are conveniently
B collected in *Roberts v Church Commissioners for England* [1972] 1 QB 278. As he contended, there are indeed some limitations on the effect that can be given to a term expressed to run from a date prior to the execution of the lease, and I must consider these.

First, it is well settled that a lease cannot retrospectively vest an estate in the lessee. If today a lease is granted for seven years from this day a year ago, no term of seven years is brought into being, but only a term of six years from today. A lease, of course, is more than a mere contract, for it operates by way of grant to create an estate or interest in the land; and you cannot grant today that X shall have had a term
C of years vested in him a year ago. Whatever contractual obligations there may be between the parties, no actual term of years can be created until the lease has been executed and so the grant has been made. Thus where the question is what term has actually been created, as where statute refers to terms of a particular length, the commencement of the term cannot be earlier than the date of the grant of the lease. There is, of course, no objection to a lease defining the term by reference to some past date, as in the grant of a term of seven years from this day a year ago; but this merely creates a term of six years from today.

Second, a lease for a term from some past date will not, at
D any rate normally, make into a breach of covenant that which was not a breach when it was committed. An act or omission by a prospective lessee which would be a contravention of the proposed lease is not turned into a breach of covenant by him merely because he subsequently accepts the grant of the lease for a term which is expressed to run from some date anterior to the act or omission. That, I think, follows from *Shaw v Kay* (1847) 1 Exch 412. There, the landlord sued on a repairing covenant in respect of acts done by the tenant before the lease was executed, basing the claim on the lease having been expressed to run from an earlier date. The claim failed, though no reasoned judgments were delivered.

E It may be tempting to infer from these two propositions that when a term is expressed to run from some date prior to the actual grant of the lease, that prior date can have no function save as a unit of calculation in ascertaining the date on which the term will expire; and in *Shaw v Kay* at p 413 there is a dictum of Parke B which can be read as indicating this. Whatever the temptation, I think that it would be wrong to draw any such inference. The dictum on which Parke B relied, uttered by Eyre CB in *Wyburd v Tuck* (1799) 1 B & P 458 at p 464, must be understood in relation to the facts of the case: and what the court was concerned with there was a lease of tithes as giving the plaintiff a title to sue for the tithes. As
F the lease did not vest the term in the tenant before it was executed, he could not sue for tithes due before the date of execution. However, when what is in issue is not title but obligation or liability, I do not see why the parties should not, if they so choose, make the obligation or liability enforceable in respect of such anterior date as they wish. Thus if a lease is granted for a term of 21 years from a date two years prior to the grant of the lease, with a provision for determination at the end of the seventh or fourteenth years of the term, then in the words of Clauson J " it is perfectly easy as a matter of construction of such a document to say that the seven years according to the obvious intention of the parties is not to run as from the date of the execution of the lease but from the moment spoken G
of, though inaccurately, as the beginning of the term in the document ": *Earl Cadogan v Guinness* [1936] Ch 515 at p 518. *Bird v Baker* (1858) 1 E&E 12 provides an example of this. There, the lease was granted in July 1851 for 14 years from Christmas 1849, with power to break " at the expiration of the first seven years thereof ": and the Court of Queen's Bench held that the break came at Christmas 1856 and not in July 1858.

I cannot see what there is to stultify an agreement in a lease to make payments in respect of past periods, or to require the court to construe a lease so as to prevent any agreement from relating to past periods unless compelled to it. It is by no H
means unknown for a lease not to be executed until after the prospective lessee has entered (usually with safeguards for the lessor) and for the lessee then to pay rent and observe the terms of the lease as from a date prior to the execution of the lease. If as a matter of construction the obligation is to pay rent at a specified rate from some date earlier than that of the execution of the lease, why should that not be enforceable?

I think that some support for this view is provided by a dictum of Pollock CB in *Shaw v Kay*. The report in (1847) 1 Exch 412 that was cited to me omits this, but it is to be found in the report of the case in (1847) 17 LJEx 17 at p 18. The Chief Baron said: " A party may covenant to indemnify J
another from what has passed. I may demise premises today, and covenant to save my tenant harmless from what has happened six months before." On the same footing I do not see why, by suitable wording, a lease should not impose on one of the parties some liability for things past.

Mr Moxon Browne stressed that rent was something that had to be paid periodically, and so an obligation to pay nearly four years' rent at a blow indicated that what was being paid could not truly be called rent. However, I do not think that rent becomes denatured when a sufficiency of gales remain unpaid. Nor do I see why an obligation to pay rent in respect of periods of occupation by the tenant prior to the grant of K
the lease should not be regarded as rent. What must matter, surely, is what the payments are due for, rather than how and when the payments are in fact made. I would follow Foa's *Landlord and Tenant* (8th ed 1957) p 101 in regarding rent as being *prima facie* " the monetary compensation payable by the tenant in consideration for the grant, however it be described or allocated," though I think that I would insert the word " periodical " before the word " monetary." I would also pay more attention to modern than medieval concepts in considering the nature of rent. If the parties to a lease choose to agree that the lessee shall pay rent to the lessor in respect of a period prior to the grant of the lease, I do not see why this L
should not be valid and enforceable. So far as it goes, I think that *M'Leish v Tate* (1778) 2 Cowp 781 at p 784 supports the view that rent may still be rent even though it is reserved in respect of the occupation by a tenant at some time prior to the execution of the instrument which reserves the rent. Even if this is wrong, and the payments in dispute, though described as " rent," are not in law " rent," strictly so called, I cannot see why the lessors should not enforce payment. There is nothing to stop a man from being liable on a covenant to pay a sum of money in respect of some past period.

Mr Moxon Browne relied to some extent on the tenant holding on the terms of the old lease, as extended by the Act, M
from March 24 1974 until May 23 1977: this is the second period that I mentioned earlier. As during this period the tenant held at a rent of £312 a year, why should he now pay more? The answer, I think, is that he would be under no obligation to pay more unless he had bound himself to do so; and on the true construction of the new lease that is just what he has done. The fact that he need not have paid more if he had not entered into a new lease, or if instead he had entered into a differently worded lease, cannot alter the consequences of what he did do. I know of nothing at common law to prevent a tenant under an existing tenancy from agreeing to pay

A a higher rent, or from entering into a new tenancy at a higher rent, in place of his existing tenancy. Nor can I see any magic in the Act which would make the terms of the extended tenancy prevail over those of the new lease. The position of the tenant during the third period is *a fortiori*; for his old tenancy had ended and until the new lease was executed the tenant's only right of occupation was by virtue of the agreement evidenced by the consent order of January 11 1977. That agreement, of course, was for the same rent as the rent under the new lease, namely, £1,750 a year, and so, whatever the road, the terminus is £1,750 a year.

I do not think that any difficulty arises from the payment
B and acceptance of rent at the rate of £312 a year after March 25 1974; and, indeed, no difficulty has been suggested. Obviously credit must be given for the payments made against the payments due at the rate of £1,750 a year, and the landlord's claim is, of course, properly made on that basis. Nor, I think, is there any difficulty in the fact that the payments at the higher rate have not been made punctually each quarter day since March 25 1974. No obligation under the new lease to make these payments could arise until the lease had been executed, and so the omission to make those payments at any earlier date could not constitute a breach of covenant.

In the result, I think that where a lease creates a term
C of years which is expressed to run from some date earlier than that of the execution of the lease, the relevant law may be summarised as follows.

(1) The term created will be a term which commences on the date when the lease is executed, and not the earlier date.

(2) No act or omission prior to the date on which the lease is executed will normally constitute a breach of the obligations of the lease.

(3) These principles do not prevent the parties from defining the expiration of the term by reference to a date prior to that of the execution of the lease, or from making contractual
D provisions which take effect by reference to such a date, as by defining the period for the operation of a break clause or an increase of rent.

(4) There is nothing in these principles to prevent the lease from creating obligations in respect of any period prior to the execution of the lease.

(5) Whether in fact any such obligations have been created depends on the construction of the lease; and there is nothing which requires the lease to be construed in such a way as to avoid, if possible, the creation of such obligations.

On the facts of the case before me, these principles and the other matters that I have discussed point to only one
E conclusion, namely, that the landlords' claim succeeds. The precise terms of the order are for discussion.

CHANCERY DIVISION

October 12 1979

(Before His Honour Judge John FINLAY QC, sitting as a judge of the High Court)

JANES (GOWNS) LTD v HARLOW DEVELOPMENT CORPORATION

Estates Gazette February 23 1980
F (1979) 253 EG 799

Landlord and Tenant Act 1954, Part II—Determination of rent and other outstanding terms of a new lease—Zoning method of rental valuation explained—Properties which should properly be treated as comparables—Additions to rental value for return frontages and rising rents and deductions for quantity—No deduction in the circumstances for lack of depth in comparison with frontage—Rent review clause to provide for movement downwards as well as upwards, but with no consequential adjustment of rent on that account—Factors to be considered in determining interim rent under section 24A—Application of English Exporters (London) Ltd v Eldonwall Ltd—Term of 20 years ordered by agreement—Rent under new lease of £12,500 per annum (previous rent £2,100) and interim rent £9,200—Useful judgment for valuers concerned with business premises and the 1954 Act G

This was an application by the plaintiff company, Janes (Gowns) Ltd, under Part II of the Landlord and Tenant Act 1954 for the grant of a new lease. The grant was not opposed by the landlords, Harlow Development Corporation, but there were certain disputed matters outstanding concerning the terms of the new lease. The premises in question were 5 and 6 Market House, The High, Harlow, Essex.

G N Huskinson (instructed by Stanley Tee & Co, of Bishop's H
Stortford) appeared on behalf of the plaintiffs; D R Hands (instructed by the solicitor to Harlow Development Corporation) represented the defendants.

Giving judgment, JUDGE FINLAY said: On September 28 1955 the Harlow Development Corporation granted to the plaintiff company, Janes (Gowns) Ltd, a lease of premises known as 5 and 6 Market House, The High, Harlow, Essex, for a term of just over 21 years to commence on September 14 1955 and to end on September 29 1976 at a rent of £2,100 per annum. On March 2 1976 the landlords, who are the defendant corporation, served upon the tenants, who were at that time still in occupation under the lease which I have mentioned, a J
notice under section 25 of the Landlord and Tenant Act 1954 specifying the date of termination of the tenancy as September 29 1976, that is to say the date upon which in fact the tenancy would but for the statutory provisions in any event have come to an end. By a notice dated April 2 1976 the plaintiff company notified the defendants that they would not be willing at the date of termination to give up possession of the property, and on June 25 1976 the plaintiffs applied for the grant of a new lease. The landlords do not oppose the grant of a new lease, and indeed but for three matters the terms of the new lease are agreed.

The outstanding matters on which the parties are not agreed are the amount of rent payable under the lease; and, secondly, K
whether, as it is agreed that there should be a rent review provision in the lease, that review provision should contemplate that the rent could be varied downwards as well as upwards, or alternatively, that there should, in the event that the review clause provided only for an upward movement of the rent, be a provision entitling the tenant to give notice breaking the lease, that is to say, notice terminating the tenancy. The third matter which arises for determination is whether any order should be made for payment of interim rent, and if so in what amount, and that arises under and by reason of an application made in the proceedings by the landlords dated October 5 1976.

At the commencement of the proceedings there were also in L
issue two other matters. One was that there was a difference between the parties as to whether the rent review should be at intervals of five or seven years, but in the course of the hearing the plaintiffs accepted that the rent review should be at five-yearly intervals. And furthermore, in the initial stages the plaintiffs' claim for a rent review provision to provide for a downwards review was coupled with and not alternative to the contention that the lease should contain a provision entitling the tenants to bring it to an end at the rent review date.

The major problem, on which the others, in my judgment, to a large extent turn, is what rent should be paid under the lease? Here again there is a certain measure of agreement between M
these parties as to this problem. Both experts—and I say both, not for the moment taking account of a third expert who gave evidence on behalf of the plaintiffs, because he did not make an initial but merely certain corroborating evaluations—both experts who gave evidence before me, that is to say P F Jones [MA FRICS], who was called by the plaintiffs, and D J Green [BSc (Est Man) ARICS], who was called by the defendants, were agreed that an appropriate approach is that which is referred to as a zoning method. This is a process whereby a notional superficial area of the premises is determined with a view to ascertaining the proper rent by applying

A to that notional superficial area the rent per sq ft appropriate to the part of the premises which is designated as zone A. Zone A is that part of any premises which is bounded by the front of the premises and is to a depth of 15 ft therefrom. If zone A is to be let at the rate of £X per sq ft, then the next zone is the next 25 ft in depth, which is, it is agreed, to be charged at the rate of half of £X per sq ft. And the next zone, which is the remainder of the shop depth (apart from the two zones which I am about to mention), is to be charged at one quarter of the rent per sq ft appropriate to zone A, ie £¼X per sq ft. There are therefore three zones, all differentiated according to the depth from the front of the shop. A fourth zone is that B comprising ancillary accommodation within the shop, the interior ancillary accommodation of offices, lavatories and the like, and that is charged at the rate of one-sixth of the rent appropriate to one sq ft of zone A. And finally the fifth zone is that of outside ancillary accommodation, and that is charged at the rate of one-tenth of the rent appropriate to zone A. The notional superficial area is accordingly ascertained by adding together the area of zone A, that is the part of the premises within 15 ft of the frontage, to one-half of the actual area of the part which falls within 15 and 25 ft in depth from the front and one-quarter of such area, if any, as falls more than 40 ft from the front, and further adding one-sixth of the area of any C ancillary interior accommodation and one-tenth of the area of any outside ancillary accommodation. When all these various areas are added together one has a notional area which, when multiplied by the ascertained amount appropriate to one sq ft of zone A will yield the basic rent. It is further agreed that although this is a proper approach particular circumstances may require modification either upwards or downwards of the basic rent thus arrived at. In determining whether any such modification should be made it is appropriate, in my judgment, and I do not think that there was disagreement with this by any of the experts, that regard should be had to any user covenants which there may be in the lease.

D The covenant as to user contained in the lease which I have mentioned and which is, it is agreed, to be maintained in the new lease was a covenant by the tenant to use the demised premises for the purposes only of a retail lock-up shop and offices in connection therewith for the sale of fashions, ladies' and children's wear and accessories "provided that the main characteristic of the shop is fashion" and subject to a further proviso that the user should be conducted so as not to contravene the provisions of a covenant contained in the lease, in common form, against the use of the premises in such a manner as to cause nuisance and noise or the like.

Applying these principles as to how the area of the premises E in terms of zone A is to be ascertained, both parties arrived at the conclusion that there are 1,628 sq ft in terms of zone A. Mr Jones for the plaintiffs took the view that the appropriate rent per sq ft of zone A was £7. Mr Green for the defendants said the appropriate rent was £8.59 per sq ft. The difference of £1.59 gives rise to a basic difference of £2,588.52 per annum for the area of 1,628 sq ft. These differences, which I should say are not the only differences between the two experts, arise from four matters. The first is this, that both experts arrive at their conclusion as to what the appropriate rent for one sq ft of zone A should be by comparing the rents of other premises in the neighbourhood. But the comparables, if I can so call F them, to which Mr Jones has had regard include four properties, nos 2, 3, 4 and 5 The Rows, which are not included among those considered by Mr Green. The second point is that Mr Jones did and Mr Green did not make an allowance for the return frontage which one finds in two properties, namely nos 1 and 2 Market House and nos 6 and 14 The Rows. Both Mr Jones and Mr Green made allowance for the return frontage of no 11 and no 15 Adams House. The point about a return frontage allowance as I understand it is this, that if there is a return frontage it is regarded as advantageous to the tenant and accordingly should result in some increase in the rent. If therefore one is considering the rent that has been G agreed in respect of premises which have a return frontage and endeavouring to arrive at what is the rent per sq ft of zone A of the premises, it is necessary first to deduct from the rent payable the allowance for the return frontage and then to divide what is left by the notional zone A area to arrive at the rent per zone A sq ft.

The third point of difference is that Mr Green has treated one of the properties which is regarded as comparable by both himself and Mr Jones, namely 10 The Rows, as being notionally two properties and not one only; and he has done so because at some date since the new lease of 10 The Rows was granted, that lease being one dated May 3 1976, there has H been an assignment of the lease, and on the assignment a premium was paid. Mr Green took the view that it would be appropriate to regard that premium as indicating that as from the date of the assignment the premises should be treated as let at a rent equal to that reserved by the lease plus an addition representing the annual value throughout the residue of the term of the premium that had been paid. Accordingly he treated 10 The Rows as furnishing evidence not only of the rent at which it was let by the lease dated May 3 1976 but also as let by the joint effect of that lease and the assignment upon which the premium was paid. Mr Jones did not view the matter in that way, and accordingly there is in effect an additional J comparable in Mr Green's calculations that does not occur in Mr Jones's.

The fourth point of difference was that Mr Jones made no allowance in his calculations for the fact that rents have risen since the dates when the comparable rents selected for comparison were arrived at, that being at the end of 1976 or early in 1977, although the exact dates when the various agreements were made was not in evidence before me and some may have been made at dates later than 1977. Mr Green on the contrary made an allowance of what appears to have been a little over 15 per cent because his K basic rent in terms of zone A is £7.44 and he increased that, as I indicated, by an addition to allow for a rise in rents to £8.59. There is a further point that I should mention as to the manner in which Mr Green arrived at the rent of the basic £7.44 per sq ft of zone A. He did so by taking (i) an average of the comparable properties in Market House betwixt and between, that average being £7.53; (ii) an average of the four properties (treated as five) in The Rows, the average being £7.36; and (iii) an average of the five properties in Adams House, that average being £7.46. He then added together the three averages and divided the total by three to arrive at £7.44. I am not satisfied that that is an appropriate way to arrive at L the average. It appears to me that the proper way would be to take the rent per sq ft in terms of zone A for each of the 12 properties, or, if it be right—and I have not yet come to this point—to treat one of them as two, the 13 properties, and then divide the total by 12 or 13. That would produce a slightly different answer.

As to the points of difference, it is said that it is wrong to include the properties 2, 3, 4 and 5 The Rows. They are in a side street that leads off Market Square. It is said that they are properties let for quite different purposes than those that we find in the Market Square, and in any event they are not part of the enclave of properties that front on to Market Square. M There is I think some force in that argument and I have come to the conclusion that the proper course is to exclude from consideration two of these four properties, namely 2 and 3, but to include 4 and 5 The Rows. I do so for this reason. The two properties I have last mentioned front on to The Rows and do so opposite the return frontage window and the return frontage of the premises with which I am concerned, 4 and 5 Market House. They are in my judgment properly to be regarded in comparing rents in the neighbourhood because in a very real sense they are adjacent to 4 and 5 Market House. I do not think, however, that the same can be said of 2 and 3 The Rows, which are removed from it not only by being on

A the other side of the road but by being in effect beyond the areas on which the side windows of 4 and 5 Market House look.

As to the second point, in my judgment it is proper to make allowance, as Mr Jones has done, for the return frontages, not only to the premises 6 to 14 The Rows, occupied by McHarris Motorcycles, but also the return frontage of the premises 1 and 2 Market House occupied by Eastern Electricity Board. So far as the McHarris Motorcycles property is concerned, that return frontage appears to me to have no features that would entitle one to disregard it. It is said in relation to 1 and 2 Market House that the return frontage looks only upon an alley way. I do not regard that as a reason for disregarding
B the return frontage and making no allowance for it, because a return frontage may be of utility to the occupier and an attraction to his customers, not merely because the paths along the thoroughfare pass the return frontage but also because the customers are attracted to the side window by looking first at the window in front.

As to the third point, in my judgment it is not right to treat 10 The Rows as two properties and to introduce a second comparable by reason of the consideration of the terms of the assignment on which a premium was taken. I have no evidence before me as to the reasons or basis on which that premium was paid or calculated. And in my judgment the proper course
C is to have regard to 10 The Rows as furnishing evidence of the state of the market so far as rents are concerned by considering the terms of the original letting for the lease dated May 3 1976 and not speculating upon the reasons for the payment of the premium at some later stage.

As to the fourth point, it is in my judgment proper to make some allowance for rising rents. I have come to the conclusion, however, on the evidence which I have heard that the allowance of 15 per cent is too much. Mr Jones accepted that some allowance should be made for rising rents, and, having regard to the evidence he gave and also the evidence of Mr Green as to this matter, I have come to the conclusion that the appropriate
D allowance for the rise in rents is an allowance of 10 per cent.

Bearing in mind what I have said about making allowances for return frontage, a matter which affects not only 6 to 14 The Rows and makes the appropriate sq ft rate in respect of that property a rate of £6.72 and not the £6.78 shown in Mr Green's schedule, an allowance should also be made for the return frontage in respect of 1 and 2 Market House, which means that the appropriate rate per sq ft in terms of zone A is £7.50 and not £7.84 as shown in the schedule. Further, the appropriate terms to consider in relation to 10 The Rows are those in the original letting. That results in my view in a rate per sq ft in terms of zone A of £6.68 and not a rate of £6.89
E as shown in Mr Green's schedule, because I calculate, as apparently does Mr Jones. that to take the average of the rents reserved over the first five years results in a figure of £6.68. Bearing these various circumstances in mind, I find that the rent per sq ft indicated by the comparables that can properly be regarded, that is to say the properties shown on Mr Green's schedule and also 4 and 5 The Rows, and taking the average of the various rents per sq ft in respect of all these properties, is in terms of zone A £7.23. 1,628 sq ft at such a rent gives a figure of £11,770.

I now come to the further matter of disagreement which
F arises not in arriving at the basic rent per sq ft but in the view taken as to whether any deductions should be made from the figure thus arrived at in respect of two matters. Mr Jones gave evidence that he thought deductions should be made in respect of the quantity of the property demised and also in respect of its shape or configuration. So far as the latter point is concerned, his evidence is not supported by that of the other expert called on behalf of the plaintiffs, R A Childs [FRICS], who did agree that some deduction should be made for quantity. I do not think, although Mr Green contended that no such deduction should be made, that he did so with quite the enthusiasm with which he differed from Mr Jones's view on certain other matters. I find that it is appropriate to make some deduction for quantity. It is a matter of advantage to a landlord to let G
property in larger units rather than in a number of small units, and the tenant is in my judgment entitled to some allowance for that advantage which the landlord has. I accept Mr Jones's evidence that an appropriate deduction for quantity is 5 per cent.

So far as the question of the shape of the premises is concerned, I am unable to accept the evidence that a deduction of 15 per cent should be made for that. It is I think to be noted that there is no suggestion that, in the course of a period of more than 20 years that the plaintiff company has occupied these premises, they have made any attempt to alter the internal arrangements in any radical way. The shape of the premises, having, as it has, a large frontage and not perhaps a corre- H
spondingly large depth, may well be apt for the nature of the trade carried on by the plaintiff company. The trade of selling ladies' and children's wear and fashion goods is no doubt one where facilities for display play a prominent part in the success of the venture, and I am not satisfied that it is appropriate to make any deduction by reason of the fact that there is a large frontage to this property and not a relatively large depth to go with it. I do not overlook the fact that the result of the large frontage is that a large part of the premises falls to be charged at the zone A rate, or rather I should say that a large part of the premises falls within zone A. That is no doubt disadvantageous to the tenant, but the disadvantage is J
of course compensated by the amenity produced by the large display area.

Accordingly I consider that there should be deducted from the figure of £11,770, which I have stated is arrived at in the way I have indicated, 5 per cent for quantity. That amount is £588, leaving a balance of £11,182. Furthermore it is appropriate in my judgment to add the figures for the return frontage, and I think the appropriate figure is £190, being £10 per ft for a return frontage of 19 ft. So far that produces a figure of £11,372. However, I have already indicated that there should be an addition to that of 10 per cent to allow for the rise in rents in the period between the time when the compar- K
able rents were arrived at and the date when the new lease will commence, which is in February 1980. That produces a figure of £12,509, which I propose to call £12,500.

In my judgment, therefore, the appropriate rent under the new lease is £12,500.

I turn now to the question whether there should be provision, not only for a rent review, which it is agreed there should be at five-yearly intervals, but for a review down as well as up. It was decided by Cross J, as he then was, in *Stylo Shoes Ltd v Manchester Royal Exchange Ltd,* reported in (1967) 204 EG 803, that in determining the terms appropriate under a new lease granted under this jurisdiction the court could in its L
discretion provide for a rent review clause which provided for review of the rent downwards as well as upwards. The nub of the decision is stated by Cross J in these terms: having said that he did not think that the courts had ever hitherto addressed themselves to the question, he said it was a rather academic matter because he saw no reason why sauce for the goose should not be sauce for the gander. It is to be borne in mind that under section 34(3) of the Landlord and Tenant Act 1954 (as amended by the Law of Property Act 1969) it is provided that: "Where the rent is determined by the court, the court may, if it thinks fit, further determine that the terms of the tenancy shall include such provisions for varying the rent M
as may be specified in the determination."

In my judgment it is appropriate that the rent review clause should deal with the possibility of rent being varied downwards as well as upwards. And I reach that conclusion because of the evidence which I have heard about the development of what is called the Harvey Centre in the near neighbourhood of the demised premises, a development which may well have the result of rendering a rent calculated prior to that development taking place inappropriate in future years. In short the rents may in the neighbourhood of the Harvey Centre go down; equally they may go up. Having regard to

A that possibility, I think it appropriate that the rent review clause should provide for the review in either direction. It is submitted, however, that if such provision is inserted in the lease then some *quid pro quo* by way of increase in the rent should be allowed, because the provision of such a clause is an advantage to the tenant. It is said that the comparable rents which have been considered in arriving at the appropriate rent for these premises were all contained in leases which contained rent review clauses providing for revision of the rent upwards, and that no doubt was an advantage to the landlords which should be taken into account in arriving at these comparable rents. I am not satisfied that that is the B right approach to take to this matter. In my judgment the only way to find out what is the appropriate rent is to have regard to the rents of neighbouring property and in doing so to arrive at a view as to what rent the holding might reasonably be expected to be let at in the open market. In conditions where it is expected that rents will over the period of the lease continue to rise it is no doubt common to insert rent review clauses to provide for an upgrading of the rent at appropriate intervals. I am by no means satisfied that the insertion of such a provision leads to any reductions in the rent paid under the lease. It is in my view more likely that in a particular state of the market leases are simply only granted on C the basis that the rent review clause is inserted. Twenty-five years ago it was a provision that was rarely if ever come across. In the last 15 years it has become so common as to be a matter of common form. I am not satisfied on the evidence before me that the insertion of rent review clauses for review in one direction only have affected the rents which have been payable, and accordingly I think that the appropriate course to take is to insert a rent review clause in this lease to provide for variation in either direction but to make no consequential adjustment to the rent which I have already indicated is the appropriate rent.

I come now to the third and last matter in dispute, which D is whether any interim rent should be payable and, if so, what amount. I have been referred to the decision of Megarry V-C in *English Exporters (London) Ltd v Eldonwall Ltd* [1973] Ch 415. That decision makes it clear that it is a matter for the discretion of the court as to whether any interim rent should be ordered between the dates when the application by a landlord is made and the date when the new lease takes effect. Further, I collect from that judgment the conclusion that a case can in appropriate circumstances be made out for tempering the market rent by some amount. That is so because the exercise that has to be carried out by the court, if it determines that an interim rent is appropriate, is that indicated E in the 1954 Act as amended by the Law of Property Act 1969, in section 24A which is inserted in the 1954 Act by the Act of 1969. That provides in subsection (1) that: "The landlord of a tenancy to which this Part of this Act applies may—(a) if he has given notice under section 25 of this Act to terminate the tenancy," which is the case here, ". . . apply to the court to determine a rent which it would be reasonable for the tenant to pay while the tenancy continues by virtue of section 24 of this Act, and the court may determine a rent accordingly." Subsection (2) provides that: "A rent determined in proceedings under this section shall be deemed to be the rent payable under the tenancy from the F date on which the proceedings were commenced or the date specified in the landlord's notice or the tenant's request, whichever is the later." It is common ground that the effect of that provision is that the interim rent, if any is awarded, will be payable as from October 5 1976. Subsection (3) provides that: "In determining a rent under this section the court shall have regard to the rent payable under the terms of the tenancy, but otherwise subsections (1) and (2) of section 34 of this Act shall apply to the determination as they would apply to the determination of a rent under that section if a new tenancy from year to year of the whole of the property comprised in the tenancy were granted to the tenant by order of the court."

Section 34 is the provision under which the rent payable G under a tenancy granted by order of the court falls to be determined, and subsection (1) of that section provides that it "shall be such as may be agreed between the landlord and the tenant or as, in default of such agreement, may be determined by the court to be that at which, having regard to the terms of the tenancy (other than those relating to rent), the holding might reasonably be expected to be let in the open market by a willing lessor, there being disregarded" certain matters set out in the subsection.

I have already held that the appropriate rent under the new tenancy is the rent of £12,500 per annum, but what has to be determined for the purposes of a determination of H interim rent is a rent determined in the same manner as is provided for by section 34 but on the footing that the new tenancy is a tenancy from year to year. There is, however, this point to be borne in mind, that in arriving at the figure of £12,500 per annum I did so having first ascertained that the appropriate rent payable without regard to the circumstance that rents have risen in the meantime would have been £11,372. The date from which the interim rent should be payable will be the date on which application is made by the landlord for its determination, that is to say October 5 1976. If the matter had been dealt with at that date and there had been then available the information as to comparable rents J which has been before me, I consider that the conclusion arrived at would have been that the rent for a new tenancy for a fixed term of years falling to be paid at that time would have been £11,372. The rent is, however, to be determined as that appropriate to a yearly tenancy, not a tenancy for a term of years. It is appropriate, in my judgment, to make a deduction to give effect to that circumstance, and the appropriate deduction I find is one of 10 per cent. Accordingly the rent which I consider would have been payable for a yearly tenancy ordered to be granted in October 1976 would have been £10,235.

The court is, however, directed in determining a rent under K section 24A to have regard to the rent payable under the terms of the tenancy, that is to say the rent payable under the terms of the tenancy granted in 1955. In my judgment the effect of my having regard to the rent payable under that tenancy, which it will be recalled is a rent of £2,100 per annum, is that there must be a substantial difference between the rent payable under the original tenancy and the rent payable under any new tenancy granted under the statutory provision. That, I think, operates in two ways. It indicates, I think, that some reduction should be made from the figure that would be arrived at by ascertaining what is the appropriate rent payable in respect of a yearly tenancy. But simply L because the difference is so large the deduction cannot in my judgment be of a very substantial amount. If it were, then the effect would be that the tenants would be continuing to occupy the premises under a rent more nearly comparable to the rent appropriate some 20 or more years ago than that appropriate for the present time. Nevertheless, the process does appear to be, and I am fortified in this view by what is said by the Vice-Chancellor in the case which I have mentioned, to be one of tempering the effect of the statutory provision. An appropriate mode of tempering in my judgment would be to make a further reduction of 10 per cent. If that is made from £10,235, one arrives at £9,212; thus in my M judgment the appropriate amount of the interim rent would be £9,200.

I indicated at the beginning of this judgment that there was a large measure of agreement between the parties on many matters. I have dealt, I think, with those which were outstanding. One matter on which the parties were agreed was that the new term should be a term of 20 years. I had some doubts as to whether there was jurisdiction to make an order for the grant of a term of 20 years, but both counsel have pressed me to do so and bearing in mind that if I were to make an order for the grant of a term of 14 years, which is the maximum term that can be ordered under section 33,

A it would be open to the parties forthwith to agree that that 14-year term should take effect and be brought into operation as one of 20 years, I have come to the conclusion that there is no reason why, provided that the order recites that the parties by their counsel agree that the term to be ordered should be one of 20 years, an order for a 20-year term should not be made.

Accordingly I will order that there should be a new lease on the terms of the existing lease but with the addition of a rent review clause which I understand has already been agreed between the parties to provide for revision of the rent at five-yearly intervals either in an upwards or in a downwards
B direction as may be appropriate and that there be no break clause in the lease, because having directed a rent review clause in that form the plaintiffs do not apply for any provision for the term to cease at their option. The rent payable under the lease will be, as I have already indicated, £12,500 and I further order that the interim rent as from October 5 1976 until the new tenancy takes effect shall be £9,200 per annum.

I will order a new lease, terms as in 1955 with deletion of clause 2(1) and the addition of an up-and-down rent review clause at five-yearly intervals, at a rent of £12,500 per annum; and an interim rent from October 5 1976 at a rate of £9,200.

C

COURT OF APPEAL

November 7 1979

(Before Lord Justice MEGAW, Lord Justice SHAW and Lord Justice WALLER)

BELL v ALFRED FRANKS & BARTLETT CO LTD AND ANOTHER

Estates Gazette March 1 1980

(1979) 253 EG 903

Landlord and Tenant Act 1954, Part II—Question as to construction and application of section 23(4)—Covenant in
D **lease that a garage was to be used for a private car only—Evidence of business use in breach of covenant—Whether immediate landlord or his predecessor in title consented to breach or immediate landlord acquiesced in it—Part II of Act not applicable where business use is in breach of covenant in the absence of such consent or acquiescence—Issue relevant to validity of notice to quit which was not in the form required by Part II—Difference between "consent" and "acquiescence"—For Part II to apply it is sufficient if immediate landlord "acquiesces" in the breach, but predecessor in title must "consent"—Immediate landlord's attitude not relied on, so that question was whether predecessor had consented—While**
E **"acquiescence" can be passive, with merely an absence of objection, "consent" means something of a positive affirmative kind—No evidence of such consent—Part II of Act not therefore applicable—Notice to quit accordingly valid and effective—Landlord's appeal in respect of possession of garage allowed**

This was an appeal by the plaintiff landlord, John Arnaud Bell, from that part of a decision of Judge Leslie at Bloomsbury and Marylebone County Court which dismissed the plaintiff's claim for possession of a garage at 9 Devonshire Close in West London. The action in the county court had included a claim in respect of a flat against the second defendant, David Blank. The present appeal, however, against
F the first defendants, Alfred Franks & Bartlett Co Ltd, related only to the garage. There was no appeal against the dismissal of the plaintiff's claim for possession of the flat.

Andrew Pugh (instructed by Manches & Co) appeared on behalf of the appellant; J J Davis (instructed by Franks, Charlesly & Co) represented the respondents.

Giving the first judgment at the invitation of Megaw LJ, SHAW LJ said: This is an appeal from that part of a judgment of His Honour Judge Leslie given on January 11 1979 at the Bloomsbury and Marylebone County Court which dismissed the plaintiff's claim for possession of a garage at 9 Devonshire Close in the west of London. At the hearing, a flat in proximity to the garage was also the subject-matter of the litigation, but G
this appeal is concerned only with the garage. The matter in contention was whether that garage had been occupied by the first defendants for business purposes. That issue arises because the plaintiff purported to give a notice to quit which was not in the form required by the Landlord and Tenant Act 1954 in relation to a business tenancy and, therefore, would have been abortive if the garage had in fact been occupied for business purposes.

The history which was dealt with by the county court judge is a long one and it is not necessary to recount it in the detail with which it was necessary for Mr Pugh to present it to this court. As long ago as the end of 1949 a gentleman named Ward, H
who occupied 94 Harley Street (which is adjacent to 9 Devonshire Close) let the garage there to the first defendants (whom I shall call "the defendant company"). In 1954 Mr Ward assigned, or agreed to assign, the lease to a gentleman named Allen, who was a dental surgeon practising at no 94, and in 1964 Mr Allen, by a written agreement, let the garage in question together with the flat to which I have referred to the first defendants. The term was for three and a quarter years beginning in September 1964 and therefore it expired by effluxion of time somewhere at the end of December 1967. The agreement, by clause 2, paragraph (4), setting out the tenant's covenants, provided that the tenant undertook to "use J
the said garage for standing a private car only" and then, since the agreement comprised the flat also, it went on to say "and the flat only as and for a private dwelling-house." It is important to note that the same word "private" is used in relation to the car and also in relation to the dwelling-house, because the question arises as to whether "*private* car" was merely a description of the character of the vehicle or was a reference to the nature of the use to which it would normally be put.

In September 1975 there was an assignment of Mr Allen's lease to the plaintiff. By that time the covenant to which I have just referred had been relaxed because it was amended so as to K
provide that two private motor cars might stand there instead of one. For some long time before the assignment to the plaintiff there is no question but that a Mr Franks—who was a moving spirit in the defendant company—had used the garage from time to time to take in and store samples collected for his company's business. Two cars which were put there—one a Bentley and another a Rolls-Royce—were used, so the evidence went, in order to assist the objects of the company by carrying prospective customers to and fro and also for the purpose of conveying samples. That was the business use which was relied upon in order to establish the character of the occupation of the garage as being for business purposes within the provisions of section 23 of the Landlord and Tenant Act L
1954.

The plaintiffs having been given the notice to quit, discontinued the action which they had started. It was not until December 15 1977 that the action under which the present appeal arises was begun. In that action the plaintiff claimed possession of the flat (which is not now in issue) and also of the garage. This was on the ground that the agreement under which they held, made in 1964, had expired at the end of December 1977. The defence, in general terms, was that the notice was ineffective because it was not such a notice as is required in the case of a business tenancy by the Landlord and M
Tenant Act 1954. In that way the critical issue from the standpoint of this appeal is as to whether the learned judge was right in finding (as he did) that the occupancy of the garage was within the Act of 1954.

Section 23, which is the first section in Part II of the Act, which is headed "Tenancies to which Part II applies," by subsection (1) provides:

> Subject to the provisions of this Act, this Part of the Act applies to any tenancy where the property comprised in the tenancy is or includes premises which are occupied by the tenant and are so occupied for the purposes of a business carried on by him or for those and other purposes.

A If a business is carried on in contravention of a covenant excluding such user, subsection (4) provides as follows:

> Where the tenant is carrying on a business, in all or any part of the property comprised in a tenancy, in breach of a prohibition (however expressed) of use for business purposes which subsists under the terms of the tenancy and extends to the whole of that property, this Part of this Act shall not apply to the tenancy unless the immediate landlord or his predecessor in title has consented to the breach or the immediate landlord has acquiesced therein.

Thus if a breach of the prohibition has occurred but it can be said to have been acquiesced in by the immediate landlord the activity constituting the breach will nonetheless give rise to a business tenancy. If the acquiescence of the immediate
B landlord is not relied upon—and it was not in this case, as is conceded—one has to look back to his predecessor in title (which in the present case was Mr Allen) and whether he consented to that use. Mere acquiescence by a predecessor in title would not be sufficient to confer upon the tenant the protection of the Act of 1954 to a tenant of business premises.

Mr Allen—who is unfortunately now deceased—had a long and friendly association with Mr Franks. Mr Allen carried on his practice as a dentist at 94 Harley Street and he very often was in the vicinity of the garage; he was able to observe what was going on there; he was able to see that from time to time there were parcels containing samples left there; and he
C also saw from time to time one or other of the vehicles which the company used, namely the Rolls-Royce or the Bentley, but not a van, or a lorry, or anything of the nature that might be called a commercial vehicle. It is contended that, as over this long history Mr Allen did not require that activities of that kind in the garage should stop as being in contravention of the tenant's covenants in the agreement, that amounts to consent.

That is really the short point which the learned judge had to decide so far as the occupancy of the garage was concerned. Two matters were relied upon before him by counsel on behalf of the defendants to show that there was a business user of the premises and that there had been consent to them. One was
D the presence of cartons containing samples for the business, fortified by conversations—which were part of the evidence—between Mr Allen and Mr Franks and which had reference to the contents of the cartons, such as sunglasses and things of that kind. The other was the fact that the cars themselves were used in order to convey customers of the defendant company as well as goods belonging to it.

The first question is did either of those activities constitute a business use? There can be no doubt that the bringing in of samples in cartons, and so forth, might well constitute such a business use even though it was intermittent and not continuous. The evidence suggested that there was nearly always something
E of that kind there—and the learned judge held that that was in this respect such a use. He made no finding as to the effect of the presence of these motor cars from time to time. That prompted Mr Davis to submit, on the basis of his cross-notice, that he should have found that the motor cars were there for business purposes, so that upon that foundation also the premises were occupied for such purposes and, therefore, there existed a business tenancy which was protected under the Landlord and Tenant Act 1954.

It is convenient to deal with that part of his submission first because it raises the narrow question of interpretation of the meaning of the word "private" car in the context of the
F agreement under which the defendant company held the garage. It seems to me that in conjunction with the sentence which follows it, which refers to the use of the flat as a "private" dwelling-house, "private" there must mean used for some personal or domestic purpose; not merely a car which is constructed for such purpose, but which is indeed used for such purpose. A dwelling-house, of course, is a house which people normally dwell in. A "private" dwelling-house is one actually used for domestic purposes. So construing that agreement, one has got to take note of the conjunction of the use of that word in two different aspects, one relating to the garage and the other relating to the flat, no 9. It would follow that use of the word "private" there means for private purposes. The fact G
that it used the phrase "for standing two private cars" did not mean two private cars used for business purposes thus introducing the element of business occupancy into the letting of the garage any more than it did into the letting of the flat at the same time. So I would hold against Mr Davis's submission on that matter under his cross-notice.

That leaves for decision only the question whether there had been at some stage in the history a consent to the use of the garage against the prohibition contained in the agreement. This is a semantic and philosophical question which requires definition of the distinction between acquiescence and consent. It is quite clear that what section 23(4) intended was to ensure H
that the immediate landlord should not be bound by mere—and I use that word deliberately—"mere" acquiescence on the part of the immediate predecessor in title, because that goes far to giving to the tenant a protection and exposing the immediate landlord to an undue risk to which he ought not to be exposed. What is meant by acquiescence? It may involve no more than a merely passive attitude, doing nothing at all. It requires as an essential factor that there was knowledge of what was acquiesced in. In this case it is not in controversy that there was such knowledge on the part of the plaintiff's predecessor in title that the garage was used in the way in which it was. J

If acquiescence is something passive in the face of knowledge, what does "consent" mean? In the context of the contrast implicit in the subsection, the only practical and sensible distinction that can be drawn is that if acquiescence can arise out of passive failure to do anything, consent must involve a positive demonstrative act, something of an affirmative kind. It is not to be implied, because the resort to implication betokens an absence of express affirmation. The only sense in which there can be implied consent is where a consent is demonstrated, not by language but by some positive act other than words which amounts to an affirmation of what is being done and goes beyond mere acquiescence in it. It may K
lead, in this context, to a false conclusion to speak of "implied consent," which is what the learned judge said was the proper inference to be drawn from the long history of acquiescence. I would prefer for myself to say "consent" involves something which is of a positive affirmative kind and that is what is required by section 23(4) if the immediate landlord is to be deprived of the opportunity of taking advantage of a breach of a prohibition contained in the terms of the tenancy.

The learned judge, in his very careful judgment, having examined the evidence, expressed himself in this way in regard to the garage: "The question is whether the plaintiff's notice to quit was effective to terminate the letting to the L
first defendants, and that depends on the use of the garage." He then set out what the contentions of the parties were and went on to say this: "It is true that the plaintiff and his witnesses saw no cartons, but having heard the evidence I find that the defendants have been using the premises for the purposes of a business, namely for the storage of cartons containing samples and goods, although contrary to the lease and contrary to the head lease. Mr Allen"—he was the predecessor in title of the plaintiff—"knew of this and must have waived the breach. It is possible that the plaintiff is entitled to bring proceedings for forfeiture but that is not what the court is concerned with in the present case." Then M
having set out section 23(4) in his judgment, he said, "and in my view that section applies to this case." That was following what he had previously said, that Mr Allen knew what was going on "and must have waived the breach." However, the inference from the history of Mr Allen's relationship with Mr Franks and his knowledge of the activities of the defendant company gives rise, so it seems to me, to no stronger inference than that there was acquiescence on the part of Mr Allen. An examination of the evidence as a whole reveals nowhere that any such positive consent, positive affirmation or permission was actually given at any time. So far as the immediate landlord is concerned, that appears to be fortified by the

A earlier part of the history when Mr Franks—who was himself, incidentally, a solicitor although he was engaged in business as well—had asked that the tenancy should be in the name of his company. That was not necessarily because the company wanted it. There might have been convenience, or financial advantage, in having the company as the tenant of the garage instead of Mr Franks himself.

I have not here dealt with the elaborate submissions that Mr Pugh presented before this court. That is not to say that they were without interest, but it seems to me that the only ones that really go to the heart of this case are: firstly, the issue as to whether or not the learned judge was right in B coming to the conclusion that there had been a business use; secondly, whether he was right in saying that the premises were occupied for the purposes of the business; and, thirdly, whether he was right in finding, as he did, that there had been a consent within the meaning of section 23(4). In my view, what is decisive of the appeal is that there was no acquiescence on the part of the plaintiff and no consent by Mr Allen. Accordingly, the requirements of the Act of 1954 were not called into play. The notice to quit was effective to terminate the tenancy and the plaintiffs are entitled to an order for possession of the garage premises.

I would allow the appeal.

C Agreeing, WALLER LJ said: I only add a few words on the question of consent because we are differing from a very experienced judge on that one finding which he made.

My Lord has already read section 23 of the Landlord and Tenant Act 1954, and in particular subsection (4), and I only repeat the last three lines of that section which say: ". . . this Part of this Act shall not apply to the tenancy unless the immediate landlord or his predecessor in title has consented to the breach or the immediate landlord has acquiesced therein." The learned judge, in considering that part of the case, made the finding already quoted by my Lord, namely D having dealt with the storage of cartons about which evidence was given by Mr Franks, he said: "Mr Allen knew of this and must have waived the breach. It is possible that the plaintiff is entitled to bring proceedings for forfeiture, but that is not what the court is concerned with in the present case."

So that, in effect, was the view the judge formed in that paragraph, and then he went on to consider whether or not the consent had to be formal and said: "When a landlord knows about a breach as Allen did, and the breach clearly goes on for a substantial period of time, as this one did, and the landlord can even accept a sample which should not be there, then one can only infer that he did consent. The parties E were on amicable terms and Mr Allen took no steps to cause this wrongful user to cease and he never even objected. I find, therefore, that the premises were used for business purposes notwithstanding the prohibition and that it was consented to by Mr Allen."

When one looks at the evidence on which the judge made that finding, perhaps the high-water mark is shown in a quotation from the evidence of Mr Franks. He said at the end of his evidence-in-chief "He never objected." Although there are other passages which clearly show acquiescence, that answer is the only one upon which the judge could found the findings that there was consent.

F My Lord has already mentioned—and I do not wish to repeat it, but I would emphasise—the distinction which is drawn in section 23(4) between the predecessor of the landlord and the immediate landlord. In order to avoid being taken out of the Act the tenant has to show that the immediate predecessor consented to the breach whereas it would be sufficient for him to show acquiescence on the part of the present landlord. The contrast between those words is, in my view, significant. Acquiescence clearly can cover a great many matters, and it seems to me the words of the learned judge where he says, "Mr Allen knew of this and must have waived the breach" was very clear evidence of acquiescence, but G acquiescence is something which has to be contrasted with consent and, in my judgment, consent requires some positive action on the part of the landlord or his predecessor—usually no doubt in words, perhaps in writing, possibly, if gestures were absolutely clear, it could conceivably be by gesture, but, in my view, careful proof of such an intention would be required. Normally one would look for some express statement, either in writing or oral, by the landlord. The answer which I have quoted clearly shows something far short of that, because there is the answer "He"—that is Allen—"never objected." In my judgment, having regard to the matters that I have mentioned, it is not possible to draw an inference from H that evidence that the predecessor consented to the use for business purposes. Accordingly, I also would allow the appeal.

Also agreeing, MEGAW LJ said: We did not consider it necessary to trouble Mr Davis to make submissions on a number of points which had been raised by Mr Pugh for the appellants in support of the appeal. That was because it was apparent, and accepted by counsel, that if Mr Davis, for the respondent, were wrong on the point which he wished to raise in the cross-notice and were wrong on the issue as to consent, then none of those other issues fell to be decided.

One of those issues on which we did not hear Mr Davis and on which, therefore, I do not find it necessary to express J anything in the nature of a concluded view, was the conclusion reached by the learned judge that here it was shown on the evidence that the premises were occupied for the purposes of the business carried on by the defendants. I will assume, therefore, without expressing any view on it, that the learned judge was right in deciding that question in favour of the first defendants, the tenants. On that assumption, Mr Davis put forward, by virtue of his cross-notice, as I understood it, the proposition that the case thus inevitably fell within the purview of section 23(1) and (2) of the Landlord and Tenant Act 1954, and then he went on to submit that for one simple reason section 23(4) did not apply to take it out again. Subsection (4) K depends upon the carrying on of a business having been in breach of a prohibition on use for business purposes; and, said Mr Davis, here there was no covenant in the relevant tenancy agreement which in any way precluded the first defendant from using the garage for business purposes. That was a point which depended upon the construction of subclause (4) of the agreement on September 23 1964, which imposed an obligation on the tenant "To use the said garage for standing a private car only and the flat only as and for a private dwelling-house." Whatever may be the meaning of "private" or "private car" in other contexts, I am satisfied that on the true construction of the words of that clause, L first the keeping of cartons of goods in the garage was in breach of that covenant to use the garage for standing a private car only; and, secondly, that the using of the garage in order to keep one or more cars there which were used to a substantial degree for the business of the defendants (they being a limited liability company) was an infringement of that covenant. If a car is used to a substantial degree for business purposes it is not a private car for the purposes of this covenant any more than the use of a house for purposes of business would be the using of the house as a private dwelling-house.

That point having gone, unless the learned judge was right in the view which he has taken that there was here consent on M the part of the plaintiff's predecessor in title, Mr Allen, then the plaintiff was entitled to succeed. I agree with my Lords that in the context of section 23(4) of the Act, whatever consent or acquiescence may mean in different contexts, in that context "consent" is put in plain antithesis to "acquiescence"; and that, therefore, if something falls within the description "acquiescence," it is not consent. The difference which is pointed out between the two in this context is that "consent" involves some affirmative acceptance, not merely a standing by and absence of objection. The affirmative acceptance may be in writing, which is the clearest obviously; it may be oral; it may conceivably even be by conduct, such as nodding the head in a

A specific way in response to an express request for consent. But it must be something more than merely standing by and not objecting.

I agree that the appeal falls to be allowed.

The appeal was allowed with costs in the Court of Appeal. The judge's order in the court below was varied to provide that no costs should be paid as between the plaintiff and first defendants. The order stood for the plaintiff to pay the second defendant's costs in the court below. The order in favour of the appellant for possession of the garage was stayed for 28 days.

B

CHANCERY DIVISION

February 4 1980

(Before Mr Justice WALTON)

SAFEWAY FOOD STORES LTD v MORRIS

Estates Gazette June 21 1980

(1980) 254 EG 1091

Landlord and tenant—Lease of business premises—Landlord and Tenant Act 1954—"Very curious clause"—Power to either lessors or lessees to determine term of 42 years at end C of 22nd or 32nd year, but with proviso that lessors should not exercise such power except for the purpose of revising the rent—Contractual and statutory notices given by lessors to determine both the contractual term and the business tenancy at the end of the 22nd year—Question as to validity—Both notices held to be valid despite absence from statutory notice of reference to a garage included in demise—Effect of notices in the light of above clause in lease—Lessors had in effect agreed not to plead the provisions of section 30 of the Act—Lessees entitled to grant of new tenancy under Part II of Act to be decided in the usual way

This was an originating summons taken out by the plaintiffs, D Safeway Food Stores Ltd, the present lessees under a lease of premises in Kensington High Street, London, dated February 28 1958, seeking a declaration as to the true construction of provisions in the lease in the events which had happened. The defendant to the summons was the present lessor, Sidney Morris. The plaintiffs sought a declaration that they were entitled to a new lease of the premises for a term equal to the residue of the term created by the lease and otherwise on the same provisions save as to rent. The premises comprised in the lease were ground-floor and basement premises, together with a garage, at 150, 152, 154, 156 and 158 Kensington High Street.

E J S Colyer QC and K M J Lewison (instructed by Lawrence Jones & Co) appeared on behalf of the plaintiffs; Ronald Bernstein QC and Christopher Priday (instructed by Brecher & Co) represented the defendant.

Giving judgment, WALTON J said: I have in front of me an originating summons which arises out of a lease dated February 28 1958 and made between the London County Freehold & Leasehold Properties Ltd as lessor and John Gardner (London) Ltd as lessee. The plaintiffs, Safeway Food Stores Ltd, have succeeded to the position of original tenant, John Gardner (London) Ltd, and Sidney Morris has succeeded to the position of London County Freehold & Leasehold Prop-F erties Ltd.

The lease, which was in fact an underlease (but nothing turns on that), was a demise of all those ground-floor and basement premises known as nos 150, 152, 154, 156 and 158 Kensington High Street, together with a garage at the rear (all of which premises are described as "the demised premises"). Then there were various rights, which I need not worry about, and exceptions and reservations. The reddendum was "to hold the demised premises unto the lessees for the term of 42 years from September 29 1957 (determinable nevertheless as hereinafter provided). . . . "Yielding and paying for the demised premises (excluding the garage)" a rising rent which rose to £11,000, "and also yielding and paying in respect of the G garage" £135 per annum, both payable by equal quarterly payments. Then there was a further rent, which I am not concerned with, in respect of the supply of hot water.

The lessees covenanted that they would not assign, transfer, underlet or part with the possession of any part or parts of the demised premises but so that that prohibition would not prevent the lessees underletting with the consent of the lessors the whole of one or more of the ground-floor shops thereby demised together with the full basement beneath such ground-floor shop so underlet and should not assign, transfer or underlet or part with the possession of the whole of the demised premises without the consent of the lessors which consent H should not be unreasonably withheld.

The lease, as I have already said, was dated February 28 1958, when rent review clauses were not as well known and frequently adopted as they are now, and in place thereof there was a very curious clause, clause 9, the whole of which I think I must read.

> If either party shall desire to determine the term hereby granted at the end of the 22nd or 32nd year thereof and shall give to the other party two quarters' previous notice in writing of such its desire and in case of a notice given by the lessees they shall up to the time of such determination pay the rent hereinbefore reserved then immediately on the expiration of such 22nd or 32nd year J as the case may be the present demise and everything herein contained shall cease and be void but without prejudice to the rights and remedies of either party against the other in respect of any antecedent claim or breach of covenant provided always that the lessors shall not exercise their option to determine at the 22nd or 32nd years of the term except for the purpose of revising the principal rent herein reserved.

On March 19 1979 two notices were served upon the plaintiffs, the first notice being a notice pursuant to clause 9 determining contractually in accordance with the terms of that clause the term granted by the lease at the end of the 22nd year, that is September 29 1979. There was also served at the same time, as was necessary to be served, a notice K under the Landlord and Tenant Act 1954 terminating the business tenancy, and it is really upon that notice that the first of the questions with which I am concerned arises. The plaintiffs in fact did challenge that the defendant had given the contractual notice pursuant to clause 9 in accordance with its terms, because although when writing his letter of March 19 1979 Mr Morris wrote that he confirmed that the notice was given for the purpose only of revising the principal rent reserved by the underlease, the plaintiffs did not, as they were fully entitled, accept that, with the result that they gave notice for production of Mr Morris for cross-examination upon his affidavit herein. Mr Morris has been produced. Mr Morris L has been cross-examined. Mr Morris has convincingly and triumphantly asserted, confirmed and established that his notice was given for the purpose only of revising the principal rent reserved by the underlease, and that head of attack by the plaintiffs on that notice is abandoned.

But that leaves a very curious point which arises on the second notice. The second notice says, "To Safeways Stores Ltd, of 16 Eastcheap, London EC3," that being, I take it, their registered office. "Tenant of premises known as ground floor and basement of numbers 150, 152, 154, 156 and 158 Kensington High Street, London W8," and then the usual matters, "I Sidney Morris," and then his address is given, M "landlord of the above mentioned premises hereby give you notice terminating your tenancy on September 29 1979," and the notice goes on to contain the statement, in conformity with the provisions of clause 9, that Mr Morris would not oppose an application to the court under Part II of the Act for the grant of a new tenancy.

What is said is that the description of the premises which I have read out is not a full description of the premises and that in order to be valid there should have been some reference to the garage. The garage in fact over the years has had its front and back walls demolished and now all it serves is as a partially covered way into the rear of the supermarket. I say partially covered because between the back wall of what used to be the garage and the supermarket itself there is a service road, a narrow service road, but still a service road. What is said is that because of the omission of any reference to the garage the notice is invalid.

This sort of point has been before the courts on a few occasions. It is interesting that in a case called *Kaiser Engineers & Constructors Incorporated v E R Squibb & Sons Ltd* [1971] Estates Gazette Digest 553 the two premises which were said to be comprised in the one notice relating to the fourth floor were some premises on the fourth floor and some premises on the ninth floor. If I had been faced with that situation uninstructed I do not think I would have taken more than two minutes or two seconds to decide it, but it did in fact go to the Court of Appeal and indeed produced a dissent in the Court of Appeal, which suggests that the matter is not quite as simple as it always looks at first blush. But it is interesting to note that in that case Ungoed-Thomas J in the Chancery Division said, "What premises are these"—that is premises known as part of the fourth floor—"Premises described as 'known as A' might refer to premises other than those to which the name A would strictly apply, for example possibly a garage let and used in conjunction with A, but to be so described the other premises would have to be known as A or as part of A," and he then went on to say that the ninth floor would not under any circumstances be described as part of the fourth floor.

In the Court of Appeal, although the decision of the learned judge was upheld by a majority of two to one, Russell and Phillimore LJJ against Lord Cairns LJ, I think that Mr Bernstein was quite right in submitting that they were not really all that far apart in their reasons as to the tests which ought to be applied. Russell LJ, for example, said this when distinguishing a case which I certainly do not find helpful in the present circumstances, *Doe d Rodd v Archer* (1811) 14 East 245: "I do not find that case of assistance, the facts being very different. Here we have no et cetera and nothing in the facts to suggest that the relevant part of the ninth floor was at the time of the notice associated in any way with the fourth floor. Indeed, the facts were exactly opposite." Cairns LJ, who dissented, said this:

> Matters that may be taken into consideration when deciding whether the notice should be given a liberal construction include the fact that the landlord must have intended to give a valid notice and that on a strict construction it would have no operation at all. In my view the critical question in the present case is whether the words typed into the notice clearly confined it to part of the fourth floor or whether the language is such to make it possible to regard it as not being so confined.

In the present case on that basis I think the question is first of all—is this curious building, the site of the former garage and now a frontless and backless wonder, pointing to the supermarket and nothing else, so closely connected with the supermarket that anybody would think that it was part of that supermarket known as the ground floor and basement of the numbers I have already read out? And it would seem to me that one has only to postulate that question to get the answer that of course it has no real connection with anything else and could not be used as part of anything else at all. It might have been much more difficult if it had remained as an actual garage which could have had some independent use and validity. At the moment the building has no independent use or validity whatsoever except as a way through from the other side of the road across the service road into the supermarket.

The matter does not rest there, because there is a reported case, *Germax Securities Ltd v Spiegal* in the Court of Appeal reported in (1978) 37 P & CR 204 in which the Court of Appeal approved the following formulation by Goulding J in *Carradine Properties Ltd v Aslam* [1976] 1 WLR 442 of a test to decide whether a notice was valid:

> I would put the test generally applicable as being this. Is the notice quite clear to a reasonable tenant reading it? Is it plain that he cannot be misled by it? Applying that test as applicable to the present case I think the notice would be so, because the tenant receiving that notice and knowing the terms of the lease must have seen there was a mistake as it would not say 1973 in 1974. Once that is accepted it is obvious the notice is for 1975 and not 1973.

That test was blessed by Buckley LJ in that case, Roskill LJ entirely agreeing with Buckley LJ, as also did Goff LJ. I was cited, of course, the case of very great weight and standing, *Hankey v Clavering* [1942] 2 KB 326, where there was a notice to determine a lease on December 21 1941 when it should have been in fact December 25 1941. But that was a prospective notice, and the tenant would undoubtedly have been in reasonable doubt as to whether the landlord had made a mistake or whether the landlord was not really trying to bring the lease to an end on a date on which he had no right to bring it to an end at all.

So I return to the section 25 notice in the present case and it appears to me quite clear that no tenant of the supermarket, for such the premises are in fact, would have the slightest hesitation in knowing that the notice which he had received was a notice to terminate the whole of his tenancy. Indeed, as Mr Bernstein for the defendant pointed out, it is not really anywhere suggested that the lessees were in any doubt, and the nearest that one gets to it is a solicitors' letter of May 15 1979 in which it is said, "Our clients expressly reserve their rights as to the validity and effect of the two notices given by you and dated March 19 1979," which being interpreted is "We think we have found a jolly good technical point which we hope to be able to run later," and it certainly does not mean, "We are in any real genuine doubt" whatsoever.

There was another matter pointed out by Mr Bernstein on which I do not for present purposes intend to place any reliance, but the section 25 form is a statutory form which is prescribed by the Landlord and Tenant (Notices) Regulations 1957 as amended and therefore must have full force, validity and effect. In it what has to be inserted as the premises of which the tenant is tenant are premises known as, and then there is an instruction "insert full address." Now Mr Bernstein points out that nobody from first to last has suggested that there is any other address for this curious building than 150 and so on Kensington High Street; in other words if you wanted to complain to the owner of that building that it was, shall we say, in a ruinous state and liable to do damage to passers-by, it is not to some other address that one would write but to 150-158 Kensington High Street. If that is so it may be that on that ground alone the notice is a valid and good notice. As I say, I record the point. I do not find it necessary to rely on it.

I now have to return to the effect of clause 9, having now decided that the relevant notices have been given, both the statutory and contractual notices, to bring the tenants' tenancy to an end. Now what is said in the originating summons is that in consequence of those findings the tenants are entitled to a declaration that on the true construction of the said underlease and in the events which have happened they are entitled to a new lease of the above-mentioned property, the term being equal to the residue of the term created by the said underlease, and otherwise on the same terms as the said underlease save as to rent. Clause 9, as I have already remarked, is a very odd clause, because it contains two quite separate matters. There is the right of the tenant to put an end to the term and there is the right of the landlord to put an end to the term thereunder, and what is provided is that in either case, and therefore it must work as much if the landlord gives the notice as the tenant gives the notice, "everything herein contained shall cease and be void but without prejudice to" accrued rights. There is then tagged on to clause 9 the proviso that the lessors shall not exercise their option to determine "except for the purpose of revising the principal rent herein reserved."

A The difficulty that I feel with the declaration that the plaintiffs seek, and it is a declaration with which I have a great deal of sympathy, is that in order to achieve such a declaration in the first instance they must, I think, read clause 9, and indeed Mr Colyer for the plaintiffs did not shrink from reading clause 9, as if it conferred a contractual option upon the tenant to take a new lease. With the best will in the world I cannot so read it. The landlords are given the option of bringing the underlease to a complete end, full stop. They can only do so when they wish to take advantage of that situation for the purpose of revising the principal rent, as it is conceded now they wish to do in the present case. But that
B does not seem to me in any way to give the plaintiffs an option to take a new lease of the premises for a term equal to the residue of the term created by the said underlease, and otherwise upon the same terms as the said underlease save as to rent. It seems to me that if that had been intended it could very easily have been put into clause 9.

Then when one comes to consider how clause 9 operates, it must be, because so far as I can see there is no other way of working it, that the first step will be with the landlord, who will submit to the tenants the lease which they are willing to grant containing terms that they are willing to have contained in the lease, such, for example, as rent review clauses and
C otherwise, and then the tenant will say either that he accepts them or that he does not accept them. How then are disputes as to the contents of the offered lease (and more particularly because the tenant would, as regards that, say that all the terms must be the same as before and you must not put in anything else) to be settled, and any dispute as to whether rent revision clauses which are only a method of revising the rent upwards should be included or not? There is no machinery whatsoever for the determination of those matters. That suggests to me, because the lease is, in so far as I have been shown any of its parts, a well-drawn lease (clause 9 is unusual, but it otherwise appears to be a well-drawn lease), that that
D was probably not what the draftsman was getting at at all. It seems to me quite clear that what the draftsman was getting at was simply the ordinary statutory rights of the tenants, which by 1958 when the lease was granted were very well known, to have a lease under the provisions of the Landlord and Tenant Act 1954; and that in a flash explains the proviso. What the proviso is in effect saying is that the lessors are agreeing not to plead the provisions of section 30 of the Landlord and Tenant Act 1954, which would enable them in certain cases to obtain possession. That they contracted themselves out of, and it may very well be that the reason that the lease does not contain, as Mr Colyer suggested it might, a simple agreement by the landlord not to plead section 30 is that E
there might well be well-justified doubts as to whether that is a good and valid covenant, or whether the general welfare represented by section 30 does not override any attempt to contract out of it.

If one once approaches it along those lines it appears to me that all is plain sailing. Mr Colyer attempted to say that that was not a fair and proper way of reading clause 9 because, he says, it might be, for example, that the tenants had subdemised the whole of the property and then the contractual notice would just put them out in the cold. The landlord could doubtless serve then a proper notice on the underlessees' subtenants and deal direct with them and in that F
way get a revised principal rent, but the undertenants themselves, the original parties to the underlease, would be out in the cold. That conclusion does not trouble me in the slightest. Once one fails to regard clause 9 of the underlease as conferring any sort of option on the tenants, and I do not think it does confer any sort of option on them, then what it is saying is that the lessors can use it and use it only for the purpose of obtaining a revised principal rent. That being so, I see no inconsistency whatsoever between the fact that the net result of operating clause 9 might very well be that the tenants were out without any profit rental accruing to them at all. It is not designed in any way to preserve the G
tenants' profit rental. What it is designed to do is ensure that the lessor shall receive a proper principal rent—that and nothing else. Therefore it appears to me that, looking at the matter in that light, the situation now simply is that the tenants are entitled to the grant of a new tenancy pursuant to Part II of the Landlord and Tenant Act 1954, which will be decided in the usual way. In those proceedings the landlords have estopped themselves completely from pleading any of the grounds upon which they are entitled to possession. Reading clause 9 in that way it seems to me a simple and sensible and logically effective clause which raises no real difficulties of construction. H

That being so, it appears to me that I must refuse both the first and second declarations which are sought by the originating summons. I gather that in any event the grant of a new tenancy pursuant to Part II of the Landlord and Tenant Act 1954 is not before me, not having been adjourned before me, and therefore I presume—and I will hear argument on the point—that I ought now to adjourn the remainder of the summons back to chambers.

The plaintiffs were ordered to pay the costs of this adjournment into court.

LANDLORD AND TENANT

RENT ACTS

A **COURT OF APPEAL**

November 14 1979

(Before Lord Justice ORMROD, Lord Justice CUMMING-BRUCE and Sir David CAIRNS)

EVANS v ENGELSON

Estates Gazette February 9 1980

(1980) 253 EG 577

Rent Act 1977—Appeal by tenant against possession order—House purchased by respondent in name of company in which he held all the shares—Possession sought by respondent under B **Case 9 in Schedule 15 to 1977 Act—Complicated series of transactions between respondent and company ending in deed of gift of house from company to respondent—Object to enable respondent to claim that he was not a "landlord by purchase"—Court of Appeal brushed aside these transactions and decided case on reality of the matter and commonsense—Corporate veil could be pierced—Company a mere nominee of respondent—Premises in effect purchased by respondent in 1964—Accordingly respondent did not become landlord by purchase after March 23 1965 and application of Case 9 was not excluded—Warning that circumstances were special—Tenant's appeal dismissed**

C This was an appeal by the tenant, Miss Kari Engelson, from a decision given in the Cambridge County Court in favour of the present respondent, Mr Gordon Page Evans. The judge made a possession order against Miss Engelson in respect of the house occupied by her as a statutory tenant at 2 Lingholme Close, Cambridge. Mr Evans claimed possession on the ground that he required the house for occupation as a residence for himself and relied on Case 9 in Schedule 15 to the Rent Act 1977.

P Salsford (instructed by Royds Barfield, agents for Crossman & Co, of Cambridge) appeared on behalf of the appellant; J Malius (instructed by Mole, Metters & Forster, of Reigate) D represented the respondent.

Giving the first judgment at the invitation of Ormrod LJ CUMMING-BRUCE LJ said: In about 1952 Mr Gordon Page Evans acquired a leasehold interest in a flat in Gerald Road, London SW1, the landlord being the Grosvenor Estate. At the time of that transaction the lease on the premises was held by a corporate body, F C Property Co, which had been formed I think at the instance of executors. The technique whereby Mr Evans acquired his leasehold interest in Gerald Road was that, instead of himself buying the lease, he bought all the shares in F C Property Co, whose only asset was the leasehold interest in the flat in Gerald Road. So the E situation then was that he acquired the lease but acquired it in the name of F C Property Co, being a company in which he and his wife held all the shares.

The next stage is that in 1964 he wanted to buy another place, 2 Lingholme Close in Cambridge, because he wanted to have it available for himself and his wife when he retired and left London. He paid £3,000 to the vendor, but arranged that 2 Lingholme Close should be conveyed into the name of F C Property Co. In 1973 unfortunately his wife died whereupon, as her successor, he held all the shares in F C Property Co. On November 20 1973 2 Lingholme Close, F Cambridge, was let to Miss Engelson and, as that house when Mr Evans bought it had been put into the name of F C Property Co, the agreement in writing which gave rise to the lease was an agreement between F C Property Co and Miss Engelson. It is perfectly clear, however, that she knew that she was dealing with Mr Evans at all material times.

In 1976 Mr Evans surrendered his remaining interest in his lease in Gerald Road to the Grosvenor Estate in consideration fall upon him at the end of his term. On the surrender of that lease he arranged with Grosvenor Estate to become their tenant in a little flat in Pimlico for a term of five years. That tenant in a little flat in Pimlico for a term of five years. That G is where he has been living ever since. In 1978 he decided to give Miss Engelson notice to quit and to go and live in 2 Lingholme Close, Cambridge, which was his house, the registered owner whereof, pursuant to the conveyance, was F C Property Co. The lady did not go and so became a statutory tenant under the Rent Act. There was then a negotiation between Mr Evans and Miss Engelson with a view to Mr Evans buying her out by paying £1,000 in consideration of which she would surrender her interest. But although those negotiations reached an advanced stage, they never culminated in an enforceable agreement. Thus he found there was a tenant who was claiming the protection of the Rent Act. H Against that background he started these proceedings in the Cambridge County Court, claiming possession of the house at 2 Lingholme Close and putting as the ground on which he claimed possession Case 9 of Schedule 15 to the Rent Act 1977, that the dwelling-house was reasonably required by him, the plaintiff, for occupation as a residence for himself and his future wife and stepson.

The particulars of claim which were initially put on the court file dated November 24 1978 were amended on April 2 1979. By the amended particulars of claim a series of transactions are pleaded which occurred in 1978, being transactions between Mr Evans personally and the F C Property J Co. In the particulars he pleaded that, after Miss Engelson on October 12 1978 decided to go ahead with an agreement to leave the premises in consideration of the receipt of £1,000, on the very next day, October 13 1978, there was a transaction at a time when Mr Evans still expected Miss Engelson to agree to leave voluntarily if he paid her a specified sum. The F C Property Co then purported to sell 2 Lingholme Close to Mr Evans for the sum of £3,500. Of course that was a sum which was then far below the market price of the property. So at that stage, as a result of that transaction, he appears to have purchased Lingholme Close from F C Property Co. Then Miss Engelson decided not to go on with her K agreement to depart voluntarily from the house. As a consequence of that, on November 17 Mr Evans purported to sell the premises back to F C Property Co for the same sum as he had originally bought it from them. In the final stage, the last November transaction was on November 20 1978 when F C Property transferred the premises to Mr Evans by way of deed of gift. The object of that final exercise of a deed of gift was to enable Mr Evans to claim that, for the purposes of the Rent Act, he did not become a land-

A lord by purchase on any date after March 23 1965, because he had acquired the premises as a gift.

So the stage was set for the proceedings in the county court, which took the form of the question of possession arising under section 98 of the Rent Act 1977 and in particular the relevant application of Case 9 of Schedule 15 thereto. The learned judge in a judgment of some length (which I regret to say I find far from easy to follow) appears to have decided the case in Mr Evans' favour on the ground that the transactions in November were all genuine and the result was that on November 20 1978 Mr Evans acquired 2 Lingholme Close from the F C Property Co by way of gift. B Therefore, the judge held on that ground that Mr Evans did not become landlord of Miss Engelson by purchasing the dwelling-house after March 23 1965, that the dwelling-house was reasonably required by Mr Evans for occupation as a residence, and it was reasonable to make an order for possession.

Many arguments have been addressed to this court and a substantial number of authorities have been cited in order to assist the court with the analysis of the transactions between Mr Evans and the F C Property Co in the month of November 1978. If it were necessary to decide the case on such an analysis, it would be right for me to review the interesting C submissions made on both sides and to examine the authorities. From the view of this case that I have formed, the answer to the question posed by the proceedings, namely, whether Mr Evans is or is not a landlord who became a landlord by purchasing the dwelling-house after March 23 1965, should be decided by an analysis of the events years before when 2 Lingholme Close was acquired in a transaction the ingredients of which were these. He paid £3,000, which came out of his own pocket, and it is relevant to observe that that purchase money did not represent the proceeds of sale of the flat in Gerald Road which had previously been acquired on the purchase of all the shares in F C Property Co by Mr Evans D and his wife. The purchase price for 2 Lingholme Close was put up by Mr Evans as cash which he possessed without any reference to the interest of F C Property Co or himself in the Gerald Road flat. When he put up the money, he asked the vendor to convey Lingholme Close to F C Property Co, all the shares in which were held by himself and his wife. So much for the first question.

In my view the last question is what was the effect, for the purposes of the Rent Act, of that acquisition of 2 Lingholme Close? The contention of the parties can be simply stated. Counsel for the appellant, Miss Engelson, submits that for the purposes of the Rent Act F C Property Co E became the owner of 2 Lingholme Close. When Miss Engelson became tenant of that hereditament, her landlord was F C Property Co and F C Property Co was the only party entitled to collect rent from her; and for the purposes of the Rent Act the court should not look behind the legal character of the conveyancing transaction and the legal form of the lease on which Miss Engelson held from F C Property Co. Therefore, on behalf of the appellant it is submitted that when the Cambridge house was acquired, it was acquired by F C Property Co and counsel goes on to make his submissions about the transactions in November 1978, which he submits make no difference at all.

F For the respondent it is contended that for the purposes of the Rent Act it is right for the court to look at the reality of the transaction when Mr Evans put up £3,000 and bought 2 Lingholme Close as a nest to which he and his wife would retire in due course when they left London. The conveyance of those premises into the name of F C Property Co, a company in which Mr Evans and his wife held all the shares, was an accidental event which did not, for the purposes of the Rent Act, divest him and his wife of the Cambridge house. He put up the money. If the matter is considered in terms of equity, he and his wife were the only beneficiaries. There was a trust held by F C Property Co but, without invoking equity, it is contended on behalf of Mr Evans that the true G nature of the transaction was that F C Property Co was simply a nominee for himself and his wife. After the death of his wife, when her interest and her shares in F C Property Co devolved on Mr Evans, from that moment F C Property Co was merely a nominee for himself, Mr Evans, so that the reality of the matter is that Mr Evans acquired the property with his wife when he bought it and that the company has never been anything except a mere shell, irrelevant for the purposes of analysis of the Rent Act rights and of litigation. In my view that contention is right. I take the view that the court ought to analyse the transaction when 2 Lingholme Close was bought and discover the reality of the H transaction, which can be discovered by looking at the form of the conveyancing in the context of the purchase and the payment of the purchase price. For my part I regard F C Property Co in relation to 2 Lingholme Close as a mere nominee. The landlord, for the purposes of the Rent Act and Miss Engelson, was—and has been at all material times —Mr Evans. He acquired the dwelling-house before March 23 1965 because he acquired it in December 1964. When he became landlord through the medium of F C Property Co of Miss Engelson, he did not become her landlord by purchasing the dwelling-house, because he had it already. If that is right, then it is unnecessary for me to analyse the rather J tortuous goings-on about the house from November 1978, and it becomes unnecessary for me to decide what the consequence, if any, of those transactions is.

I permit myself, however, to say that after hearing the argument, if one accepts the learned judge's finding of fact that those transactions were real, then there was a stage in those transactions when Mr Evans did acquire the house from F C Property Co by purchase. Having done so, what has been called in one of the cases the "clog" on his interest remains a "clog" on the subsequent successor in title. However, it is unnecessary, on my analysis of the events in 1964, to decide the case on that ground and I would expressly refrain K from doing so. I would dismiss this appeal on a ground very different from those which appear to be the grounds on which the learned county court judge founded his decision, and the ground on which I would decide the case and dismiss the appeal is that Mr Evans in 1964 acquired the premises by purchase, but did not become landlord of Miss Engelson by purchase after March 23 1965.

I would add that the facts of this case are very unusual. The unusual feature is that F C Property Co was from Mr Evans' point of view a mere accident which he accepted because the vendors of his leasehold interest in Gerald Road wanted for some reason of their own to sell him the shares L in the company which owned the lease instead of the lease simpliciter. This accident seems to explain his decision to put 2 Lingholme Close into the name of the same company instead of into his own name or the joint names of himself and his wife. It is this circumstance which in my view makes it commonsense to pierce the corporate veil when answering the question whether he as owner was seeking possession of the house as a residence for himself within the meaning of Case 9. So this decision is certainly no authority for any proposition that ownership by a corporate body is commonly to be disregarded for the purposes of Case 9 situations. It is also important to take note that the tenant never raised any M suggestion of estoppel at the time when the contract was formed; she relied upon the fact that her landlord was a corporate body as compared to Mr Evans personally. Had the facts given rise to such a defence Mr Evans could not have succeeded.

For these reasons I would dismiss the appeal.

Agreeing, SIR DAVID CAIRNS said: I too would dismiss this appeal. Two lines of argument were put forward here on behalf of the appellant. One might be described as the legalistic one and the other as the commonsense one. If the legalistic

A approach is adopted on the construction of Case 9 of Schedule 15 to the Rent Act 1977, then one gets into the realm where attempts are made by divers transactions to escape from the part of the case which restricts the right of the landlord to get an order for possession. That was what was attempted to be done here. Whether it could have succeeded it is unnecessary, as my Lord has already said, to decide, so like him I decline to do so. In my view, however, the commonsense approach is the right approach. Support for that is to be found in the judgment of Lord Donovan sitting in the Court of Appeal in *Thomas v Fryer* [1970] 1 WLR 845 at p 852, where Lord Donovan approved of the question being put in this form:

> B Would the ordinary and reasonable person call this transaction a purchase and sale of the house?

It appears to me that the learned judge who heard this action did in fact adopt this approach in one part of his judgment. I quote from it:

> The question is as to whether the landlord became a landlord by purchase or is a landlord by purchase for the purpose of this act, would be to ask him "When did you purchase the premises?" Well, if that were put in the wide application of the matter "When at any time did you purchase the premises?" I have no doubt that Mr Evan's reply would be "Well, I first really purchased it in 1964 and I purchased it then with my money C and had it put into the name of F C Property Co."

Whether or not that was the ultimate ground (or one of them) on which the learned judge reached his conclusion, it is unnecessary to say. If it was, I think that it was the right ground and on that ground I, too, would dismiss this appeal.

Also agreeing, ORMROD LJ said: The only point on which I wish to comment is that the pleadings in this case are not satisfactory in any way. Mr Malins is not in any way responsible for them, but the particulars of claim as originally filed were manifestly inaccurate and inadequate and when they were amended in April this year the contention that the pro- D perty was originally purchased in fact by Mr Evans and not by F C Property Co was not pleaded. No objection seems to have been taken in the court below and it is clear that the learned judge had this very point directly before him because, as my Lord pointed out, it was the first point to which he referred in his judgment. He continued to discuss the situation in 1964 over the first two or three pages of his judgment and it plainly formed an important part of his consideration in this case. No question of the pleadings was raised in this court. I merely mention this in case someone hereafter says that the point was not pleaded. We have not asked Mr Malins to amend the particulars of claim because there was no real E need to do so. The case as put in argument below and here is sufficiently plain.

I agree with the reasoning expressed in both the judgments which have just been given and do not wish to add anything further of my own. I, too, would dismiss this appeal.

The appeal was dismissed with costs. The appellant's liability was determined at nil. An order nisi was made against the legal aid fund. The appellant was ordered to vacate the premises by January 1 1980. Leave to appeal to the House of Lords was refused.

F

COURT OF APPEAL

October 12 1979

(Before Lord Justice MEGAW and Lord Justice SHAW)

RHODES TRUST v KHAN

Estates Gazette February 16 1980

(1979) 253 EG 689

Principles applicable to exercise of judge's discretion to set aside order for possession against tenant protected under Rent Act—Prolonged absence abroad of protected tenant who made no arrangements to get in touch with landlords or pay rent—County court judge made orders for possession and payment of rent arrears while tenant still abroad—Tenant's flat subsequently G included in long lease of landlords' property to third parties with vacant possession—Tenant on return from abroad, finding himself dispossessed, sought to set aside judgment on ground that he was unaware of proceedings—Judge in exercise of discretion under Order 37 of County Court Rules set orders aside—Held by Court of Appeal that discretion had been exercised on wrong principle—Judge failed to take into account the prejudice to the third parties, who in good faith had acquired the landlords' property with vacant possession of the flat—Appeal allowed

This was an appeal from the Rhodes Trust from a decision of Judge McNair at the Bloomsbury and Marylebone County H Court setting aside orders which he had previously made in their favour for possession and arrears of rent in an action brought by them as landlords against Allah Bax Khan as tenant of a flat at 111 Inverness Terrace, London. It appears from the judgment of Shaw LJ that Mr Khan had originally been a subtenant of the flat, but had subsequently become the tenant of the appellants on the termination of the intermediate tenancy.

D A Wood QC and R J Furber (instructed by Coward Chance) appeared on behalf of the appellants; S Goldblatt QC and Richard Llewellyn Davies (instructed by G Lebor & Co) represented the respondent; T R F Jennings (instructed by J Cowan, Lipson, Rumney) held a watching brief on behalf of Mobax Securities Ltd (not parties to the proceedings).

Giving the first judgment at the invitation of Megaw LJ, SHAW LJ said: This is an appeal by the plaintiff in the action from an order of His Honour Judge McNair made at the Bloomsbury and Marylebone County Court on August 8 of this year whereby he set aside two judgments which had been obtained by the plaintiffs on May 22 and July 3 of this year respectively. The first of those judgments took the form of an order in favour of the plaintiffs for possession of a flat known as 111 Inverness Terrace in the west of London, which the defendant had occupied as a subtenant since July 1973, and K where his status was that of a protected tenant. The second of those judgments was a written claim in respect of arrears of rent amounting to some £637. On August 17 the learned judge gave leave to appeal against his order and so the matter came before this court, it having been expedited in the interests of all now concerned in this matter.

The history, so far as it is material to this appeal, is that the plaintiffs acquired the freehold of property which included 111 Inverness Terrace in 1969 and thereafter, by lawful subletting, the defendant became a tenant of 111. The contractual term out of which his subletting was derived would have ended in March 1978 and, before that date had arrived, the defen- L dant, who is the respondent in this appeal, had sought to negotiate an extension of his tenancy with the plaintiffs, but these negotiations had not produced any result.

In the month of February 1978 the defendant, who was a man of business, went to Nigeria in connection with that business. He had announced that he expected to be away for two months, but it might be, he thought, as short as two weeks; and he made no arrangements, whatever the length of his absence might be, to see that the rent due in respect of his tenancy was paid to the plaintiffs or to his intermediate landlord, whose name was Rahim. In the event, the defendant remained absent from this country until July of this year and, in the M meantime, so far as the history of the matter has revealed, he made no attempt whatever to get in touch with the plaintiffs as his landlords, or to take any steps to see that the rent was paid as it fell due. What he says he did was to put a notice on the door of his flat, giving the address of persons who might be referred to in case of need. So far as the evidence goes—and it includes the evidence of somebody who acted at material times as a caretaker of the premises, and another gentleman who acted as a manager of those premises—in the course of their not infrequent visits to the premises they saw no such notice on the door. That is the most the defendant may have done to see that the relationship of landlord and tenant between the plaintiffs and himself was preserved in the normal way with the rent being paid at the intervals at which it fell due.

The plaintiff appellants appear to have been very forbearing indeed, for they took no steps for over a year. In the end and, no doubt, because there was in prospect a sale in its entirety of the property of which 111 Inverness Terrace formed part, they took steps in the spring of 1979 to obtain an order for possession. Such an order was procured from the court on May 22 of this year and it was an order for possession forthwith. The money claim was not pursued at that time because, no doubt, the plaintiffs thought it would have been an empty judgment. Later, however, when they came to execute their order for possession—which they did some time in June 1979—it was found that there was a good deal of furniture inside the premises. Thus, there was something upon which execution might be levied, and so the judgment for the rent was also sought and was obtained.

Then, on June 26 1979—and the date is a critical one—the plaintiffs granted a long lease of the whole of the residential part of the property in which 111 Inverness Terrace was situated to a company called Mobax Securities Ltd. The purchase consideration was £127,000. The term granted was a 75-year lease and, in a schedule which formed part of that lease, there were set out the properties comprised in it, namely, residential flats with an indication against each of them as to which were being sold with vacant possession and which were already occupied. Against no 111 was the indication that vacant possession was available. That formed part, therefore, of the bargain between the parties.

On July 8 of this year Mr Kahn, the defendant, came back to England and, when he went back to what had been his abode and sought to enter it, he found that his furniture had been removed and that the locks had been changed so that he was unable to enter the premises. He says in his affidavit that he was able, by looking through the letterbox, to see that certain changes had been effected during his absence. He made an application to the court under Order 37 of the County Court Rules that the judgment for possession should be set aside, and he complained that he had been unaware of the proceedings against him. Through his counsel it was submitted to this court that no effort had been made by the landlords to make contact with him so as to apprise him of the fact that they were seeking an order for possession. The learned judge, in the exercise of what is undoubtedly a discretion which he had, took the view that, in all the circumstances which were disclosed before him, it would be right to set the judgment aside on the general principle that it is wrong to maintain a judgment against somebody who has had no knowledge of the proceedings, so that he was not present when the matter was before the court, and who knew nothing of the judgment at the time when it was given against him.

That, as a general principle, is absolutely right. The interests of justice require that somebody against whom proceedings are being pursued should be accorded the opportunity of resisting them, for he may have some reason for which he can effectively resist them or, at least, ensure that, if the judgment goes against him, it will not be in the full terms sought by the plaintiff but with such modifications as justice may require.

The learned judge, dealing with the matter, said this:

> It is a difficult case. After the plaintiffs obtained the order for possession they executed a new lease of the flat on the basis of vacant possession. They sold the flat on a long lease to Mobax Securities who paid a capital sum for this and other properties. Obviously the plaintiffs received a capital sum which was substantially more, because of the vacant possession, than they would have received if Mr Khan was the statutory tenant.
>
> It is submitted on these facts by the plaintiffs that I ought not to exercise my discretion to set aside the order which would be the normal step to take when a defendant is out of the country at the time of service, especially in a non-payment of rent case where there is no history of bad delay in payment. The normal procedure in a non-payment of rent case is that if the defendant pays money into court no order for immediate possession is made.

As a matter of history the proceedings were started by way of substituted service, the necessary order having been obtained from the court. The learned judge went on to say that the plaintiffs claimed that they had been prejudiced. He referred in the course of his judgment to the decision of the Court of Appeal in *Grimshaw v Dunbar* [1953] 1 QB 408 and, in particular, to a passage in the judgment of Jenkins LJ specifying the matters which, in general, ought to be taken into account where a court is called upon to exercise its discretion whether to make an order setting aside a judgment. Jenkins LJ at p 415 dealing with a delay said:

> ... I venture to think that delay in itself would not be important, but delay prejudicing the other party, or delay enabling rights of third parties to intervene, would be most material. How does the matter stand here as regards delay ... Therefore, in this case I should not say that there had not been any undue delay.

Mr Goldblatt on behalf of the respondent has urged that the delay there referred to is the delay which arises as between the date when the judgment has been given and the time when an application is made to set it aside. It appears to me that delay overall is a factor which ought to be taken into account in deciding whether or not the discretion to set aside a judgment ought to be exercised in favour of a defendant. Here, the major and effective delay resulted from the fact that the defendant absented himself for a long period; and that during the whole of that period, knowing that rent was accruing and falling due, knowing that he had instructed nobody to make contact with the landlords either to tender the rent to them or else to explain his absence and give them some assurance that he proposed to return and take up his tenancy and provide for the payment of any rent in arrears, he had created a situation in which any reasonable landlord would have supposed that, for one reason or another, the premises had been abandoned by the tenant and that he did not propose to come back at all. So, in those circumstances, they could take steps to obtain an order for possession and to re-enter and deal with the premises in whatever way they desired.

It seems to me that, when the learned judge came to deal with this matter, he took the view that, if the plaintiffs got the arrears of rent and if they were to adjust the situation between themselves and Mobax (who had bought the long lease), they really would not be under any serious prejudice at all. What the learned judge appears not to have kept in the forefront of his mind—as I think he should have done—is the position of Mobax, who had in good faith purchased the whole property with vacant possession of flat 111. They had in consequence removed the contents of that flat and then had gone on to effect improvements with a view to disposing of the flat on a long lease. The learned judge makes no reference to their position and does not appear to have taken it into account.

Mr Goldblatt conceded that their position is a relevant and material consideration. In a case where the rights of third parties have intervened and are threatened with prejudice, it is something which the court, when asked to exercise discretion, should weigh very heavily in deciding as to how to exercise that discretion.

Mr Goldblatt says it is premature to consider the effect on the third party. That is a matter which can still be effectively dealt with. If this appeal is dismissed and there is a new trial, then at that trial the third parties can come in and it will then be appropriate for their situation to be canvassed and methods of redressing any possible loss to them can be gone into. With all respect to Mr Goldblatt's argument—which he put as persuasively as it could be put—that very much begs the question of what is the immediate situation confronting the judge when asked to exercise his discretion under Order 37. He can of course—not that I for one moment suggest that it was so in

A this case—shelve the responsibility and say: "Well, this is to be left to be dealt with at the trial. Nobody will be hurt if that happens." On the other hand, for so long as a new trial is pending, the third party's interests are precarious. They cannot be sure what will happen when the trial takes place.

In my view it is right at the earliest available opportunity—which is when the learned judge is being asked to set aside the judgment—for him to weigh the interests of the third parties and to see whether there is a real threat to them or whether that threat can be dismissed as illusory and insubstantial. In the present case, it is quite obvious that the threat is a substantial and a real one. Where such a situation exists,
B the interests of the third party ought not to be exposed to such a degree of risk in a matter for which they have no responsibility whatever, and with regard to which they have acted at all times in good faith. To find that the situation so far as they are concerned is turned upside down because, as in the present case, the defendant has brought this situation about by his own dilatoriness and by his own neglect to take any step to look after his own interests, would be, it seems to me, utterly to ignore the impact upon innocent third parties.

While this is a matter in which this court would not usually interfere with the exercise of the discretion of the judge in the court below, it does seem to me that the learned judge
C had not applied his mind sufficiently to the effect upon the position of a third party. In the result, he made an order which could not reasonably be made if reasonable consideration had been given to all the relevant factors. For those reasons, notwithstanding the elaborate and careful argument of Mr Goldblatt, I would allow the appeal.

I would also add that, at a very late stage, Mr Goldblatt suggested that the appellants had themselves acted with inclement haste, having got an order for possession forthwith. They did not seek to execute that order through the court, which might have involved more time, and they also sought an order forthwith instead of the usual order, which would allow 28
D days so that there might be an application for a stay of execution. As a result they prevented Mr Khan by only a few days from approaching the court earlier or from appearing at the court. They it was who made it necessary for him to go back and put himself in mercy and ask for the judgments against him to be set aside.

I find those arguments unimpressive in the light of the defendant's own conduct. I consider that the learned judge erred in principle in the exercise of his discretion and I would allow this appeal.

Agreeing, MEGAW LJ said: This is a matter that was for the discretion of the learned judge and, for myself, I would not
E begin to think it right to intervene merely because of any view which I might take that I would have been disposed to exercise the discretion in a different way. But it seems to me in this case that the learned judge, as appears from the careful note of reasons for judgment approved by him which has been placed before us, has indeed acted upon a wrong principle in arriving at his conclusion. That being so, it is the right and the duty of this court to give effect to its own discretion based upon the material before the county court judge, but upon what we conceive to be the correct principle.

There are various unsatisfactory matters about this case. Our attention was called, for example, to discrepancies between,
F on the one hand, some of the contents of the affidavit put in on behalf of the defendant on the day before the hearing of the county court application resulting in the order to set aside the previous order, and, on the other hand, notes, apparently accepted by the parties and approved by the learned judge, of oral evidence given by the defendant at the hearing of the application. But, however much they might be matters of which one would wish explanation, they would not, in my view, in themselves begin to justify this court in thinking of setting aside the learned judge's order.

The principle on which the learned judge, with great respect to him, erred is that, as I read his notes of judgment, he did not take into account—or certainly did not take into account G
as being a material factor—the fact that rights of third parties had been affected during the period between the making by the learned judge of his order for possession in favour of the plaintiff on May 22 1979—that was an order which was perfectly validly and lawfully obtained and was a valid order of the court—and the date when the application came before the court to set aside that order. Between those dates third parties had obtained rights perfectly properly. No criticism can be made either of the third parties or of the plaintiffs in respect thereof. Those rights would, at any rate in some degree, be put in peril if the judgment lawfully obtained and lawfully executed were to be set aside as the learned judge decided H
should be done.

The principle is that which is set out in a sentence in a judgment of Jenkins LJ in *Grimshaw v Dunbar* [1953] 1 QB 408 at p 415. Each case, of course, must be decided on its own facts, as was clearly recognised and stated by Jenkins LJ in that judgment. He set out certain guiding principles to be followed. The two sentences which I wish to quote are these:

> Secondly, has there been any undue delay by the absent party in launching his proceedings for a new trial? In answering that question I venture to think that delay in itself would not be important, but delay prejudicing the other party, or delay enabling rights of third parties to intervene, would be most material. J

Now, it is perfectly true that the principle as there laid down is not literally applicable to the question which we have to consider. Jenkins LJ was there considering a case where it was suggested on behalf of the plaintiff, who had obtained an order and that order had been set aside at the instance of the defendant, that the defendant had shown undue delay in his application to set aside the original order. That question does not arise in this case because it cannot be suggested, and has not been suggested, that, after the defendant, Mr Khan, returned to this country on July 8 1979, there was anything other than proper expedition in the steps which he took to seek to have the order, of which he then learned for the first time, set aside. K

But the analogy with the passage which I have read appears to me to be this. By analogy with the delay which Jenkins LJ was there considering, you have in this case the conduct of the defendant in failing throughout all this time to pay rent, or to take any steps to have the rent paid or to take any steps to ascertain what was happening about the rent, which he must have realised was falling due in respect of these premises, of which he remained, so far as Rent Act law is concerned, in occupation, and for which he was liable for rent. That is one of the factors.

The other is that he absented himself from this country for
a period which, however short he originally may have thought L
it was going to be, extended to a period of some 18 months and, neither initially nor at a later stage when he realised, if it was a matter of realisation, that his stay was going to be much more prolonged, did he take any steps which could conceivably be regarded as reasonable steps—and, after all, he is a man of business—to ensure or to make it probable that matters of importance relating, for example, to any legal proceedings that might arise in connection with his rent or non-payment of rent or his right to remain in occupation of the flat should be brought to his notice. He simply did nothing that could really be described as reasonably meeting the sensible requirements in such circumstances. M

In these circumstances, it seems to me that, while these circumstances in themselves would not be sufficient to justify the refusal of the setting aside of the judgment and of the order for possession, nevertheless when, in relation to these circumstances and arising out of them, rights have accrued to third parties on the basis of an order lawfully made by the court and lawfully executed, that is a most material factor. The learned judge failed to take into account its materiality.

It may be that it would turn out that the third parties would have a right of indemnity, which might require legal proceedings to enforce, and which might or might not be

A successfully resisted, in respect of the loss which they would have suffered by reason of the contract which was made by them on the assumption, based upon a lawful order of the court, that Mr Khan was no longer interested in the property. But it does not seem to me to be material whether that loss or potential loss which the third parties would suffer, their legal rights being undoubtedly affected, is potentially large or potentially small. It is sufficient that there is here a real interference with the right of third parties and, in the absence of some strong factor to the contrary, it seems to me that justice requires that the order which had been lawfully made and executed should not be set aside.

B I agree that the appeal should be allowed.

The appeal was allowed with costs. An application by the respondent for leave to appeal to the House of Lords was refused.

QUEEN'S BENCH DIVISION

December 21 1979

(Before Sir Douglas FRANK QC, sitting as deputy judge of the division)

GUPPYS PROPERTIES LTD v KNOTT AND ANOTHER

C Estates Gazette March 1 1980
(1979) 253 EG 907

Rent Act 1977, sections 69 and 70 and Schedule 12—Certificate of fair rent—Appeal from decision of a Devon and Cornwall rent assessment committee on a reference to them of an application to a rent officer for a certificate of fair rent —Submission that in assessing the fair rent which would be chargeable under a regulated tenancy after the completion of improvements the committee were wrong in not taking into account the effect of works of repair which the applicants were also proposing to carry out—Further submissions that
D the committee applied the wrong test in finding that certain items of work were in the nature of repairs and not improvements and that the committee did not give sufficient reasons for their decision—Held by judge that the committee were correct in not taking account of proposed repairs in assessing the fair rent and that they had applied the correct test in distinguishing repairs from improvements—Held, however, that the committee had failed in their statutory obligation to give proper reasons and that the case must be remitted to them for further consideration on this ground—Judge cites with approval the strict Scottish doctrine as to the kind of reasons which should be given by committees, as formulated by the
E Court of Session in Albyn Properties Ltd v Knox—Leave given to appeal to Court of Appeal

In this case Guppys Properties Ltd, of Bridport, Dorset, appealed against a decision of a rent assessment committee of the Devon and Cornwall Panel determining the fair rent on the reference to the committee of an application for a certificate of fair rent. The premises concerned consisted of a cottage known as Chez Nous, Trehaddle, Cusgarne, Truro, Cornwall. The tenants, who were in occupation of the cottage under a regulated tenancy, were Mr and Mrs Knott. The application for a certificate of fair rent was made by Guppys Properties Ltd under section 69(1)(a) of the Rent Act 1977. The rent
F officer for the Cornwall registration area decided that the information supplied to him by the applicants was insufficient to enable him to issue a certificate of fair rent and that, accordingly, he could not entertain the application. The application was then referred, at the request of the applicants, to a rent assessment committee in accordance with Schedule 12 to the 1977 Act. The committee were satisfied that they could proceed and decided that in determining the fair rent for the purpose of the certificate only those works which constituted "improvements" within the meaning of section 75 of the Act should be taken into account and that works of repair should be excluded. They also decided that of the list of works pro-posed by the applicants the only items which constituted G
improvements as defined were the provision of an inside wc, an insulating quilt to the first-floor ceilings, and a hot water system and linen cupboard.

Robert C Pryor (instructed by Turner, Peacock) appeared on behalf of the appellant landlords; there was no appearance by or on behalf of the tenants.

Giving judgment, SIR DOUGLAS FRANK QC said: This is an application by the landlords to set aside and remit a decision by a rent assessment committee relating to a house near Truro in Cornwall. The landlords, intending to carry out works to the property, applied for a certificate of fair rent, but the rent
officer declined to entertain it, for it appeared to him that the H
information supplied was insufficient. Subsequently the application was referred to the committee. The committee considered the list of proposed works, which were estimated to cost £4,500, and which of the items constituted improvements and which repairs. They came to the conclusion that only certain items constituted improvements and only those would they take into account in assessing the fair rent. They then went on to determine the fair rent to be £9.75 per week after completion of the works of improvement. It is apparent that, apart from the question of what works should be taken into account, the methods of arriving at the fair rent were canvassed and,
in particular, assessments of capital value and the application J
of rents for comparable properties. The landlords' case was conducted by a director of the landlord company, who put in a long statement of evidence to be given.

When the matter came before me Mr Pryor appeared for the landlords, but there was no appearance for either the tenants or the rent assessment committee. Mr Pryor first argued that the committee should have taken into account the repairs to be carried out, but that submission was supported on grounds of logic rather than the provisions of the Act. Indeed, I can find nothing in the Act which empowered the committee to take into account repairs to be carried out in the future, and
in my judgment the committee cannot be faulted in having K
regard only to proposed improvements. I further think that the committee applied the correct test in deciding what constituted improvements and to that extent their decision is unassailable.

I turn now to the main ground upon which the decision is attacked, and that is that it is said that it failed to give reasons. The decision deals thoroughly with the state of the property as the committee assumed it to be, but when they come to the valuation of the property in that state they said:

> Having considered all the evidence put before them, and having regard to the provisions of the Rent Act 1977 and to their
> inspection, the committee are of the opinion that the only L
> alterations to qualify as admissible improvements are the provision of an inside wc, of an insulating quilt to the first-floor ceilings, and of a hot water system and linen cupboard, and accordingly determine that after the completion of those improvements a fair rent will be £9.75 per week exclusive of rates.

Thus all they say about the evidence is that they have considered it and, without giving any reason beyond that, they determine the fair rent to be £9.75. There have been numerous decisions concerning requirements under the Tribunals and Inquiries Act 1958 to give reasons for decisions under town planning and compulsory purchase legislation and the Rent
Acts. However, I think for the purpose of this case it is M
sufficient to refer to *Mountview Properties Ltd v Devlin* (1970) 21 P&CR 689. There Lord Parker CJ said (at p 692):

> What reasons are sufficient in any particular case must, of course, depend upon the facts of the case. I approach the matter in this way: that reasons are not deficient merely because every process of reasoning is not set out. I further think that reasons are not insufficient merely because they fail to deal with every point raised before the committee at the hearing. Indeed, I would adopt the words used by Megaw J in *Re Poyser and Mills' Arbitration* [1964] 2 QB 467. That was dealing with an arbitrator's award, but Megaw J said: "The whole purpose of section 12 of the Tribunals and Inquiries Act 1958 was to

enable persons whose property or whose interests were being affected by some administrative decision or some statutory arbitration to know, if the decision was against them, what the reasons for it were. Up to then, people's property and other interests might be gravely affected by a decision of some official. The decision might be perfectly right, but the person against whom it was made was left with the real grievance that he was not told why the decision had been made. The purpose of section 12 was to remedy that, and to remedy it in relation to arbitrations under this Act. Parliament provided that reasons shall be given, and in my view that must be read as meaning that proper, adequate reasons must be given. The reasons that are set out must be reasons which will not only be intelligible, but which deal with the substantial points that have been raised." A little farther down, he said: "I do not say that any minor or trivial error, or failure to give reasons in relation to every particular point that has been raised at the hearing"—and he was dealing with an error of law on the fact of an award—"would be sufficient ground for invoking the jurisdiction of this court."

Here, as it seems to me, while the tribunal had before them evidence of comparable flats in the vicinity, they had in addition a schedule of all the contractual rents and past histories including fair rents fixed by the rent officer and by the committee itself with regard to other flats in Mountview Court, yet the striking bit of evidence which they had before them was no 64, and, beyond saying, in the first reasons which they gave, that they had, to use their own words, "considered the evidence adduced," the decision in objection no LON/31/551 and their inspection, there is no explanation at all to show why they rejected what one would think would have to be a greater market value, a greater fair rent in respect, at any rate, of the two larger flats, no 43 and 73. In my judgment, that failure does result in the fact that they have here given insufficient reasons; they have not complied fully with their duty under section 12 of the Tribunals and Inquiries Act 1958.

Accordingly, the next question is what flows from that. There is, in my judgment, undoubtedly power to compel any tribunal to which section 12 of the Act applies to give sufficient reasons; they could be ordered by mandamus to do so; in an appeal they could be required to do so by remission of the case to them. Indeed, as Lord Denning MR said in *Earl of Iveagh v Minister of Housing and Local Government* [1964] 1 QB 395: "The whole purpose of the enactment is to enable the parties and the courts to see what matters he"—and that was the Minister of Housing and Local Government in that case—"has taken into consideration and what view he has reached on the points of fact and law which arise. If he does not deal with the points that arise, he fails in his duty: and the court can order him to make good the omission."

I have quoted at length from that judgment because what is said there seems very pertinent to the instant case. However, I appreciate that valuation is not an exact science and at the end of the day, having considered the evidence, the committee are entitled to use their own knowledge and experience: see the cases cited by Lord Denning MR in *Metropolitan Properties Co (FGC) Ltd v Lannon* [1969] 1 QB 577. Lord Denning however pointed out that that does not mean they should throw over the evidence altogether. At any rate, they should not throw over the evidence without saying why. It is perhaps not inappropriate to point out that the Lands Tribunal, in dealing with rating assessments of dwelling-houses, have to carry out an exercise comparable to that carried out by rent assessment committees. Although the Lands Tribunal has for the 30 years of its existence been required to give reasons for its decisions, the Court of Appeal has never found it wanting in that respect and, although I would not expect rent assessment committees to set out the facts, evidence and arguments as extensively as is done by the Lands Tribunal, I suggest that their decisions are a useful guide.

Further assistance can be gained from the judgment of the court in *Albyn Properties Limited v Knox* 1977 SLT 41 at p 43 where it was said:

> The statutory obligation to give reasons is designed not merely to inform the parties of the result of the committee's deliberations but to make clear to them and to this court the basis on which their decision was reached, and that they have reached their result in conformity with the requirements of the statutory provisions and the principles of natural justice. In order to make clear the basis of their decision a committee must state: (i) what facts they found to be admitted or proved; (ii) whether and to what extent the submissions of the parties were accepted as convincing or not; and (iii) by what method or methods of valuation applied to the facts found their determination was arrived at. In short they must explain how their figures of fair rent were fixed.

In the instant case it is quite impossible to discern from the decision what reasons lay behind the assessment of £9.75 per week, other than that the committee disregarded works which were not improvements. In particular, they have failed to state whether and to what extent the submissions concerning capital value and comparables were accepted as convincing or not. Consequently, the applicant knows not whether any consideration was given to his submissions and thus is left in a state of mind of dissatisfaction, which the requirement to give reasons was enacted to avoid. Accordingly, in my judgment, the committee have failed to fulfil their statutory obligation to give reasons and, accordingly, the decision must be sent back to them for further consideration.

The case was remitted to the rent assessment committee to consider further in the light of the judgment. Leave was given to appeal to the Court of Appeal. No order was made as to costs.

CHANCERY DIVISION

October 31 1979

(Before Mr Justice BALCOMBE)

FONTANA NV v MAUTNER

Estates Gazette April 19 1980

(1979) 254 EG 199

Promissory estoppel—Action for possession of flat not subject to the Rent Acts—Tenant in occupation of flat under a succession of terms since 1952—Tenant's wife suffering from severe mental and physical illness likely to be aggravated if she had to move out of the familiar surroundings of the flat—After a change of landlords, tenant was warned that he would have to give up possession of the flat—At a meeting with a representative of the landlord company who had authority to arrange the lettings of the flats the tenant explained his wife's serious condition—Landlord's representative said "it is quite all right, you can stay here as long as you wish"—Subsequently tenant was given a notice to quit—In the possession proceedings tenant relied on this statement as constituting a promissory estoppel—Judge found that tenant had all along been convinced that his possession of the flat could not be disturbed and that he did not do anything, either by action or inaction, in reliance on this representation—Doctrine of promissory estoppel considered at length—Whether necessary that promisee should suffer "detriment" and, if so, at what point of time—Lord Denning's views in Alan & Co Ltd v El Nasr Export and Import Co questioned—In present case, as tenant had not in fact relied on the representation made on behalf of landlord company, an essential ingredient in the equitable doctrine of promissory estoppel was lacking—Judgment in favour of landlord company for possession and other relief

In these proceedings the plaintiff company, Fontana NV, registered in the Netherlands Antilles, sought possession, arrears of rent (a claim subsequently abandoned) and mesne profits against the defendant, Jack Mautner, in respect of a flat, no 55, at Fountain House, Park Lane, London W1, which the defendant had occupied as a tenant since 1952.

Robert Pryor (instructed by Stephenson, Harwood) appeared on behalf of the plaintiff company; G Godfrey QC and J Littman (instructed by Bernard Sheridan & Co) represented the defendant.

Giving judgment, BALCOMBE J said: Although this case, in the way that it has been presented and argued before me (and

A may I say at once very well argued) raises a point of law of some interest, since I have formed a very clear view of what is the answer here I think it is not right that I should delay to reserve my judgment merely to put it in perhaps more elegant form than I can do if I give an extempore judgment, and therefore I propose to give judgment straight away.

The facts are as follows. The defendant, Mr Jack Mautner (who told me in evidence he was 29 at the outbreak of the war and therefore must now be a man of some 69 years of age) has been a tenant of Flat 55, Fountain House, Park Lane, since 1952. He took that tenancy shortly after the war, during which he had, although in reserved occupation, volunteered for the B RAF and served as aircrew. During the war, in 1941, he married his present wife. They have had no children.

As I said, they moved into this flat in Fountain House, Park Lane, which, as I suppose will be apparent to anybody from the address, was a flat at a considerable rental, and a material fact for the purposes of this case is that it was not subject to any statutory protection under the Rent Acts. At the time he moved in the owners of the block were the Prudential Assurance Co Ltd, and they had a policy of granting five-year leases, which at the end of each term they renewed to people who, like himself, were satisfactory tenants. That position continued with renewed five-year terms at intervals, with a slight increase C in rent on each occasion, until 1973, when the last of the five-year terms expired. In the meantime, in 1970 or 1971, the Prudential had sold the reversion to Mr Mautner's lease to a company, the Lyon Group Ltd, and at the time of that sale there was an exchange of letters, both by the Prudential and the Lyon Group to the tenants, saying that the Lyon Group would propose to continue a similar policy to that of the Prudential. Whether or not they would have done so is not absolutely clear. Certainly at one stage they were reluctant to renew a five-year lease and were prepared to grant only a shorter-term tenancy, but events overtook them because, as I have heard in evidence and I think is quite well known, the Lyon Group was one of D the companies that got caught in the property difficulties of late 1973, and in due course a receiver was appointed by, as I understand it, a mortgagee of the building and eventually, in 1977, that property was sold to a company, the plaintiff in these proceedings, registered in the Netherlands Antilles, Fontana NV.

In the meantime, as I have said, Mr Mautner's last five-year term had expired in March 1973. He held over. He had a one-year term eventually granted until March 25 1974 at £2,200 per annum. That was extended until December 31 1974. In May of 1975 the receiver of the Lyon Group offered him a one-year term at an increased rent of £6,750. He did not accept that; E but in May 1976 (and this is perhaps the vital fact underlying this case) Mr Mautner's wife regrettably and tragically became seriously ill. Her illness, as has been described in a medical report which has been put before me and which is not contested, was cerebral atrophy aggravated by senile arteriosclerosis. Early in 1979, which is after the date I am speaking of now, this was further complicated by a minor cerebrovascular accident involving the right side of the brain. I will at a later stage deal with the effect that that illness has on Mrs Mautner. The symptoms of her illness became apparent in May 1976 and were apparently aggravated by a fire in the flat below, and she had to be rescued by firemen from her own F flat. It does not require much imagination to realise that to a person who was already suffering from illness that could have a traumatic effect. I should have said that Mrs Mautner was herself a talented dress designer and had, at any rate until the onset of her illness, practised in partnership with her sister. In any event, in May 1976 the effect of her illness aggravated by the fire became apparent, and in June 1976 Mr Mautner accepted an agreement from the receiver for a further one-year term, from June 25 1976 until June 24 1977, at a rent of £5,000 per annum.

A large bundle of correspondence has been put in and I do not propose to go through it in any detail, but I can summarise the effect of the letters which were passing between Mr Mautner G and the various representatives of the landlords, because after Keith Cardale Groves & Co, the agents for the Lyon Group, there were J Trevor & Sons, who came in when Fontana became the owners of the building. It was Mr Mautner's contention, which indeed he has maintained in evidence, that as a model tenant, as he put it, for a period of 25 years upwards, of a flat which was his home, and where a move could be disastrous in relation to his wife's health, he had a right to stay on; and, as he put it in his letters, he could not believe that any court before whom the matter might come could make an order against him, evicting him, and also of course inevitably his wife, from their flat. H

In April 1977 the building was sold to this company Fontana, and there was a suggestion (I am not sure how far it was proved in evidence, but it matters not for this purpose) that the owner of the share capital of Fontana at this time was a Mr Sassoon Sofair, and his nephew, a Mr Isaac Sofair, was, through the medium of his company, Al Gazar Properties (London) Ltd, given certain powers of management over the block. Although on the pleadings the extent of Mr Sofair's powers of management was put in issue, Mr Pryor for the plaintiffs fairly conceded on the evidence which I have heard (and that included not merely Mr Mautner himself but the evidence of Mr Carter, who was the house manager of the block for a considerable J time) that Mr Isaac Sofair was held out by the owners as having ostensible management authority, and certainly for the purposes of this judgment I find as a fact that, to Mr Mautner, Mr Isaac Sofair was held out as having authority to make arrangements connected with the block on behalf of the landlords. This authority continued up to the relevant date, which as will be apparent in a moment was July 2 1977, although in June of 1977 it appears that Mr Sassoon Sofair sold his shares in Fontana to a Sheik Khojah, who, it is again asserted, though I have heard no direct evidence on this, is an Arab, and so far as I am aware that may be where the ownership of the shares in Fontana still rests. Again it does not seem to be material. K

On July 2 1977 there was a vital conversation between Mr Mautner and Mr Sofair. Before I deal with that I should mention the position of Mrs Mautner's health, because it is that which has throughout been the prime consideration of Mr Mautner. As I have said, she suffers from chronic illness of the brain. In October-November 1976 she had a period at the Priory Hospital in Roehampton for psychiatric treatment. But it then became apparent that her illness was neurological in nature and she was subsequently examined by a neurologist, who diagnosed the cerebral atrophy which I have already mentioned. In late 1976 she came back to the flat, Fountain House, where she has since lived and where she lives to this day. L I quote from one of the medical reports which I have already mentioned, of Dr Myrddin-Evans, who is her medical practitioner. He says, "Her present medical condition is such that she is delusional, hallucinatory and almost completely bedridden. Added to this she is incontinent. It requires the assistance of two nurses to get her out of bed and stand her up. To move her to any other room bar the bedroom a wheelchair has to be used. She can, with great difficulty, manage to feed herself, but it is obviously much easier and quicker if the nursing staff undertake this. Drinking is a physical impossibility as she cannot control a cup." Mr Mautner gave evidence, describing in graphic terms the effect of his wife's illness; she M has very occasional, very brief, moments when she is aware of her surroundings, and, as he puts it, she relies very much in those moments on the familiarity of her surroundings. She would be distraught if she had to move, so he tells me, and this again is borne out by what the doctor says, in the sense that he says he does not know how much her surroundings are actually taken in by a person in her condition, but he is certain that, if she were moved to somewhere completely strange and without her personal possessions in their usual places, that would be extremely bad for her. So there we have the position from late 1976 onwards to date of Mrs Mautner, living in the flat with the full-time attention of day and night nurses, ill in the way that I have described and in her occasional brief moments of lucidity recognising and relying on her surroundings. She occasionally gets taken out into the park by the nurses and, perhaps as a detail I should mention, Mr Mautner told me that in order that his wife should get the maximum feeling of being at home rather than in an institution even the nurses are not allowed to wear nurses' uniform; they have to wear their civilian clothes.

There had been from time to time, prior to this date in 1977, notices served on Mr Mautner by the landlord—I say notices, it would be more accurate to say letters—saying that his tenancy would not be extended. On one occasion there was in fact a three months' notice to quit, but it was never in fact carried into effect; and in early 1977 Mr Mautner was negotiating with Trevors for the possible purchase of this flat. He had put forward an offer on behalf of himself and some of the other tenants, but that offer was never accepted, and those negotiations never came to anything. Mr Mautner had learned that Mr Isaac Sofair was in fact the person who appeared to be "the boss," as he put it—the owner's representative in relation to these flats—and in the early summer of 1977 Mr Mautner had been attempting to arrange a meeting with him to discuss his problems, but had been unable to do so because Mr Sofair's secretary had said—and for all I know this may be true—that Mr Sofair was busy, and no meeting had in fact come about. But on July 2 1977—which was a Saturday—Mr Mautner met Mr Sofair in the following circumstances. Mr Mautner had two tickets for the men's singles final which was due to be held at the All England Tennis Club at Wimbledon on that day, and because of his wife's health (I think that was the reason, but it does not matter) he was unable to go himself. He telephoned his nephew, Roger Mautner, to see whether he could take advantage of those tickets. Mr Roger Mautner was delighted to take advantage of them, and it was arranged that he should come by to Fountain House to pick them up on his way to Wimbledon. He arrived there at about 1.20 or 1.25 that afternoon. When he came into the entrance hall of the block (and Mr Carter who was there telephoned up for Mr Jack Mautner to come down and hand over the tickets), Mr Roger Mautner happened to meet in the foyer Mr Isaac Sofair, whom he had known as a social acquaintance for a number of years. Mr Isaac Sofair had with him a young lady. When Mr Jack Mautner came down to hand the tickets over to Mr Roger Mautner, Mr Roger Mautner took the opportunity of introducing Mr Isaac Sofair to his uncle, as he knew that Mr Jack Mautner was anxious to meet Mr Isaac Sofair, and thereafter a comparatively brief conversation took place. But in the course of that conversation Mr Jack Mautner told Mr Sofair of his problem, that he had been a model tenant for 25 years, and he emphasised his wife's condition. Mr Sofair, he said, seemed a reasonable man, and Mr Roger Mautner, who also gave evidence before me, stressed that as far as he knew Mr Sofair was an honourable man. And it was the evidence of both Mr Jack Mautner and Mr Roger Mautner (which was not contradicted in any way whatsoever, because in fact Mr Sofair was not called to give evidence, and indeed no evidence was called on behalf of the plaintiff landlords) that Mr Sofair said, "Mr Mautner, it is quite all right, you can stay here as long as you wish." Mr Roger Mautner soon after that left to go to Wimbledon, but Mr Jack Mautner invited Mr Sofair upstairs to his flat. He and the girl with him came upstairs. Mrs Mautner was in the front room with her nurse, and in order to avoid distressing her Mr Jack Mautner took Mr Isaac Sofair into an adjoining room from which at any rate he could see Mrs Mautner and could therefore realise Mr Mautner had not exaggerated the state of her health. And again Mr Mautner described his wife's condition: the nature of her brain damage and the fact that she had to have day and night nurses. Mr Sofair expressed sympathy. He asked Mr Mautner what rent he was paying and he was told £5,000 a year exclusive of rates. Mr Sofair said that that was very satisfactory and he again said, according to Mr Mautner —and this was not challenged—that Mr Mautner need have no concern, he could stay as long as he wished. It is on the legal effect of those remarks that this case turns.

That was on July 2 1977, although on June 2 1977 Mr Mautner had had a letter from J Trevor & Sons saying: "As you know we act for the managing agents for Fontana. Under the terms of your lease your tenancy expires on June 24 1977. We confirm that our clients will require full vacant possession of the flat on that date." As I have said, Mr Mautner had had letters of that kind before at intervals and in the event had not had to leave. Then the letter proceeds, "Of course if your negotiations with our residential department proceed with the purchase of a 999-year lease of the above flat then this will not apply."

Then there was the meeting of July 2 and then nothing happened until August 2, when the solicitors for Fontana, Malkin, Cullis & Sumption, through their partner Mr Bluston, wrote a letter to Mr Mautner in the following terms:

> Dear Sir—Flat 55, Fountain House—Would you kindly note we act for Fontana NV, the freeholders of Fountain House, and we have been asked to write to you in particular following your meeting with Mr Isaac Sofair.

And I pause there to say that it is common ground that there was only the one meeting between Mr Mautner and Mr Isaac Sofair, namely that of July 2 1977.

> Following this meeting it has in fact been decided that it is not our clients' policy to sell flats at this time, and indeed our clients wish to pursue the possibilities of taking possession of those flats which are available for this purpose in order to carry out their refurbishment programme. We have looked through your correspondence with our clients' managing agents, J Trevor & Sons, including the correspondence with their predecessors, Keith Cardale Groves & Co. It is our understanding that following your letter of June 2 1976 to the latter firm a one-year's tenancy which would therefore have expired in June last—

there is an obvious omission there but the sense is clear

> This, of course, would entitle our clients to pursue the possibility of obtaining possession of the flat, but our clients do not wish to take any precipitate action and we look forward to your reply.

Although Mr Mautner said that when he read that letter he did not see it was in any way inconsistent with what Mr Sofair had told him—and he has said more than once that he is not a lawyer; he is, however, a very intelligent man, and indeed is a property developer—it does seem to me that anybody reading that letter would realise that at the very least it contains the seeds of an inconsistency, because if Mr Sofair were to be taken entirely at face value Fontana could not be even considering pursuing the possibilities of taking possession as against Mr Mautner, even though they might not wish to take any precipitate action. And indeed, Mr Mautner realised this. There was some correspondence between him and Malkin, Cullis & Sumption, to which I need not refer in detail until one gets to a letter of September 2, when he wrote back to them. There had been intervening letters. Mr Bluston of Malkin, Cullis & Sumption was, according to Mr Godfrey in his opening, speaking with forked tongue— I would rather put it that he was, perhaps very reasonably in the circumstances, trying not to exacerbate a delicate position, and he had written a letter which was not an outright demand for possession, but simply saying his clients' policy was that set out in the letter of August 2 and there were these plans for refurbishment, and would a meeting be of some assistance. To which Mr Mautner replied on September 2 1977:

> I am not quite sure what this is all about as I thought I had made the position clear in my previous letter of August 8 that I wished to continue to live here in my home of the last 25 years. Either you have misread that letter or misinterpreted Mr Sofair's instructions, as regardless of any other aspect of the matter he has already informed me that I can continue to live here as long as I wish.

If before that date Mr Mautner had not realised that there was a potential inconsistency between what Mr Sofair had

A told him on the one hand and what the landlords' solicitors were saying on the other, by that date he clearly was aware of that inconsistency, because he makes the point, "Either you have misread my letter or misinterpreted Mr Sofair's instructions." But the fact is—and I stress this, and it comes out quite clearly from Mr Mautner's evidence—that nothing I have heard in the evidence persuades me that Mr Mautner either did or did not do something in reliance upon what Mr Sofair had told him. He had all along taken the line, and he continued to take it after the conversation with Mr Sofair, that Flat 55 was his home, that that gave him if not some sort of right at any rate a position which no judge of an B English court would disturb, and he neither considered looking for alternative accommodation nor did so. That, I think, is an important question when I come to consider the effect of estoppel. I find as a fact, and I am quite clear on the evidence of Mr Mautner himself (which was not contested on this point; indeed it was given very fairly), that his view all along was that he could stay there. He was going to stay, and although he was delighted to hear from Mr Sofair that he would be able to stay he in no way changed his position, either by action or inaction, in reliance upon what Mr Sofair had told him.

Matters then went ahead. There were further letters. There C was a notice to quit given by solicitors on behalf of Fontana certainly on June 6 1978, and that was a six months' notice to quit. But the point was considered that that might not terminate on the right date, and a further notice to quit was given on December 20 1978, terminating the tenancy on June 24 1979. Mr Mautner made it quite clear that he was not proposing to quit on that date, and in due course the action 1979 F No 1 was issued in which the landlords Fontana set out the relevant facts and simply claimed a declaration that the defendant's tenancy had been determined and would expire on June 24 1979, and a declaration that they were entitled to possession of the flat on June 24 1979. Subsequently, D after the second notice had expired, they started a second action, F No 895 between the same parties, claiming possession, arrears of rent and damages for trespass—and at the beginning of this hearing I consolidated those two actions, so that the claim for the declaration and the claim for possession are both before me.

To that action the defendant, Mr Mautner, has put in a defence and counterclaim, relying on the equitable doctrine which is known generally by the term promissory estoppel. After setting out the uncontroversial facts he sets out in paragraph 4 an allegation of his wife's illness, and then in paragraph 5 sets out the effect of the conversation of July 2, 1977, E with Mr Isaac Sofair. Then in paragraph 6 it is said:

> The Defendant in reliance upon the said statement of the said Sofair has forborne to seek other accommodation for himself, and other accommodation and treatment for his wife, than in the said flat; has altered his position to his and his wife's detriment or prejudice; and has incurred expense. As a result of the Defendant's said acts, the Defendant's wife has become more dependent for her health upon continued residence in the said flat than would have been the case had she been moved to other accommodation during or about July 1977.

As I have already found as a fact, I do not accept that the defendant has altered his position in any way in reliance upon Mr Sofair's statement.

F Then the defence and counterclaim as amended goes on:

> In the premises, the Defendant avers that by his said statement the said Sofair intended the Defendant to believe and act upon the assurance that the Plaintiffs would forbear to seek to recover possession of the said flat, so long as the Defendant was prepared to pay a commercial rent therefor, until the happening of the earliest of the recovery of health by the Defendant's wife, the death of the Defendant's wife, or the death of the Defendant; and the Plaintiffs are estopped from alleging that they will be entitled to possession of the said flat on June 24 1979.
>
> Further or alternatively, in the premises it would be inequitable to allow possession of the said flat to the Plaintiffs on the 24th day of June, 1979.

Then the relief counterclaimed in its original form was, "A G declaration that the Defendant's said yearly tenancy might not be determined by the Plaintiffs until the happening of the earliest of the recovery of health of his wife, her death, or his death." At an early stage in the course of the proceedings I pointed out to Mr Godfrey that a declaration in that form was not entirely consistent with the allegation contained in the statement of claim, that the defendant should be allowed to remain a yearly tenant at a commercial rent, because if they could never determine the existing tenancy it would follow from that that the landlords could never put up the rent. So that claim was amended to claim

> A declaration that the Plaintiffs are not entitled to seek to H recover possession of the said flat, so long as the Defendant is prepared to pay a commercial rent therefor, until the happening of the earliest of the recovery of health of his wife, her death, or his death.

Before getting on to the law there are two problems there which I should deal with. First, Mr Mautner said in his evidence that as far as he was concerned what he understood from Mr Sofair's statement that "he could stay there for as long as he liked" was that his wife would be secure. It certainly never entered his head that should he die first his wife would not have security of tenure. Mr Pryor, for the landlords, relied on that difference between what Mr Mautner believed as a matter J of his understanding of the assurance and how the claim was pleaded for a submission that therefore the representation by Mr Sofair was not sufficiently clear or unequivocal to found the basis of promissory estoppel. I do not accept that. It seems to me that Mr Godfrey was right in his submission, that provided the person who relies on the representation establishes the fact of the representation sufficiently clearly, and if the court is satisfied that the representation was clear and unequivocal, then the legal consequences to be drawn from that statement are a matter for the judge, and the fact that in this case Mr Mautner may have believed that Mr Sofair's representation had a wider ambit than his lawyers have felt able to put before the court K does not, as it seems to me, defeat his claim.

The other point about the amendment of the claim to pay a commercial rent I will deal with at a later stage, because it does seem to me that that introduces an additional difficulty in the way of Mr Mautner.

However, having said that, and that being the legal issue I have to decide, I now turn to the question of the relevant law. It has been said more than once that the doctrine of promissory estoppel is in a state of development, and so it is. I think all lawyers will know that from the late 19th century when there were the two cases, one in the House of Lords, *Hughes v Metropolitan Railway Co* (1877) 2 App Cas 439, and the L Court of Appeal case of *Birmingham and District Land Co v London and North Western Railway* (1888) 40 Ch D 286, the doctrine seems to have lain dormant until it was revived by Denning J (as he then was) in the *Central London Property Trust Ltd v High Trees House Ltd* case just after the war [1947] KB 130. And since then it has been the subject of much development. But although I know that Lord Denning himself would not accept the definition which I am about to cite as a definitive definition (and Mr Godfrey submits that I should not so accept it), for the purposes of this case I accept as a correct statement of the law the judgment of Lord Hodson in the Privy Council (and I realise, of course, that Privy Council M decisions are only of persuasive authority, but of the very highest persuasive authority) in the case of *Ajayi v R T Briscoe (Nigeria) Ltd* [1964] 1 WLR 1326, where Lord Hodson says this at p 1330:

> Their Lordships are of opinion that the principle of law as defined by Bowen LJ—

and pausing there to interpolate that is in the *Birmingham and District Land Co* case—

> has been confirmed by the House of Lords in the case of *Tool Metal Manufacturing Co Ltd v Tungsten Electric Co Ltd* [1955] 1 WLR 761, where the authorities were reviewed and no

A encouragement was given to the view that the principle was capable of extension so as to create rights in the promisee for which he had given no consideration. The principle, which has been described as quasi estoppel and perhaps more aptly as promissory estoppel, is that when one party to a contract in the absence of fresh consideration agrees not to enforce his rights an equity will be raised in favour of the other party. This equity is, however, subject to the qualifications (1) that the other party has altered his position, (2) that the promisor can resile from his promise on giving reasonable notice, which need not be a formal notice, giving the promisee a reasonable opportunity of resuming his position, (3) the promise only becomes final and irrevocable if the promisee cannot resume his position.

B If I might say so, with respect, that, as I understand it, and subject to the various references which I shall deal with in a moment, correctly states the law, and certainly accurately summarises what Lord Simonds said in *Tool Metal Manufacturing Co Ltd v Tungsten Electric Co Ltd (supra)*. In that case at the bottom of p 763 Viscount Simonds sets out the passage from the judgment of Bowen LJ in the *Birmingham and District Land* case, which perhaps I should read as it is referred to so frequently.

> It seems to me to amount to this, that if persons who have contractual rights against others induce by their conduct those against whom they have such rights to believe that such rights will either
> C not be enforced or will be kept in suspense or abeyance for some particular time, those persons will not be allowed by a court of equity to enforce the rights until such time has elapsed, without at all events placing the parties in the same position as they were before.

Then Viscount Simonds goes on:

> These last words are important, for they emphasise that the gist of the equity lies in the fact that one party has by his conduct led the other to alter his position.

I stress those words, because the doctrine is an equitable doctrine and is based entirely, as I understand it, on the premise that where one person has made a representation to another
D in reliance upon which the other alters his position, then it would be inequitable to allow the first person to go back on what he had said. This appears, and I am very much obliged for the citation, in a helpful passage from the leading textbook on the subject, *Spencer Bower & Turner on Estoppel by Representation*, 3rd ed at p 109, where the learned author, Sir Alexander Kingcome Turner, who was formerly President of the Court of Appeal in New Zealand; says this:

> Whether the representee has "altered his position" by way of positive change, or merely by refraining from some action which he was otherwise at liberty to take

E and I pause there to interpolate that I of course accept that you can alter your position by inaction just as much as you can by action

> he must be able to show "detriment," which as we have seen is some prejudicial effect upon his temporal interests. But it is of the utmost importance to notice that the "detriment" which the representee must be shown to have suffered is judged only *at the moment when the representor proposes to resile from his representation*. It has been pointed out by Dixon J

who of course later became Chief Justice Dixon of the High Court of Australia

> F in the course of his enunciation of the principles governing this part of the subject contained in his judgment in *Grundt v Great Boulder Proprietary Gold Mines Ltd* [1948] Ch 145 that in measuring the detriment, or demonstrating its existence, one does not compare the position of the representee, before and after acting upon the representation, upon the assumption that the representation is to be regarded as true. So long as the assumption continues to be regarded by the parties as true, the question of estoppel does not arise. It is only when the representor wishes to disavow the assumption contained in his representation that an estoppel arises and the question of detriment is considered, accordingly, in the light of the position which the representee would be in if the representor were allowed to disavow the truth of the representation.

and then perhaps somewhat ironically G

> As Lord Denning has observed, the principles governing this branch of the subject cannot be better put than in the words of this great Australian jurist, which are accordingly reproduced in full in the following quotation.

I do not think it would help and would probably merely lengthen what is already turning out to be an unduly long judgment if I were to cite the quotation in full, but it is quite clear that what the learned editor said there follows from what Dixon J says: the detriment is determined at the moment when the representor otherwise seeks to resile from the representation. The final paragraph in the text is:

> The test of detriment, in a word, is whether it appears unjust or H
> inequitable that the representor should now be allowed to resile from his representation, having regard to what the representee has done, or refrained from doing, in reliance on the representation.

Now in deference to the careful and attractive argument of Mr Godfrey it is only right that I should deal with the cases which he has cited in which Lord Denning has suggested that detriment is no longer necessary. The cases are first, *Alan & Co Ltd v El Nasr Export and Import Co* [1972] 2 QB 189. That was a commercial case, and the question essentially was one of waiver. Lord Denning sought to equate waiver with the doctrine of promissory estoppel, and then said at p 213:

> The principle of waiver is simply this. If one party by his J
> conduct leads another to believe that the strict rights arising out of the contract will not be insisted upon, intending that the other should act on that belief and he does act on it, then the first party will not afterwards be allowed to insist on strict legal rights when it would be inequitable for him to do so.

I pause there to say that there is the question of equity arising again. He goes on:

> There may be no consideration moving from him who benefits by the waiver. There may be no detriment to him by acting on it. There may be nothing in writing. Nevertheless the one who waives his strict rights cannot afterwards insist on them. His strict rights are at any rate suspended so long as the waiver lasts. He may on K
> occasion be able to revert to his strict legal rights for the future by giving reasonable notice in that behalf or otherwise making it plain by his conduct that he will thereafter insist upon them (*Tool Metal Manufacturing Co Ltd v Tungsten Electric Co Ltd*). But there are cases where no withdrawal is possible. It may be too late to withdraw or it cannot be done without injustice to the other party. In that event he is bound by his waiver. He will not be allowed to revert to his strict legal rights. He can only enforce them subject to the waiver he has made.

With the greatest respect to Lord Denning I do not believe anyone would dissent from the propositions there cited. He then sets out certain cases by way of example, and then says:

> In none of those cases does the party who acts on the belief L
> suffer any detriment. It is not a detriment but a benefit to him to have an extension of time or to pay less, or as the case may be. Nevertheless he has conducted his affairs on the basis that he has that benefit and it would not be equitable now to deprive him of it.

It does seem to me, and I do not propose to go into this further because at the end of the day I believe it is not necessary for me to deal with it in my judgment, that Mr Pryor was right when he said that with the greatest of respect to Lord Denning he was considering the question of detriment at the wrong point in time. As was said by Dixon J as set out in *Spencer Bower*, almost in every case at the moment of acting on the representation the representee is far from suffering a det- M
riment, he is getting a benefit. And the only question of considering whether he has altered his position, whether you say it as a matter to his detriment or whether you say because it would be inequitable to hold him to the original bargain, must be at the point of time at which the original representor says, "I didn't mean it, I want to go back to what I said." If that is right then it seems to me that with the greatest of respect the judgment of the Master of the Rolls in *Alan & Co Ltd v El Nasr Export and Import Co* is not directly helpful to me in this case. I would add in any event that the other two learned Lord Justices in that case, Megaw and Stephenson

A LJJ, did not accept this approach to the matter at all; they simply said it was a straight question of waiver, and they held there was a waiver and that was that.

Lord Denning has made a similar suggestion about the lack of necessity for detriment in other cases. The point was touched on in *Crabb v Arun District Council* [1975] 3 WLR 847, which again was a case not of the type of equitable or promissory estoppel we are concerned with here but a question of proprietary estoppel, and again with respect it does seem to me that certainly for a judge of first instance it makes it difficult if one has to consider the separate doctrines all under the one head. But there again it was not dealt with in the same way
B by the other members of the Court of Appeal. Equally, in the most recent case, *Brikom Investments Ltd v Carr* [1979] 2 WLR 737, again Lord Denning says that detriment is unnecessary. But the other two members of the Court of Appeal in the case, Roskill LJ and Cumming-Bruce LJ, treat the matter on a totally different basis. I was also referred to another waiver case, *Bremer v Vanden*, in the first instance before Mocatta J and reported in [1977] 1 Lloyd's Rep 133, where he cites (p 165) Lord Denning's remarks in *Alan v Nasr*, that proof of actual detriment was unnecessary; and in the same case in the House of Lords, Lord Wilberforce says he prefers Mocatta J's judgment to that of the Court of Appeal. These are cases of
C straightforward waiver. Mr Godfrey submitted to me that this really was all this case was about; that there had been a once-and-for-all waiver by the landlords here of their right to serve a notice to quit. The difficulty it seems to me in Mr Godfrey's way on that point is that if it is simply a mere waiver and nothing else, then the sort of relief which he was originally claiming would follow; namely if it was a mere waiver of a right to serve a notice to quit it would follow, it seems to me, that Mr Mautner's existing tenancy would continue until it could be determined, which might be never. But once you accept, as Mr Godfrey rightly it seems to me was forced to accept, first that the existing tenancy could not continue
D indefinitely because it needed in equity to be at any rate varied to the extent that the rent should keep pace with the commercial levels of inflation, and secondly that there could come a point of time, whether it be only at the death of Mrs Mautner or whether it be at the first to happen of the death of Mr Mautner or the death of Mrs Mautner or her recovery of health, when the tenancy would end, then it is not just a question of mere waiver; it is, as Mr Godfrey rightly submitted, a question of the doctrine of promissory estoppel. That is why, with the greatest respect, I hope, to Lord Denning, I do not find these cases which were cases essentially of waiver of any assistance. I prefer to go back to what I believe myself to be
E the true statement of the doctrine of promissory estoppel as first propounded in *Hughes v Metropolitan Railway Co* and in the *Birmingham and District Land Company* case, and set out in the Privy Council case of *Ajayi v Briscoe*. And, as I hope I have made clear, I am quite positive in my mind on the evidence I have heard that Mr Mautner did nothing, either by action or inaction, in reliance on the representation made by Mr Isaac Sofair. He did what he was going to do anyway, and that was to sit tight for as long as he possibly could. I do not say I blame him for that, but it does seem to me that that is not an example of the doctrine of promissory estoppel.

It follows from everything I have said that the plaintiffs here
F must succeed in their action. I will hear submissions in a moment as to the form of the relief I should grant, but I think I ought to add one thing. Mr Mautner has in the course of this dispute (if I can call it that) written a number of intemperate letters. Mr Godfrey suggested, and I accept the suggestion, that he should be forgiven those because he was under a great and understandable strain arising from his wife's health. But at one point in the course of the correspondence he seems to have taken the view that in an action between a company incorporated in the Netherlands Antilles, as the plaintiff company is here, with Arab shareholders, and a person like himself, a British subject who served his country well during
the war and having the humanitarian claims in regard to his G
wife's health that he has, there could only be one answer. I feel in fairness to him that I should say this. A judge in this country has to do justice according to the law, and it is freedom under the law which is one of the freedoms that Mr Mautner, among others, fought for during the last war. It would not be, in my view, justice according to the law to weigh in the balance on the one hand the personalities of the plaintiff company and the personalities of the defendant and therefore say that the defendant must have judgment. I have to give judgment according to what I believe to be my duty under the law, and though I have, and I believe everybody in this court who has
heard Mr Mautner's evidence has, the greatest sympathy for H
him, my duty requires me to give judgment according to what I believe to be the law, and that is what I have tried to do.

Judgment was given in favour of the plaintiffs for possession and mesne profits, possession to be given on or before March 25 1980 and the mesne profits to be determined by the master. The defendant was ordered to pay the plaintiffs' costs of the action except in respect of two items as to which the plaintiffs were to pay the defendant's costs. A stay of execution pending a possible appeal was refused.

J

COURT OF APPEAL

January 22 1980

(Before Lord Justice ORR, Lord Justice ORMROD and Sir Robert MEGARRY, Vice-Chancellor)

MANN v CORNELLA

Estates Gazette May 10 1980

(1980) 254 EG 403

**Rent Acts—Appeal from decision of county court judge—Tenancy granted on January 24 1974, ie before Rent Act 1974
came into operation on August 14 1974—Flat let at a rent K
which included payment for some furniture—Whether tenancy was now a restricted contract within section 19 of Rent Act 1977—Answer depended on whether the amount fairly attributable to the use of furniture in the letting of January 24 1974, having regard to the value of the use to the tenant, formed a substantial part of the whole rent—Evidence of tenant's valuer accepted by county court judge—Depreciation over a period of 12 years with interest at 14 per cent—Resulting rental value of furniture about 9 per cent of whole rent—Not a "substantial part"—The tenancy was not therefore a pre-1974 Act furnished letting and was not now a restricted
contract—Tenancy protected—Appeal from judge's decision L
in favour of tenant dismissed—Woodward v Docherty discussed**

This was an appeal by the landlord, Anthony Gerald Mann, from a decision of Deputy Judge Battenbury at Westminster County Court in favour of the tenant, Luigi Cornella. The landlord had applied as plaintiff for a declaration that a tenancy agreement in respect of a flat at 2 Lupus Street, Westminster, London SW1, occupied by the tenant, constituted a restricted contract within the meaning of section 19 of the Rent Act 1977.

Miss J M Livesey (instructed by Geoffrey B Gush & Co)
appeared on behalf of the appellant; R St Barbe Methuen M
(instructed by Hextall, Erskine & Co) represented the respondent.

Giving the first judgment at the invitation of Orr LJ, ORMROD LJ said: This is an appeal from a judgment of Deputy Judge Battenbury given on November 9 last year sitting at the Westminster County Court. He had before him in form an originating application by the landlord as plaintiff, Mr Anthony Gerald Mann, for a declaration that a tenancy agreement dated January 24 1974 in respect of the premises situated and known as the first-floor flat, 2 Lupus Street, Westminster, London SW1, is a restricted contract within the meaning of section 19 of the 1977 Rent Act. In substance the question that he was asked to decide and did decide was whether or not the letting was a furnished letting within the meaning of the Rent Act.

So far as it is relevant the brief history of the matter is this. This particular flat was originally let by the landlord and his two brothers, who are joint owners of the property, to the tenant, Mr Cornella, in 1968 on a monthly tenancy originally for a rent of £45.50 a month. It was afterwards increased to £60.67 a month and it was a letting which included a substantial quantity of furniture. For some reason or other, on January 24 1974 the parties entered into a new tenancy agreement which was before the court and which is the relevant tenancy agreement for the purposes of this case. It provided briefly for the letting of a suite of rooms consisting of three rooms on the first floor of 2 Lupus Street with use of common parts; and, secondly, for the use of the fixtures, furniture and effects now in and upon the property and, " more particularly, specified in the inventory thereto signed by the parties." It was let for a term of one year from January 31 1974 at a rent of £75.84 per month clear of all deductions. Unfortunately, there was in fact no inventory signed by the parties pursuant to that agreement and so there was a certain amount of doubt as to what furniture had been included in that letting. By November 1979 the rent had risen to £101.50 a month.

By a notice to quit dated February 7 1978 the landlord terminated the tenancy, subject to the Rent Act, by August 31 1978. There were then two applications to the rent tribunal to defer the operation of the notice to quit: on one occasion a deferment was granted, but on the second application it was adjourned pending a declaration that the contract was a restricted contract within the meaning of section 19 of the Rent Act 1977.

The question for the learned judge was, in my judgment, whether the letting on January 24 1974 was a furnished letting within the definition of the Rent Act. I may as well now deal with one point in this appeal raised by Miss Livesey, who sought to argue that the court ought " to have regard," to use her own phrase, to the position under the original tenancy in 1968. I am unable to accept that submission. It seems to me that, whatever the position was in 1968, it is wholly irrelevant to the question the judge had to determine or that we in this court have to determine. The agreement which is the subject-matter of the litigation is the agreement of January 1974, and so the judge had to look at the situation at the time when that agreement was entered into to decide whether the value of the furniture in the property at the time could be said to represent a substantial part of the rent. He held that it did not represent a substantial part of the rent.

Looking at the photographs and the inventory of furniture it is not difficult to have some considerable sympathy for the landlord, because my impression from the photographs and the list of furniture is that the premises were substantially furnished: that is, the main furniture required for living in that flat was provided by the landlord. There were tables and chairs and beds and cupboards, and I doubt whether anybody looking at it would form the impression that it was otherwise than a furnished letting.

But, unfortunately, that is not the test because, under the 1968 Rent Act, which, I think, re-enacted an earlier section, it was provided as follows:

> 2(3): . . . a dwelling-house shall not be taken to be bona fide let at a rent which includes payments in respect of attendance or the use of furniture unless the amount of rent which is fairly attributable to attendance or use of furniture, having regard to the value of the attendance or the use to the tenant, forms a substantial part of the whole rent.

What has happened in fact is that rents of any kind of privately rented property have risen owing to inflation and the shortage of accommodation, but such rise has not been followed by any equivalent rise in the value of second-hand furniture, which probably, if anything, has depreciated; so that, as the market rises so far as rents are concerned, it becomes progressively more and more difficult to show that the payments for the use of the furniture amount to a substantial part of the whole rent. In fact, when this court decided *Woodward v Docherty* [1974] 1 WLR 966 and decided in terms that the right approach was to look closely at the value of the furniture, it was apparent to most people that it was going to be exceedingly difficult to establish that any letting was a furnished letting within the meaning of section 2 of the Rent Act 1968 and the present one. Thus, the situation confronting Miss Livesey in this court is that she has to show, following *Woodward v Docherty*, that the value of this furniture and the payments made in respect of it represent a substantial part of the rent.

The learned judge conducted a most careful and detailed investigation into this furniture. When one looks at it in the photographs it looks quite reasonable furniture, but it turns out that its monetary value, as one might perhaps expect, is not at all high. The learned judge, first of all, had to determine what furniture there was on the premises in January 1974 when the agreement was made. For this purpose the only evidence he had before him was an inventory which had been prepared by Miss Barnes, a secretary of the landlord, originally for the 1968 letting. She had, some time in 1975, gone over the flat and checked the inventory to see what was there then in 1975 and, by making ticks and crosses against each item in her original inventory—the ticks and crosses in question being on the right-hand side of the page—she provided the judge with a list of the articles on the premises at the relevant time.

There then had to be a valuation of this furniture and the evidence as to values was clearly more than a little confused. The learned judge in fact determined which of the items in the inventory were present and which were not in 1974 as best he could. He had evidence on behalf of the landlord from a Mr Dalrymple, who was a qualified valuer working as the manager of Harrods Valuation Department. He had originally prepared a valuation for the purposes of insurance in September 1979, which of course is much later than the relevant date. The figure in 1979 pounds was £1,412. That obviously was very much too high for the figure for this furniture in 1974, but he ultimately adapted his figures and produced a valuation of £403, which he said was the nearest he could get to auction prices. He said that, if one were to try to find the corresponding cost of buying this furniture in second-hand shops, he thought that it would probably be marked up by 100 per cent, making £806 altogether.

The tenant called a Mr Ray, who was not a qualified valuer, but who was the sole principal of a firm of auctioneers operating in the Pimlico area. He put an entirely different value on it and he thought the value, allowing for the mark-up, was £371. His auction value was £310 and he increased it by 20 per cent, I think, but at any rate arrived at the figure of £371 as the value of this furniture.

The learned judge accepted Mr Ray's evidence; and, in spite of Miss Livesey's criticism that these were the sort of prices Mr Ray said Pimlico landladies, if any still exist, would have paid for this furniture—and Miss Livesey suggested that was a false basis of valuation—which it seems to me impossible for this court to give effect to, the learned judge accepted the figure of £371. Then he had to decide the question of depreciation and there was evidence that wood furniture would last 20 years, possibly longer, and other evidence was given about the mattresses and carpets and so on. The judge decided to take a mean figure for the life of these various sorts of furniture and he took it at 12 years; so, converting the value of £371 into an annual value, he came to the figure of £31. Then he took a figure for the interest on the capital investment in this furniture. He was prepared to take a figure of 14 per cent—which Mr Methuen says may well be high, having regard to interest rates in 1974—and the

A interest figure worked out at £52; so his calculations showed that the value of the furniture to the tenant was £83 a year, which gave a percentage of the rent of 9.12 per cent. He took the rent at £910 and he held, trying to carry out the exercise as indicated in *Woodward* v *Docherty*, that a percentage of 9 per cent was not a substantial proportion of the total rent and that, therefore, the property was not a furnished letting within the meaning of the Rent Act.

It seems to me, with respect to Miss Livesey's argument, that it is almost impossible for her to challenge the findings of the judge, who heard the witnesses, heard the valuers and who made a clear choice between them and came down in
B favour of the tenant's valuer and, therefore, the tenant's valuation. It is not seriously contested in this court that 9 per cent is enough to amount to a substantial proportion of the rent. In fact, the figure in *Woodward* v *Docherty*, which this court operated on, works out, if my arithmetic is right, at about 8 per cent, which this court held was not a substantial proportion of the rent. In my judgment, it is not possible for this court to interfere with this finding, and I repeat what I said at the outset: the way in which the market has moved has made it virtually impossible to establish that any property falls now within the definition of a furnished letting. In fact, soon after *Woodward* v *Docherty* was decided, Parliament
C abolished the protection given to furnished lettings and brought them within the protection of the Rent Acts to a very large extent.

I think I need add only one thing to explain the nature of the proceedings. It was the landlord's intention in this case to seek possession of this flat, relying upon section 12 of the Rent Act 1977, which means that he has to show various things, the relevant ones being:

> (a) The dwelling-house forms part only of a building and that building is not a purpose-built block of flats; and (b) the tenancy was granted by a person who, at the time that he granted it, occupied as his residence another dwelling-house which also forms
> D part of that building.

He also has to show, if he is to take advantage of section 12, that at the time when the tenancy was granted to the tenant he was not a protected tenant. In order to satisfy that condition the landlord has to show in this case that the letting was a furnished letting. Having failed to do that he does not qualify and cannot qualify to take advantage of section 12, and that is really all there is to it.

I find it quite impossible to say on the material before us, giving full weight to Miss Livesey's argument, that there would be any possible justification for this court interfering with the learned judge's conclusion. I would dismiss the appeal.

E THE VICE-CHANCELLOR and ORR LJ agreed.

The appeal was dismissed with costs.

COURT OF APPEAL

November 21 1979

(Before Lord Justice MEGAW, Lord Justice BROWNE and Lord Justice BRIGHTMAN)

GRANGE LANE SOUTH FLATS LTD v COOK AND ANOTHER

F
Estates Gazette May 17 1980

(1979) 254 EG 499

Rent Act—A somewhat involved procedural matter, but illustrating the substantive point decided in Penn v Dunn —Husband, tenant of flat which was the matrimonial home, left it leaving wife in occupation—County court registrar in possession proceedings struck out tenant's defence and made order for possession against him only—Wife's solicitors were present, but did not appreciate significance of registrar's order in relation to wife—Decision in Penn v Dunn had made it clear that wife's rights under Matrimonial Homes Act 1967 did not assist her after possession order against husband had G **effectively terminated his statutory tenancy—Application for leave to appeal against order not filed until some 8 months after it was made—Deputy county court judge decided reluctantly that he must refuse leave—Appeal from refusal to Court of Appeal—Landlords' solicitors aware of Penn v Dunn when order made by registrar—Deputy county court judge held to have exercised discretion incorrectly in refusing leave —Registrar's order which, contrary to his intention, deprived wife of any defence to possession claim, was a bad order which could have been instantly upset on appeal by wife— Justice required that leave to appeal out of time should be given—Deputy county court judge's decision reversed** H

This was an appeal against the refusal of Mr Recorder M J W Marsh, sitting as a deputy county court judge, to grant leave to appeal out of time from an order of the registrar of the Scunthorpe County Court, against Steven Cook, who was the tenant of a flat at 105 Staindale Road, Scunthorpe. The registrar made no order for possession against Ann Cook, the tenant's wife, although she was in fact in occupation of the flat, her husband having left the matrimonial home.

S A Allison (instructed by Winocour & Williamson, of Scunthorpe) appeared on behalf of the appellant, Mrs Ann Cook; Douglas Hogg (instructed by R A C Symes & Co, of Scunthorpe) represented the respondents. J

Giving the first judgment at the invitation of Megaw LJ, BRIGHTMAN LJ said: This is an appeal from a decision of Mr Recorder Marsh sitting as a deputy county court judge. He refused leave to appeal out of time from an order of the registrar of the Scunthorpe County Court which had been made on October 10 1978. The appeal from the deputy county court judge was out of time, but leave was granted by this court on October 22 for this application to be made. So that today's application is for leave to appeal out of time from the registrar's order, contrary to the decision of Mr Recorder Marsh.

The facts are as follows. Mr and Mrs Cook are husband and wife. The matrimonial home was a flat, 105 Staindale K Road at Scunthorpe. This was held under a tenancy agreement dated in August 1973 granted in favour of the husband. It was a monthly tenancy. At the end of 1977 Mr and Mrs Cook separated. Mr Cook left the matrimonial home leaving his wife in occupation. In May 1978 a notice to quit was given. On July 5 1978 the landlords, Grange Lane South Flats Ltd, the respondents to this appeal, began proceedings in the county court for possession against Mr and Mrs Cook. The claim to possession was based on allegations that Mr Cook, the tenant, was no longer in residence and that breaches of covenant, to do with the keeping, I think, of a caravan on the premises, had occurred. I should say that the premises fall within the Rent L Act legislation.

On July 21 1978 Mrs Cook filed a defence to the claim. She relied on the statutory tenancy of her husband and on her own rights under the Matrimonial Homes Act 1967. It may be recalled that under section 1(5) of that Act it was declared that a spouse's occupation by virtue of that section should, for the purposes of the Rent Act, be treated as possession by the other spouse. We were told by counsel, and I have no reason to dispute it, that that was probably declaratory of the existing law. The wife also alleged in her pleading that a contractual tenancy had arisen by reason of payment of rent by her. Mr Cook purported to defend the claim on the basis that he was M not in possession of the premises and that, therefore, no order to give up possession could properly be made against him.

I will leave the narrative at this point to attempt a brief examination of the law which is applicable to a statutory tenancy of a husband where the wife is also resident on the premises. Putting the matter shortly, I understand the position to be as follows, according to the decision of this court in *Middleton* v *Baldock* [1950] 1 KB 657. Where the husband is the statutory tenant and the wife resides on the premises, as in the instant case, the landlords cannot properly be granted an order for possession against the husband unless there are available grounds, under the Rent Act, for possession against both husband and wife, the wife's residence being equated with the husband's residence. If, however, an order is made against the husband alone—and it may well be that in the circumstances indicated this would occur only by mistake—the wife will thereby lose her statutory protection. This was decided in *Penn v Dunn* [1970] 2 QB 686.

On September 15 1978 the landlords made an application to the registrar of the Scunthorpe County Court. The landlords applied, first, for leave to amend the particulars of claim by alleging certain breaches of covenant on which I need not dwell; secondly, the landlords sought an order that the defence of Mr Cook should be struck out under the County Court Rules on the ground that it disclosed no reasonable ground of defence; and thirdly, and consequential upon the making of such an order, an order for possession against the first defendant. This came before the registrar on October 10 in the presence of the solicitor for the landlords and the solicitor for the second defendant, Mrs Cook. Mr Cook was not present or represented. The registrar gave leave to amend the particulars of claim. He then directed that the defence of Mr Cook be struck out, because it disclosed no reasonable ground of defence. He made an order for possession against Mr Cook. No reference was made to any order for possession against Mrs Cook.

The action came on for trial as between the plaintiffs and Mrs Cook on June 19 of this year. The question of a contractual tenancy was apparently debated first and it became apparent, we are told, that that submission was likely to fail. So Mrs Cook would be thrown back on her defence arising under section 1 of the Matrimonial Homes Act 1967. It then became apparent to all concerned that, if the registrar's order for possession against Mr Cook stood, Mrs Cook would lose her defence dependent upon her occupation, because the Matrimonial Homes Act does not confer on a wife any greater right to a statutory tenancy than was afforded by the possession of her husband, and her husband had lost his right to possession when the registrar's order was made.

In those circumstances, His Honour Judge Geoffrey Jones, who was the county court judge trying the case, made an order that the action should be stood over on three undertakings: first, that an application for leave to appeal out of time against the registrar's order be made within 14 days; secondly, and presumably on the hypothesis that that application was granted, an undertaking that an appeal against the registrar's order be made within 14 days; and thirdly, that Mrs Cook pay to the landlords' solicitors all mesne profits and arrears of rent to date within 14 days and all future mesne profits as they arise. Without intending to be in any way critical of the course taken by the learned county court judge, it does seem to me that it would have been preferable, had it been possible so to arrange matters, that the county court judge himself should, there and then, have considered an application for leave to appeal out of time against the registrar's order, then (if leave were granted) have dealt with that appeal and proceeded to try the action before him to its final conclusion. I do not know how exactly the matter proceeded before the county court judge and possibly that simple and convenient course was not available. I should add this, that in fact mesne profits or arrears of rent have not been fully paid to date, but no point is taken by Mr Hogg, who appears for the plaintiff landlords. Apparently the money has been paid by Mrs Cook, who is still in occupation, to her solicitors and they have it available for this purpose.

An application to enlarge the time for appealing from the registrar's order was issued on June 25 and it came before Mr Recorder Marsh on September 3. The learned recorder, with manifest reluctance, refused the application. He said this:

> In my opinion the learned registrar was wrong in making an order for possession. No criticism, however, whatsoever could be levelled at the learned registrar in view of the omission to draw his attention to *Penn v Dunn* and the consequences of the making of such an order. However, the second defendant, Mrs Cook, did not appeal against the learned registrar's order until after the time allowed for doing so had elapsed. The principles governing leave to appeal out of time seem to have been made quite clear by the various authorities on that point, namely, the rules of the court have to be observed. It is not sufficient to show that there would be no expense or injustice to the other side. If the rules are not observed the reason must be explained. Then having regard to the explanation or excuse and all the circumstances I have to decide whether it would be unjust to refuse the application.
>
> The learned registrar's order was made on October 10 1978. The application for leave to appeal out of time was filed on June 26 1979—some 8 months later. The second defendant's explanation was in effect as follows: It was not my fault; my solicitors did not realise the implication of the registrar's order. I should not be penalised by his mistake. Had I realised the position I would have appealed in time. The existence of *Penn v Dunn* and the implications arising therefrom was known to the second defendant's solicitors in March 1979. Had an application for leave to appeal out of time been made as soon as possible thereafter I would have had no hesitation in allowing that application. That, however, was not the case. Leave to appeal was not filed by the second defendant until August 1979.

—I interpose to say that that is a mistake, either in the judgment or in the typing of the note, for June 1979—

> By that time the case had come before the court and was subsequently adjourned. However, before the case was adjourned it became clear that the second defendant's first barrel of her defence, that a new contractual tenancy had been granted to her, was bound to fail; and it was only when that position became clear that the second defendant tried to bring her "second barrel" into operation by appealing against the learned registrar's order. In my opinion, that was too late, and with the utmost reluctance I have to refuse the application to appeal out of time.

A little before the passage I have read there appears this finding: "Apparently the solicitors for the plaintiffs realised the implications of the making of an order for possession." It may not perhaps be absolutely clear from that finding what the solicitors knew. What is certain that they knew is the existence of the authority of *Penn v Dunn*. They knew that the order being sought against the first defendant, which apparently the solicitors for the second defendant permitted to be made without any representations, would put the second defendant in an impossible position as regards her claim under the Matrimonial Homes Act. It is not absolutely clear to me whether they actually knew that the order was one which ought not to have been made and whether they knew that the order, if made, could have been instantly upset on appeal without further ado.

I seek to summarise the arguments presented to us by Mr Allison on behalf of the appellant. He said that an appeal from the registrar's order is bound to succeed, and I do not think that that is challenged. The order, he said, was therefore wrong. It was a decisive factor on what has turned out to be the most important part of the action, which has yet to be tried. If the order cannot be appealed from at this distance of time it can cause severe hardship to the second defendant. He said that the plaintiffs' solicitors knew that the point that they took before the registrar was a bad one and that, in the circumstances, notwithstanding the stringency of the rules for appealing out of time, this case falls within that class of case in which leniency should be exercised.

Mr Hogg, for the landlords, in a powerful submission, referred to the three authorities of *Ratnam v Cumarasamy* [1964] 3 All ER 933, *Revici v Prentice Hall Incorporated* [1969] 1 All ER 772 and *Warinco AG v Samor PA* [1979] 3 All ER 64 to underline the well-known propositions that rules of court are made to be observed and that a good and cogent reason must be advanced to explain and excuse a default by a party in complying with the time-limits for appealing. Otherwise, of course, the finality which is so important in litigation will be at stake. No matter, said Mr Hogg, that the intending appellant is willing to pay the costs which would be thrown away if the time were enlarged; no matter that no injustice would be done to the intending respondent. As he pointed out, the real question before us is whether the learned recorder exercised his discretion wrongly: that is to say, did he direct his mind to the wrong considerations or did he fail to take into account those considerations to which he ought to have directed his mind? Mr Hogg said that the relevant authorities were before the learned recorder. He was reluctant not to extend the time for appealing but, as found by the recorder, the delay which existed after March 1979, when Mrs Cook's solicitors knew of *Penn v Dunn*, was what Mr Hogg aptly described as a "knowing delay." Therefore, the order of the learned recorder was the only order that could properly be made in the circumstances; it could not possibly be said to be manifestly bad. For these reasons, Mr Hogg urged upon us that it would be quite wrong to interfere with the discretion of the recorder.

I myself take the view that this appeal ought to be allowed. With all respect to the learned recorder, I do not think that he fully directed his mind to the correct considerations. I accept that cogent reasons are required to justify an appeal out of time from a final order and no doubt from an interlocutory order in the normal case. But in the instant case the interlocutory order was not made against the intending appellant but against a co-defendant and that order only affected the intending appellant by a side-wind. It was not a decision against the intending appellant upon an issue which was being consciously raised between the plaintiffs and the intending appellant and of course it was never argued. After the registrar's interlocutory order the action was intended to continue against the appellant and, having regard to the state of the pleadings, it must, in my judgment, be taken to have been in the contemplation of the parties that it would proceed on the basis of the pleadings as they existed between the plaintiffs and the second defendant; and that the defences of the second defendant would in due course be dealt with on their merits, including paragraph 4, which is the reference to the Matrimonial Homes Act. There would, I think by concession, be no conceivable injustice to the plaintiffs in allowing Mrs Cook to appeal out of time against the order of the registrar, so as to enable the defence of the second defendant to be fully raised and argued. The trial of the action is not yet concluded and I think that, to do justice to the parties, it is important that the second defendant should be at liberty to raise her defence under paragraph 4 and should not be shut out by what I described as a side-wind and to my mind is at any rate somewhat of a technicality.

I do not think that the matter rests there. There is also this consideration to which I have already adverted. The registrar's order was clearly one which ought not to have been made. The plaintiffs' solicitors knew all about *Penn v Dunn*. They may or may not have appreciated that the order, if made against the first defendant alone. was one which could be instantly upset on appeal by the wife. I myself take the view that it was the duty of an advocate in these circumstances to bring *Penn v Dunn* to the attention of the registrar. That I think would inevitably have opened up a debate which can hardly have failed to touch on *Middleton v Baldock* and almost inevitably the order would never have been made. That I think was the duty of an advocate. I do not in any way wish to criticise the solicitors of the plaintiffs professionally, but they had cast themselves in the role of advocate for that occasion and I do not think that their duty to the court can have fallen short of the standard which I have indicated, although they certainly did not appreciate it. To my mind that is a most important consideration in this case and enables, and I think requires, this court to interfere with the manner in which the learned recorder exercised his discretion and requires us to qualify, for the purposes of this case, the application of the authorities to which I have referred.

For the reasons which I have attempted to express, I think that this appeal ought to be allowed.

Agreeing, BROWNE LJ said: This appeal should be allowed for the reasons given by my Lord, with which I entirely agree. The first two of the three authorities cited to us by Mr Hogg, and to which my Lord has referred, seem to me to have been directed to exploding what was supposed to have been a principle based on some 19th-century authorities that a party in default on a timetable was entitled to an extension on payment of costs unless irreparable mischief to the other side would be caused by the extension. I do not think anybody now would try to maintain that proposition. In the third case the Appeal Committee of the House of Lords emphasised the importance of compliance with time provisions in the relevant rules. Of course this court, and any court, is very conscious of the vital importance of keeping to timetables laid down by the rules. But there is no doubt that the court has a discretion to extend time if, in all the circumstances of some particular case, it is just to do so. On an application for leave to extend time, relevant factors in the decision whether or not to exercise this discretion include both the merits of the decision from which it is sought to appeal and the reasons for the delay. The weight to be attached to these various factors will, of course, vary in each case.

In the present case the recorder said that if the application for leave to appeal out of time had been made as soon as possible after he found that the existence of *Penn v Dunn*, and the implications arising therefrom, became known to the appellant's solicitors in March 1979, he would have had no hesitation in allowing the application. But his decision was that the further delay from March until June, during which, as I understand his finding, he found that the appellant's solicitors knew of the implications but did nothing about it, was fatal, and he accordingly refused the application.

It seems to me that in the present case the overriding consideration must be the nature of the registrar's order from which leave is sought to appeal out of time. It is quite plain, and indeed Mr Hogg rightly concedes, that that order was a bad order in the sense that it should never have been made and was bound to be reversed on appeal. There is also no doubt that its effect was to deprive the appellant, the wife, of any defence to the plaintiffs' claim for possession. As my Lord has said, it is quite plain that the registrar did not intend that result. What he intended by giving judgment only against the husband was that the action should go on and the wife should have the opportunity of raising her defences, in particular the defence based on the Matrimonial Homes Act.

It seems to me that the injustice of giving effect to an order of that nature must be the overriding consideration in the present case in deciding whether or not to give leave to appeal out of time. In my judgment, the recorder was plainly wrong in not giving decisive effect to that consideration. Accordingly, for the reasons given by my Lord and such additional reasons as I have tried to give, I agree that the appeal should be allowed.

Also agreeing, MEGAW LJ said: I place particular stress upon the last reason given by Brightman LJ in his judgment. This, as appears to me, is a case which falls within the words used by Atkin LJ in *Maxwell v Keun* [1928] 1 KB 645, where, at p 653, with regard to an appeal to this court asking the court to reverse a trial judge's order refusing a plaintiff an adjournment, Atkin LJ had this to say: "I quite agree the Court of Appeal ought to be very slow indeed to interfere with the discretion of the learned judge on such a question as an adjournment of a trial, and it very seldom does do so." In my view, precisely the same observation applies in relation to an application for leave to appeal out of time. Atkin LJ goes on: "but, on the other hand, if it appears that the result of the order made below is to defeat the rights of the parties altogether, and to do that which the Court of Appeal is satisfied would be an injustice to one or other of the parties, then the court has power to review such an order, and it is, to my mind, its duty to do so."

I agree that the appeal should be allowed.

The appeal was allowed with costs in the Court of Appeal. Legal aid taxation of appellant's costs was ordered. Liberty to apply was given in regard to certain other matters affecting costs.

COURT OF APPEAL

March 5 1980

(Before Lord Justice MEGAW, Lord Justice BRANDON and Lord Justice OLIVER)

YEWBRIGHT PROPERTIES LTD v STONE AND OTHERS

Estates Gazette June 7 1980

(1980) 254 EG 863

Rent Acts—Appeal from decision of county court judge refusing order for possession—Main issue suitability of alternative accommodation offered by landlords—Whether accommodation was reasonably suitable to the needs of the tenant as regards proximity to place of work—Appellant a freelance clothes designer who, apart from short visits abroad, carried out her work at the customers' premises or at home in the subject premises, both her customers' premises and her home being in the London SW6 area—Alternative accommodation offered in East Dulwich—County court judge held that her place of work was the SW6 area and that the accommodation offered at East Dulwich was not reasonably suitable as regards proximity to place of work—Judge entitled to come to this conclusion on the evidence—Various criticisms of decision rejected—Observations on state of disrepair of the East Dulwich premises which, even if appellants had succeeded otherwise, would have prevented the Court of Appeal from doing more than remitting case to county court judge—Appeal dismissed

This was an appeal by the plaintiffs, Yewbright Properties Ltd, the landlords, from a decision by Judge Hayman at the West London County Court refusing the plaintiffs a possession order in respect of a maisonette at 35 Mirabel Road, London SW6. The respondent was Miss Jean Stone, who had been held by the county court judge to be the tenant of the maisonette. Two other ladies, Miss Mills and Miss Checkter, were found by the judge to be licensees of the tenant. Miss Stone had been described in the action as the first defendant and the other two ladies collectively as the second defendant. There was no appeal against the judge's finding that the latter were licensees.

Richard Fernyhough (instructed by Friedman, Fredman & Co) appeared on behalf of the appellants; Simon Levene (instructed by Russell-Cooke, Potter & Chapman) represented the respondent.

Giving judgment, MEGAW LJ said: This appeal from the judgment and order of Judge Hayman in the West London County Court arises out of an action by the plaintiffs, Yewbright Properties Ltd, claiming possession of a maisonette at 35 Mirabel Road, London SW6. The order of the learned judge was made on December 4 1979. A written reserved judgment giving the reasons for the decision was, at the request of the plaintiffs, delivered by the learned judge on January 7 1980.

The judge refused the claim for possession on two grounds. The first defendant is Miss Jean Stone who, at the time when the action for possession began, had become a statutory tenant of the premises. She had lived there as a tenant since 1967, the landlords in earlier years being the plaintiffs' predecessors in title. There are two other defendants, Miss Mills and Miss Checkter, who are described in the title to the proceedings collectively as the second defendant. The judge held, contrary to the submission of the plaintiffs, that those two ladies were licensees of the tenant, Miss Stone, the first defendant. That finding defeated certain of the grounds on which the plaintiffs claimed possession. There is no appeal against that finding or the judge's rejection of those grounds.

Because of the view which I take on one of the several issues raised in the appeal—being an issue on which, unless the plaintiffs show that the judge was wrong, the appeal must fail irrespective of the answer on the other issues sought to be raised—I do not think it necessary to go into the facts in any great detail. Many of the matters canvassed in evidence at the trial are not relevant to the issue on which I think this appeal can and should be decided.

Because of the poor state of repair of the premises at 35 Mirabel Road and because of repeated notices from the local authority requiring repairs to be carried out, at some time in the summer of 1979 the defendants agreed to the plaintiffs' request that they should temporarily move out from 35 Mirabel Road in order to enable the required work to be done. That was agreed on the basis, as I understand it, that, for the purposes of the then contemplated legal proceedings by the plaintiffs, the defendants should be treated as though they were actually residing in the premises. If the contemplated claim for possession should fail they would return on the completion of the repairs. If an order for possession were to be made meanwhile they would not return to the maisonette.

On July 20 1979 the plaintiffs served a formal notice to quit. On its expiry Miss Stone became, as I have said, a statutory tenant. The proceedings for possession were begun in the county court on August 29 1979. The grounds on which the claim was originally made are no longer relevant. They have all fallen by the wayside but, on November 14 1979, which was the first date of the hearing in the county court, the plaintiffs amended their particulars of claim to add a further ground: the availability of suitable alternative accommodation.

The hearing before the judge took two days, November 14 and December 4 1979. At the end of the hearing the judge announced that the claim for possession would be dismissed and, as I understand it, he gave brief oral reasons. He fairly and courteously said that, if either party were to wish for fuller reasons for the judgment, he would be prepared to give them at a later date. The plaintiffs took advantage of that offer, hence the full and careful written judgment actually delivered on January 7 1980. The judge held against the plaintiffs on other grounds which are not relevant to this appeal. He also held against them on the two issues sought to be raised in this appeal: first, that the alternative accommodation offered was not suitable and, second, that it would not be reasonable to make the order for possession.

The relevant statutory provisions are contained in the Rent Act 1977. Section 98, so as far as it is relevant, reads as follows:

> (1) Subject to this Part of this Act, a court shall not make an order for possession of a dwelling-house which is for the time being let on a protected tenancy or subject to a statutory tenancy unless the court considers it reasonable to make such an order and either—
>
> (a) the court is satisfied that suitable alternative accommodation is available for the tenant or will be available for him when the order in question takes effect

and then I need not trouble with (b). Subsection (4) reads:

> Part IV of Schedule 15 shall have effect for determining whether, for the purposes of subsection (1)(a) above, suitable alternative accommodation is or will be available for a tenant.

Part IV of Schedule 15 provides, so far as is relevant, in paragraph 4 as follows:

> . . . accommodation shall be deemed to be suitable for the purposes of section 98(1)(a) of this Act if . . . in the opinion of the court, the accommodation fulfils the relevant conditions as defined in paragraph 5 below.

Paragraph 5, so far as it is relevant, reads:

> (1) For the purposes of paragraph 4 above, the relevant conditions are that the accommodation is reasonably suitable to the needs of the tenant and his family as regards proximity to place of work, and either—

and then it goes on to make provision for other matters which I need not read, having regard to the view that I have formed in this case.

It follows that, if the plaintiff landlord (the onus being on him) fails to satisfy the court that the alternative accommodation offered is reasonably suitable to the needs of the tenant as regards proximity to place of work, the claim to possession must fail. So also, even if the judge were satisfied as to that, because of the express provisions of section

98(1) the judge would have to consider whether it was reasonable to make the order and, if the answer to that were "No," then the claim also would fail.

The plaintiffs put forward two flats or maisonettes as being suitable alternative accommodation. The judge rejected both of them. As regards one of them, in Harrowby Street, W1, we are not concerned in this appeal. The plaintiffs do not ask us to say that the judge was wrong as respects that accommodation. The other alternative accommodation offered was a maisonette at 28 Friern Road, East Dulwich. Counsel for the plaintiffs submits that, on paragraphs 4 and 5 of Part IV of Schedule 15, the plaintiffs have to show that the alternative accommodation offered satisfies each of seven tests. Those include tests prescribed in a part of paragraph 5 which I do not find it necessary to read. The judge held that the Friern Road maisonette passed four of those tests. One of the three which it did not pass in the judge's view on the evidence was that it should be reasonably suitable to the needs of a tenant as regards place of work. Unless the judge was wrong as to that, the claim for possession, as I have said, necessarily failed.

At the end of the submissions of counsel for the plaintiffs on that point we were satisfied that the plaintiffs had failed to show that the judge was wrong. We did not find it necessary, nor did we think it would be profitable, to hear submissions on the other issues, the other hurdles which the plaintiffs would have had to surmount.

The relevant evidence as to suitability for the needs of the first defendant as regards place of work is summarised in the written judgment of the learned judge. As we have been shown all the relevant parts of the notes of evidence, it is right that I should say that it is, as one would expect, admirably accurately summarised by the learned judge. I look first at p 5 of the judgment:

> With regard to the first defendant personally, her evidence was as follows. She is a freelance clothes designer by occupation. Most of her customers are in the Fulham-Chelsea-Kensington area (described, during the course of the hearing, as "the SW6 area," and I shall hereafter refer to it as such). The first defendant has lived in that area all her life. She does not have any work in the Oxford Street area nor does she like the West End. She does some work abroad, mainly in Spain, but does not spend more than about two months a year out of the United Kingdom. When commissioned to do a job, the first defendant carries out the work either at the premises of the customer in question or at home.

As the judge has found that most of her customers are in the SW6 area, then, when the learned judge says "When commissioned to do a job, the first defendant carries out the work either at the premises of the customer in question or at home," the premises of the customer in question would, in the majority of cases, be in the SW6 area.

The further relevant passage in the judgment is at p 6. The judge says:

> I turn to the conditions laid down in Part IV of Schedule 15 to the 1977 Act. First, *proximity to place of work*. On the first defendant's evidence, I find that the SW6 area is her place of work. The fact that she also works at home—and, indeed, abroad —does not, in my judgment, affect the position. I further find that Friern Road is not reasonably suitable to her needs as regards proximity to her place of work.

I should say that there was evidence which the learned judge was entitled to take into account and did, no doubt, take into account both from the first defendant herself and, I think, another witness as to the length of time that it would take to go from East Dulwich to the SW6 area.

For the plaintiffs, counsel submits that the judge was wrong for a number of reasons which are, in some degree at any rate, interrelated. First, as I understood his submissions as he made them yesterday in the course of his address, he says that, as a matter of construction of the wording of paragraph 5(1) of Schedule 15, Part IV—the needs of the tenant and his family as regards proximity to place of work—a place of work has to be a place such as one factory or office. It cannot be an area. If that proposition is indeed put forward, that the place of work has to be one building, I must say that I see no basis for accepting it. The only authority cited to us was the case of *Dakyns v Pace* [1948] 1 KB 22. So far as that judgment of Hilbery J is relevant, in my view it tells against the proposition and in favour of the view that, where one's work requires one to leave, as it were, one's base and, in the course of the work, to travel to a number of places in a surrounding area, that surrounding area can be a place of work for the purposes of this statutory provision. I think that is right. It is in each case a question of fact and degree.

Then it is contended that, at least, the fact that the work is carried out not in one particular building but at various places makes it less easy to find that there is a place of work consisting of an area. It may be so. It is a question of fact and degree and I see no reason to find that the learned judge has erred in any way here in the conclusion that he reached on the question of fact and degree on the evidence in this case.

There were two other factors on which counsel for the plaintiffs particularly relied supporting his submission that, in this case, the judge was wrong in holding that there was a place of work and that, if there was a place of work, that place was what he called the SW6 area. One of those factors was that the evidence of the first defendant showed that she worked also in other places outside the SW6 area; for example, she worked abroad in Spain when she went on holiday for perhaps two months of the year. When she was there she did some work, combining business with pleasure. Also, her evidence showed, when one looks at the notes of the evidence, that her work took her to a shop in Oxford Circus outside the SW6 area and perhaps elsewhere outside that area. That, in my judgment, is a factor on the question of fact and degree. It is for the judge to assess on the evidence whether and to what extent the carrying out of work at a place or places which are outside the claimed area of place of work prevents the claimed area from being the place of work for the purposes of the Act.

The other of the two factors relied upon by counsel for the plaintiffs in his submission on this aspect was that Miss Stone, as she said in her evidence, did some of her work at home. Counsel submitted that, in so far as the work was done at home, it could equally well be done at home in Friern Road in East Dulwich as at Mirabel Road in West Fulham. Again, that, I should have thought, was a relevant factor and might, in some cases, be an important factor depending, no doubt, in some degree at any rate on the extent, or proportion, to which the work was done at home as compared with doing it, using the home as a base and going to other places outside the home. The fact that some work is done at home is no more than one possible relevant factor, and I see no reason to doubt that the learned judge had his mind directed by counsel's argument to that factor so far as it ought to be thought relevant on the evidence here. I see no reason to think that the judge misdirected himself or to fault his conclusion on that question of fact.

Then it was said by counsel for the plaintiffs that, on the evidence here, the judge ought to have taken the view, and that this court ought to take the view, that, if there could be more than one place of work, it ought to be regarded as the defendant having had no less than four places of work. One was the place where she gets her commissions, that is where her customers are to whom she goes, the majority of whom, according to the evidence, were in SW6, but some may have been elsewhere. Any place outside the SW6 area where she happened to see a customer and get a commission, counsel suggests, would be another place of work for the purposes of this statutory provision. Another was at home, because she did some of her work at home and another place of work was abroad, because when she went on holiday she did some work in Spain.

I am quite unable to accept that, on the ordinary and sensible construction of this Act and on the evidence as it was in this case, it can properly be regarded as there being four places of work which are relevant for the provisions of the Act. The relevant place of work, the judge found and was entitled to find for the purposes of this provision, was the SW6 area. It would follow that the various submissions which counsel elaborated, referable to the proposition that there were here four places of work, do not arise. His submission was that the judge misdirected himself because, there being four places of work, the judge ought to have taken into account the question of the proximity of the alternative accommodation to each one of those alternative places of work.

Whether the learned judge was invited by counsel in submissions made to him to consider that matter I do not know. The fact that, in this careful judgment, that submission has not been referred to would make me wonder whether it had been made clear to the learned judge. If it were not, it would not be open in this court properly to criticise the judge for having misdirected himself in respect of a way of putting the case which had not been put to the judge. But let me assume in favour of the plaintiffs that it was put to the judge: he was, I think, right and entitled and, indeed, on the evidence before him, bound to reject it, because there was not here more than one relevant place of work for the purposes of this statute on the evidence as it stood.

The final submission which counsel for the plaintiffs made was that, if he is wrong on all the other matters, nevertheless accommodation in East Dulwich could not properly have been held by the judge not to have been reasonably suitable to the needs of the first defendant as regards her place of work, the place of work being accepted for this purpose as the SW6 area. On that submission I would simply say that I have no hesitation in agreeing with the judge's view. I therefore think that the plaintiffs fail to show that the judge was wrong in holding that the offered alternative accommodation in Friern Road did not fulfil the requirements of the Act. Therefore, the appeal must fail on that ground.

I should add, however, that, as I see it, whether or not either of the parties desired to raise the point, it would have been impossible for this court to do as the plaintiffs asked in their notice of appeal: that is, to make an order for possession on the ground that suitable accommodation had been offered. The judge did not have to consider whether he could make such an order or, if he could, what form such order would take. It may be that he could have adjourned the proceedings. In the circumstances I do not desire to go into this matter at length, but the Friern Road accommodation was unquestionably not suitable in the condition in which it was when the judge at the hearing was asked to make an order for possession on the basis of an offer of suitable alternative accommodation. Quite apart from any other objections the accommodation offered in Friern Road had been in a state of disrepair. This is what the judge said about it at p 4 of the judgment:

> With regard to the flat in Friern Road, I think it was accepted by the plaintiffs—if not, I find—that that accommodation is not fit for occupation until certain work has been carried out to it. In other words, the flat is not available for occupation at the present time. I will have more to say about this at a later stage.

Then at p 5 of the judgment, the judge said:

> The flat has been unoccupied for at least a year and probably a good deal longer than this. It is damp; fungus is present; and there is a large structural crack in one of the walls. I accept that the plaintiffs have offered to carry out such work to the premises as is reasonably required by the first defendant. She expressed the view, when giving evidence, that work more extensive than mere decorative work is required to render the premises fit to live in. I accept that. I was not told how long such work would take to carry out, but I infer from the evidence that it would be quite a considerable time.

The most that the judge could have done, if he could have done so, would have been to make a future and conditional order, conditional upon the repairs being done presumably within some stated time, and the maisonette having by then been put into a condition in which it would objectively have been regarded as suitable. There are, no doubt, cases where, although the offered accommodation is not immediately available, the judge can properly make an order to take effect at a future date. Whether the judge could or should have found it possible to take such a course on the facts as they were before him, and the facts that remain unknown, on the evidence in this case I do not need to speculate. But, so far as this court is concerned, it is clear that, even if the plaintiffs had succeeded on every other issue in the appeal, we could not properly have made the order asked in the notice of motion and the notice of appeal—a simple order for possession. I doubt whether we could properly have done more than, if we could have done so much as, to remit the action to the judge for a further hearing, leaving it to him to decide, possibly on further evidence, what order, if any, he could make. But, as the appeal, in my judgment, fails on the ground which I have mentioned, I do not find it necessary to pursue that matter further.

Agreeing BRANDON LJ said: I would like to add a few words with regard to the true meaning and effect of paragraph 5(1) of Part IV of Schedule 15 to the Rent Act 1977. That paragraph provides:

> For the purposes of paragraph 4 above, the relevant conditions are that the accommodation is reasonably suitable to the needs of the tenant and his family as regards proximity to place of work . . . .

It was argued by Mr Fernyhough, on behalf of the appellants, that, in considering reasonable suitability to the needs of the tenant as regards proximity to place of work, the court should look only at the distance between the proposed alternative accommodation and the tenant's place of work, and that it was irrelevant that, although the distance might be such-and-such, the journey from the one to the other involved very considerable time and possibly the use of more than one form of transport.

In my view, to interpret this provision as referring only to distance as the crow flies would be absurd. The proper way to interpret the provision is to construe proximity and suitability in a commonsense way, taking into account not only what the distance is along such routes as are available but the means of transport available to the tenant concerned and the amount of time and the degree of inconvenience involved in making the journey. So, in the present case, it would be right for the judge to consider, as anyone who knows this area at all well would know, that it would be impossible for the first defendant to travel from the proposed alternative accommodation to her place of work or the other way without either using two buses, or a train and a bus, or two trains, and occupying a time certainly as much as an hour and perhaps considerably more than an hour on occasions.

It seems to me that these are all matters which have to be taken into account when applying this provision in a commonsense way and that any judge who did not approach the matter in that way would be in error. It seems to me that the judge in this case must have taken into account the reality of the matter: that is to say, how the first defendant would have to go from the one place to the other, what means she would have to use, how long it would take her and how inconvenient it might be. In arriving at his conclusion he took those matters into account and it seems to me that he was fully justified in reaching the conclusion that he did.

OLIVER LJ also agreed.

The appeal was dismissed with costs. The respondent being legally aided, legal aid taxation was ordered.

COURT OF APPEAL

March 6 1980

(Before Lord Justice STEPHENSON, Lord Justice BRIDGE and Lord Justice TEMPLEMAN)

BEEBE v MASON

Estates Gazette June 14 1980

(1980) 254 EG 987

Rent Act 1977, section 12(1)(c) and Schedule 2—Position on death of resident landlord—Possession action by deceased's brother who was both sole executor and sole beneficiary under the will—Brother went into occupation as a resident in deceased's flat about 10 months after her death—No vesting assent, however, made by brother as personal representative in favour of himself as devisee during the 12 months "disregard" period allowed by paragraph 1(c) of Schedule 2—Submission on behalf of tenant that the deceased landlord's interest remained vested in the brother as personal representative until after expiry of the 12 months' period, so that the tenant became fully protected—Tenant's argument rejected—Landlord's interest belonged to brother both before and after any assent as the sole executor and also sole beneficiary of a solvent estate—The only period which required reliance on Schedule 2 was the 12 months during which he was not in occupation—County court judge's order for possession upheld—Cautionary note that this was a case where the executor was absolutely entitled to the property and the estate was solvent.

In this appeal the tenant, Mrs Christina Mason, of the upper part of a house at 45 Park Road, Hounslow, Middlesex, challenged the decision of Judge Bernard Lewis at Brentford County Court granting possession of this part to Ernest Beebe. Mr Beebe was the brother and personal representative of Mrs Ethel Keen, the owner of the house, who had occupied the lower part of it as a resident landlord. Mrs Keen had let the upper part to Mrs Mason on July 16 1976, ie after the resident landlord provisions of the Rent Act 1974, now contained in the Rent Act 1977, section 12 and Schedule 2, had come into operation on August 14 1974.

D R Watkinson (instructed by Bensons) appeared on behalf of the appellant; A M Hughes-Chamberlain (instructed by Bonnett, Son & Turner) represented the respondent.

Giving the first judgment at the invitation of Stephenson LJ, TEMPLEMAN LJ said: This is an appeal against an order of His Honour Judge Bernard Lewis made in the Brentford County Court on December 3 1979 whereby he granted possession to the plaintiff respondent, Mr Beebe, of the upper part of 45 Park Road, Hounslow, Middlesex. The house, 45 Park Road, was owned by the late Ethel Keen, and she lived there. On about July 16 1976 Mrs Keen let the upper part to the appellant defendant, Mrs Mason. Mrs Keen continued to occupy the lower part herself.

By her will dated December 17 1973 Mrs Keen gave all her property to her brother, Mr Beebe, and appointed him sole executor. Mrs Keen died on September 7 1978 and her will was proved by Mr Beebe in the Winchester District Probate Registry on October 30 1978.

On July 2 1979 Mr Beebe took up residence in the lower part of 45 Park Road, and on July 18 1979 solicitors acting for Mr Beebe gave notice to Mrs Mason requiring her to quit the upper part on August 25. Mrs Mason remained in possession and Mr Beebe brought proceedings against her in Brentford County Court, proceedings which were served on September 7 1979, and in those proceedings he set out that he was the executor and residuary devisee and he set out the other facts to which I have referred, and claimed to be a resident landlord within section 12 of the Rent Act 1977 and entitled to possession.

It is common ground, and vital for the argument which has been adduced to us on behalf of Mrs Mason, that Mr Beebe did not before any relevant period make a written assent vesting the estate in himself as absolute owner, as distinct from the legal estate being vested in him in his capacity as personal representative of Mrs Keen.

On December 3 1979 the learned judge made an order for possession on January 14 1980, and subsequently on January 7 1980 he granted a stay until further order. In these days any order for possession is bound to create a disturbance and hardship on the tenant, and Mr Watkinson, on behalf of Mrs Mason, has very properly sought to analyse the relevant provisions of the Rent Act in such a way as will enable her to claim the protection from the Act and remain in possession. We are bound by the provisions of the Act and must give them their true and proper construction. The tenancy of Mrs Mason would be a protected tenancy under the Rent Act were it not for the provisions of section 12 of the Rent Act 1977, and the dispute in this appeal is whether the provisions of that section apply so as to enable Mr Beebe to obtain possession.

Section 12 of the Rent Act 1977 provides, with an exception not immaterial:

> . . . a tenancy of a dwelling-house granted on or after August 14 1974 shall not be a protected tenancy at any time if—

three specified conditions are satisfied. The tenancy in favour of Mrs Mason was granted after August 14 1974 and it remains to be seen whether the conditions were, and continue to be, satisfied. Condition (a) stipulates that the dwelling-house (in this case the upper part of 45 Park Road) forms part only of a building (in this case the whole of 45 Park Road) and that the building is not a purpose-built block of flats. This condition is satisfied. Condition (b) requires that the tenancy was granted by a person (in this case, Mrs Keen) who, at the time she granted it, occupied as her residence another dwelling-house (that is the lower part of 45 Park Road) which also forms part of that building, namely 45 Park Road, as a whole. That condition is also satisfied. Condition (c) stipulates:

> (c) subject to paragraph 1 of Schedule 2 to this Act, at all times since the tenancy was granted the interest of the landlord under the tenancy has belonged to a person who, at the time he owned that interest, occupied as his residence another dwelling-house which also formed part of that building.

When the tenancy was first granted, namely in July 1976, the interest of the landlord under the tenancy did belong to Mrs Keen, who also occupied the lower part of 45 Park Road, so the condition was satisfied so long as Mrs Keen was alive. It was also, it seemed to me, satisfied when Mr Beebe went into possession on July 2 1979, because on July 2 1979 the interest of the landlord belonged to him and to nobody else. He had proved the will, so he had the legal estate, and he was the sole beneficiary under the will. There is no question of any insolvency. The one person to whom, in my judgment, the interest of Mrs Keen then belonged, was Mr Beebe and no one else. But, of course, on the words of section 12(1)(c) there was a time during which the interest of the landlord under the tenancy did not belong to a person who occupied part of 45 Park Road and that time was between September 7 1978 when Mrs Keen died, and July 2 1979, a period of nearly 10 months which elapsed before Mr Beebe himself moved into 45 Park Road; so that unless Mr Beebe can bring himself within paragraph 1 of Schedule 2 he is not entitled to possession.

I turn, therefore, to Schedule 2 paragraph 1. That provides certain periods which are to be disregarded for the purposes of section 12(1)(c). The question is whether the period between September 7 1978 when Mrs Keen died and July 2 1979, when Mr Beebe went into possession, is to be disregarded as a result of Schedule 2.

The relevant period of disregard is (c) which is as follows:

> (c) any period of not more than 12 months beginning with the date on which the interest of the landlord under the tenancy becomes, and during which it remains, vested—
>
> (i) in the personal representatives of a deceased person acting in that capacity;

The period with which we are dealing is the 10-month period beginning with the death of Mrs Keen and the entry into

A possession of Mr Beebe. During that period of 10 months the interest of the landlord was vested in Mr Beebe as personal representative of Mrs Keen, and therefore it seems to me that is a period which we must disregard. We have to strike it out of the calendar and therefore we strike out the one period which does not fall squarely within section 12(1)(c) and on striking that period out it is true to say that this is a tenancy which is not protected because there has been a resident landlord, or landlords, throughout the tenancy.

Mr Watkinson says that because there was no written assent this property remains vested in Mr Beebe as personal representative and the period of sanctity granted by the relevant B provisions of Schedule 2 is limited to 12 months and those 12 months expired. Therefore, he says, Mr Beebe can no longer claim the protection of paragraph 1; but looking at the language of section 12(1)(c) there is no need to rely on Schedule 2 during any time when the interest of the landlord belonged to Mr Beebe and he occupied the property. One is only driven to Schedule 2 if either the interest of the landlord did not belong to Mr Beebe or he was not in occupation.

In the circumstances of this case, and I stress that we are dealing with a case where the executor was absolutely entitled to the property in question, it seems to me that both before and after any assent the property belonged to Mr Beebe, and C that once he went into occupation he satisfied the requirements of 12(1)(c) and therefore only relies on the Schedule for that period during which he was not in occupation. That period was not more than 12 months. During that period the property was vested in him as personal representative of Mrs Keen and therefore that period must be ignored.

Mr Watkinson referred us to a statement in Megarry's *Law of Real Property*, 4th ed, p 539 and to several cases which establish that in the case of an unadministered estate a beneficiary under a will cannot gain any equitable interest in D the property, and that the property belongs to the personal representative until he makes an assent. One can perceive of different circumstances in which one would have to look again at the section and at the Schedule to see what the results are, but in the present case, where the facts are simple, it seems to me that there are no grounds for saying that when Mr Beebe was in occupation the interest of Mrs Keen did not belong to him and that the only period which he must invite the court, or require the court, to disregard is the period of 10 months during which he was not in residence and to which subparagraph (c) paragraph 1 of Schedule 2, in my judgment, plainly applies. Accordingly it seems to me that on the true E construction of section 12 and Schedule 2 there is no answer to the claim for possession, that Mrs Mason's tenancy never was a protected tenancy provided one ignores, as one must, that period of 10 months during which the interest of the landlord was vested in a personal representative who was not in possession. For these reasons I come to the same conclusions as the learned judge and I would dismiss the appeal.

BRIDGE LJ agreed.

Also agreeing STEPHENSON LJ said: Mr Watkinson has not persuaded me that by not availing himself of the power he has under section 36 of the Administration of Estates Act 1925 the plaintiff has broken the duty imposed on him by section F 25(a) of that Act, as substituted by section 9 of the Administration of Estates Act 1971, and thereby failed to satisfy the requirements of section 12(1)(c) of the Rent Act 1977. For the reasons given by Templeman LJ he seems to me to have satisfied those requirements and I therefore agree that the appeal should be dismissed.

The appeal was dismissed with costs limited to £272 and legal aid taxation was ordered.

LANDLORD AND TENANT

RENT REVIEW

A **QUEEN'S BENCH DIVISION**

July 31 1979

(Before Mr Justice PHILLIPS)

PIVOT PROPERTIES LTD v SECRETARY OF STATE FOR THE ENVIRONMENT

Estates Gazette January 26 1980

(1979) 253 EG 373

Rent review clause in lease of premises to the Secretary of State—Award by arbitrator in the form of a special case for decision of the High Court under section 21(1)(b) of the Arbitration Act 1950—Rent from review date to be the "rack
B **rental market value" as defined in the lease—Question of law as to whether account had to be taken of the possibilities of continuation or renewal of tenancy under Part II of the Landlord and Tenant Act 1954—Held that the hypothetical landlord and tenant competing in the market would bear in mind the possibilities of continuation or renewal—Hence the rent should be the higher of the two rents mentioned in the award—Tenant's contrary submissions rejected**

This was a special case stated by the arbitrator, who framed his award in the form of a special case under section 21 of the Arbitration Act 1950 for the decision of the court. The arbitration concerned the true construction of the rent review
C clause in an underlease between Pivot Properties Ltd as lessors and the Secretary of State for the Environment as lessee. The properties subject to the lease in question were Lunar House and Apollo House, Wellesley Road, Croydon.

J S Colyer QC, G W Seward and K M J Lewison (instructed by Goodman, Derrick & Co) appeared on behalf of the claimants, Pivot Properties Ltd; M A B Burke-Gaffney QC (instructed by the Treasury Solicitor) represented the respondent Secretary of State.

Giving judgment, PHILLIPS J said: This is an award by an arbitrator stated in the form of a special case under section 21(1)(b) of the Arbitration Act 1950 in an arbitration between
D Pivot Properties Ltd as claimants and the Secretary of State for the Environment as respondent. The dispute between the parties concerns the construction of a rent review clause contained in an underlease made on August 4 1971 between Pivot Properties Ltd as lessors and the Secretary of State for the Environment as lessee, whereby land and buildings known as Lunar House and Apollo House in the County Borough of Croydon were let to the lessee for a term of 42 years from June 24 1971 at a rent of £785,000 for the period from August 5 1971 to August 5 1972 and thereafter at a yearly rent of £1,570,000, subject to the provisions as to rent review.

The buildings are described in the underlease as being
E "lately erected." I have been shown photographs of them and, as the rent suggests, they are very large.

Clause 7 of the underlease provides as follows:

IT IS HEREBY AGREED AND DECLARED as follows:

(1) The Lessors shall have the right to call for a review of the yearly rent for the time being payable hereunder at each of the following dates (hereinafter referred to as "the review date") namely

Twenty-fourth day of June 1978
Twenty-fourth day of June 1985
Twenty-fifth day of December 1990
Twenty-fourth day of June 1996 F
Twenty-fifth day of December 2001
Twenty-fourth day of June 2005
Twenty-fifth day of December 2010

(2) The said right shall be exercised by the Lessors giving to the Lessee not less than six months' notice in writing prior to a review date.

(3) If and so often as the Lessors shall give to the Lessee such notice as aforesaid then from and after a review date this Lease shall be read and construed and shall take effect in all respects as if the yearly rent for the time being payable hereunder had from such review date been the rack rental market value (as hereinafter defined) of the demised premises but without prejudice to any of the other terms and conditions contained in this Lease and so that G
in no event shall the rent payable by the Lessee to the Lessors after a review date be less than the rent payable by the Lessee to the Lessors immediately before a review date.

(4) For the purpose of this clause the rack rental market value of the demised premises at any review date shall (subject as hereinafter provided) be the amount which shall be agreed between the Lessors and the Lessee (or in default of agreement be determined by arbitration as hereinafter provided) to be the best rent at which the demised premises might reasonably be expected to be let in the open market as a whole for a term not exceeding five years and one half of another year and subject to similar covenants and conditions (other than the amount of the rent and the provisions of this present clause for reviewing the rent) to those H
contained in this Lease there being disregarded:

(a) Any effect on rent of the fact that the Lessee has or his predecessors in title or by any undertenant or undertenants of the Lessee or his predecessors in title have been in occupation of the demised premises or any part thereof.

(b) Any goodwill then attaching to the demised premises or any part thereof by reason of the carrying on on the demised premises or on some part thereof whether by the Lessee or by his predecessors in title or by any undertenant or undertenants of the Lessee or his predecessors in title of any business or businesses therein.

(c) Any effect on rent of any improvement carried out by the Lessee or any person deriving title under the Lessee with the consent of the Lessors otherwise than in pursuance of an obligation to the Lessors. J

(5) If by three calendar months prior to any review date the Lessors and the Lessee have failed to agree upon the rack rental market value of the demised premises as at any review date then and in such case the question shall forthwith be referred upon the request of either party to the decision of some competent person to be nominated by the President for the time being of the Royal Institution of Chartered Surveyors and these presents shall be deemed to be a submission to Arbitration within the meaning of the Arbitration Act 1950 or any statutory modification or re-enactment thereof for the time being in force and the decision of such person (including any decision as to the costs of such arbitration) K
shall be binding on both the Lessors and the Lessee.

(6) Until such time as the amount of any increase in the rent reserved by this Lease shall have been settled the Lessee shall continue to pay such rent as is payable by this Lease immediately prior to any review date on the dates and in manner hereinbefore provided and when the amount of any increase has been settled the Lessors will serve upon the Lessee a notice in writing specifying the amount and demanding payment thereof and the first payment of the rent to which such notice relates shall immediately become due and payable in respect of the period from the review date to the following quarter day.

(7) SUBJECT as hereinbefore provided to the contrary and to any award as to costs made by an Arbitrator the Lessors and the Lessee shall each bear their own costs of ascertaining the rack rental market value of the demised premises at any time for the purpose of this present Clause.

The lessors having called for a review of the rent as from June 24 1978 and the parties having failed to agree, the arbitrator was duly appointed to determine the rent in accordance with clause 7 and, at the request of the parties, stated his award in the form of a special case for the decision of the court. The award states the question of law as follows:

Whether upon the true construction of the underlease made on August 4 1971 the rack rental market value of the premises (which is to be assessed on the assumption that they are to be let as a whole for a term of 5½ years) is also to be assessed upon the assumption that no account is to be taken of any possibility of the tenancy being continued or renewed under the provisions of Part II of the Landlord and Tenant Act 1954.

The award continues as follows:

(1) In the event of the Court deciding the question of law in the affirmative I award as follows: (i) That the rent for the period 24 June 1978 to 23 June 1985 shall be at the annual rate of Two Million One Hundred Thousand Pounds (£2,100,000). . . . (2) In the event of the Court deciding the question of law in the negative I award as follows: (i) That the rent for the period 24 June 1978 to 23 June 1985 shall be at the annual rate of Two Million Nine Hundred and Twenty-Five Thousand Pounds (£2,925,000). . . .

It will be observed that those paragraphs are expressed in the negative. The first of the two alternatives takes no account of the matter stated; the latter takes account of the matter stated.

The parties were agreed that the length of the term to be considered by the arbitrator for the purposes of clause 7(4) was five and a half years.

Thus the question for my decision is whether in assessing under clause 7(4) the best rent at which the premises might reasonably be expected to be let in the open market as a whole for a term of five and a half years and subject to similar covenants and conditions (other than the amount of the rent and the provisions of clause 7(4) for reviewing the rent) to those contained in the underlease, account is to be taken of any possibility of the tenancy being continued or renewed under the provisions of Part II of the Landlord and Tenant Act 1954. Mr Burke-Gaffney for the lessee submitted that no account should be taken of this possibility; Mr Colyer for the lessors submitted that it should, and the arbitrator was of opinion that it should.

As will have been seen from the extracts from the award set out above, a lot hangs on the answer to the question posed for my decision: £825,000 a year until 1985; and over the whole term £28,000,000, according to Mr Burke-Gaffney, or £14,000,000, according to Mr Colyer. The question is one of general importance in other cases also.

The provisions of Part II of the Landlord and Tenant Act 1954 relating to business tenancies are well-known to property owners and especially their professional advisers. Thus we are not here concerned with some obscure or little-known part of the law. There is no doubt that the existence of the Act of 1954 has an effect on the level of rents of business premises. Thus, assuming business premises to be offered for a term of five years, a bidder in the market, bearing in mind the possibilities of a continuation or renewal of the tenancy, will tend to offer more than he would do if the Act did not exist and there was no such possibility. Mr Burke-Gaffney agreed that this is so. His argument was different, namely that to take account of the possibilities of continuation or renewal of the term postulated by clause 7(4) is inconsistent with and repugnant to the terms of the underlease and its tenor. Apart from this dispute there is agreement between the parties as to how the rack rental market value ought to be determined in accordance with clause 7(4). Thus the lessor and the lessee are hypothetical, it being assumed that the hypothetical lessor and the hypothetical lessee are willing negotiators in the open market for a term of five and a half years of the actual land and buildings on the covenants and conditions of the underlease, save where otherwise stated in clause 7(4).

In *Priestman Collieries Ltd v Northern District Valuation Board* [1950] 2 KB 398, Morris J at p 407, delivering the judgment of the court, said this:

In the opinion of the court, the phrase "open market" in section 13(4) of the Act (that is the Coal Industry Nationalisation Act 1946) does not contemplate a purely hypothetical market which is to be regarded as exempt from any restrictions imposed by law. The section does not postulate conditions wholly divorced from reality. The directed inquiry in effect is: "what would the sale of the unit have realised on January 1 1947?" In pursuing that inquiry it must be assumed that on that date the seller was willing to sell and that the buyer was willing to buy; any notion of a compulsory sale must be ignored. The seller must be deemed to have done as well as a willing seller would reasonably have been expected to do on January 1 1947. But that does not mean that he should be regarded as a person not bound by the law. It must be assumed that he would deal on the market on January 1 1947, subject to the law of the land. Any restriction or limitation imposed on him by law or imposed on any willing buyer must be recognised and taken into account. The sale to be regarded for the purposes of section 13(4) must be a notional one, but a notional sale in the market in which and subject to the conditions under which willing buyers and sellers could legitimately have operated on January 1 1947. It is enjoined by section 13(5) that regard should be had "to all relevant circumstances." What is contemplated, therefore, is a sale that might have taken place in the actual market on the prescribed date—*rebus sic stantibus.*

Although the terms of that Act, of course, were particular, it seems to me that similar considerations apply in the present case. See also *East End Dwellings Co Ltd v Finsbury Borough Council* [1952] AC 109.

So the parties and the term are hypothetical, but the figure at which the parties would settle has to be determined according to the facts as they are in the real world, *rebus sic stantibus,* save in so far as reality is contradicted by the hypothesis. In the real world, for reasons given earlier in this judgment, I consider that the hypothetical tenant in making his bid and the hypothetical landlord in deciding whether to accept it would bear in mind the possibilities of a continuation or a renewal of the term in accordance with the provisions of the Act of 1954. Uncertainties, for example due to a contention that the provisions did not apply, or to a possibly successful opposition by the landlord under section 30, or a possible repeal of the Act, would be allowed for in the process of negotiation, in the case of agreement, or of valuation in the case of reference to arbitration. Such uncertainties would not lead to the possibilities being disregarded. There is here a real possibility to be assessed and the situation seems to me to be quite different from that which obtained in *Plinth Property Investments Ltd v Mott, Hay & Anderson* (1979) 249 ESTATES GAZETTE 1167, a decision of the Court of Appeal, upholding a decision by Slynn J given on May 30 1977, relied upon by Mr Burke-Gaffney. There the arbitrator had to determine the rent of premises let upon terms containing a restriction as to user, the effect of which would be to depress the rent. The question was whether the arbitrator must take into account the possibility that the landlord would relax the restriction. The arbitrator, Slynn J, and the Court of Appeal all said: "No." Two reasons were relied upon for this conclusion. First, that the possibility of relaxation was so intangible that it could not be assessed; secondly, that to assume such a possibility would be to depart from the terms of the lease under which the rent was being assessed. The first reason is that given by Lord Denning MR and Shaw LJ in the Court of Appeal and to some extent it was relied upon by Slynn J and Brandon LJ. The second reason appears to have been that principally relied upon by Slynn J and Brandon LJ.

Perhaps Mr Burke-Gaffney may derive some support from the principle behind the second reason, but, as I have said, it seems to me that the situation in the present case is quite different. The possibility of the hypothetical tenant enjoying a continuation or renewal of the term at the date of its expiry seems to me to be a real one, and just as much a part of the real world as the state of the market, the rate of inflation or the merits of building. Experienced surveyors when making valuations are well accustomed to taking into account possibilities of this character. The arbitrator in the present case seems to have had no difficulty. The parties have clearly been able to make their submissions upon value allowing for this possibility. Nor do I think there is any reason to suppose that the hypothetical tenant would—whether for technical reasons, see *Compton Group Ltd v Estates Gazette Ltd* (1978) 36 P&CR 148 at p 157; (1977) 244 EG 799 at p 805—or otherwise not enjoy the *possibility* of benefiting under the Act of 1954. In any event, others competing with him in the market certainly would and their competition would help to set the market rate.

It struck me at the outset of the case, and I still think, that clause 7(4) is oddly worded, and I cannot help wondering whether there has not been some slip in the drafting. In giving directions to the parties or the arbitrator as to the basis upon which the rack rental market value of the premises is to be agreed or to be determined, one would have expected an exact formula to be prescribed, yet the phrase, "a term not exceeding five years and one half of another year" leaves it to the parties to agree or determine one of the principal features by which the rent is to be fixed. Nor is it apparent what criteria should be taken into account in so deciding the length of the term. It is, it seems to me, an odd way of proceeding to leave it to the arbitrator, subject to the upper limit, to determine for himself the very measure which he is to apply. This peculiarity perhaps makes it difficult to know what weight to attach to the words "not exceeding" on which Mr Burke-Gaffney relied. His principal submission, put in a variety of ways, was that taking into account the possibility of a continuation or a renewal of the hypothetical term could be inconsistent with and repugnant to the terms of the underlease and in particular clause 7(4). The directions in that clause, he said, are perfectly clear and base the assessment on a term not exceeding five and a half years. The express words could not be departed from, or any implication relied upon, and that to take into account the possibilities under the Act of 1954 would be to substitute a term in excess of that prescribed by the underlease. He relied on *Affréteurs Réunis Société Anonyme v Leopold Walford (London) Ltd* [1919] AC 801 and *London Export Corporation Ltd v Jubilee Coffee Roasting Co Ltd* [1958] 1 WLR 661. It was also inconsistent with those cases, he said, for the arbitrator to have to go into the realms of conjecture and speculation which taking into account the possibilities under the Act of 1954 would involve.

The principle of those cases is not in doubt, but I do not think that there is any express inconsistency or repugnancy here. In the first place, I cannot see that the words "not exceeding" can directly assist Mr Burke-Gaffney's submission to the contrary, though they may perhaps contribute to the atmosphere, and thus to the tenor of the underlease upon which he relies. Those words are merely a direction to the parties or the arbitrator to be used in determining the length of the hypothetical term or, as Mr Colyer put it, they are mere words of limitation and are not words describing the attributes or consequences of that term. In this sense, said Mr Colyer, they are commonly used in statutes: see, for example, section 33 of the Landlord and Tenant Act 1954, section 54 of the Law of Property Act 1925, section 41 of the Settled Land Act 1925 and section 123 of the Local Government Act 1972. I think that this is correct. So Mr Burke-Gaffney's submission of inconsistency in terms is no stronger than it would be if clause 7(4) had specified the length of the term precisely, say five and a half years. Secondly, and in that event, I do not think that there would be any such express inconsistency or repugnancy. To take account of the possibilities under the Act of 1954 is not to assess the rent for a term longer than five and a half years, as Mr Burke-Gaffney contended, but to assess the rent for a term of five and a half years, one of the potentialities of which is that it may be continued or renewed. This potentiality is as much an ingredient contributing to its value as other characteristics such as the terms and conditions of the lease.

So I reject this part of Mr Burke-Gaffney's submissions. But he put the case more broadly and submitted that to take into account the possibilities under the Act of 1954 is inconsistent with the tenor or the substance of the underlease. I agree that in a case of this kind it will not do to be too technical and that I should be wary of the skilful logic of Mr Colyer's argument which, as logically-riveted chains of argument sometimes do, might lead to a foolish result; in other words, keep an eye on the commonsense of it. The difficulty in this is that if I am right in saying that in the ordinary way a bidder for a term of business premises would take into account the possibilities under the Act of 1954 (and this was not disputed) there is little on which to found the view that to do so in this case is contrary to the tenor, substance or spirit of the underlease. All that can be said on this is that the underlease directs the assessment of the rent for a term of five and a half years and that it is therefore forbidden to look beyond the end of that period. By itself this seems to me to carry the argument no further. None the less, I can see that this could be a peg on which to hang Mr Burke-Gaffney's contention, if it could be shown that the claimants' contention leads to an absurd result. Accordingly, I thought it right to inquire during the course of the hearing whether the result arrived at by the arbitrator, that is taking into account the possibilities under the Act of 1954—I do not mean the mere quantum of the assessment—was in principle unreasonable or absurd. While no concession was made, Mr Burke-Gaffney disclaimed any contention to the effect that the claimants' contention leads to an absurdity.

Indeed, it seems to me, it may lead to a more sensible result than the respondent's. It is common ground that the reasons for the large difference between the alternative awards (£825,000) is connected with the huge size of the buildings. The cost of occupation of such buildings would be so large that there would be great reluctance to take them except for a period which would make it worthwhile to incur such costs. A renewal at intervals of between five and seven years throughout a term of 42 years, at which the rent was assessed on each occasion on the basis of a letting for a term not exceeding five and a half years, would not, except at the end of the term, seem to reflect the reality of the situation. In a sense, this is an aside and I merely pray it in aid negatively as not supporting the view that the claimants' contention here leads to absurdity.

Finally, it should be noted that clause 7(4) lists a number of matters to be disregarded in determining the rent and that the provisions of the Landlord and Tenant Act 1954 are not among them; and that the subclauses to clause 7(4) seem to be borrowed or adapted from section 34 of that Act.

So, at the end of the day, the position is this: I accept Mr Colyer's arguments to the effect that the hypothetical tenant and the hypothetical landlord competing in the market would take into account the possibilities of the tenant obtaining a continuation or renewal of the term. I do not find any express inconsistency here with the terms of the underlease, nor can I say that to take into account such possibilities is inconsistent with the tenor or substance of the underlease. It is not contended that the claimants' case leads to an absurdity. Accordingly, I find in favour of the claimants and answer the question of law posed for my decision: No.

Judgment was given for the claimants in the arbitration, Pivot Properties Ltd, with costs. Leave was given to the respondent, the Secretary of State for the Environment, to appeal. The answer given to the specific question raised in the special case was "No," the effect being that the possibilities

of the tenancy being continued or renewed under Part II of the Landlord and Tenant Act 1954 were to be taken into account in determining the rent under the review clause.

COURT OF APPEAL

January 18 1980

(Before Lord Justice MEGAW, Lord Justice EVELEIGH and Sir David CAIRNS)

AMHERST v JAMES WALKER GOLDSMITH & SILVERSMITH LTD

Estates Gazette April 12 1980

(1980) 254 EG 123

Rent review clause in lease—Whether provision in clause for submission by lessor of a "trigger" notice by a fixed date, for the purpose of bringing the review machinery into operation, made time "of the essence"—Time expressly stated to be of the essence for subsequent procedures under the clause, but not for the "trigger" notice—Lessor failed to submit the notice by the specified date—Deputy High Court judge's decision that time was not of the essence for the purpose of this provision in review clause upheld by Court of Appeal—United Scientific Holdings Ltd v Burnley Borough Council followed

This was an appeal by the defendants, James Walker Goldsmith & Silversmith Ltd, from part of a decision by Judge Mervyn Davies QC, sitting as a deputy High Court judge in the Chancery Division, which was in favour of the plaintiff, William John Amherst. The proceedings before Judge Mervyn Davies took the form of an originating summons in which the plaintiff claimed declarations in regard to a rent review clause in a lease dated August 29 1961. The judge did not make the declarations claimed, but did declare that time was not of the essence so far as concerned a provision in the rent review clause that an assessment of the rent for the second period under the lease should be submitted by the lessor to the lessee for approval in writing on or before December 25 1974. Subject to this the judge dismissed the originating summons.

Richard Moshi (instructed by Bulcraig & Davis) appeared on behalf of the appellants; Roger Gray QC and John Hamilton (instructed by Macdonald, Stacey & Co) represented the respondent but were not called on.

Giving the first judgment at the invitation of Megaw LJ, SIR DAVID CAIRNS said: This is an appeal from a decision of His Honour Judge Mervyn Davies, sitting as a deputy judge of the Chancery Division. He had before him an originating summons raising questions about a rent review clause in a lease. At the material time the plaintiff was the lessor and the defendant company was the lessee. The plaintiff claimed by his originating summons a declaration that, under the rent review clause, he was entitled to receive increased rent from June 24 1975 and he claimed a further declaration in support of that entitlement. At the hearing, it became apparent that one question which was relevant was whether the time specified in the lease for the submission by the plaintiff to the defendant of an assessment of the increased rent that he wished to operate was of the essence or not. The plaintiff was arguing that it was not; the defendant was arguing that it was. The learned judge decided this issue in favour of the plaintiff and made a declaration to the effect that time was not of the essence. Save as to that, he dismissed the originating summons and awarded the defendant company its costs of the proceedings. The plaintiff has not appealed against that decision, but the defendant appeals from that part of the order made by the learned judge containing the declaration that I have referred to.

The lease was a lease made on August 29 1961 of the premises, 449 High Road, Wembley, and was a lease for 28 years from June 24 1961 at a rent defined in this way, "during the first fourteen years the yearly rent of two thousand five hundred pounds and during the remainder of the said term either the yearly rent of two thousand five hundred pounds or such higher yearly rent as shall be ascertained in accordance with the proviso next hereinafter contained." This was the proviso:

> Provided always and it is hereby agreed that the yearly rent payable by the lessee during the second fourteen years of the term hereby granted (hereinafter called "the second period") shall be the yearly sum of two thousand five hundred pounds aforesaid or such yearly sum representing the rental value of the demised premises in the open market on a lease for fourteen years certain on the assumption that the premises are available for letting with vacant possession and without the payment of a premium whichever shall be the higher such assessment of the second period rent to be made in the following manner that is to say: (a) Such assessment shall be made in the first instance by the lessor and submitted to the lessee for approval in writing on or before the twenty-fifth day of December one thousand nine hundred and seventy four; (b) In the event of the parties hereto failing to reach such agreement as aforesaid on or before the date appointed (in respect of which time is to be deemed to be of the essence of the contract) then the yearly rent for the second period shall be fixed or assessed by an independent surveyor appointed for that purpose by the parties hereto failing agreement as to such appointment by the twenty-fifth day of January one thousand nine hundred and seventy-five (time in this respect to be deemed to be of the essence of the contract) then by an independent surveyor appointed for that purpose by the President for the time being of the Royal Institution of Chartered Surveyors. The assessment fixed by the independent surveyor shall be communicated to the lessor and lessee in writing and immediately upon such communication the yearly rent so assessed as a reasonable rent for the second period or two thousand five hundred pounds per annum whichever shall be the higher shall be the yearly rent payable for the second period under the terms hereof.

I need not read paragraph (c) of the proviso.

> (d) It is hereby agreed that the fees payable to the independent surveyor hereinbefore mentioned in respect of the assessment to be made herein shall be borne by the parties hereto in equal shares.

The present defendant company became lessee by having the lease assigned to it on December 16 1970. The lessor did not, before December 25 1974, submit any such assessment as is mentioned in clause (a) of the proviso that I have read. Soon after Christmas 1974 the solicitors acting for the lessor realised that the time provided for giving such notice had arrived and had indeed been passed. They accordingly wrote a letter to the defendant company on January 25 1975 in these terms:

> Dear Sir, Re: 449 High Street, Wembley. We act for Mr W J Amherst the lessor of the above premises. We observe from the lease dated

—they give the wrong date, October 29 1961: it ought to have been August—

> that there is a rent review in relation to the premises to come into effect on June 24 1975. Under the terms of the lease the landlord was required to assess the rent by submitting a notice in writing before December 25 1974 but if for any reason the rent was not agreed before that date then the rent should be fixed or assessed by an independent surveyor appointed by the landlord and tenant by January 25 1975, and on failure to agree such independent surveyor an independent surveyor should be nominated by the President for the time being of the Royal Institution of Chartered Surveyors. As no notice had been given to you of the proposed increase of rent it seems to us that prior to January 25 1975 steps should be taken to endeavour to agree the name of an independent surveyor in accordance with the terms of the lease. We are instructed to inform you that our client is prepared to nominate Mr A C W McKenna of Messrs Healey & Baker to act as such independent surveyor. If you are not able to agree to his appointment or are prepared to nominate someone else we would be prepared to consider any person whom you are prepared to nominate. We appreciate that the notice given is rather late and we would be prepared to agree to the time specified in the lease being extended so that no application need be made to the President for the time being of the Royal Institution of Chartered Surveyors until February 25 1975. If we do not hear from you by January 25, in order to protect our client's position we will make application to the President of the Institution without prejudice to any agreement which may be reached as to the parties deciding on an

> A independent surveyor. We are also instructed to inform you that entirely without prejudice to the procedure which has been laid down in the lease, our client is prepared to negotiate with you as to the amount of rent

and then suggesting communicating with a Mr Allan in that connection.

The defendant company did not agree to that course. On February 10 the lessor's solicitors wrote to the president of the institution asking for a nomination of an independent surveyor. The president did not comply with that request, having been informed by the lessees' solicitors that the lessees were of opinion that the lessor's application was out of time.
B Nothing further happened after that until June 2 1978 when the solicitors for the lessor wrote to the lessees' solicitors, recalling the earlier correspondence and mentioning the decision which had taken place in the meanwhile in the leading case in this field of rent review clauses, *United Scientific Holdings Ltd v Burnley Borough Council* [1978] AC 904. The lessees were called upon by that letter to withdraw their objection to the appointment of an independent surveyor. They refused to do so and the lessor's solicitors thereupon issued the originating summons in this case.

The declarations that they asked for, as appear from the summons, were:

> C (1) A declaration that on the true construction of the lease and in the events which have happened the plaintiff is entitled to receive from the defendant as from the 24th day of June 1975, rent at such rate as may be determined by an independent surveyor appointed by the President for the time being of the Royal Institution of Chartered Surveyors notwithstanding the fact that the plaintiff did not submit to the defendant a written assessment of rent for approval on or before the 25th day of December 1974 in pursuance of the terms of the lease. (2) A declaration that the defendant is not entitled to object to the said appointment of an independent surveyor by reason only of the plaintiff's failure to submit to the defendant a written assessment of rent as aforesaid.

The learned judge held that the time for the plaintiff to
D submit an assessment of rent by December 25 1974 was not of the essence, but he did hold that the assessment (what has been called in some of the earlier cases a "trigger" notice) must be submitted before any rent review could operate and accordingly the plaintiff was not entitled to the main declaration which he was claiming asking for the court to say that he was entitled immediately to an increase in rent. The plaintiff attempted in the course of the hearing to get over the difficulty of there having been no "trigger" notice served at all by serving one while the hearing was in progress—I think Mr Moshi has told us that while he was actually on his feet in court that notice was served. The learned judge, however,
E said that he could not, in those proceedings, give any effect to a notice which had been served so long after the issue of the originating summons.

Looking simply at the wording of this proviso in the rent review clause, it seems to me, on the face of it, that the provision for the submission of the "trigger" notice by December 25 1974 is not of the essence of the contract. I say that for two reasons. First, where you have an agreement which specifies the time for doing three things and two of the three are stated to have a time-limit which is of the essence of the contract and the other one is not, then it follows that, in all probability, where it is not stated that it is to be of the essence,
F it is not intended to be of the essence. If the parties, when they wish something to be made of the essence, say so in terms, then obviously it is of the essence if they do say so and the probability is that, where they do not say so, it is not intended to be of the essence.

That view is, to my mind, strongly supported by the decision of the House of Lords in the *United Scientific* case, to which I have referred, where the unanimous decision of the House, overruling the previous decision of the Court of Appeal, was to the effect that in these timetables in rent review clauses *prima facie* the times given are not of the essence and they are to be regarded as being of the essence only if they are expressly, by some clear indication in the document itself, G
made to be of the essence. Therefore, where, in relation to this particular part of the procedure, there is no indication that it is to be of the essence, it seems to me that the conclusion is that it cannot be so regarded and that it would be contrary to the actual decision in the House of Lords to hold that it was.

Mr Moshi, on behalf of the defendant company here, has sought to rely upon a passage in the speech of Lord Dilhorne in that case, in which he took a view which had already been taken in the Court of Appeal, that where a step to be taken was something which the landlord, and the landlord only, was, under the provisions of the agreement, in a position to take, H
time should be regarded as of the essence. Mr Moshi asks us to accept the decision of the Court of Appeal and the language of Lord Dilhorne as sufficient authority for that proposition. The first thing that has to be said is that the expression of opinion, both in the Court of Appeal and by Lord Dilhorne, in relation to that matter was all *obiter dictum*. The case in question is not the *United Scientific* case itself but another case which the House of Lords heard along with it, *Cheapside Land Development Co Ltd v Messels Service Co*. In that case, although these opinions were expressed, they were *obiter*, because the notice that had been given in that case was in fact in time within the terms of the contract. Not only that, J
but while Mr Moshi is, of course, entitled to rely on the expression of opinion (*obiter* though it be) in this court and by one member of the House of Lords in support of his argument, the great weakness in his argument is that three members of the House of Lords took the opposite view. Lord Diplock, Lord Salmon and Lord Fraser, in their speeches, took the view that it made no difference whether it was a notice which the landlord alone was entitled to give or whether it was one that could be given on either side. So that one must take the view, when the majority view in the House of Lords is to that effect, that, with the respect that we must, of course, pay to any such expression of opinion by K
Lord Dilhorne, the weight of authority must be deemed to be that of the three members of the House of Lords who said that that was not a material consideration.

So far as other matters are concerned, I think that the most effective part of Mr Moshi's argument before us was to this effect, that, unless we hold that the taking of the first step was one which had to be taken by a specified date, the lessee was put in difficulty in relation to the later steps, because there was a fixed date which could not be departed from, because it was provided to be of the essence of the contract, up to which he was entitled to agree either as to the amount put forward by the landlord for the rent or as to the surveyor L
who was to be appointed to fix that rent. The difficulty of that argument appears to me to be this, that, putting it at its highest, assuming that it could be said that the period for the "trigger" notice was subject to an absolute limit of December 25, there is certainly nothing in the agreement to say that it must be given before December 25. If the date is to be of the essence of the contract at all, it is on or before December 25; it is quite impossible to put it at a reasonable time before December 25. So that, in effect, even if it were regarded as being of the essence of the contract, the position is that, if the notice were given by the landlord at the last moment, one would immediately be into what may be called M
the second stage.

It seems to me that the true meaning of this lease is that, if the notice is not given by December 25, it can still be given at a later date. It may well be that the later date would have to be within a reasonable time, but as to that I express no more than a tentative opinion. But if notice is given thereafter and if it is not given so late that it can be said it was not given within a reasonable time and that that makes it invalid, then the position would be that, indeed, in effect, the second and third stages are concertinaed and we come straight on to the position where either party can go straight to the president of the RICS and ask for the appointment of an independent surveyor.

That I think is the meaning and effect of the agreement between the parties and, in my view, the learned judge came to the right conclusion in a very careful judgment and the conclusion which he reached, for reasons which I would have been prepared to adopt as my own. I would dismiss the appeal.

Agreeing, EVELEIGH LJ said: I find some difficulty in seeing that, in a case like the present, a "trigger" notice has any real part to play, certainly once the date specified in the lease has passed. But we are not concerned to determine the effect of that; we are concerned simply with the question whether time is of the essence in the making of the assessment in the first instance by the lessor and submission by him "to the lessee for approval in writing on or before the twenty-fifth day of December one thousand nine hundred and seventy four."

For the reasons stated by my Lord, I agree that time is not of the essence for that.

Also agreeing, MEGAW LJ said: I think that it emerges from the speeches in the *United Scientific* case that there was some measure of confusion in the development of that case through the courts by reason of submissions having been made to the effect that the cases there in question—rent review cases and the law of landlord and tenant—were to be treated as being matters of a commercial character. In one sense of the word, of course, letting contracts and the law of landlord and tenant are of a commercial character, but they are not contracts which fall within the narrower meaning of "commercial" which usually is accepted in the law. No one, for example, would dream of expecting to find in a textbook on commercial law discussion of the law of landlord and tenant.

Lord Diplock, in his speech in the *United Scientific* case *supra*, p 924 at B, said this:

> I shall have to examine rather more closely what are the legal consequences of "time being of the essence" and time not being of the essence; but I do not think that the question of principle involved in these appeals can be solved by classifying the contract of tenancy as being of a commercial character. In some stipulations in commercial contracts as to the time when something must be done by one of the parties or some event must occur, time is of the essence; in others it is not. In commercial contracts for the sale of goods *prima facie* a stipulated time of delivery is of the essence, but *prima facie* a stipulated time of payment is not (Sale of Goods Act 1893, section 10(1)); in a charterparty a stipulated time of payment of hire is of the essence. Moreover a contract of tenancy of business premises would not appear to be more of a commercial character than a contract for sale of those premises. Nevertheless, the latter provides a classic example of a contract in which stipulations as to the time when the various steps to complete the purchase are to be taken are not regarded as of the essence of the contract.

I refer to that passage in Lord Diplock's speech in order to emphasise that, in my view, the considerations which apply in deciding whether or not time is of the essence in a contract of the nature with which we are here concerned are by no means necessarily the same as are the considerations affecting contracts which are more strictly of a commercial nature. The decision of the House of Lords in the *United Scientific* case has shown that, in contracts of the nature with which we are concerned here, the general bias is against treating time as being of the essence.

Subclause (a) provides that: "Such assessment shall be made in the first instance by the lessor and submitted to the lessee for approval in writing on or before the twenty-fifth day of December one thousand nine hundred and seventy four." In my view, as a matter of construction—though I do not think it matters whether this is right or wrong—those words mean that the date December 25 1974 is the final date for the approval in writing and not for the submission by the lessor of the assessment. If that be right, then it would follow that the intention of the parties in that clause must have been that the assessment by the lessor should be submitted to the lessee at some date before December 25, so as to enable the lessee to have the opportunity to express his approval or his disapproval on or before December 25. That is to say, this clause does not purport to lay down any specific date on which the lessor's assessment is to be sent, save, of course, that it would be implied that it had to be sent at some date before December 25.

Subclause (b) provides that: "In the event of the parties hereto failing to reach such agreement as aforesaid on or before the date appointed (in respect of which time is to be deemed to be of the essence of the contract). . . ." I should accept that "the date appointed" must be taken (because I do not see what other meaning can be given to it) to be December 25 1974 and that the subclause is dealing with a case in which the approval of the lessee has not been given before that date. It would, of course, be absurd to suggest that if the lessor had given his assessment well before December 25 and the lessee had simply done nothing by way of acknowledging or considering or replying to it, it could then be said that time, being of the essence, had expired on December 25 and that the lessor had thereby, through no fault of his own, lost his right to have a higher rent during the remaining 14 years of the contract. That is not, and could not be, suggested. It follows, however, that what the subclause is dealing with is the time of the approval and not the time of the giving of the notice by the lessor. In those circumstances, it seems to me that it would be extremely difficult, in any event, to read the words used here as imposing upon the lessor, by implication, the consequence that, if he fails to give his assessment in whatever is the reasonable time before December 25, the lease thereafter has to continue for the remaining 14 years at the original rent.

I refer, simply in order to emphasise that point, to a passage in a judgment in a commercial case, properly so called. It is, therefore, the sort of case in which the law, as I have already suggested, would be more ready to hold time as being of the essence than in a case of the nature with which we are concerned. It is a sale of goods case, *Bremer Handelsgesellschaft v Vanden Avenne-Izegem PVBA* [1978] 2 Lloyd's Rep 109. One of the questions which the House of Lords had to consider was whether, in an exemption clause in a contract, a provision that notice should be given without delay if a party sought to rely on the exemption, was a condition of the contract: so that, if the notice was not given without delay, the person who sought to rely on it would be held, without more ado, not to be entitled to the protection of that clause. The House of Lords held that the clause was not to be so treated, even though it was a clause in a commercial contract. There were passages in the speeches of each of their Lordships to the same effect. I simply cite that of Lord Salmon. In the report at p 128 in the left-hand column he said this:

> In the event of shipment proving impossible during the contract period, the second sentence of cl 21 requires the sellers to advise the buyers without delay of the impossibility and the reasons for it. It has been argued by buyers that this is a condition precedent to the sellers' rights under that clause. I do not accept this argument. Had it been intended as a condition precedent, I should have expected the clause to state the precise time within which the notice was to be served, and to have made plain by express language that unless the notice was served within that time, the sellers would lose their rights under the clause.

So also in the present case, *a fortiori,* I should have expected that, if the intention had been that for which Mr Moshi contends, the latest date by which the lessor was required to give his assessment would have been expressly stated and not left to be, by implication, a reasonable time before December 25; and the clause would, by its express terms, have made it clear that, if that action was not taken within that time, the lessor would lose his rights.

I agree that this appeal falls to be dismissed.

The appeal was dismissed with costs. Leave to appeal to the House of Lords was refused.

RATING

COURT OF APPEAL

October 2 1979

(Before Lord Justice MEGAW, Lord Justice SHAW and Lord Justice WALLER)

LIVERPOOL CITY COUNCIL v BRITISH TRANSPORT HOTELS LTD AND ANOTHER

Estates Gazette January 5 1980

(1979) 253 EG 53

Rating—Case stated by Lands Tribunal—Rating of former railway hotel owned by British Transport Hotels Ltd—Proposal by ratepayers that the assessment of the hotel (gross value £18,000, rateable value £14,972) should be deleted from the list, the grounds being that the assessment was "incorrect, excessive and bad in law"—Local valuation court decided not to direct that the assessment be deleted, but that it be reduced to gross value £1,000 and rateable value £805—Submission by rating authority that local valuation court had no power on the basis of the proposal to reduce the assessment, but only to decide whether the assessment should be removed or remain in the list—Submission accepted by Lands Tribunal, but rejected by Court of Appeal—The correct interpretation of the proposal and the grounds was that, if the assessment was not deleted, at any rate it should be reduced—Criticism by court of rating authority's action in appealing to Lands Tribunal against "rational" and "just and fair" conclusion by local valuation court

This was an appeal by case stated from a decision of the Lands Tribunal (Victor Wellings QC) (1977) 246 EG 491 reversing a decision of a Liverpool local valuation court. The appellants before the Court of Appeal were the ratepayers, British Transport Hotels Ltd, and Alan Antrobus Gibbons, the valuation officer. The respondent in the Court of Appeal was the rating authority, Liverpool City Council. The property which was the subject of dispute was known as the Exchange Hotel, Tithebarn Street, Liverpool, which had in fact ceased to be used as an hotel from 1971 onwards.

J A R Grove (instructed by the solicitor to the British Railways Board) appeared on behalf of the first appellant (British Transport Hotels Ltd); A P Fletcher (instructed by the Solicitor of Inland Revenue) represented the second appellant (the valuation officer); and P H Goldsmith (instructed by the solicitor to the Liverpool City Council, represented the respondent.

Giving judgment, MEGAW LJ said: There is before us a case stated by the Lands Tribunal at the request of British Transport Hotels Ltd. It relates to a decision made by one of the legal members of the Lands Tribunal, Victor Wellings QC, on December 19 1977. The questions of law raised by the special case stated to this court are in effect whether the member of the Lands Tribunal was correct in law on a matter which he decided.

The relevant facts are set out in a statement of facts which was agreed between the parties for the purpose of the hearing before the Lands Tribunal, which itself was an appeal from a local valuation court at Liverpool given on September 29 1976. I summarise, as briefly as I think is possible, the statement of facts as it appears.

The appellant before the Lands Tribunal, the respondent before this court, is the Liverpool City Council, who were the rating authority in respect of the property in question. The first respondent before the Lands Tribunal and the first appellant before us is British Transport Hotels Ltd. It changed its identity at one stage but nothing turns on that. The second respondent before the Lands Tribunal and the second appellant before us is Alan Antrobus Gibbons, who is the valuation officer concerned in the valuation in question. The dispute relates to the value for the purposes of rating in the Liverpool Metropolitan District Rating Area of property which is described in the valuation list as "Hotel and Premises, Exchange Hotel, Tithebarn Street: Gross Value £18.000 Rateable Value £14,972."

The Exchange Hotel had been occupied as a railway hotel up to the year 1971, but from 1971 onwards it had ceased to be used as an hotel. Whether or not it had been wholly unused or whether it had been used for some kind of storage is not clear, but I do not think it matters for the purposes of this appeal. British Transport Hotels Ltd decided in the year 1975 that they would make a proposal in respect of the valuation list in relation to the Exchange Hotel to apply to the rating year 1974-75. They accordingly made a proposal which is set out in document 1 of the bundle before us. As everything in this appeal turns on the wording of that document it is desirable that I should read it, if not in full, at any rate fairly comprehensively.

It is headed "General Rate Act 1967, Proposal for alteration of Valuation List." Then the address of the property to which the proposal relates is given as "Exchange Hotel, Exchange Station, Tithebarn Street, Liverpool 2." The rating area is Liverpool and the name of the person or body by whom or on whose behalf the proposal is made is British Railways Board/British Transport Hotels Ltd. Then in print on the form are the words "Alteration which that person or body hereby proposes should be made in the valuation list" and there is typed in the words "That the Assessment should be deleted from the Valuation List." Then there are the printed words "Grounds on which that alteration is proposed" and typed in the words "That the Assessment is incorrect, excessive and bad in law." The name of the owner is given as British Railways Board, it is dated March 21 1975 and it is signed by a gentleman called R C J Bray FRICS, estate surveyor and manager, as agent for the owner (ie British Railways Board or British Transport Hotels Ltd).

On June 23 1975 the valuation officer, Mr Gibbons, duly objected to the proposal of March 21 1975 and in due course he transmitted to the clerk of the local valuation panel copies of the proposal and his notes of objection. On September 24 1976 the local valuation court, duly constituted from the Merseyside local valuation panel, heard the appeal of the valuation officer against the proposal. On September 29 1976 the court decided—a decision for which it gave reasons clearly stated in the document—that it was competent for the court to direct that an alteration should be made on the valuation list in pursuance of the proposal, ascribing an assessment of gross value £1,000 and rateable value £805. The court gave directions for the valuation list to be altered accordingly.

At the same time, as I understand it, there was also before the local valuation court a proposal by the same ratepayers in respect of the rating list for the same property for the following year 1975-76. The form of the proposal, which was submitted on March 12 1976, had been altered. It simply asked that the assessment should be substantially reduced and gave as its grounds that the assessment was incorrect and excessive. An agreement was made in writing between all concerned—the potential ratepayer, the rating authority and the valuation officer—under which the proposal of March 12 1976 was agreed and, in accordance with the provisions of section 72 of the General Rate Act 1967, the valuation list was altered accordingly to a gross value of £1,000 and a rateable value of £805. In other words, it was accepted that that was a fair and appropriate rateable value for the years 1975-76. There is no suggestion that there is any material difference which would have made the appropriate rateable value different in the previous year 1974-75.

Nevertheless, the rating authority, the Liverpool City Council, saw fit to contend—on what I am bound to say was the most technical of technical grounds—that, in spite of the obvious injustice resulting therefrom, the decision of the local valuation court ought to be held to have been wrong in respect of the March 1975 proposal because, in view of the form of that proposal, the local valuation court had no authority merely to reduce the value of the assessment; that they could either decide that the entry of the hereditament should be removed altogether from the valuation list or that it should remain in the valuation list, but that they had no jurisdiction to alter the amount if the decision was that it should remain in the valuation list.

Now, the basis of that contention is really quite simply stated, having regard to the document which I read at the outset, the proposal form of March 21 1975. It will be remembered that, against the printed words " Alteration which that person or body hereby proposes should be made in the valuation list " were the words " That the Assessment should be deleted from the Valuation List." It is said that that was the contention of the ratepayer and that the ratepayer is, as a matter of law and on the basis of statute and authority, not entitled to go beyond that contention, and that the local valuation court was not entitled to give them any relief beyond that contention. The words " the Assessment should be deleted," it is submitted, should be taken to mean that the whole entry ought to be deleted. If it meant that the value of the assessment should be deleted, then it would have left a gap, because it did not make any proposal that anything should be substituted for the value instead. That was the submission of the rating authority, and so they still contend before us, having succeeded on that submission before the Lands Tribunal. On the true meaning of this proposal, said the rating authority, the only " contention " of the ratepayers was that there should be a complete removal of the entry from the valuation list; hence the local valuation court had no jurisdiction to reduce the amount to what we now know was indeed the fair and appropriate amount for the year in question. This submission ignores, or fails to give proper effect to, the fact that the proposal went on in its printed form to say " Grounds on which that alteration is proposed." It is wrong, as an ordinary matter of construction, to segregate the two and to treat the first part as decisive of the meaning, without taking into account the second part—the grounds on which that alteration is proposed. The grounds on which that alteration is proposed were that the assessment was incorrect, excessive and bad in law. There is there, plainly, an inference. The inference is that the ratepayer is contending that, if the assessment is not " deleted "—is not reduced to a nil assessment—at any rate it should be a lesser assessment as currently appears in the valuation list. Since it is asserted that the assessment is " excessive," that plainly involves the " contention " that it should be less, and the acceptance that a lesser value may be correct.

I am wholly unable to see, despite the careful and ingenious and elaborate argument which was presented to us, that that document, on its true construction and in relation to any of the surrounding facts of which we can properly take notice, involves any ambiguity. If the ratepayer was claiming that the amount that appeared as the value of the assessment was excessive and, on the true construction of that document, was asking that, if it was not eliminated altogether, it should be reduced, I can see no reason why the local valuation court were not entitled to do that which they themselves decided that they were entitled to do. With great respect to the learned member of the Lands Tribunal who gave his decision to the contrary, I find great difficulty in following the basis on which he arrived at that decision.

I refer to one or two passages in the report of *R v Winchester Area Assessment Committee, ex parte Wright* [1948] 2 KB 455. That is a decision of this court. At p 459 Scott LJ, beginning the leading judgment, said: " It has been judicially recognised that, in passing the Rating and Valuation Act 1925 it was the intention of Parliament that an aggrieved person should be able to act upon many of its provisions, including section 37, without legal assistance." Greer LJ said in *R v West Norfolk Assessment Committee ex parte Ward* (1930) 94 JP 201 at p 203: ". . . it is not intended by the statute that his difficulties "—ie the difficulties of the person making the proposal—" should be such that he would have to go to an expert or to a lawyer to draw up his proposal."

Asquith LJ in his judgment in *R v Winchester Area Assessment Committee ex parte Wright* at the bottom of p 463 said: " But it has been laid down emphatically that a proposal need not be a strict or formal document and I think statements contained in parts of the proposal other than the space opposite ' grounds of proposal ' can be prayed in aid to eke out insufficiency at that point."

I respectfully agree and I would add that I think that, in the document with which we are here concerned, there is no basis on which it can justifiably be said that the words inserted against " Grounds on which that alteration is proposed " cannot be prayed in aid to eke out any insufficiency or any ambiguity (if there were such) in the wording of the part of the document in which the " alteration proposed " is stated. In my opinion, with great respect to the learned member of the Lands Tribunal, the decision at which he arrived was wrong and I would deal with the case in the appropriate way by answering the questions in paragraph 5 of the case stated by the Lands Tribunal, the effect of which will be to restore the decision of the local valuation court.

Agreeing, Shaw LJ said: On what is a narrow question of construction of particular words used in a particular context there is nothing to be added to the judgment of my Lord.

I cannot, however, forbear to say that it is a great pity that, when the Merseyside Valuation Panel had given its decision—which was not only, if I may say so, rational but also one which led to a just and fair result—the Liverpool City Council should have thought it necessary to pursue the matter to the Lands Tribunal. That has made it necessary for the ratepayer to come to this court so that the position might be put right. It is not as though some general proposition of law, which might in the future affect the interests of the rating authority, had to be considered. It is simply the meaning of certain words which appear to me, as they did to my Lord, to bear only one sensible meaning.

I would answer the questions posed accordingly and allow the appeals.

Waller LJ, after expressing agreement with both the judgments which had been delivered, said: I can understand that, if the words " that the assessment should be deleted from the Valuation List " stood entirely alone, it would be possible to interpret them as referring to the whole entry, as was contended by Mr Goldsmith, particularly as the ratepayers were connected with the British Transport Commission. But the words do not stand alone.

Firstly, one year later the same ratepayer put in a similar proposal using the word " assessment " and Mr Wellings, in his judgment of the Lands Tribunal, accepted that, in that notice, the words were clearly intended to refer to the values and not the whole entry. One would expect, in the normal

A case, that words used by the same authority, signed on behalf of the same man (albeit maybe by different individuals), would be intended to mean the same thing.

Secondly, there are the grounds to which my Lord has already referred "that the assessment is incorrect, excessive and bad in law." It is true that the words "incorrect and bad in law" would cover the meaning contended for by the respondents before this court (that is to say the rating authority), but the word "excessive" can only mean that the assessment should be smaller. In my opinion "excessive" clearly shows that the assessment refers only to the valuation figure, as my Lord has said, and deleting the figure would B be the best that could be achieved by the ratepayer.

Like my Lord, Shaw LJ, I express surprise at this argument. I can understand the attempt to adopt the strained meaning of the words if to do so would be to achieve justice, but in this case the adoption of the strained meaning of the words would be to achieve a valuation some £14,000 higher than what all parties agreed was the correct valuation in the following year and would obviously have been the correct valuation for that year if the documentation had been more clearly or differently set out. I find it surprising that a local authority, having had a decision from the local valuation court, should have seen fit to upset it on such grounds.

C The appeal was allowed with costs in the Court of Appeal and below.

QUEEN'S BENCH DIVISIONAL COURT

October 18 1979

(Before Lord WIDGERY CJ, Lord Justice EVELEIGH and Mr Justice KILNER BROWN)

ENFIELD LONDON BOROUGH v WOOLLS

Estates Gazette January 12 1980

(1979) 253 EG 149

D **Rating—Dispute as to which of two divorced parties was liable to pay rates on premises formerly the matrimonial home —Premises consisted of dwelling-house and garage—After divorce the court ordered wife's rights of occupation to continue, the husband having left—Subsequently a maintenance order was made and an order requiring ex-husband to transfer the premises to his ex-wife—Various undertakings were to be given by wife which were not implemented—Ex-husband continued to occupy the garage for storage of industrial equipment and occasionally used garage as a workshop—Some two years after order for transfer of premises the actual transfer was made—The magistrates held that the ex-husband had remained E in beneficial occupation of the premises during this period of approximately two years, which was the period covered by the complaint—The ex-husband and not the ex-wife was therefore liable for rates during this period—Held by the Divisional Court that on the evidence the magistrates were entitled to come to this conclusion and the rating authority's appeal was accordingly dismissed**

This was an appeal by case stated by the rating authority, the London Borough of Enfield, against the decision of Middlesex magistrates holding that the respondent, Elaine Ann Woolls, was not liable for payment of rates, for the period which was the subject of the complaint, in respect of F premises at 214 Wellington Road, Enfield.

R J Mitchell (instructed by Wilfrid D Day, town clerk, Enfield) appeared on behalf of the appellant authority; Brian Ash (instructed by Edward Kramer & Co) represented the respondent.

Giving the first judgment at the invitation of Lord Widgery CJ, EVELEIGH LJ said: This is an appeal by way of case stated from an adjudication of the justices for the Middlesex Area of Greater London for the petty sessional division of Edmonton.

On July 12 1977 the respondent answered a summons before the court pursuant to section 97 of the General Rate Act 1967 to show why she had not paid the rate specified in the com- G plaint in respect of premises at 214 Wellington Road, Enfield. The amount and how it was made up is not relevant for present purposes. Suffice it to say in so far as this appeal is concerned that it covered a period from April 5 1974 until June 24 1976. There was a period subsequent to that date included as to which liability was not denied.

The facts found were as follows. The premises comprised a dwelling-house and a garage which were not separately rated. They were registered in the Land Registry in the name of the husband. The respondent and her husband thereafter lived in the premises and for a time before June in 1972 when the husband left. On December 8 1971 a caution had already been H registered at the Land Registry pursuant to the Matrimonial Homes Act 1967 in respect of the premises in favour of the respondent wife. On June 6 1972 there was a decree nisi. The court then ordered that the wife's rights of occupation of the premises were to continue. The decree nisi was made absolute on September 8 1972. Thereafter the wife and children of the marriage remained in the former matrimonial home.

On January 29 1973 the High Court made a maintenance order in respect of the children of the marriage. The maintenance order was varied on April 5 1974 as to the amount of money payable, but otherwise it remained in force during the relevant period. On that same day the court ordered that the respond- J ent's former husband should forthwith transfer the premises to the respondent. The order which is now being placed before this court was made on the basis of undertakings being given to the court by the wife petitioner through her counsel to pay various sums of money to her former father-in-law and also to the husband. The first payment of £3,000 was to be on April 5 1975 and a further sum of £3,000 on April 5 1979. There was another undertaking that she should pay or cause to be paid to the husband a lump sum of £4,500 by April 5 1975. From the next finding it would appear that those undertakings were not complied with, for the case continues: "That order was not in fact complied with until June 24 1976 K and then upon payment by the respondent to her former husband the sum of £10,500."

The rates paid up to March 1974 had been demanded of and paid by the respondent's former husband.

After the date of the order for transfer of the property the husband continued to occupy the garage portion of the premises for storing a considerable amount of industrial equipment, including a lathe, heavy-duty bench with tools, transformers and bandsaws used by him and his father in connection with a bandsaw repair business run by them, and he occasionally used the garage as a workshop. The magistrates accepted the respondent's contention to this effect that from L April 5 1974 until June 24 1976 the husband was in beneficial occupation of the premises and therefore liable to pay the rates.

The arguments were advanced on this basis, that the husband continued to occupy the garage for his own purposes and that he was during this period discharging his duty in accordance with the order of the High Court to provide accommodation for his former wife and children by allowing them to remain in the dwelling-house. He paid no maintenance to the wife, it being taken into account by the judge that the wife was residing in the matrimonial home with the two children free of expenses.

Various authorities were quoted to the magistrates, including M the case of *Cardiff Corporation v Robinson* [1957] 1 QB 39 to which it is unnecessary to refer at the moment. Before this court counsel for the appellant has relied strongly upon the case of *Mourton v London Borough of Hounslow* [1970] 2 QB 362. In that case a maintenance order between husband and wife contained a recital of an agreement between the parties the effect of which was that so long as it subsisted the husband was constituted a licensor of the house and the former wife a licensee with an exclusive right of occupation so that she was in actual and exclusive occupation which was beneficial and had the necessary degree of permanence for rateable occu-

A pation. Those words are taken from the headnote on that case at p 362.

The court was also referred to the words of Bridge J, as he then was, at p 371 where he said:

> It follows that if the wife acquires any right to continue in occupation of the former matrimonial home after the marriage has been dissolved, it is a right of a new and different character. It may arise under an order of the court or by agreement. Superficially it may appear to resemble the former right, especially if the dissolution of the marriage has been preceded by a period of separation during which a similar order or agreement may have been operative. But to extend the principle of the *Cardiff Corporation* [1957] 1 QB 39 line of cases across the watershed of divorce, B although logically plausible on the wider view of what is meant by the obligation to maintain to which those cases refer, would in our judgment be to introduce a large and unjustifiable element of fiction into this branch of the law of rating.

It is submitted on behalf of the appellant in this case that an analogy should be drawn with the case of *Mourton v Hounslow Borough Council* because here there was an order of the court referred to by Bridge J in that passage. In consequence it is submitted that the reasoning of that passage applies to the facts of this case. On behalf of the respondent it is argued that the magistrates' decision was one of fact and that it is incumbent upon the appellant to show that the magistrates C arrived at a conclusion that no reasonable bench could have arrived at on the material before them. One therefore turns to examine the facts of this case with those competing arguments in mind.

The linchpin of the appellant's arguments rests upon the making of that court order, for it is said that, the court order having been made, the wife thereupon obtained a right of her own which superseded all that had gone before, and she was then in sole beneficial occupation by virtue of that order and she alone would be the rateable occupant.

Attractive as that argument is, it seems to me that it does not fit in with the actual facts of the case, because that court D order was not in fact complied with in its terms, and it certainly does not appear that the husband was the one who was responsible for its not being complied with, so that the position could well be at this stage that there was a continuation of the wife's former status in relation to that property. She had, it is conceded, as I understood it, up to the order of April 5 1974 been there as a licensee of the husband. It is conceded that at that point the husband did have a beneficial interest in her occupation in pursuance of his obligation to provide accommodation for the wife and the children. It is quite clear that some negotiations of some kind took place between the parties in spite of the order that had been made by the learned E judge, because, as I have said, that order was not only not immediately implemented but it was never implemented at all in its precise terms. There was a transfer of ownership some two years later.

That being so, it seems to me that there was material there from which various inferences could properly be drawn. If the magistrates drew the inference from that situation and the evidence taken as a whole, including the fact that the husband continued to occupy the garage and operated a lathe there and a heavy-duty bench with tools, it seems to me that they were entitled to come to the conclusion which they did, namely, that the husband had the degree of beneficial occupation necessary for them to come to the conclusion that he was F the rateable occupier.

Finally, I would add that this court was referred to the decision in *Wirral Borough Council v Lane* [1979] RA 261 in which it is emphasised that a finding of this nature is essentially one of fact. I would accordingly dismiss this appeal.

LORD WIDGERY CJ and KILNER BROWN J agreed.

The appeal was dismissed with costs.

COURT OF APPEAL G

November 9 1979

(Before Lord Justice MEGAW, Lord Justice SHAW and Lord Justice WALLER)

ELLESMERE PORT AND NESTON BOROUGH COUNCIL v SHELL UK LTD AND ANOTHER
SAME v UKF FERTILISERS LTD AND ANOTHER
SAME v BURMAH OIL TRADING LTD AND ANOTHER
CITY OF CHESTER COUNCIL v SHELL UK LTD AND ANOTHER
SAME v BURMAH OIL TRADING LTD AND ANOTHER

Estates Gazette January 19 1980

(1979) 253 EG 259 H

Rating—Five appeals by ratepayers to Court of Appeal from Lands Tribunal decisions—Very large sums involved—Jurisdiction of Lands Tribunal to increase assessments challenged—Figures agreed by ratepayers and valuation officer put before local valuation court—Rating authority appeared and sought to dispute agreed figures—Rating authority had not objected to any of the proposals—Local valuation court decided that figures were agreed between ratepayers and valuation officer and that rating authority had no further right to be heard—On appeal by authority to Lands Tribunal the assessments were increased—Court of Appeal invited to decide that tribunal had no jurisdiction to do so—Submissions that "contention" J of valuation officer was not proposal figure but agreed figure and that there was no justiciable issue after agreement—Submissions rejected—Rating authority's powers and position discussed—Lands Tribunal's task is to decide by considering validity of objection rather than propriety of proposal—Tribunal can arrive at a higher figure than that determined by local valuation court, but not higher than the proposal—Ratepayers' appeals dismissed—Megaw LJ's observations on procedure and admission by court of new evidence

In these five appeals the ratepayers, Shell UK Ltd, UKF Fertilisers Ltd and Burmah Oil Trading Ltd, appealed against decisions of the Lands Tribunal in favour of the rating K authorities, Ellesmere Port and Neston Borough Council and City of Chester Council. In each case the valuation officer supported the decision of the Lands Tribunal. The history of the proceedings is set out in the judgment of Waller LJ.

D G Widdicombe QC and M B Horton (instructed by the Solicitor to Shell UK Ltd) appeared on behalf of Shell UK Ltd; M B Horton (instructed by the Solicitor to Shell UK Ltd) appeared on behalf of UKF Fertilisers Ltd; C S Fay (instructed by Allen & Overy) appeared on behalf of Burmah Oil Trading Ltd; W J Glover QC and Miss S Hamilton (instructed by the Solicitor to Ellesmere Port and Neston Borough Council) represented that council; L Viscount Colville of Culross QC and Miss S Hamilton (instructed by the Solicitor to Chester City Council) represented the city council; and A P Fletcher (instructed by the Solicitor of Inland Revenue) represented the valuation officer.

Giving the first judgment at the invitation of Megaw LJ, WALLER LJ said: These are five appeals against decisions of the Lands Tribunal called on and heard together at the request of all parties. The case stated by the Lands Tribunal is in identical terms in each case and raises the same point for the decision of this court. In each case it is the ratepayer who is appealing and the rating authority and the valuation officer M who are supporting the decision of the Lands Tribunal. The ratepayers' own pipelines and oil refineries and, in the case of UKF Fertilisers Ltd, manufacturing plant and, although in reality in each case one hereditament, each is divided into a number of items which are divided between two rating authorities. The sums involved are very great; in the case of Shell UK Ltd, for instance, a total amount amounting to some £2,600,000 odd. Some of the proposals were by the ratepayers for reductions and some were by the valuation officer for new items with a stated valuation.

We were told that, over a period of three years, the rate-

A payers and the valuation officer were in negotiation and that, at the end of that time, agreement was reached between them at figures which, in the case of Shell, produced a total of some £2,150,000. Since in each case there had been an objection to the proposal by the valuation officer, or the ratepayer as the case might be, it was necessary to go to the local valuation court to effect any alteration to the valuation list. Before the local valuation court neither the valuation officer nor the ratepayer called evidence, but each put forward their agreed figures. The rating authority, however, appeared and claimed the right to ask questions and call evidence. The local valuation court, having heard argument, decided that the figures were
B agreed and that the rating authority had no further right to be heard. The rating authority appealed to the Lands Tribunal and the ratepayers took the preliminary point that the Lands Tribunal had no jurisdiction to increase the figures found by the local valuation court.

Counsel on behalf of each of the appellants submit that it is unfair and contrary to the spirit of rating legislation since 1948 that the rating authority should stand by for three years without even objecting to the proposals and then be allowed to go before the local valuation court and seek to argue and prove that the figures should be higher than that which had been agreed.

C I do not propose to refer at this stage to the earlier legislation, namely the Local Government Act 1948 and the Rating and Valuation (Miscellaneous Provisions) Act 1955, because they can only be relevant if there is an ambiguity in the current statute, namely the General Rate Act 1967 and, in particular, Part V. Section 67 deals with the publication of a new valuation list and then there follow a number of sections dealing with alterations to the current valuation list. Section 69 deals with proposals for alteration of the current valuation and enables anyone aggrieved to make a proposal for amendment either by adding a hereditament or altering its description or value.

Section 69 reads:

D (1) Subject to subsection (6) of this section, any person (including a rating authority) who is aggrieved—

(a) by the inclusion of any hereditament in the valuation list; or
(b) by any value ascribed in the list to a hereditament or by any other statement made or omitted to be made in the list with respect to a hereditament; or
(c) in the case of a building or portion of a building occupied in parts, by the valuation in the list of that building or portion of a building as a single hereditament,

may at any time make a proposal for the alteration of the list so far as it relates to that hereditament.

(2) Subject to subsection (6) of this section, the valuation officer may at any time make a proposal for any alteration of a valuation
E list and in particular, in addition to the proposals authorised or required by, or by virtue of, the following provisions of this Act, namely, paragraph 6(1) of Schedule 1, paragraph 4(1) of Schedule 4, paragraph 8(4) of Schedule 5, paragraph 13 of Schedule 6 and paragraph 15 of Schedule 7, shall from time to time make such proposals as may be requisite—

(a) for deleting from the list any premises exempted from rating by virtue of section 33(1)(b) of this Act;
(b) for excluding from the list any premises which form part of a hereditament shown in the list and which, by virtue of section 33(1)(a) or (b) of this Act, are not liable to be rated, and for including in the list, as one or more separate hereditaments, so much of any such hereditament as remains liable to be rated;
F (c) for altering the list in consequence of any event whereby premises cease to be within the exemption from rating conferred by section 32(3), 33(1) or 34(1) of this Act.

(3) Without prejudice to any right exercisable by rating authorities by virtue of subsection (1) of this section, where—

(a) it appears to a rating authority that a hereditament in their rating area which is not included in the list ought to be included therein; and
(b) the valuation officer gives notice in writing to the rating authority that he does not intend to make a proposal for inserting that hereditament in the list,

the rating authority, at any time within twenty-eight days after the date on which that notice was given, may make a proposal for the alteration of the list by the insertion of that hereditament therein. G

....

(5) Every proposal under this section must—

(a) be made in writing; and
(b) specify the grounds on which the proposed alteration is supported; and
(c) comply with any requirements of any regulations made by the Minister with respect to the form of such proposals and otherwise with respect to the making thereof,

and every such proposal made otherwise than by the valuation officer must be served on the valuation officer.

Section 70 makes provision for objections to proposals:

(1) The valuation officer shall, within twenty-eight days after the H date on which a proposal under section 69 of this Act is served on him, or within seven days after the date on which such a proposal is made by him, as the case may be, transmit a copy thereof together with a statement in writing of the right of objection conferred by subsection (2) of this section, to each of the following persons, not being the maker of the proposal, that is to say—

(a) the occupier of the hereditament to which the proposal relates; and
(b) the rating authority for the area in which the hereditament in question is situated.

And by section 70(2) the owner or occupier or the rating authority may within 28 days of service serve notice of objection on the valuation officer. Where the proposal was J made by the ratepayer the valuation officer must give him notice of objection by the rating authority. The valuation officer's duty is under section 74(1) and he has to object within three months. In this case the proposals for alterations were made by both the ratepayer (in all cases) and the valuation officer in the cases of proposed new entries in the valuation list. The ratepayer objected to the valuation officer's proposals and the valuation officer in due course objected to the ratepayer's proposals but no objection was made by the rating authority.

Section 71 deals with unopposed proposals: K

(1) Where in the case of any proposal under section 69 of this Act—

(a) no notice of objection is served within the time limited by section 70(2) of this Act, or every such notice is unconditionally withdrawn; and
(b) either—
   (i) the proposal was made by the valuation officer; or
   (ii) the valuation officer is satisfied that the proposal is well-founded; or
   (iii) at the end of the period of four months beginning with the date on which the proposal was served on the valuation officer, that officer has not given a notice under section 74(1) of this Act,

the valuation officer shall cause the valuation list to be altered so L as to give effect to the proposal.

There then comes section 72 which deals with agreed alterations after proposals but before determination by the local valuation court. If the requirements of the section are fulfilled it becomes the duty of the valuation officer to cause the alteration to be made in the valuation list.

(1) Where, in the case of any proposal under section 69 of this Act, the requirements of section 71 of this Act are not satisfied, but—

(a) all the persons referred to in subsection (2) of this section agree on an alteration of the valuation list (whether the alteration is that specified in the proposal or alteration); and M
(b) the agreement is reached without, or before the determination of, any appeal to a local valuation court, or reference to arbitration, with respect to an objection to the proposal,

the valuation officer shall cause that alteration to be made in the valuation list.

(2) The persons referred to in subsection (1)(a) of this section are—

(a) the valuation officer;
(b) the person who made the proposal, where the proposal was not made by the valuation officer;
(c) any person who has served and who has not unconditionally withdrawn a notice of objection to the proposal;

(d) the occupier of the hereditament to which the proposal relates, if he is not included by virtue of paragraph (b) or (c) of this subsection;

(e) the rating authority (if not included by virtue of paragraph (b), (c) or (d) of this subsection), unless they have notified the valuation officer that they do not desire to be included by virtue of this paragraph either generally or as respects a class of hereditament which includes the hereditament to which the proposal relates.

It is not suggested that in this case section 72 had been complied with. Under section 73 the valuation officer is under a duty to transmit proposals to which there has been an objection to the clerk to the local valuation court and section 74 makes provision for proposals to which the valuation officer objects to be so transmitted. Under either section the transmission has effect as an appeal to the local valuation court by virtue of section 76(1):

(1) Where a copy of a proposal is transmitted to the clerk to a local valuation panel and by virtue of section 73(2), 74(3) or 75 of this Act that transmission has effect as an appeal to a local valuation court against an objection to the proposal, it shall be the duty of the chairman or a deputy chairman of that panel to arrange for the convening of such a court.

(2) The procedure of a local valuation court shall, subject to any regulations made in that behalf by the Minister, and subject to subsection (3) of this section, be such as the court may determine; and the court—

(a) shall sit in public, unless the court otherwise order on the application of any party to the appeal and upon being satisfied that the interests of one or more parties to the appeal would be prejudicially affected; and

(b) may take evidence on oath and shall have power for that purpose to administer oaths.

....

(4) On the hearing of an appeal to a local valuation court—

(a) the appellant; and

(b) the valuation officer, when he is not the appellant; and

(c) the owner or occupier of the hereditament to which the appeal relates, when he is not the appellant; and

(d) the rating authority for the rating area in which the hereditament in question is situated, when that authority are not the appellant; and

(e) the objector, where he is not one of the persons aforesaid,

shall be entitled to appear and be heard as parties to the appeal and examine any witness before the court and to call witnesses.

(5) Subject to the provisions of this Act, after hearing the persons mentioned in subsection (4) of this section, or such of them as desire to be heard, the local valuation court shall give such directions with respect to the manner in which the hereditament in question is to be treated in the valuation list as appear to them to be necessary to give effect to the contention of the appellant if and so far as that contention appears to the court to be well founded; and the valuation officer shall cause the valuation list to be altered accordingly.

As I have already indicated, the rating authority exercised their right to appear and be heard as parties to the appeal, but the local valuation court, having heard argument, decided that the figures were agreed. In effect, the question which is posed for the opinion of this court is: were the local valuation court right in deciding that the agreed figures stood?

The case for the appellants is put in two ways: (1) in the case of proposals for new entries and valuations made by the valuation officer the "contention" of the valuation officer was not the proposal figure but was the agreed figure or some lower figure. This argument is based on the wording of section 76(5) where it says ". . . the local valuation court shall give such directions with respect to the manner in which the hereditament in question is to be treated in the valuation list as appear to them to be necessary to give effect to the contention of the appellant if and so far as that contention appears to the court to be well founded; . . ." It was submitted that, once the valuation officer reduced his own figure, his "contention" was the new figure as a maximum; and it was not open to the local valuation court to go above the new figure because its jurisdiction was limited to "so far as that contention appears to the court to be well founded"; (2) it was submitted, in the case of proposals by the occupiers for a lower figure than in the valuation list, once there was agreement with the valuation officer there was no justiciable issue, ie no dispute. This argument did not depend on the meaning of the word "contention," but on the simple submission that there was no issue for the local valuation court to try.

On the meaning of "contention" it was submitted that, once the valuation officer agreed a lower figure than that in the valuation list or in his proposal, as the case might be, this lower figure would be the limit of his contention and it would not be possible to say that the higher figure in the proposal was his contention. The case of *Morecambe and Heysham Corporation v Robinson (VO)* [1961] WLR 373, when carefully considered, does not support this limited view of contention. Furthermore, the decision is really on the question of whether contention refers to the figures or refers to the arguments. I do not find any help from other authorities cited to us. In this case the valuation officer published a valuation list with valuations in it. These valuations were communicated to all concerned including the rating authority. The occupier made a proposal of a nominal figure. In my judgment, the limits of contention were contained in the valuation list figure on the one hand and the figure in the proposal on the other and, unless all parties agreed, no agreement between the two parties could alter that. In other instances, where the valuation officer made proposals, the limits of contention would be that figure on the one hand and that made in any objection on the other.

We were referred also to the case of *Ellerby v March* [1954] 2 QB 357, a decision which was concerned with appearances before the Lands Tribunal. There the argument is on appeal from the local valuation court which is the court of first instance. In my opinion the decision does not apply to proceedings before the local valuation court.

Subsidiary arguments by the ratepayers that the words of section 76(4) entitling the rating authority to be present at the local valuation court do not give the rating authority the right to attack and overrule an agreement between the valuation officer and the ratepayer were to the effect that there were other possible ways of achieving an alteration without the rating authority's concurrence. One argument was based on making a number of proposals for one hereditament and then withdrawing all but one. Another was founded on section 81(1) where there could be a partial withdrawal of the proposal. In my opinion these arguments do not assist the appellants. The fact that it is possible to point to one anomaly is no argument for ignoring or restricting the meaning of plain words in the statute. It is implicit throughout the part of the Act that I have set out above that no conclusion can be reached without the co-operation of the rating authority. The rating authority could adopt one of three possible courses of action: firstly, it could object in accordance with the provisions of section 70 and thereafter it would be an active party to the dispute; secondly, it could notify the valuation officer that it did not desire to be included (see 72(2)(e)); thirdly, it could do nothing until it could see whether things were going wrong, ie whether their rateable values were to be reduced to figures which they regard as being too low. If they were, then the rating authority could attack the figure and call evidence before the local valuation court. The sections I have mentioned above require that at every stage the rating authority must be involved unless it has notified the valuation officer that it does not desire to be included (see section 72(2)(e)).

I said earlier that I was not looking at the statutes which preceded the 1967 Act unless there was an ambiguity about the General Rate Act 1967. Although there may be some anomalies, I do not find any ambiguity about the words which say that the rating authority "shall be entitled to appear and be heard as parties to the appeal and examine any witness before the court and to call witnesses" and I therefore do not need to examine these earlier statutes.

I am of the opinion that the president of the Lands Tribunal came to a correct decision in law and that the Lands Tribunal has jurisdiction to increase the assessments which are the subject of these appeals.

Since preparing this judgment I have had the advantage of reading in draft the judgments which Megaw and Shaw LJJ are about to deliver and I agree with them both.

Agreeing, SHAW LJ said: The scheme embodied in Part V of the General Rate Act 1967, for the maintenance and alteration of valuation lists, recognises that, apart from other possible parties, the valuation officer, the ratepayer and the rating authority must always have an interest in the compilation and alteration of a valuation list which relates to the area of that authority and is a hereditament for which the ratepayer has a liability. This concurrence of interest is made clear from a cursory examination of the relevant provisions of the Act beginning with section 69.

Where a proposal has been made and has been met with a notice of objection which is persisted in, the valuation officer is required, within the limits of time prescribed by section 73, to transmit a copy of the proposal and of every notice of objection to the local valuation panel, and by section 76(1) this step in the procedure prescribed under the Act "has effect as an appeal to a local valuation court against an objection to the proposal."

By section 76(5), after due hearing (a matter which gives rise to one aspect of this appeal and to which I shall later refer) "the local valuation court shall give such directions . . . as appear to them to be necessary to give effect to the contention of the appellant if and so far as that contention appears to the court to be well founded."

As I understand these provisions they not only determine and delimit the function and jurisdiction of the valuation court but they also define the role of the party responsible for the proposal. That role is to challenge and refute the objections and, as necessary corollary, to support the proposal. The function of the valuation court is to assess the impact of the contentions of the appellant (and in particular where the appellant is the valuation officer) upon the force of the objections. The court may reject the objections or accept them in whole or in part with the consequence that the proposal may be left intact or may be reduced or obliterated. Even if the appellant does not seek to support his proposal in its entirety it is still for the court to decide to what extent the objections to it have been refuted by facts proved before them. The appellant cannot modify his proposal (unless he withdraws it altogether) save as provided by section 72; and the court cannot determine upon a more onerous proposal than that which was objected to since it is the objections to that proposal which constitute the matter in contention rather than the primary validity of the proposal itself.

Thus the jurisdiction of the court is measured by the area of controversy lying between the proposal and the matters of objection. The court will refuse to give effect to these last to the extent that the appellant's contentions appear to the court to be well founded. *Prima facie* the contention of the appellant must be that the objections are ill-founded. Otherwise he is virtually putting forward a new proposal and must follow the procedural requirements appertaining to a fresh proposal unless there is an agreement to which all interested bodies or persons are parties in accordance with the requirements of section 72(2) of the Act. It is pertinent to observe that section 72(1)(b) provides that such an agreement may be reached without, or before the determination of, any appeal to a local valuation court with respect to an objection to the proposal.

No such agreement to which the rating authority was a party was made in the present case. The agreement between the valuation officer and the ratepayer could be no more than persuasive before the valuation court. Their jurisdiction remained what it had been from the outset, namely, to allow the appeal in whole or in part and by rejecting or upholding the objections in whole or in part correspondingly to uphold the proposal in its entirety or partially.

As to the operation of subsection (4) of section 76 it seems to me plain that any of the persons or bodies enumerated in paragraphs (a), (b), (c), (d) and (e) are entitled to intervene in the appeal and to be heard as parties. Whether they support or oppose the objections will depend on their interest; but the valuation court will take account of the results of their intervention in deciding whether and in what measure the contention of the appellant that the objections are untenable is well founded.

I see no substance in the suggestion made by the appellants that it is unfair on the part of the rating authority to come in as a party to the appeal when they have taken no earlier action. There was neither need nor obligation to do so. When the proposal has been made and met with an objection the issue is defined and all parties whose interests may be affected have a status to intervene at the hearing.

When the matter is taken to the Lands Tribunal its jurisdiction corresponds to that which the valuation court had when the appeal before it was constituted; this may be stated shortly in the formula that it is to decide the outcome by consideration of the validity of the objections rather than by addressing itself to the propriety of the proposal. It may therefore, in the result, arrive where the amount is in issue at a higher figure than that determined by the valuation court. What it cannot do is to find a figure higher than the proposal itself, for it, like the valuation court, has no jurisdiction to entertain a different proposal from that to which objection has been made.

It follows that the answer to the question posed by the case in each of the appeals is in the affirmative and that the Lands Tribunal can make a higher assessment than that determined by the valuation court so long as their assessment so found is not greater than that contained in the proposal.

I would accordingly dismiss these appeals.

Also agreeing, MEGAW LJ said: There are two procedural matters to which I should refer. The responsibility of formulating, for consideration by the president of the Lands Tribunal, the question or questions of law on which they sought to contend before this court that the decisions were wrong rested with the respective appellants. It is unfortunate that the appellants formulated one general question, instead of formulating two separate questions. This is particularly to be regretted since the appellants put in the forefront of their criticism of the president's decision the submission that he failed to recognise that there were two separate points of law. Before us it emerged that the appellants' submissions involved that if we were to decide one of the two points of law against them, but were to uphold their submission on the other point of law, it would be impossible for the court to answer the question asked other than by some such formula as "the Lands Tribunal has jurisdiction unless . . ." or "the Lands Tribunal does not have jurisdiction except if and in so far as. . . ."

We were asked on behalf of one of the appellants, Shell UK Ltd, by original motion, for leave to adduce additional evidence in the form of a lengthy affidavit and exhibits. Since an appeal to this court from the Lands Tribunal is only on questions of law, and since questions of fact are for the tribunal itself, it is obviously in the highest degree undesirable that, in the absence of very exceptional circumstances, this court should admit evidence not given before the tribunal. The court has jurisdiction to do so; but it should be very sparingly exercised. If evidence tendered in the Lands Tribunal has been rejected, and it is desired to contend that the rejection was wrong, that should be dealt with as a question of law: not as a ground for simply resubmitting the evidence in this court.

We decided that the only course open to us was to look at the proposed evidence provisionally. Two of the parties to these appeals, the valuation officer and one of the rating authorities, said that, if the evidence were to be admitted, they would wish to put in affidavits in reply on certain matters of fact. In the end, they did not do so, perhaps because it became

A evident that the evidence was, at best, of little real significance to the issues of law which we had to decide.

The proposed evidence turned out to be, to a considerable extent, argument, which, if relevant, should not have been put forward in an affidavit. The affidavit dealt also with facts and figures concerned with the particular assessment. Such facts and figures, it was, I think, accepted on behalf of the party seeking to adduce the evidence, could not directly affect a decision of general principle to be reached on the construction of statutory provisions: which principle, when ascertained, must be equally applicable to all assessments, large or small, complex or simple. It was contended that some of the facts set out in B the affidavit would be helpful to the court as what was called "background" in reaching its decision on the issues of law. In the end, I think it is right to say that the affidavit was used only for the purpose of drawing illustrations, or examples, from an exhibit thereto.

We were not asked to give a definite ruling, following our decision to look at the evidence provisionally. I think we ought now to give such a ruling. In all the circumstances, it seems to me to be better to treat the evidence as admitted. But if I had thought that anything in it was material to the decision of this court, as putting in evidence facts which were not before the Lands Tribunal, I should have regarded it as necessary to C consider further, with the assistance of counsel, whether the proper course would not have been to remit the relevant appeal to the Lands Tribunal for further consideration on the basis of the evidence now adduced.

The point which, logically, I should have thought, was the appellants' first point was dealt with second in the arguments for the appellants and has thus been dealt with as the second point in the judgment of Waller LJ. If it were right, the answer to the question of law posed would be that the Lands Tribunal have no jurisdiction to increase any of the figures found by the local valuation court in any of the appeals. I have been unable to understand how, in the absence of some D express provision in section 76(4) of the General Rate Act 1967, affecting the position in cases such as those with which we are concerned, it can be said that the words of that subsection are consistent with the appellants' contention that the rating authority is precluded from being heard as parties to the appeal: that is, from saying, if they see fit, "we require the appellant ratepayer to prove his case on this appeal." It was suggested, if I understood the submission correctly, that the rating authority could, indeed, "be heard as parties to the appeal," to the extent of challenging the validity of the agreement as to figures between the ratepayer and the valuation officer. I do not understand either what sort of "invalidity" E is here contemplated, or why the words of the statute should be treated as being so limited in effect. In substance, also, as I see it, the appellants' argument on this point cannot stand in the light of the provisions of section 72 of the Act. They are intended to provide for agreed alterations of the proposals, and not to leave it open for the rights of a rating authority under section 76(4) to be circumvented by a purported "agreed alteration of the proposals" which does not comply with the requirements of section 72 as to such agreement.

As regards the appellants' other points, it would, if right, as I have already said, not require or entitle this court to say that the Lands Tribunal had no jurisdiction to increase the values F accepted by the local valuation court. The court would have to answer the question with a qualification. If I understood correctly, it was accepted at the end of the argument before us that the question was whether "the contention" of the appellant in section 76(5) is to be treated as being the proposal for the alteration of the list as set out in what I may call the relevant "pleading," or is the figure (where it is an appeal as to the proper figure) which the appellant is putting forward before the local valuation court at the final moment before the local valuation court has to consider its decision. I think that both reason and convenience support the view that it is the former.

The appeals were dismissed. G

The ratepayers were ordered to pay the costs of the rating authorities and the valuation officer. An application on behalf of the ratepayers for leave to appeal to the House of Lords was refused.

QUEEN'S BENCH DIVISIONAL COURT

March 26 1979

(Before Mr Justice MICHAEL DAVIES and Mr Justice NEILL)

THE POST OFFICE v OXFORD CITY COUNCIL H

Estates Gazette February 2 1980

(1979) 253 EG 479

General Rate Act 1967 as amended—Whether rating surcharge in respect of unused "commercial building" under section 17A was payable on premises owned by Post Office—Exemption from surcharge claimed on ground that premises constituted a hereditament constructed or adapted for use wholly or mainly for industrial purposes—Premises had been used as garage premises for sale and repair of cars, petrol and motor accessories—Over 70 per cent of the floor area had been devoted to repair of motor vehicles—Submission that J the exemption referred to premises on which manufacture had been carried on rejected—Held that the premises were exempt from the rating surcharge imposed by section 17A—Appeal from decision of magistrates to issue a distress warrant allowed

This was an appeal by case stated from a decision of Oxford magistrates to issue a distress warrant against the Post Office on the application of Oxford City Council for failure to pay a rating surcharge alleged to be due on premises known as Morris Garage, St Aldates, Oxford, owned by the Post Office and vacant since December 1975.

G Seward (instructed by the Solicitor to the Post Office) K appeared on behalf of the appellant; D Micklethwait (instructed by the City Solicitor, Oxford City Council) represented the respondent.

Giving the first judgment at the invitation of Michael Davies J, NEILL J said: This is an appeal by way of case stated from a decision of the Oxford City Justices given on March 3 1977 on a complaint by the Oxford City Council against the Post Office for failure to pay a rating surcharge for the period December 25 1975 to March 31 1977, amounting in total to the sum of £26,086.79. The Post Office disputed liability on the ground that the premises in question, Morris L Garage, 69-74 St Aldates, Oxford, were not subject to the surcharge. The justices upheld the claim by the city council and agreed to the issue of a distress warrant. The Post Office appeal to this court.

The issue between the city council and the Post Office can be shortly stated. Section 17A of the General Rate Act 1967 imposed a rating surcharge on certain unused commercial buildings. An exception from this surcharge was given, however, by regulation 3 of the Rating Surcharge (Exemption) Regulations 1974 (SI 1974 no 1563) in respect of hereditaments which have been and remain constructed or adapted as factories, mills or other premises of a similar character used wholly or mainly for industrial purposes. The question M is whether these premises come within that exemption. It is common ground that if the premises are not covered by the exemption this surcharge is payable.

Before I turn to the facts, I should refer in more detail to the provisions of the General Rate Act 1967, which is the principal Act which governs the rating of property. This Act was a consolidating Act which brought into a single statute provisions which had previously been contained in a number of statutes which date back to the Poor Relief Act 1601. The 1967 Act came into force on April 1 1967. For the purpose of ascertaining the net annual value, hereditaments were

A divided into two categories. The two categories were set out in the General Rate Act 1967, section 19(2) and section 19(3). The first category (to be found in section 19(2)) was as follows: "a hereditament consisting of one or more houses or other non-industrial buildings, with or without any garden, yard, court, forecourt, outhouse or other appurtenance belonging thereto, but without other land." The second category comprised "any other hereditament": see section 19(3).

Section 19(6) of the 1967 Act contained a definition of the words "non-industrial building." For the purposes of this case it is sufficient to read part only of that definition. Section 19(6) provided: "In this section, the following expressions have the following meanings, respectively, that is to say B —'non-industrial building' means a building, or part of a building, of any description other than—(a) factories, mills and other premises of a similar character used wholly or mainly for industrial purposes . . ."—it is not necessary for me to read the rest of that subsection.

In 1974 two new sections were added to the General Rate Act 1967. By section 16 of the Local Government Act 1974 a new section, 17A, was added. Subsection (1) of that section was in these terms:

> If for a continuous period exceeding six months a commercial building is not used for the purpose for which it was constructed C or has been adapted, its owner shall pay in respect of that period (the "period of non-use") a surcharge additional to the rates (if any) payable apart from this section.

The words "commercial building" in that subsection were defined in the new section 17B(2) as follows:

> References in section 17A of this Act and this section to a commercial building are references to a hereditament (not being a dwellinghouse, or a hereditament having a floor space not exceeding 240 square feet and used as a lock-up garage) whose net annual value falls to be ascertained under section 19(2) of this Act. . . .

It will be remembered that factories, mills and other premises of a similar character used wholly or mainly for D industrial purposes are not "non-industrial" buildings within section 19(2) and are therefore within the scope of "any other hereditaments" in section 19(3). But the exclusion from the definition of non-industrial buildings is limited to factories, mills and other premises which are in active use. Accordingly a hereditament which in active use would fall outside the definition of non-industrial building might, when disused, be within the definition of a commercial building in section 17A, and therefore, in appropriate circumstances, attract liability for the surcharge.

Later in 1974, however, after the passing of the Local Government Act 1974, regulations were made, pursuant to E powers conferred by the General Rate Act 1967, which contained exemptions from the section 17A surcharge. Regulation 3 of the Surcharge (Exemption) Regulations 1974 was in the following terms:

> No surcharge shall be payable under section 17A of the General Rate Act 1967 in respect of—(a) hereditaments which have been and remain constructed or adapted as factories, mills or other premises of a similar character for use wholly or mainly for industrial purposes, or (b) . . . .

This appeal is concerned with the scope of paragraph (a) in regulation 3.

F The relevant facts are set out in the case stated and they can be summarised as follows. The premises were built in or about 1932 for Morris Garage Ltd and were used from that date until December 1975 for the preparation for sale and the sale of new and used cars to members of the public, for the repair of vehicles belonging to members of the public and the sale to the public of petrol and accessories. At no time were vehicles manufactured on the premises. In 1971 the premises were sold to the Post Office and were then leased back to Morris Garage Ltd until the lease expired in December 1975. The premises have been vacant since then.

The premises were on three floors. The ground-floor area consisted of a car showroom, used-car area, administration offices, repairs and servicing area, parts store and petrol filling area. The first-floor area consisted of a store area immediately above the ground-floor parts store. The second-floor G area consisted of a paint-spray shop and a panel-repair shop. Those areas are depicted on a plan which is attached to the case. The total area of the premises amounted to 39,297 sq ft. That was divided—as shown in an agreed document—into 28,510 sq ft of industrial and subsidiary accommodation and 10,787 sq ft of non-industrial accommodation. In percentage terms the proportion of the total floor area consisting of industrial and subsidiary space amounted to 72.5 per cent.

The staff at the premises was as follows: 12 staff in the showroom area, 9 staff in the paint-spray area and about 20 H staff plus labourers in the workshop area.

The premises were registered under the Factories Act 1961. Thus the premises fell within the definition of a factory in the Factories Act, which provided in section 175(1): "Subject to the provisions of this section, the expression 'factory' means any premises in which, or within the close or curtilage or precincts of which, persons are employed in manual labour in any process for or incidental to any of the following purposes . . .", and then in subparagraph (b) was set out, among other things, "the repairing of any article." Though the premises were so registered, they were also at all material times assessed under section 19(2) of the General Rate Act 1967—that is J the subsection which deals with non-industrial premises.

It was argued on behalf of the city council that the premises were not within the scope of the exemption in paragraph (a) of regulation 3. There is no definition of the word "factory" in the 1967 Act, and the ordinary meaning of the word, said Mr Micklethwait on behalf of the city council, imports the carrying on of some manufacturing process. A factory is a place where things are made. Secondly, he submitted that the second characteristic of a factory is that it is not a place where the public go. Moreover in ordinary parlance, a garage mechanic is not a factory worker.

In the course of the submissions which were made to the K court reference was made to the decisions in the House of Lords in *Turpin v Middlesbrough Assessment Committee and Bailey* [1931] AC 451 and *Kaye (Barnsley Revenue Officer) v Eyre Brothers Ltd* [1931] AC 475. In both these cases the House was concerned with garage premises of a type similar to those involved in this case. The decisions in those cases were based on the wording of section 3 of the Rating and Valuation (Apportionment) Act 1928.

At p 471 of the *Turpin* case Lord Dunedin said "The hereditament is registered as and is clearly a factory within the Factory and Workshop Acts 1901 to 1920. It is accordingly an 'industrial hereditament' within section 3, subsection (1) L of the Act of 1928, unless it is primarily occupied and used for any of the purposes enumerated in the proviso or for any combination of such purposes." The House decided that the premises in question were outside the definition of "industrial hereditament" because they were used for the purpose of a retail shop.

I do not consider that those decisions assist us in the matter which we have to decide in this case. It is to be noted that section 3 of the 1928 Act was repealed by the Rating and Valuation Act 1961.

The question for our consideration is whether, in the absence M of any statutory definition, the premises with which we are concerned fall within the scope of paragraph (a) of regulation 3. I am satisfied that they do. I do not, in the context of this regulation, accept the narrow definition of the word "factory" which Mr Micklethwait has submitted was the appropriate meaning. In my judgment the premises, which were used to the extent of 72.5 per cent for the repair of motor vehicles and which included a repair workshop of some 16,000 sq ft and a paint-spray shop and a panel workshop of a combined area of 7,000 sq ft, were properly to be regarded as a factory for use wholly or mainly for industrial purposes. Moreover, even if the premises were not a "factory," I am satisfied that

A they constituted "other premises of a similar character for use wholly or mainly for industrial purposes." I am persuaded, therefore, that the premises with which we are concerned do properly come within the scope of paragraph (a) of regulation 3.

I would allow this appeal and would answer the question posed in the case stated by the justices: "The question for the court is whether we came to a correct determination in law on the facts found" in the negative.

MICHAEL DAVIES J agreed.

The appeal was allowed with costs in the Divisional Court. The matter was remitted to the magistrates with a direction to refuse the distress warrant and with the court's opinion
B that the order for costs made by the magistrates in favour of the respondent Post Office should be reversed. Leave was given to appeal to the Court of Appeal.

CHANCERY DIVISION

October 18 1979

(Before Mr Justice DILLON)

C WESTMINSTER CITY COUNCIL v HAYMARKET PUBLISHING LTD

Estates Gazette February 23 1980

(1979) 253 EG 805

General Rate Act 1967 as amended, sections 17A and B—Incidence of liability for rates on unused commercial premises —Rating surcharge registered by plaintiff rating authority as a local land charge in accordance with the Land Charges Act 1925—Owners of premises subject to charge mortgaged them to a bank and the bank in exercise of its power of sale as mortgagee subsequently sold them to the defendant company
D —Was the defendant company's interest bound by the charge? —Question depended on whether the charge was binding on all interests in the land or only on the mortgagor's equity of redemption—Held, following authorities on charges under Public Health Acts and similar statutes, that the charge was binding on all interests—As defendant company derived title through mortgagee bank, it was bound by charge

In this action Westminster City Council sought a declaration that the council's charge for £16,940.93, being the amount of the rating surcharge in respect of unused commercial properties at 17 Lancaster Gate and 12 and 13 Lancaster Mews, London W2, was binding on the defendant company,
E Haymarket Publishing Ltd, as the registered proprietors of the premises.

W J Mowbray QC and C Braham (instructed by the solicitor to Westminster City Council) appeared on behalf of the plaintiffs; P Millet QC and G Hill (instructed by Wilde, Sapte & Co) represented the defendants.

Giving judgment, DILLON J said: I am concerned in this case with the scope and priority of the charge in respect of the rating surcharge on unused commercial property, which is imposed by sections 17A and B of the General Rate Act 1967. These sections were introduced into the 1967 Act by the Local Government Act 1974 and have effect from February
F 8 1974. Their relevant provisions are as follows:

Section 17A(1):

If for a continuous period exceeding six months a commercial building is not used for the purpose for which it was constructed or has been adapted, its owner shall pay in respect of that period—

the "period of non-use"—

a surcharge additional to the rates (if any) payable apart from this section.

(2) Subsection (1) of this section will not apply where—
(a) the owner has tried his best to let the building, or
(b) . . .

another condition which is not relevant in the present case.

In subsections (3) and (4) there are alternative conditions. G
Subsection (3) applies where the owner is in occupation of the building throughout the period of non-use, and subsection (4) applies where the owner is not in occupation of the building throughout the period of non-use.

Section 17B(3) says:

A surcharge imposed under section 17A of this Act in respect of a hereditament shall until recovered be a charge on the land comprised in the hereditament; and for the purposes of the application to such a charge of section 15 (registration of local land charges) of the Land Charges Act 1925 this Act shall be deemed to be a similar statute to the Acts mentioned in subsection (1) of that section. H

Section 17B(6) makes Schedule 1 to the General Rate Act 1967 apply for the purposes of section 17A with minor modifications, and then section 17B(7) provides that in sections 17A and B "owner" means "the person entitled to possession, and where different persons are entitled to possession of a hereditament during different parts of a period of non-use, a surcharge in respect of that period shall be apportioned between them according to the length of each part and levied accordingly."

In Schedule 1 to the General Rate Act 1967, paragraph 13 provides:

Any amount due . . . shall, without prejudice to the operation J
of any other enactment under which it is recoverable, be recoverable as a simple contract debt in any court of competent jurisdiction.

The premises with which I am concerned in these proceedings are 17 Lancaster Gate and 12 and 13 Lancaster Mews, London W2. These premises were acquired in fee simple by Shop Investments Ltd, which I shall call "the company," on January 3 1974 and on the next day they were charged by the company by way of legal mortgage with the payment to National Westminster Bank Ltd, which I shall call "the bank," of all moneys and indebtedness due from the company to the
bank. The legal charge to the bank contains nothing to cut K
down the ordinary immediate right of a legal mortgagee to possession of the mortgaged property.

On February 8 1974, when sections 17A and B came into effect, the premises, which were commercial premises, were empty and unused and it is common ground that they remained empty and unused until October 24 1975 and that the rating surcharge is payable by the owner in respect of the premises for that period. The total amount is £16,940.93. The premises remained empty after October 24 1975, but it is common ground that no surcharge is payable in respect of time after that date because of the efforts then being made to let the
premises. On March 19 1976 the bank served a demand on L
the company for the payment of the amount due under the legal charge. This amount was at all material times in excess of £6m because the company had guaranteed its parent company's obligations to the bank. On April 21 1976 the local authority served a demand on the company for the payment of the £16,940.93 rating surcharge, and on the following day the local authority's charge in respect of the surcharge was duly registered in the Land Charges Registry as a local land charge. The amount has not yet been paid.

On July 15 1977 the bank in exercise of its powers as legal mortgagee contracted to sell the premises to the present
defendants, Haymarket Publishing Ltd, for a sum of £220,000. M
A few weeks later this contract was duly completed and the defendants are now the registered proprietors of the premises at H M Land Registry. In the July 1977 contract the bank agreed to indemnify the defendants against the local authority's charge for the rating surcharge if that charge is binding on the defendants.

By virtue of section 17B(3) of the General Rate Act 1967 the surcharge while unpaid is declared to be a charge on the land. It is common ground between Mr Mowbray for the local authority and Mr Millett for the defendants that this wording can only mean either (1) a charge on all interests,

A legal or equitable, in the premises including the bank's interest as mortgagee, or (2) merely a charge on the interest, such as it was at the relevant time, of the owner in the premises. Mr Mowbray's first contention is that the charge is a charge on all interests in the premises including the bank's. It is common ground that if this is correct the charge binds the defendants because the charge on this contention had priority to the bank's interest and the defendants' claim under the bank. Mr Mowbray's second contention is that, even if the charge is only a charge on the interest of the owner in the premises, the bank was at all relevant times an owner or the owner of the premises whether or not the company was also an owner. B It is common ground that, if this be right, again the charge binds the defendants.

Mr Millett by contrast contends that the charge is only a charge on the interest of the owner in the premises, that under the scheme of the Act only one interest at any one time can be the interest of the owner and that the owner at the relevant times was the company, whose interest was merely the equity of redemption subject to the bank's charge. Mr Mowbray concedes that, if that be right, the defendants are freed from the local authority's charge.

In support of his second contention Mr Mowbray relies on the well-established rule that by virtue of his legal estate C a legal mortgagee is entitled to possession of the mortgaged property even before the ink is dry on the mortgage—certainly before demand and before any default on the part of the mortgagor. I prefer Mr Millett's view that it is inconsistent with the scheme of the Act for there to be several owners of a property at the same time, not being mere joint owners of the same legal estate like partners or trustees.

Subsections (3) and (4) of section 17A are concerned in the alternative with the situation where the owner is or is not in occupation of the premises, and I do not find these readily workable if there is one owner who is in occupation and another owner in respect of a different interest who is not in D occupation. Subsection (2) refers to the situation where the owner had tried his best to let the building and this hardly fits the situation where there is one owner, the mortgagor, who is trying his best to let and another owner, the mortgagee, who has not been concerned to do anything because the interest under his mortgage had been duly paid. Moreover, I cannot conceive that Parliament intended to impose a personal liability to pay the surcharge recoverable as a simple contract debt on a mortgagee, let alone a second or subsequent mortgagee, who was not at the relevant time a mortgagee in possession in the accepted sense of that term. However, it is unnecessary for me to express any final view on this since E I have reached a clear conclusion that Mr Mowbray is right on his first contention.

In support of this contention Mr Mowbray has referred me to a line of established authorities on earlier Acts, such as Public Health Acts and Highways Acts, where charges on land created by statute in favour of local authorities had been held to be charges on all the interests in the land, including the interests of chargees under subsisting mortgages, and not merely charges on the equity of redemption subject to all prior mortgages. See particularly *Corporation of Birmingham v Baker* (1881) 17 Ch 782. The terms of the relevant statute are set out at p 783 as follows:

F

> the Public Health Act 1875, enacts that "Where any local authority have incurred expenses for the repayment whereof the owner of the premises for or in respect of which the same are incurred is made liable under this Act or by any agreement with the local authority, such expenses may be recovered, together with interest at a rate not exceeding £5 per centum per annum, from the date of service of a demand for the same till payment thereof, from any person who is the owner of such premises when the works are completed for which such expenses have been incurred, and until recovery of such expenses and interest the same shall be a charge on the premises in respect of which they were incurred. . . .

Jessel MR said at p 786: G

> Now, the houses, *per se* being inanimate, they cannot bear the burden. If there be a charge on the houses it is a charge on the total ownership—if I may call it so, on the proprietorship; not on any particular section or portion of the proprietorship, but on the whole.

See also the judgment of Fry LJ in *Guardians of Tendring Union v Dowton* (1891) 3 Ch 265 p 269, where he said:

> All the Act does is to create a charge on the premises—that is, on the land—that is, on all the interests of the owners of the land.

He went on to explain that the charge did not override a restrictive covenant in favour of the owners of other property H for reasons which are not relevant to this present case.

Mr Millett has sought to distinguish these authorities. He has said, correctly, that they were decisions under other Acts, and under those other Acts the definition of the term "the owner" was somewhat different. He has pointed out that under recent Public Health Acts and Highways Acts the charge is expressly stated in the Act to be on the land and each and every interest in the land: see section 291(1) of the Public Health Act 1936 and section 264(1) of the Highways Act 1959. But this does not, in my judgment, invalidate the reasoning of the earlier decisions under the earlier Acts. Mr Millett has contended in particular as a ground for distinguishing the J earlier decisions that in all those cases moneys had been laid out by the local authorities in carrying out works which were calculated to benefit the owners of all interests in the relevant premises including mortgagees. It is quite true that in the reported cases the judges have pointed out that the works concerned would benefit the owners of all interests in the premises. But I think they have done so not as the ratio of their decisions but as an indication why the decisions they have reached are intelligible or in accordance with commercial sense: see Lord Halsbury LC in *Guardians of Tendring Union v Dowton* (1891) 3 Ch 265 at p 268. The real ratio is that the words in the relevant statute "charge on the land" K naturally import a charge on all interests in the land, that is to say, a charge on the proprietorship or total ownership. The emphasis of the judgments is on the interests rather than on the benefits. It would be impossible for the court, in applying such sections which create statutory charges, to enter on an inquiry whether the matters which have brought about the creation of the statutory charge were or were not beneficial to the owners of all or any of the interests in the land concerned. The true beneficiaries may well be other sections of the community, users of the public highway or occupiers of neighbouring premises, and any benefit to any owner of an interest in the subject premises may on any view L be incidental. It would be equally impossible as a matter of construction for the meaning of such words as "a charge on the land" in a public statute to vary from case to case, depending on whether or not the matters which had given rise to the imposition of the charge were or were not in some sense beneficial to the owners of particular interests in the land.

I conclude, therefore, that the plaintiffs' charge for £16,940.93 under section 17B has priority over and binds the defendants' interest in the land and all other interests in the land and I will so declare.

M

COURT OF APPEAL

March 21 1980

(Before Lord Justice STEPHENSON, Lord Justice BRIDGE and Lord Justice TEMPLEMAN)

POST OFFICE v OXFORD CITY COUNCIL

Estates Gazette June 7 1980

(1980) 254 EG 867

Rating—Liability to surcharge on unused commercial building under section 17A of General Rate Act 1967 inserted by section

16 of Local Government Act 1974—Unused garage premises in Oxford owned by the Post Office—Whether exempt from surcharge under Rating Surcharge (Exemption) Regulations 1974 because constructed as a factory or other premises of a similar character for use wholly or mainly for industrial purposes—Hereditament in this case was constructed as a garage which inter alia undertook some repair work—Although a single hereditament constructed exclusively for repair work might be a factory, premises such as this garage were similar to a shop—Garage was constructed for purpose of supplying goods and services directly to the public—Accordingly the premises were commercial premises liable to the surcharge—Decision of Divisional Court reversed and that of magistrates restored

This was an appeal to the Court of Appeal from a decision of the Divisional Court reversing a decision of Oxford City magistrates who had held that the Post Office was liable to a rating surcharge under section 17A of the General Rate Act 1967. The premises in question consisted of an unused garage owned by the Post Office at 69 to 74 St Aldates, Oxford.

D Micklethwaite (instructed by the city solicitor, Oxford City Council) appeared on behalf of the appellants; Guy Seward (instructed by the solicitor to the Post Office) represented the respondents.

Giving the first judgment at the invitation of Stephenson LJ, TEMPLEMAN LJ said: In the City of Oxford, only a Cambridge man expects to find a factory or premises similar to a factory. No one was surprised to discover at 69 to 74 St Aldates a garage which dealt in motor cars and accessories, in the servicing and repair of motor cars and in the supply of petrol. The garage premises comprised a car showroom, repair shops, stores, a petrol filling area and offices and were rated as commercial premises.

The premises have not been used since December 1975 and the present appellants, the Oxford City Council, imposed a rates surcharge on the respondent Post Office, the present owners of the premises. The Post Office claims that the premises are exempt from surcharge because they were constructed as a factory, or as premises similar to a factory for use mainly for industrial purposes.

The Oxford City magistrates crossed the road to inspect the premises and decided that the premises were not a factory or similar to a factory and that they were commercial premises.

On March 26 1979 the Divisional Court of the Queen's Bench Division, Michael Davies J and Neill J, reversed the magistrates. The city council appeal to this court.

The rates surcharge is an invention of section 16 of the Local Government Act 1974, which enables a surcharge to be imposed on a commercial building during periods of non-use. The objects of the invention were to encourage owners of property to make use of commercial property, to protect the rating authorities against loss of revenue and to discourage the manoeuvre illustrated by Centre Point whereby a developer generated more security and more cash by keeping vacant commercial property in a town centre and benefiting from the increasing sums raisable by mortgage or sale based on vacant possession value than he would have received after taxation from leasing or using the property. Regulation 3 of the Rating Surcharge (Exemption) Regulations 1974 (SI 1974 No 1563) exempted from the rates surcharge "Hereditaments which have been and remain constructed or adapted as factories, mills or other premises of a similar character for use wholly or mainly for industrial purposes."

I apprehend that the exemption was made because factories and mills and similar premises are unlikely to remain idle for long periods if they can in fact be usefully occupied whereas commercial premises may fall into disuse for a variety of reasons, particularly in city centres, to moulder awaiting inspiration and inflation unless the owners are spurred into action. No doubt there is a respectable or plausible reason in the present case for the melancholy void now exceeding four years, but it is not surprising that the Oxford City Council seek to recover a rates surcharge in respect of these premises.

The Post Office claims that the garage at St Aldates was constructed as a factory or similar to a factory because maintenance and repair works were carried out on the premises.

The legislative history to which we were referred does not shed light on the problem whether a garage is a factory or similar to a factory or a shop or similar to a shop.

Section 22 of the Rating and Valuation Act 1925 provided for the ascertainment of the rateable value of houses and buildings outside London excluding "Mills, manufactories or premises of a similar character used wholly or mainly for industrial purposes"; see the note to Part I of Schedule 2.

Section 3 of the Rating and Valuation (Apportionment) Act 1928 defined an "Industrial hereditament" so as to include a factory or workshop unless it was primarily occupied and used for the purposes of a retail shop and premises of a similar character where retail trade or business (including repair work) was carried on. This section was only relevant to the derating of industrial hereditaments and was repealed by the Rating and Valuation Act 1961.

Section 5(3) of the Rating and Valuation (Miscellaneous Provisions) Act 1955, amending section 22 of the 1925 Act and applying that section with modifications for London, defined the expression "non industrial building" as "a building, or part of a building, of any description, with the exception of factories, mills and other premises of a similar character used wholly or mainly for industrial purposes."

The General Rate Act 1967 replaced all the earlier legislation, and section 19(2) provided for the ascertainment of the rateable value of "non industrial buildings" by deductions from gross annual value. By section 19(6) "non industrial building" was defined as a building of any description other than "(a) factories, mills and other premises of a similar character used wholly or mainly for industrial purposes." Until the present dispute arose the garage premises now under discussion were rated under section 19(2) without demur.

The Local Government Act 1974 amended the General Rate Act 1967. Section 17A(1) of the General Rate Act 1967 now provides that if for a continuous period exceeding six months a commercial building is not used for the purpose for which it was constructed or has been adapted, then, subject to certain exceptions not material, the surcharge is payable. A commercial building is defined by section 17B(2) as a hereditament whose net annual value falls to be ascertained under section 19(2), in other words a "non industrial building."

Finally, by the Rating Surcharge (Exemption) Regulations 1974, made in exercise of powers conferred by the General Rate Act 1967, it was provided in regulation 3 that no surcharge shall be payable *inter alia* in respect of "(a) hereditaments which have been and remain constructed or adapted as factories, mills or other premises of a similar character for use wholly or mainly for industrial purposes." This regulation was necessary because non industrial buildings included any building not in use.

We were not referred to any authority which shed light on the question whether a garage falls within the exemption accorded by the 1974 regulation.

For present purposes I am prepared to assume that a single hereditament constructed exclusively for repair work may be a factory or similar to a factory. In the present case, however, the hereditament was constructed as a garage whose facilities included repair work. Garage premises of the kind now in question are constructed or adapted in order to attract and supply the requirements of individual members of the public who seek to buy or sell motor cars and accessories, spare parts and fuel and to obtain repairs and servicing of motor cars. Such premises in my judgment are not a factory but are a retail establishment; they are not similar to a factory, they are similar to a shop. It matters not to the customer or the garage whether some or all of the repair and servicing works are carried out on the premises; depending on the facilities of the garage, the extent of the required repairs and the skill of

A the particular garage employees repairs may be carried out either in the garage or elsewhere. Alternatively the customer may be persuaded to buy another motor car. The hereditament in the present case was constructed for the purpose of enabling the occupiers to supply goods and services directly to the public. In my judgment such a hereditament is used for commercial purposes. A garage is the name which identifies a particular class of goods and services which are offered and supplied by the occupiers directly to the public. A retail shop which supplies and repairs goods is not similar to a factory. A garage is similar to a shop.

Mr Seward, who appeared for the Post Office, pressed the B arguments which convinced the Divisional Court. The garage was registered under the Factories Acts; but those Acts were intended to protect " the safety, health and welfare of employed persons " and apply to many premises which are not factories or similar to factories in common parlance. The garage was a large garage; the proportion of the floor space attributable to servicing and repairs was 72.5 per cent; the premises included a repair shop, a paint spray shop and a panel work shop. In addition Mr Seward relied on the fact that the garage employed mechanics and engineers. In my judgment a garage which supplies goods and services to the public does not cease to resemble a shop or begin to resemble a factory C according to size or distribution. The size of the garage, the precise proportion of the floor space, or the turnover or income attributable to repairs and the presence or absence of facilities or employees for particular kinds of repairs are all matters of degree and not of kind. For the reasons I have endeavoured to indicate, premises constructed for use as a garage of the kind now in question are similar in character to a retail shop and not similar in character to a factory. A garage is a commercial enterprise offering sales and repair services and not an industrial enterprise albeit that on the garage premises some works of repair similar to works which might be carried out by an industrial enterprise in D a factory or in premises similar to a factory are performed.

In the result I would allow the appeal and restore the decision of the magistrates.

Agreeing, STEPHENSON LJ said: Is this garage in St Aldates, Oxford, a hereditament constructed (or adapted) as a factory, mill or other premises of a similar character for use (wholly or) mainly for industrial purposes? If it is, it is exempted by regulation 3 of the Rating Surcharge (Exemption) Regulations 1974 from the rates surcharge provided for by section 16 of the Local Government Act 1974, as the Divisional Court have held. If it is not, it was rightly surcharged by the Oxford City justices.

E I agree that the justices were right for the reasons so cogently given in the judgment of Templeman LJ. Although the premises are registered as a factory under the Factories Act 1961, I do not agree with the Divisional Court that they are a factory. No one suggests that they are a mill. But I have doubted whether the Divisional Court may not have been right in holding that if they are not a factory, they are similar to a factory. Have they not something in common with factories and indeed mills? If garage premises, or at least F garage premises of this kind, are not similar to factories and mills, what premises are? Or do garage premises lack the characteristic of being used for industrial purposes and does that deficiency keep them out of the exempted class?

Templeman LJ has traced the legislative history of this enactment. Its language goes, I think, further than he went, right back to the dawn of legislation for the protection of persons employed in mills and factories at the beginning of the 19th century with the Act of 42 Geo 3 c 73 "for the Preservation of the Health and Morals of Apprentices and others employed in Cotton and other Mills, and Cotton and other Factories." Later legislation extended protection to non- G textile factories and workshops; Factory and Workshop Acts continued to be passed until 1937, the precursor of the Act of 1961. Shops have been the subject of separate legislation culminating in the Shops Act of 1950 and the Offices, Shops and Railway Premises Act of 1963. But, as has been pointed out, those Acts relate to the safety, health and welfare of employed persons and do not decide what I agree is the right question here, namely whether a garage is a factory or similar to a factory and for use mainly for industrial purposes—or a shop or similar to a shop—and for use mainly for commercial purposes.

The difficulty is that it is similar to both. It is similar to H a factory, as Mr Seward pointed out, in being constructed for manual work and for engineering work; in having shops or workshops for panel-beating, paint-spraying and repairing. And the repair workshop of a garage is classified as light industry for the purposes of planning legislation. On the other side, Mr Micklethwaite called attention to the characteristics of a factory (or mill) which a garage lacks: production of some product; dealing with that product in bulk and not with one car at a time for one individual customer, in which it resembles a retail shop. And this garage has been rated as non-industrial since 1925.

Faced with these conflicting characteristics and treatments J I would give the words in the regulation their ordinary meaning which, in the words of the second ground of appeal, "would not (then) include a motor car showroom and repair workshop." I do not find it possible to conclude that because these premises are a factory as defined in section 175 of the Factories Act 1961, they are also a factory constructed for use mainly for industrial purposes within the General Rate Act 1967 and the 1974 regulations, or at the very least bear a strong enough resemblance to a factory so constructed as to come within the words of the statutory exemption. Looking, as the justices I think did, at the real nature of these premises and at the object of the statutory exemption, I agree that they K are more similar to a shop than to a factory and are constructed mainly for commercial purposes, and that we should, therefore, allow the appeal and restore the justices' decision.

I am authorised by Bridge LJ, who is unable to be present, that he agrees with both judgments.

Appeal allowed with costs in the Court of Appeal and below. Leave to appeal to the House of Lords was refused.

REAL PROPERTY AND CONVEYANCING

A **CHANCERY DIVISION**

April 3 1979

(Before Mr Justice WALTON)

JOHN TRENBERTH LTD v NATIONAL WESTMINSTER BANK LTD AND ANOTHER

Estates Gazette January 12 1980

(1979) 253 EG 151

Trespass—Bank, owners of a building which had become dangerous, sought permission of neighbouring owners to go on their land in order to carry out remedial works—Permission refused—Bank and bank's building contractors B **eventually went ahead and trespassed on neighbouring property—Works partially carried out—"Flagrant invasion of another's rights of property"—Although damage likely to be slight it was a fit case for a prohibitory injunction—No ground for suspending injunction—Interlocutory orders made for prohibitory injunctions and also for mandatory injunctions to remove scaffolding and rubble and debris from plaintiffs' land**

In these proceedings the plaintiffs, John Trenberth Ltd, moved for interlocutory injunctions, both prohibitory and mandatory, until the trial of the action for trespass, against National Westminster Bank Ltd, the first defendants, and E C Turner & Sons Ltd, the second defendants, who were the bank's building contractors. The trespass alleged concerned the plaintiffs' property at 117 Bute Street, Cardiff. In earlier ex parte proceedings the judge had granted the plaintiffs ex-parte relief in the form of negative injunctions.

James Munby (instructed by **Gilbert Robertson & Co, of** Cardiff) appeared on behalf of the plaintiffs; R J Powell-Jones (instructed by Cartwrights, of Cardiff) represented both the first and second defendants.

Giving judgment, WALTON J said: In this case the plaintiffs and the defendants own adjoining property in Bute Street, D Cardiff, the plaintiffs' property being 117 Bute Street. Next door to them, the premises to the south are owned by the first defendant, the National Westminster Bank Ltd. The National Westminster Bank has found itself in a very unfortunate situation because the building which it occupies is apparently in a dangerous condition; or the bulk of it was in a dangerous condition, so far as the frontage of it was concerned at any rate, because that consisted of stone slabs with metalwork behind. Water had penetrated between the metalwork and the stone slabs, forming an accumulation of rust, which occupies a greater volume than the metal itself, and the stones have been therefore pushed forward and are E to some extent dangerous. Therefore, the building had become in a dangerous condition.

Mr Powell-Jones, who has said everything that can possibly be said on behalf of the defendants, has submitted to me that it was the defendants' duty to remedy the defective state of their building, and that I readily accept. At common law there is a duty to maintain premises adjacent to the highway, and these premises were adjacent to Bute Street, in an unsafe condition; and he also called my attention to section 58(1) of the Public Health Act 1936, whereby if one does not comply with the notice to make a defective building safe certain penalties follow. Therefore, he said, the defendants F found themselves in this dilemma, that their building is so constructed that it cannot be effectively repaired without to some extent trespassing upon the plaintiffs' land. And so, not wishing to commit a trespass, but realising that they might, in order to repair the premises cheaply and efficiently, have to do so, they in fact entered into a correspondence with the plaintiffs, starting by a letter of May 30 1978, and pursued thereafter down to December 8 1978, after which the plaintiffs did not answer any more letters.

Mr Powell-Jones has submitted that the defendants, that is to say the National Westminster Bank Ltd, the owner of the property, and their contractors for putting the building right, G E Turner & Sons Ltd, behaved throughout perfectly properly, not in flagrant disregard of the plaintiffs' rights but offering full indemnities against any accidents that might happen, full assurances that nothing would happen, and so on and so forth. And ultimately, whatever they asked for and however they put it, the reply of the plaintiffs to the request for permission was a flat "No."

It may be or it may not be, and I do not think I am really concerned with that, that the refusal of the plaintiffs to grant permission was irrational. But it certainly was made in good faith and it certainly was not made, as the evidence makes perfectly clear, as a bargaining counter in order to extract concessions from the defendants. H It was a genuine belief and desire on the part of the plaintiffs—they did not want their property to be trespassed upon by the defendants.

The situation at the end of that correspondence was that the defendants, according to Mr Powell-Jones, and I accept his formulation as being absolutely accurate, felt they were never going to get the plaintiffs' permission to carry out their works—that is to say, the plaintiffs' permission to trespass on the plaintiffs' land in order to carry out their works. Therefore, Mr Powell-Jones says, the defendants (particularly the first defendants) were in an impossible position. There was on the one hand the Scylla of the dangerous building and there was on the other hand the Charybdis of trespassing upon J the plaintiffs' land, and they chose, says Mr Powell-Jones, to go ahead and trespass. And he says that the defendants were not thereby acting unreasonably in pressing ahead and completing works, in so far as they have been completed, to make the building safe.

I can see fully and I can sympathise with the dilemma in which the defendants (particularly the first defendant) found themselves, but on Mr Powell-Jones' own summary of the evidence, and what happened, it is clear beyond peradventure that the defendants, knowing that they had not and were not likely to get permission to trespass upon the land of the plaintiffs, nevertheless went ahead and trespassed. K And if flagrant invasion of another's rights of property of that nature is not sufficient to call forth the interposition of a court of equity, I do not know what invasion ever could be said so to do.

Mr Powell-Jones has said that the invasion was not for the commercial benefit of the bank, but was merely to ensure that the property was made safe so as not to be dangerous to third parties, including the plaintiffs themselves. I do not accept that for one moment. The defendants wished to use their building as a building, possibly for aught I know as a bank. If the building was dangerous, another alternative could easily be to pull it down and start all over again. But, of course, that course would be one which cost them a great deal of money, and so they deliberately chose the less expensive course of trespassing upon the plaintiffs' land.

Therefore, it seems to me that this is a case, there being nothing else in the facts which raise the faintest shadow of a cloud upon the plaintiffs' title to relief, on which the injunction must at any rate in a normal case go. So far from acceding to Mr Powell-Jones' first real submission, that this was not a clear case for an injunction and it is inconceivable it would be granted, it appears to me the clearest possible case for an injunction and inconceivable that it would not be granted.

He then says, well, supposing his submission is not accepted, his second submission is that the plaintiffs can be compensated in damages; that to obtain an injunction would be obtaining an injunction for an injunction's sake, and that it would be a great hardship to the first defendant if the injunction were to go. I do not accept that either. Indeed, it may be that there are special circumstances here which will ultimately result in the plaintiffs obtaining aggravated damages. I must not be taken to express any opinion upon that at all. But it is perfectly clear that the actual damage, apart from any question of aggravation, caused by the mere trespass, both by oversailing the front of the plaintiffs' property and by actually resting upon the rear of the plaintiffs' property, is comparatively slight; so slight that if an action were brought for it, it would hardly command the smallest coin in the realm. But so far from that being a reason why an injunction should not be granted, it has been said in many of the cases to which Mr Munby drew my attention that the fact that any damage would be trifling is the very reason why an injunction should be granted. People are not to infringe the property rights of others and then say "And I am entitled to go on doing it because I am really doing you no tangible harm, and fivepence will amply compensate you for that harm."

Mr Powell-Jones' third point was on the question of suspension. He had, and he rightly says, that in a proper case the court can always suspend the operation of an injunction. He referred me in this connection to the case of *Stollmeyer v Petroleum Development Co Ltd* [1918] AC 498, a case in the Privy Council. Now, I would remark that that case in fact does not do very much for his proposition because that was a case of nuisance and not a case of trespass; and as has been well pointed out so far as nuisance is concerned, damage being the gist of the action it is possible, normally, to compensate the plaintiff—at any rate for a short suspension—in damages. Moreover, one has to look at the nature of the works required in that case, which were obviously going to be very very considerable indeed, dealing with property which had been put up to exploit Trinidad's then burgeoning petroleum industry. That is a totally different sort of case from the present, where half a day's work, if that, is all that will be required to remove the trespass to the plaintiffs' property.

I think I must refer in this connection to the case of ***Woollerton & Wilson Ltd v Richard Costain Ltd*** [1970] 1 WLR 411 where, on facts which in some ways bear a great resemblance to the present case, Stamp J (as Stamp LJ then was) granted an injunction, but then suspended it for a couple of years. I find that that case is not one which, on the question of suspension, has received universal approbation; and, indeed, in ***Charrington v Simons & Co Ltd*** [1971] 1 WLR 598, there occurs in the judgment of Russell LJ (as the third Lord Russell of Killowen then was) these words:

> Further, we wish to reserve our opinion whether the decision in *Woollerton & Wilson Ltd v Richard Costain Ltd* [1970] 1 WLR 411 was correct. Neither counsel seemed to think that it was.

I therefore think that this must be a case where Homer nodded, and that that case cannot be relied upon so far as suspension is concerned. And I think the reason it cannot be relied upon was well illustrated by Mr Munby in his submissions to me when he said that in a sense, when one is dealing with the direct physical invasion of a right of property by a trespass one is very close to the line of cases stemming from ***Doherty v Allman*** (1878) 3 App Cas 709, which decide that an injunction to enforce a negative stipulation in a contract goes almost as of course. The parties, having agreed that something shall not be done, the court simply says that what the parties have agreed shall not be done.

The matter can be tested, I think, very simply in this way. Supposing that instead of putting up that scaffolding, as the second defendants doubtless did, taking their time about it in the normal way, supposing they had signalled notice of their intention well in advance so that it was quite clear what was going to happen if it was not restrained, and the plaintiffs had applied to the court for an injunction *quia timet*; what conceivable ground could have been put forward by the defendants in answer to the claim for such an injunction? There is not the faintest shadow of reason they could have advanced why they should not be restrained from committing the trespass. Can it then be any better that they have actually committed it? Does the fact of their having committed it and it being, I doubt not from a structural point of view and commercially, highly advantageous to them that they should be able to go on and complete their building, make the matter better from their point of view Does that give them any right at all to have a suspension of the injunction which, if the plaintiffs had been able to come realising what was going to happen earlier, they would have got without the faintest difficulty at all? The answer must clearly be "No," there can be no conceivable reason for any such postponement at all.

Therefore, for those reasons it appears to me that the negative injunctions which are sought, and which I gave to the plaintiffs by way of ex parte relief on March 29 1979, must be continued until judgment in the action or further order.

There are also two mandatory orders sought, namely, an order that the defendants and each of them do forthwith remove or cause or procure to be removed from the plaintiffs' land all scaffolding or staging erected thereon by the defendants or either of them, or by their respective directors or servants or agents; and, in similar terms, that they remove or cause or procure to be removed from the plaintiffs' said land all rubble and debris deposited thereon by the defendants or either of them, or by their respective directors or servants or agents. Mr Powell-Jones for the defendants said quite frankly that if the negative injunctions asked for were granted, then of course the relevant bits of scaffolding on and over the plaintiffs' land were of no further use to the defendants and he could not then seriously object to the mandatory orders. But whether he objects or not, this is a case where there is no dispute as to any of the relevant facts; there is no dispute as to the law; no conceivable reason or justification in law for the defendants' acts have been forthcoming. Therefore, this is one of the admittedly very rare class of cases where, even on an interlocutory motion, a mandatory order must go.

Therefore, for those reasons I propose to make the orders sought by the notice of motion herein under the first three heads.

COURT OF APPEAL

July 19 1979

(Before Lord Justice ORMROD, Lord Justice BRANDON and Sir David CAIRNS)

E & L BERG HOMES LTD v GREY AND ANOTHER

Estates Gazette February 2 1980

(1979) 253 EG 473

Possession of land sought against persons who had occupied a caravan and other structures by licence of owners of site used as a brickworks until it became derelict—Occupation lasted over 26 years—Brick company arranged for water and

A **electricity to be supplied to caravan—After brickworks closed down occupiers installed own electric generator, subsequently purchased a replacement caravan, had main electricity laid on and telephone installed—Evidence that original owners of site were aware of these actions by occupants and acquiesced in them—Land subsequently conveyed by brick company to present plaintiffs, an associated company—Notice then given by new owners to occupiers to remove caravan and other structures and to vacate—County court judge held plaintiffs entitled to possession—Defendants invoked equitable right to remain based on proprietary estoppel—Court of Appeal applied rules laid down by Fry J in Willmott v Barber—Occupiers** B **failed to satisfy requirements that they had mistaken their legal rights and had been encouraged by the owners to spend money and take other action on the faith of such encouragement—No legal or equitable right to remain after termination of licence—Occupiers' appeal dismissed**

This was an appeal by Edwin Thomas Arthur Grey and Eadythe Eileen Grey, his wife, from a decision of Judge Peck at Slough County Court granting an order for possession of land at Pinkneys Green, Maidenhead, to the plaintiffs in the county court action, E & L Berg Homes Ltd. The land in question consisted of about 0.7 acre, forming part of a larger area owned by the plaintiffs.

C G M Godfrey QC and J Boggis (instructed by Eric Levine & Co) appeared on behalf of the plaintiffs (respondents to the appeal); T S Baker QC and D R Foskett (instructed by Field, Gamble & Kirk, of Reading) represented the defendants (appellants).

Giving the first judgment at the invitation of Ormrod LJ, BRANDON LJ said: This is an appeal by the defendants from an order made by His Honour Judge Peck in the Slough County Court on May 13 1979. The order was made in an action brought by the plaintiffs against the defendants for possession of land. It required the defendants to give up possession of the land concerned to the plaintiffs, subject to a D stay of execution till the happening of certain specified events. I shall refer to the precise terms of the order, so far as the stay is concerned, later in my judgment.

The plaintiffs in the action are E & L Berg Homes Ltd. They are successors in title to the Maidenhead Brick and Tile Co Ltd. The two companies are associated, both being controlled by the same parent company, Maidenhead Investment Holdings Ltd. They further have three directors in common. I shall refer to Maidenhead Brick and Tile Co as "the brick company." The defendants are a married couple, Mr and Mrs Grey, who occupy the land of which the plaintiffs claim possession. Mr Grey is 51 and Mrs Grey 50. I shall refer to them collectively E as the defendants, but individually as Mr and Mrs Grey.

The land concerned is a quadrangle of about 0.7 acre forming part of a much larger area of land at Pinkneys Green, Maidenhead. There are presently certain structures standing or erected on the quadrangle and occupied by the defendants as or with their home. In the history of the matter which follows I shall refer to the larger area of land, including the quadrangle, as the plaintiffs' land, and the quadrangle itself as the site.

At the trial in the Slough County Court the defendants, on whom the burden of proof lay, for reasons which will become apparent, called three witnesses. They were Mr Grey, Mrs F Grey and a surveyor, Mr Wilton, whose main evidence in chief was contained in a written statement. None of the evidence given by any of these three witnesses was challenged in cross-examination, and the plaintiffs, when their turn came, did not call any witnesses themselves. A substantial number of agreed documents, including photographs, were also put in evidence. In those circumstances, there being no conflict of evidence, the case fell to be determined in the court below, and falls to be reviewed in this court, on primary facts which are substantially agreed.

The story begins in 1949. The land was then owned by the brick company, who operated brick works on it. Mr Grey, G then aged 21, obtained employment with the brick company at those works. In 1952 Mr Grey married Mrs Grey and early in 1953 they had a daughter born to them. About that time, Mr Grey obtained permission from the then manager of the brickworks, Mr Smith, to put a caravan on the site. In March 1953, pursuant to that permission, Mr Grey put a rented caravan on the site and he and Mrs Grey, with their child, moved into it and occupied it as their home. Mr Grey was then 25 and Mrs Grey about 23. In August 1953, Mr and Mrs Grey bought a caravan on hire purchase at a cost of £469 and substituted it for the caravan which had previously been rented. Between 1953 and 1975 Mr Grey carried out work of various H kinds on the site. He built sheds, tidied and cleaned up the site and erected fences. The brick company further arranged for water to be laid on and allowed Mr Grey to bring in an electricity lead to the caravan from their own power-house. Subsequently the Grey family continued to occupy the caravan as their home for many years. In the late 1950s or early 1960s Mr Taylor became the general manager of the brick company. He was at the works most weeks. Mr Grey saw him from time to time and regarded him as the boss. In 1964 Mr Grey replaced the fences around the site, which had by then become damaged.

In the summer of 1968 Mr Taylor called a meeting of the J employees of the brick company. He told them that there was bad news. The brickworks were going to close down and they would be made redundant. Mr Grey was given a month's notice by Mr Taylor, who said nothing at that time about Mr Grey's continued use of the site. Mr Grey fortunately obtained other employment at Knowl Hill and he and his family continued to live in the caravan as before. Some time afterwards, just after the brickworks had begun, at any rate, to close down, there was a chance meeting between Mr Taylor and Mr Grey, and a conversation took place between them in which Mr Taylor raised the subject of the continued occupation of the site by Mr Grey. The latter gave an account K of that conversation in his evidence, which was, as I indicated earlier, in no way challenged or contradicted on behalf of the plaintiffs.

The judge's note of Mr Grey's evidence of the conversation, reads as follows:

> Taylor said he was a bit worried in case property sold, and if I saw anything come along, it might be in my own interest to take it. As far as brickworks concerned, he did not think they would ever put me off. I said I would stay as I was, as I was alright for the moment, ie I did not intend moving. I did not want to go. I had no other discussions with Taylor about living on that site. At no time did he suggest I should look for anywhere else to live. L He may have said they would not bother to put me off.

Counsel told us that that last sentence resolved a doubt which had arisen as to whether what Mr Taylor had said earlier was that he did not think they would *ever* put Mr Grey off, or he did not think they would *bother* to put Mr Grey off.

The brickworks finally ceased all operations, after a progressive run-down, in September 1969. After that a Mr Weller came to look after the remaining stock of bricks and to act as caretaker for the brick company. The plaintiffs' land, apart from the site with the caravan on it, became derelict. During 1969 Mr Grey bought an electric generator for £170 as the electricity from the power-house of the brickworks was M no longer available. By 1972 the caravan bought in 1953 was approaching the end of its useful life. Mr Grey decided to replace it with a twin mobile home unit at a cost of over £2,600. Before doing so, he spoke to Mr Weller about it. His evidence about that conversation was as follows:

> Before I got it

that is, the unit

> I saw Weller, and asked him to tell Mr Taylor, if he saw him, that I was thinking about replacing the caravan. This was before I bought. Weller said he would see Colin Taylor, and mentioned it to him. Weller said as far as he could see, it did not make any difference as long as I did not take any more ground in.

He said he could not really see what difference it made if I did not take more ground in. I saw Weller after that, but he did not say anything more about it. I did not ask him if he had cleared it with Taylor.

Following that conversation, without hearing anything more about the matter, Mr Grey carried out the replacement. The nature of the twin mobile home which he acquired and put on the site is described in Mr Wilton's evidence and can be clearly seen from the photographs. It is supported on wheels and jacks. It is not, and has not at any time, been fixed to the ground. In 1973 Mr Grey erected certain new structures on the site. These consisted of a carport on a concrete stand, porches and concrete steps. He spent over £500 on the materials and did the work himself. He also arranged for mains electricity to be brought to the home by the Electricity Board at a cost of £50. Mr Grey arranged for a telephone to be installed by the Post Office at a cost of £40. Both lines were brought in on the same pole. Mr Grey also erected chain-link fencing around the site. Between 1973 and 1977 Mr Taylor was still visiting the plaintiffs' land from time to time. Mr Grey did not see him often; he did not visit the Greys' home and there were no discussions with him.

In 1975 the clearing of the plaintiffs' land, previously derelict, was completed. Mr Weller left and has since died. In July of that year the land, including the site, was conveyed by the brick company to the plaintiffs. In 1976 Mr Grey built a new shed, spending £30 on materials and again using his own labour.

On July 29 1977 solicitors acting for the plaintiffs wrote a letter to Mr Grey in these terms:

> Dear Sir,
>
> *Land at Pinkneys Green Maidenhead*
>
> On the instructions of our clients, E & L Berg Homes Ltd, we require you to remove the temporary building which you have erected on their land at the above site and to vacate the same within the next seven days.
>
> In the event that you fail to comply with this notice then our clients will take immediate steps to obtain vacant possession and remove all temporary structures which you have erected on the site.

The defendants had lived there for 24 years and this letter asked them to get out in seven days. It is conceded for the plaintiffs that, whatever their rights, this was a most unfortunate and indeed callous letter to have been sent on their behalf.

On receipt of that letter Mr Grey consulted solicitors and correspondence ensued between them and the solicitors of the plaintiffs. Mr Grey's solicitors claimed that he was entitled to remain on the site. In September 1978 the plaintiffs began the present action. The particulars of claim were filed on September 25 and the defence on November 1. The defence was later amended to add a counter-claim, and later still reamended. On March 12-13 the action was heard by His Honour Judge Peck, and on March 13 he made the order to which I referred earlier and against which this appeal is brought.

Before I state the cases put forward by either side, it will be convenient to get one matter out of the way. It was common ground in the court below, and again on appeal to this court, that the conveyance of the plaintiffs' land by the brick company to the plaintiffs in July 1975 did not affect whatever would have been the rights of the defendants if no such conveyance had taken place. This was on the basis that the two companies were associated with common directors, and the plaintiffs could not be in a better position in relation to the defendants than the brick company, but for the conveyance, would have been. It follows that the change of ownership of the land in July 1975 can, for all practical purposes, be treated as irrelevant.

The case put forward for the defendants in the court below was that, whatever the strict legal rights of the parties might be, the defendants had an equitable right to remain indefinitely on the land. This equitable right was based on proprietary estoppel, arising from the relationship between the parties and the conduct of the brick company, the plaintiffs and the defendants. The conduct relied on for this purpose was threefold. First, what Mr Taylor said to Mr Grey in the conversation in the summer of 1968. Secondly, what Mr Weller said to Mr Grey in the further conversation in 1972. Thirdly, the acquiescence by the brick company and the plaintiffs in the various things done by the defendants in, on or in connection with the site from 1968 onwards.

It was contended for the defendants that the conduct of the brick company and the plaintiffs gave rise to an equity in favour of the defendants, and further that this equity could only be satisfied either by conveyance by the plaintiffs to the defendants of the freehold of the site, or alternatively by the granting of a right to occupy which would subsist during the joint lives of the defendants.

The case put forward by the plaintiffs was that the defendants had no equitable right of any kind. They were bare licensees; that is to say, holders of a licence granted without any consideration. That licence had been revoked by the plaintiffs' solicitors' letter of July 29 1977. A reasonable time for removal had elapsed and the plaintiffs were accordingly entitled to possession forthwith. Alternatively, if the defendants had an equitable right at all, it was a right to a licence to occupy, terminable on reasonable notice, and no more.

With regard to the first item of conduct relied on by the defendants, the conversation between Mr Taylor and Mr Grey in 1968, the defendants pleaded that Mr Taylor had said in that conversation that the defendants could stay on the site for as long as they liked. As will be apparent from the note of Mr Grey's evidence which I read earlier, his account of the conversation did not support what had been pleaded, but fell well short of it. This necessarily reduced very much the reliance which counsel for the defendants could place upon the conversation in question.

The judge, in a most careful and sympathetic judgment, held that the defendants had not established the equity which they claimed at all and that the plaintiffs were accordingly entitled to an order for possession. He held, alternatively, that, if he were wrong about that, the extent of the equity was no more than a right to occupy for a limited period; 12 months after any planning permission for development of the plaintiffs' land, or the sale of the land at arm's length, and that the equity would be satisfied by giving the defendants that limited right. If the judge had given full effect to his decision, he would have made a simple order for possession on the claim. In fact, however, there was an agreement between the parties as a result of which he made an order for possession, subject to a stay of execution. The material terms of the order read as follows:

> AND IT IS ORDERED that the Defendants do give possession of the said land to the Plaintiffs subject to the matters hereinafter appearing
>
> AND UPON THE PLAINTIFFS consenting to the suspension of the execution of the said order on the terms hereinafter appearing
>
> IT IS FURTHER ORDERED that execution of the aforesaid order for possession be stayed—
>
> 1. Until September 13 1979; and thereafter
>
> 2. Until six months after either the grant of planning permission satisfactory to the Plaintiffs for development of property which includes the land or the entry by the Plaintiffs into a contract at arm's length for the sale of the land or any property which includes the land, whichever shall be the sooner; or
>
> 3. With the leave of the Court at a date earlier than the happening of either of the events referred to in 2 above, such leave to be sought by application on not less than two days notice to the Defendants, with liberty to apply for such leave; and in any event
>
> 4. Until the final determination of any appeal herein provided that the Defendants pursue such appeal with due diligence and expedition. . . .

The judge further dismissed the counterclaim.

On the appeal before us counsel on either side repeated substantially, with some variations and additions, the arguments which they had advanced below. In considering the case so

A raised, it will, I think, be convenient to divide the problem into two parts. First, what were the legal rights of the parties? Second, have those legal rights in effect been varied by the acquisition by the defendants of an equitable right or title, based on estoppel?

I consider first the legal rights of the parties. It seems to me that the first and second caravans were placed on the site, and occupied by the defendants as their home, pursuant to an oral licence given to Mr Grey by Mr Smith on behalf of the brick company in March 1953. The other things done subsequently by Mr Grey on the site, including fencing, replacing of the old caravan by the twin mobile home and the erection
B of the carport and other structures, were not authorised by that original oral licence, but the fact that these things were being done, or had been done, was known, or must be taken to have been known to the brick company and the plaintiffs through the agency of Mr Taylor, who continued to visit the land at all material times. No objection to what was being done, or had been done, was ever taken by the brick company or the plaintiffs. In those circumstances the original oral licence granted in 1953 was extended by conduct to cover the doing of all of those things.

On the footing that the defendants had a licence from the brick company and the plaintiffs to do all that they did in or
C in connection with the site, and that the licence was a bare licence granted without any payment or consideration, it was revocable by the brick company, or later by the plaintiffs, at any time. Further, on revocation of the licence, the defendants were entitled to a reasonable time to remove themselves and their belongings, including the twin mobile home, from the site: *Minister of Health v Bellotti* [1944] KB 298. That was a case involving a contractual licence without any express terms, but I do not see why similar principles should not apply in the present case.

I consider, secondly, whether the legal rights of the parties which I have described were varied by the acquisition by the
D defendants of an equitable right or title, based upon estoppel. The principles applicable to the acquisition of such a right or title were considered and explained in a comparatively recent case in this court, *Crabb v Arun District Council* [1976] Ch 179. That case was cited to the judge below and referred to at some length in his judgment, and cited again to us on appeal. Scarman LJ said in his judgment in that case, at pp 192 and 193:

> If the plaintiff has any right, it is an equity arising out of the conduct and relationship of the parties. In such a case I think it is now well settled law that the court, having analysed and assessed the conduct and the relationship of the parties, has to
> E answer three questions. First, is there an equity established? Secondly, what is the extent of the equity if one is established? And thirdly, what is the relief appropriate to satisfy the equity?

Then in relation to the first of the three questions, the learned Lord Justice, at pp 194-5 cited, with subsequent approval, a passage from the judgment of Fry J in *Willmott v Barber* (1880) 15 Ch D 96. That passage reads as follows:

> It has been said that the acquiescence which will deprive a man of his legal rights must amount to fraud, and in my view that is an abbreviated statement of a very true proposition. A man is not to be deprived of his legal rights unless he acted
> F in such a way as would make it fraudulent for him to set up those rights. What, then, are the elements or requisites necessary to constitute fraud of that description? In the first place the plaintiff must have made a mistake as to his legal rights. Secondly the plaintiff must have expended some money or must have done some act (not necessarily upon the defendants' land) on the faith of his mistaken belief. Thirdly, the defendant, the possessor of the legal right, must know of the existence of his own right which is inconsistent with the right claimed by the plaintiff. If he does not know of it he is in the same position as the plaintiff, and the doctrine of acquiescence is founded upon conduct with a knowledge of your legal rights. Fourthly, the defendant, the possessor of the legal right, must know of the plaintiff's mistaken belief of his rights. If he does not, there is nothing which calls
> upon him to assert his own rights. Lastly, the defendant, the G
> possessor of the legal right, must have encouraged the plaintiff in his expenditure of money or in the other acts which he has done, either directly or by abstaining from asserting his legal right.

While approving that passage, the learned Lord Justice pointed out that what Fry J called "fraud" in 1880 would today be called taking advantage of a person in a way which is unconscionable, inequitable or unjust.

On the footing that the existence of an equity is to be tested by reference to the five requirements set out by Fry J in the passage of his judgment in *Willmott v Barber* cited above, I turn to see what is the situation with regard to the fulfilment
of those five requirements in the present case. In doing so, H
I bear in mind that the case for the defendants involves an allegation of unconscionable conduct against the brick company and the plaintiffs, and that such an allegation can only be made good by clear and cogent evidence.

The first requirement is that the party claiming the equity must have made a mistake as to his legal rights. Mr Grey did not say anywhere in his evidence that he had made a mistake as to his legal rights. Mrs Grey, however, did give certain evidence which was relied on by counsel for the defendants in this connection. What she said, as recorded in the judge's
notes, was this: J

> There was a public inquiry. The planning people came to see me for a statement regarding the lorries. I was told our circumstances had already been looked into, and vetted by the council. We had been there over 20 years and we could not be got off. When new shed was to be built, I went to local authority. The little shed by back door was put up in 1956-57 about. When we decided to replace the old caravan with a mobile home, we definitely would not have bought it if we thought there was risk of being turned off site. Saw Mr Taylor now and again. Did not have conversation with him, merely passed good day.

I do not regard this as evidence that Mrs Grey, or Mr Grey,
made a mistake about their legal rights. At most it is a state- K
ment that they did not think that, if the brick company had a right to make them leave, there was any real risk of their doing so. There also appears to have been an idea, based on what Mrs Grey was told by the local authority, that, because they had occupied the site for over 20 years, they had acquired a prescriptive right to it. If they did have that idea, it was plainly wrong, because their occupation had not been adverse, but under licence. Faced with this paucity of evidence on the matter, counsel for the defendants contended that it was a necessary inference from the various things done on the site by Mr Grey from 1968 onwards that he made a mistake about
his legal rights. If the mistake sought to be inferred is a L
mistaken belief that he had a legal right to remain on the site indefinitely, I should not be prepared to draw that inference, which I do not regard as necessary at all. His conduct was equally consistent with the belief that, if he did those things without the prior permission of the brick company, they would raise no objection but would, on the contrary, acquiesce in them. Indeed, I think the latter inference far more likely to be correct than the former.

On this analysis of the evidence given by the defendants, and the inference to be drawn from Mr Grey's conduct, it seems to me that the defendants fail to satisfy the very first require-
ment laid down by Fry J. In other words, they fall at the M
first of the five hurdles. If that is right, it is sufficient to dispose of the defendants' case that an equity in their favour exists, and it is unnecessary to go on to consider whether the defendants satisfy the other four requirements. In case I should be wrong with regard to the first requirement, however, I shall consider the others.

The second requirement is that the party claiming the equity should have expended money or done some act in the faith of his mistaken belief. The third requirement is that the party against whom the equity is to be claimed should know of the existence of his own right which is inconsistent with the right claimed by the other party. On the assumption which I must make for this purpose that the first requirement is satisfied, I should have no difficulty in holding that the second and third requirements were also fulfilled. The fourth requirement is that the party against whom the equity is claimed should know of the other party's mistaken belief about his rights. Here again, on the same assumption and further on the basis, which I do not accept, that the things done by Mr Grey on the site necessarily lead to an inference that he had a mistaken belief about his rights, it would, I think, follow that the brick company and the plaintiffs, being aware of what was being done, should have drawn that inference. The fourth requirement would accordingly, on that assumption and basis, also be satisfied.

The fifth requirement is that the party against whom the equity is claimed, must have encouraged the other party in his expenditure of money or in the other acts which he has done, either directly or by abstaining from asserting his legal right. With regard to this, it was contended for the defendants that Mr Grey was encouraged to spend the money and do the various things he did on the site from 1968 onwards, first by what Mr Taylor said in 1968, secondly by what Mr Weller said to him in 1972, and thirdly by the acquiescence in all that was done.

So far as what Mr Taylor said in 1968 is concerned, this argument seems to me to turn the reality of the matters topsy-turvy. The purpose of Mr Taylor's remarks on that occasion was a warning to Mr Grey that, while he would probably be permitted to remain so long as the brick company owned the land, he might well not be allowed to do so if the land was sold, and that in those circumstances he might well wish to consider making other arrangements. Mr Grey's reaction to this warning was that he wished to stay where he was and intended to do so. Accordingly he disregarded the warning which he had been given. How the giving of that warning could be interpreted as an encouragement to Mr Grey to stay and spend the money in the belief that he would be allowed to remain on the land indefinitely is beyond me. It is true that Mr Taylor said that he thought that the brick company itself would not make Mr Grey leave, but this was an expression of opinion rather than an assurance, and I do not think that Mr Grey was entitled to regard it as anything more. A statement by way of representation or promise, in order to found an estoppel, must be clear and unambiguous. What Mr Taylor said in this respect does not satisfy those requirements.

So far as what Mr Weller said in 1972 is concerned, I do not think that it assists the defendants. Assuming without deciding that the information given to Mr Weller can be treated as being given to the brick company, without any evidence that it was passed on to Mr Taylor, the information amounted only to this; that Mr Grey was proposing to replace one kind of mobile home, namely a caravan, with another kind. Further, all Mr Weller said was that he did not think it would make any difference. I should add that Mr Grey already had a licence to keep a mobile home on the site and it was unlikely that, so long as a mobile home only was involved, the brick company would be concerned with the particular type of such home in use. So far as the acquiescence from 1968 onwards is concerned, I have already said that the things done by Mr Grey were consistent with Mr Grey believing that the brick company would not object to extending its licence to cover those additional things. On that view, I do not think that the brick company or the plaintiffs, by not objecting to them, were encouraging Mr Grey to do them.

Assuming, therefore, contrary to the view which I expressed earlier, that the first requirement was satisfied, I should still be unable to hold that the fifth requirement was also satisfied.

For the reasons which I have given, I have reached the conclusion that the answer to the first of the three questions which Scarman LJ said in *Crabb v Arun District Council* had to be answered in a case of this kind, namely whether an equity is established, must be answered in the negative. On that answer, the second and third questions do not arise.

On the footing that no equity is established, the legal rights between the parties, to which I referred earlier, must prevail. That means that the plaintiffs are entitled to the order for possession which they claim, unless the period which the defendants had for removal was insufficient. By the time of the hearing in March 1979 21 months had elapsed since the revocation of the defendant's licence by the plaintiffs' solicitors' letter dated July 29 1977.

It was argued for the defendants that, in deciding what was a reasonable time for removal, account should be taken, firstly, of the long period of occupation by the defendants and, secondly, the difficulty facing them in finding another site for their mobile home. I would accept that those matters should be taken into account, but, in deciding what is a reasonable time, the situation of both parties has to be considered and not only that of the defendants. In my judgment the maximum period for removal which could be regarded as reasonable and fair for both sides is 12 months. This period had long since expired by the time of the hearing before the judge in March of this year.

There is one other matter to which I should like to refer. It was said on behalf of the defendants that, if the judge's order were to stand, it would work a great hardship on the defendants. Like the judge below, I have much sympathy for the defendants in the predicament in which they find themselves, but there are, I think, certain factors which fall to be weighed on the other side. Firstly, the defendants have been extremely fortunate to be allowed to occupy a site, without any payment of rent or other consideration of any kind, for over 26 years. That they have been allowed so to do was due to the kindness of the brick company. Secondly, their present predicament arises in part, at least, from Mr Grey's unwillingness to face realities after the warning given to him by Mr Taylor in 1968. Thirdly, the terms for a stay of execution of the possession order agreed below, and incorporated in the judge's order, are by no means unfavourable to the defendants. Indeed, even if an equity had been established, it would, it seems to me, have only been one much more limited in extent than that claimed by the defendants, and it is doubtful whether, on the basis of such equity, the defendants would in substance have been much better off than under the existing agreed order.

For the reasons which I have given I would uphold the order of His Honour Judge Peck and dismiss the appeal.

Agreeing, Sir David Cairns said: It is clear that when the defendants were allowed to put their caravan on land to which this appeal relates, they became licensees. As there was no consideration for a licence they were bare licensees and the licence could be terminated without notice. They could not be treated as trespassers and they had a right to a reasonable opportunity to make arrangements to leave and to take their property with them. How long a time would be reasonable would, I think, depend upon the character of the individuals, but I would consider that it clearly could not exceed one year. I am not persuaded that anything which happened thereafter estopped the owners of the land from terminating the licence or created an equity enlarging the defendants' rights in respect of the land.

I am prepared to assume that the plaintiffs are in no stronger position than their predecessors in title, the Maidenhead Brick and Tile Co, because no point seems to have been taken below on this aspect of the case, but in order to establish the estoppel on which the defendants rely, it would be necessary in the first place for them to show that they believed that they were entitled to remain there for longer than the licence would entitle them to stay, and that that belief was induced by some statement or promise, or something done or left undone on behalf of the landowners which encouraged that belief. Mr Grey gave no evidence that he had any such belief. Mrs Grey, as I understand her evidence, was saying either that she merely expected to be allowed to stay without any actual right to do so, or more probably that she believed that because she and her husband had been there for 20 years, they had a prescriptive right to remain there.

There is, of course, no suggestion that anything done or left undone, by the plaintiffs or their predecessors, encouraged any belief in a prescriptive right. If, however, contrary to my view the defendants, or either of them, had any belief on any other basis that they were entitled to remain there for any extended period, I can find nothing in the language used by Mr Taylor, or Mr Weller, which would have induced such a belief. Mr Taylor gave a warning that the defendants might have to leave if this land were sold. He expressed an opinion, but no more, that the brick company would not be likely to want them to remove. Mr Weller, assuming—and it is a big assumption—that he had any sort of authority on behalf of the landowners, did no more than indicate that the defendants' position would not be changed for the worse by the bringing in of a mobile home.

Mr Baker, however, relied mainly on the contention that the landowners encouraged the defendants to believe that they had some interest in the land, entitling them to remain there for life, by failing to raise any objection to the introduction of the mobile home and the carport or to the obtaining of the supply of electricity and other services. I think it can fairly be assumed that the plaintiffs, or their predecessors, knew that this was going on, but I cannot accept that the developments were of such a character that a reasonable person would suppose that the defendants believed themselves to have any right greater than that of licensees. The home is a mobile home; the services were such as licensees would wish to have, even if there was a risk of their not being able to stay there very long. The carport is the only item of which it can possibly be said that it was of a permanent character, but I do not consider that it was of such magnitude as to raise an equity; alternatively if, contrary to my view, any equity could be said to have been raised it was, at the utmost, no more than the maximum envisaged by the learned judge.

It comes to this. In my judgment, the plaintiffs were entitled to terminate the licence as they did in July 1977 by giving notice of termination. The time elapsing between the giving of that notice and the judgment in the action was such that by the time of that judgment, the defendants had no right to remain, and the actual order made against them was in the circumstances a generous one. It may seem hard on the defendants to have to find a new home, but as Brandon LJ has said, they enjoyed for 26 years a site for which they had nothing to pay and even now they have no immediate prospect of having to leave. I would dismiss the appeal.

Also agreeing, ORMROD LJ said: Nobody listening to this story could feel anything but sympathy for Mr and Mrs Grey, who are in an extremely unfortunate position, but it would be quite wrong, as my Lords pointed out, to think that the merits in this case lie all one way. Mr Baker, on behalf of the appellants, argued the case very much on the footing that the merits were all in his clients' favour. I, like my Lords, do not accept that. It seems to me that Mr and Mrs Grey have been to some extent at an advantage, by living rent free over a period of many years. Putting it in another way, they have been able to house themselves at a very much less cost than any of their friends and neighbours have been able to, and that owing to the kindness of the brick company. Also I think it is important that this court should not do or say anything which creates the impression that people are liable to be penalised for not enforcing their strict legal rights. It is a very unfortunate state of affairs when people feel obliged to take steps which they do not wish to take, in order to preserve their legal rights and prevent the other party acquiring rights against them. So the court in using its equitable jurisdiction must, in my judgment, approach these cases with extreme care. It is vital that the courts recognise and apply the limits of this equitable jurisdiction set out in the first place by Lord Kingsdown in *Ramsden v Dyson* (1866) LR 1 HL 129 at p 170 and later by Fry J in the passages which were cited by Scarman LJ and that those criteria are applied to this case. I agree that Mr and Mrs Grey fail to satisfy the very first and, therefore, they do not succeed in bringing themselves within reach of the equitable remedy which they are claiming.

I think also it is important to recognise that the extent of the remedy to which they would be entitled if they were able to qualify for it must, in my view, depend upon the sort of legal rights they mistakenly thought they had, or to use Lord Kingsdown's words, "under an expectation created or encouraged" by the other side." That is not, as I understand it, to be invoked as a general jurisdiction in equity to relieve hardship resulting from the application of the ordinary law; it is a restricted jurisdiction, and is one which should, in my judgment, be exercised with due caution.

I agree, therefore, that this appeal should be dismissed, not least because I think the order made by the court below, finally by consent, gives Mr and Mrs Grey at least as much, if not more than they could have got by winning this appeal.

The appeal was dismissed with costs. The appellants' contribution was assessed at nil. An order was made for costs to be paid out of the legal aid fund. A legal aid certificate was granted.

HOUSE OF LORDS

November 8 1979

(Before Lord WILBERFORCE, Viscount DILHORNE, Lord SALMON, Lord RUSSELL OF KILLOWEN and Lord KEITH OF KINKEL)

ROYCO HOMES LTD v SOUTHERN WATER AUTHORITY

Estates Gazette March 15 1980
(1979) 253 EG 1119

Water Act 1945 as amended—"Leap-frog" appeal to House of Lords from decision of Forbes J—Claims by water authority against developers of residential site in respect of provision of water supply—Whether new main proposed by authority to supply the developers' housing site was a "necessary main" within the meaning of section 37 of the Act, for which the developers could be required to undertake to make contributions and to deposit a sum as security—Question as to the point from which the "necessary main" should start—Nearest distribution main to site was too small and was fully committed—Authority proposed to construct a new main with sufficient capacity but considerably further away—Developers' contention that a "necessary main" must run from nearest point where there is an existing distribution main—As this was too small and fully committed, developers argued that authority should carry out such extensions and other works required to provide a sufficient supply, however uneconomic and contrary to good water engineering practice this might be—Developers' arguments rejected and decision of Forbes J affirmed—Proposed main a "necessary main" and developers liable to pay required contributions—Speech of Lord Diplock in Cherwell District Council v Thames Water Board considered—The necessary starting point for new main is to be determined in the light of sound water engineering practice

This was an appeal to the House of Lords from a decision of Forbes J (1978) 248 EG 599 under the "leap frog" procedure authorised by section 12 of the Administration of Justice Act 1969. The appellants, Royco Homes Ltd, had sought a declaration that the respondents, Southern Water Authority, were not entitled under section 37 of the Water Act 1945 as amended to demand a deposit and guarantee of payments in respect of a proposed new main between a point on the A2 road (known as Key Street) and the appellants' site boundary. Forbes J refused the declarations sought and dis-

A missed the appellants' action. The relevant facts are set out in the speech of Lord Keith of Kinkel.

John Drinkwater QC and Barry Payton (instructed by Harold Benjamin & Collins of Harrow) appeared for the appellants; Conrad Dehn QC and William Ainger (instructed by Barlow, Lyde & Gilbert) represented the respondents.

In his speech, LORD KEITH OF KINKEL said: This appeal comes directly before the House, pursuant to section 12 of the Administration of Justice Act 1969, from a judgment of Forbes J dated June 9 1978.

The appeal is concerned with the proper construction of section 37 of the Water Act 1945 which, as amended by section
B 46 of the Housing Act 1949 and section 11 of the Water Act 1973, provides:

Duty of Water Authority to provide domestic supply for new buildings

(1) Where an owner of land proposes to erect thereon buildings for which a supply of water for domestic purposes will be needed, he may require any water authority within whose limits of supply that land is situated to construct any necessary service reservoirs, to lay the necessary mains to such point or points as will enable the buildings to be connected thereto at a reasonable cost and to bring water to that point or those points, and thereupon the water authority shall, subject as hereinafter provided, comply with that requisition:
C Provided that the water authority before complying with a requisition under this subsection—

(a) may require the owner to undertake to pay in respect of each year a sum amounting to one-eighth of the expense of providing and constructing the necessary service reservoirs and providing and laying the necessary mains (less any amounts received by the water authority in respect of water supplied, whether for domestic or non-domestic purposes, in that year from those mains) until the aggregate amount of water charges as are stated by the water authority to be payable for a supply of water for domestic purposes payable annually in respect of the buildings when erected and in respect of any other premises connected with the said mains at the rates for the time being charged by the water authority equals or exceeds such sum as
D aforesaid or until the expiration of a period of 12 years, whichever first occurs; and

(b) except where the owner is a local or public authority, may also require him to deposit with the water authority as security for payment of the said annual sums, such sum, not exceeding the total expense of constructing the service reservoirs and providing and laying the mains, as the water authority may require.

(2) The water authority shall pay interest at the prescribed rate or, if no rate is prescribed, at four per cent per annum on any sum in their hands by virtue of a requirement under paragraph (b) of the proviso to the last foregoing subsection, and shall, on the request of the owner of the land, appropriate out of that sum any amount due under the undertaking referred to in paragraph (a)
E of the said proviso and shall, when the said undertaking is finally discharged, repay to the owner any sum remaining in their hands as aforesaid.

(3) Any question arising under subsection (1) of this section as to the point or points to which mains must be taken in order to enable buildings to be connected thereto at a reasonable cost shall, in default of agreement, be determined by the Minister.

(4) If the water authority, after receipt of a requisition under subsection (1) of this section and after tender to them of any undertaking or deposit which they may require in accordance with that subsection, do not before the expiration of three months (or, where a question has, before that time, been referred to the Minister under the last foregoing subsection before the expiration
F of three months from the date when the Minister notifies the water authority of his decision, if that period expires later) comply with the requisition, they shall, unless they show that the failure was due to unavoidable accident or other unavoidable cause, be guilty of an offence against this Act.

The relevant facts as found by Forbes J are as follows. The appellants are developers of land for residential purposes. They own and propose to develop for those purposes a substantial site near Sittingbourne in Kent, known as the Milton Church Farm Estate. The site is capable of accommodating 2,200 houses, and the appellants have so far received outline planning permission for the erection of 700 houses, and detailed planning permission for the erection of 204 of those. G
The respondents are the statutory water authority within whose limits of supply the site lies. By letter, dated June 7 1976, the appellants required the respondents to provide a domestic water supply to the 204 houses, but it was common ground that this requisition was to be taken as applying to the whole of the 700 houses for which the appellants had outline planning permission. The existing position as regards water supply was that there were no mains actually on the site. The nearest distribution main to the site was in Middletune Avenue at a point to the south of it. This was a 4-in main. There were in the locality other small distribution mains somewhat further away and certain larger mains further away still. All these H
mains were fully committed. Well to the south of the site, however, was a 10-in main laid under the east-west A2 Dover-London road. This main was not fully committed. The respondents proposed, in response to the appellant's requisition, to construct a new 8-in distribution main connected to the 10-in main under the A2 road at a point (known as Key Street) near the junction of the latter with the A249 road running north-east to the Isle of Sheppey. The proposed new main was to follow the A249 for some distance and then turn eastwards into the appellants' site. Its total length was to be about 3.5 km. The total number of houses capable of being supplied by this new 8-in main was 950. The next smaller stock size of main, J
however, would not have sufficient capacity to supply as many as 700 houses, while the cost of manufacturing a special size of main capable of supplying precisely 700 houses would be prohibitive. The respondents required from the appellants, in terms of section 37(1)(b) of the 1945 Act, a deposit of £115,000 in respect of the construction of the new main. There happened to be road works in progress at the time on the A249, and the respondents took the opportunity, in order to save greater expense in the future, of laying about one kilometre's length of the new main under the road. In respect of this work the appellants paid to the respondents under protest the sum of £16,000 in part satisfaction of the required deposit of K
£115,000.

By the present action the appellants seek a declaration that the respondents are not entitled in law to demand any deposit and guarantee of revenue in respect of the proposed main from Key Street to the boundary of the appellants' site, and that the appellants are entitled to repayment of the sum of £16,000 with interest. By the judgment under appeal Forbes J refused the declarations so sought and dismissed the action in relation thereto. It is to be mentioned that by the action the appellants also sought certain declarations related to another site owned by them within the respondents' limits of supply. These declarations were granted by Forbes J, and the respondents do L
not seek to disturb the part of his judgment concerned with that aspect of the case.

The appellants accept responsibility in respect of distribution mains within the boundaries of their site, but they maintain that they cannot lawfully be required to give any undertaking or pay any deposit in respect of the proposed new main between Key Street and the site boundary, on the ground that it is not a "necessary" main within the meaning of section 37(1). Their argument, in effect, is that no main outside the boundaries of the site can be a "necessary" main other than one which runs to the site from the nearest point where there is an existing distribution main. The particular point which M
they pick on here is the location of the 4-in distribution main in Middletune Avenue. That main is, of course, fully committed, and its capacity is in any event far too small for the purpose of delivering an adequate supply of water for the appellants' development. But the appellants say that it is up to the respondents, at the expense of the general body of ratepayers, to carry out all such extensions and improvements of the existing system as may be necessary to secure that a sufficient supply of water is available at the point indicated. This is on the face of it an absurd proposition, involving as it does that the respondents might be required to carry out

works that were completely uneconomic and quite contrary to sound water engineering practice. The appellants argue, however, that this result is compelled by the decision of this House in *Cherwell District Council v Thames Water Board* [1975] 1 WLR 448 and the reasons for that decision contained in the speech of Lord Diplock, concurred in by all the other Lords of Appeal who were party to the decision.

In my opinion the decision, properly understood, does not support the appellants' proposition. The facts were that the appellant council proposed to build houses on two sites in Banbury, where there were no existing water mains. The water supply at Banbury was obtained from a service reservoir into which water was pumped from the river Cherwell. The supply was becoming inadequate, and the proposed new houses would exacerbate the situation. To supplement the supply the respondent water authority proposed to extract water from the Thames at Woodstock and convey it by a new 27-in trunk main to the existing service reservoir at Banbury. In response to an inquiry from the appellant council as to the amount which they would be required to undertake to pay under section 37(1) in respect of the supply of water for the proposed housing development, the water authority quoted a figure which comprised the cost of new mains in the immediate vicinity of the site and also a very small contribution towards the cost of the new 27-in trunk main. The matter at issue was the appellant council's liability in respect of the latter element. The case had become unnecessarily complicated in the lower courts through counsel for the parties agreeing that the issue turned on the answer to a hypothetical question as to whether a trunk main was capable in law of being a "necessary main" within the meaning of section 37(1), but there is no need to go into that aspect. The House decided that the appellant council was not liable for any contribution towards the cost of the new trunk main, on the ground that, since it was not provided solely for the purpose of conveying water to points at which the new houses could be connected up by means of service pipes, it was not a "necessary main." The ultimate ratio of the case is to be found in the speech of Lord Diplock at p 457:

> In the instant case it is common ground that the 27-in trunk main, when completed, would serve to supply water to the general body of consumers of water for domestic purposes as well as to the proposed new buildings that were the subject of the contemplated requisition. How small a part the necessity to supply water to these buildings played in creating the need to provide the new trunk main is indicated by the minute proportion of the total cost of it that the board thought it equitable to ascribe to the council as requisitioners in respect of those buildings. That the 27-in trunk main would serve such other purposes as well in my view precludes it being a "necessary main" within the meaning of section 37 of the Water Act 1945.

In the course of arriving at this ratio Lord Diplock made a number of general observations about section 37, certain parts of which, as they were strongly founded on by counsel for the appellants in the present case, it is necessary to quote in full (p 455):

> "Mains" is, by virtue of its definition in the Act, a generic term which *prima facie* includes trunk mains as well as distribution mains, unless the context in which it appears makes it clear that it was used in some more restricted meaning. In section 37 (as also in clause 29) the only express restriction upon its generality is that the mains must be "necessary." This is, in my view, the key word in the section for the purposes of deciding the actual question raised by the originating summons. It poses immediately the question: "Necessary for what?"
>
> The section presupposes, first, that there are no existing mains bringing water to points at which it would be practicable at reasonable cost to connect the proposed buildings to those mains by service pipes. If there were, there would be no need for any requisition. Once the building was erected the owner would be entitled to require the connection to be made under clause 30, without any liability to contribute to the cost previously incurred by the undertakers in laying the existing mains or to the cost of replacing them by mains of greater capacity to meet the additional demand resulting from the connection of the newly-erected building to them.
>
> Secondly, the section presupposes that there will be a supply of water in bulk capable of being brought to the new mains by the undertakers, which if it is not already available to them it is their duty to procure without demanding from the requisitioner any contribution to the cost of doing so.
>
> So the "necessary mains" with which the section deals are confined to new mains to be laid by the undertakers in a street or other place where there are no existing distribution mains and to start from a point to which a supply of water is already being brought in existing works belonging to the undertaker; and the answer to the question: "Necessary for what?" is: necessary in order to convey the water from that starting point to points at which it would be practicable to connect the proposed buildings to those mains by service pipes, and for no other purpose.
>
> The section contemplates that as well as "necessary mains" there may be "necessary service reservoirs" to be constructed. *Ex hypothesi* these must lie between the starting point to which a supply of water is already being brought in existing works belonging to the undertakers and the new distribution mains to which the proposed new buildings will be connected. The new main which will carry water from that starting point to the service reservoir may well fall within the definition of "trunk main," and if the new trunk main and the new service reservoir will serve no other purpose than to convey water from that starting point to the new distribution main or mains to which the proposed buildings in respect of which the requisition has been made will be connected it would, in my view, fall within the expression "necessary mains." For in the context of section 37 I would construe "necessary" as applicable to new mains and service reservoirs provided for that purpose only and not to mains or reservoirs which will also serve some other purpose of the undertakers in connection with the supply of water to existing consumers or to potential new consumers whose water will not be brought to them through the new distribution main or mains.
>
> To ascribe any wider meaning to the word "necessary" where it appears in section 37—or, in a similar context, in clause 29—would, in my view, conflict with the general scheme of the Act for allocating the cost of works between the proceeds of the annual water rate payable by the general body of consumers of water for domestic purposes and charges additional to the water rate imposed upon particular consumers. Once works have been constructed by the undertakers which bring a supply of water to an existing distribution main, the cost of additions or improvements to any of those works lying between the ultimate source of the supply and service pipes connected with that main which are needed to meet increased demand for water must be provided for out of the proceeds of the annual water rate or loans serviced out of the proceeds of that rate. No additional charge may be imposed upon the individual consumers who benefit from particular additions or improvements, even where they are new consumers whose exercise of their right under clause 30 to have their premises connected to an existing main has caused the increased demand. The Act thus appears to accept as a general principle that new consumers as well as old are entitled to the benefit of additions and improvements to existing works rendered necessary by increased demand for water for domestic purposes which they themselves have caused; and that they cannot be required to make any individual contribution to the cost of those additions or improvements beyond what is payable by them by way of the annual water rate in common with consumers who may derive no such benefit.
>
> It would seem to conflict with that principle that an exception should be made in respect of additions or improvements to works lying between the ultimate source of supply and the point to which water had already been brought by the undertakers merely because the supply to the new consumer involved, in addition to the laying of service pipes, the laying of a new distribution main beyond that point to a point in the vicinity of the premises for which the water was required.

In my opinion it is clear that Lord Diplock did not have present to his mind, anywhere in the course of this passage, the kind of situation which prevails here, where the water authority have already brought to a point quite close to the proposed development a distribution main of small capacity, carrying a limited supply of water for existing buildings in the immediate vicinity, and where the extension of that main so as to serve the proposed development would be impracticable, or at least uneconomic and contrary to sound water engineering practice. The passage must be read *secundum subjectum materiam*, namely a situation where the water authority proposed to execute, and charge the requisitioner with part of the cost of, works intended to enhance and improve the general supply of water in the broad locality within which the proposed development lay. It is true that one of the cases which Lord Diplock had in contemplation was that where the need for an enhanced or improved supply was brought about entirely by the requirements of the proposed development. But even so, the principle that the cost of improvements or additions to the system cannot be laid to the charge of the developers applies only, on Lord Diplock's analysis, to improvements or additions executed upon sections of the system short of the point at which any new main necessary for the connection of the proposed development takes its departure. The present appeal is concerned with the question what is the appropriate starting point of a new main laid solely for the purpose of supplying the new development, and the related question whether the choice by the water authority of a particular starting point for such a main can have the effect of eliding the developers' liability in respect of it. These questions were not in issue in the *Cherwell* case, and it would, in my opinion, be a mistake to regard anything in Lord Diplock's speech as being directed to their resolution, or as imposing any necessary restriction upon the range of the search for the correct answers.

It is evident that any "necessary main" must start from a point where water is available to be led into it. So much was pointed out by Lord Diplock in the *Cherwell* case in the course of the passage quoted. But what, if any, are the other essential characteristics of the starting point? Is it enough that there should be a certain limited supply of water there, albeit quite insufficient to charge the new main, or is it right to postulate that the existing supply of water at the point in question should be adequate for that purpose? Neither answer would, in my view, necessarily be correct in all circumstances. I am of opinion that the necessary starting point is to be determined in the light of sound water engineering practice. In some instances that might indicate that the necessary starting point lay somewhere on an existing main of adequate capacity. But in others it might involve replacing an existing main of inadequate capacity with a larger one, and connecting the new main to the replacement at a suitable point. In that event no part of the cost of the replacement main could, under the decision in the *Cherwell* case, lawfully be charged to the requisitioner. But there are no sound grounds whatever for the view that the necessary starting point must invariably be the nearest point to the proposed development at which there is an existing supply of water in an existing main, however limited.

At this stage there is to be noted an argument for the respondents to the effect that the matter of the necessary starting point for the new main is entirely for the judgment of the water authority, acting reasonably and in good faith. In my opinion this argument is to be rejected. In this context the question is one to be decided by the court, in the event of dispute, in the light of evidence led about sound water engineering practice.

Upon the facts of this case, as found by Forbes J, there can be no doubt that the starting point selected by the respondents for the proposed new main was the correct one from the point of view of sound water engineering practice, and that it did not, in all the circumstances, have the effect of depriving the main of the character of a "necessary main." The main would be laid with the sole purpose of supplying the domestic water requirements of the appellants' proposed development, and at the end of the day it was not maintained that its superfluity of capacity for that purpose involved, in the circumstances, that it was larger than it need be.

For these reasons I am of the opinion that the judgment of Forbes J was correct and soundly reasoned, and that the appeal should be dismissed.

VISCOUNT DILHORNE and LORD RUSSELL OF KILLOWEN delivered speeches agreeing that the appeal should be dismissed. LORDS WILBERFORCE and SALMON also agreed.

COURT OF APPEAL

December 7 1979

(Before Lord Justice STEPHENSON, Lord Justice WALLER and Lord Justice CUMMING-BRUCE)

KNIGHT v HEXT AND OTHERS

Estates Gazette March 22 1980

(1979) 253 EG 1227

Liability for damage caused by fall of tree on to neighbour's barn—Cause of action based on allegation of negligence against owners who had purchased and entered into occupation of adjoining land a short time before the tree fell—Evidence of conversation between the parties indicating that some of the trees on the new owner's land, including the one that came down, were old, but not that any urgent action was necessary—11 days after this conversation, a period which included two Sundays, Christmas and Boxing Day, the tree fell on the barn—Court of Appeal reverse county court judge's finding of negligence—No duty in the circumstances on the new owners of the offending tree to arrange for it to be inspected by a tree expert as a matter of urgency

This was an appeal against a decision of Judge Goodall at Launceston County Court in favour of the plaintiff, Frederick Charles Knight. The appellants, defendants in the action, were Elizabeth Priscilla Hext and Jacqueline Loveday Hext (administratrices of the Estate of J W Hext) and Elizabeth Priscilla Hext in her own capacity. Elizabeth Priscilla Hext was the widow of the deceased J W Hext who, as mentioned in the judgment, was ill at the time when the tree fell on the plaintiff's barn. Mr Hext in fact went into hospital two days after he entered upon the property, leaving his wife in occupation and control.

Christopher Moger (instructed by W R Kirk, of Manchester) appeared on behalf of the appellants; Clive Toomer (instructed by Bright, Broad and Skinnard, of Callington, Cornwall) represented the respondent.

Giving judgment, STEPHENSON LJ said: This appeal arises out of the fall of a beech tree on property owned and occupied by Mrs Hext, the second defendant, on December 27 as long ago as 1972. It fell on to a barn belonging to the plaintiff, Mr Knight, who is the owner and occupier of the adjoining property. On October 10 1978 His Honour, Judge Goodall found that that fall was due to the negligence of the defendant (and I treat Mrs Hext as "the defendant" although the first defendants are the executors of her late husband, who was ill at the time of the accident and has since died). The learned judge, on that date, in the Launceston County Court gave judgment for the plaintiff for the sum of £1,579.78.

The first question, and I think the decisive question, raised by this appeal is, ought Mrs Hext, as a prudent landowner going into occupation of this property on December 12 1972, to have inspected this tree before it fell on December 27 1972? The learned judge, after saying that he found it a difficult case and that he had altered his mind in the course of it, answered that question "Yes" and accordingly found against Mrs Hext. Differing with hesitation from a decision of this learned judge on a matter with which I at once confess he would be much better acquainted than I am, I nevertheless have come to a firm conclusion that the answer to this question ought to be "No" and it should be answered in favour of Mrs Hext and not against her.

A The facts are, as I have said, that she and her husband, who was then a sick man, went into occupation of this property, which they had recently purchased, on December 12 1972. It was a property of some dozen acres containing quite a large amount of woodland, including a spinney of 12 or more trees which contained the beech tree which fell and damaged the plaintiff's barn. On this property there were a number of old trees. Mr Knight, the plaintiff, had lived on the adjoining property for over 40 years. He had been a tenant for 10 years and he had bought the adjoining property about the same time as the Hexts had bought their property. Mr Hext went into hospital on December 14 1972 and Mrs Hext was there-
B after in control and occupation of this property as its new landowner.

On December 16, which was a Saturday, she saw the plaintiff and one of his workmen, a Mr Curtis, cutting up a tree which had fallen in the park, which had belonged to the property which she had bought but was then in the ownership of the plaintiff. It may have been the first occasion on which the two parties had met, although the plaintiff had realised that Mrs Hext had moved into the property. However that may be, it was she who approached him and had a conversation with him, beginning with her expressing regrets that his tree had come down. That led to a conversation, as to which,
C as the learned judge said, there was a dispute between the plaintiff and the defendant. It is not necessary to go into details of that dispute, except to say that the plaintiff thought that he had used the word "danger" in reference to the trees on Mrs Hext's property. There is no dispute that he said there were old trees on her property, past mature, that could come down, or that he pointed to a spinney or clump which contained the tree which did come down on the night of December 27; but he thought that he had referred to the danger of the trees in that clump coming down and said not "that they *could* come down" but "that they *should* come down."

D That account of the conversation was rejected by the learned judge in favour of Mrs Hext's account, supported, as Mrs Hext's account was, by the evidence of Mr Curtis, who did not remember the word "danger" being used. What the learned judge found about that conversation was that it contained no specific warning about the particular tree and no suggestion of urgency or any specific statement of danger. "I am left with a clear impression," said the learned judge, "that the plaintiff spoke of trees nearby, probably pointing to the clump or spinney and certainly his statement included it. He said the trees were old and could come down. This was in the context of the tree which had come down. But I accept that
E there was no suggestion of urgency or any specific statement of danger." The learned judge went on to say this: "The problem is, where a new landowner is told that there are old trees which could come down, in the context of a tree which had come down, should that put him on inquiry? I think it should. There was no question of urgency. But after that conversation with the plaintiff, I think that a reasonably prudent landowner would have said: 'Well, I've got old trees near someone else's buildings; I think I ought to take a look at them.' It was only mid-December; the bulk of winter storms were to come. She knew that the trees were near another's building—she told me 'You can see the roof of the stable'"
F —that is the barn—"'from the lawn.'"

He repeated a number of times that he thought a prudent landowner, after that conversation, should have gone to look at the trees and his final view was expressed in this sentence: "I think a prudent landowner would have looked at the trees that day or the next—it would only take half an hour—and would have picked up the telephone and asked someone to come and have a look. Mrs Hext did not do that."

Then he went on to find that, if she had done that, there was a reasonable chance that remedial action would have been taken in time, because there was evidence that there was a tree-feller working nearby, some 400 yards away, and the G
learned judge thought he could take judicial notice of the fact that there were a number of enterprising men in the country with chain saws. So he came to the conclusion that this defendant had failed (as he put it, "just failed") in her duty of care and, having failed in her duty of care, could not really be heard to say, without some further evidence, that, if she had discharged her duty of care, it would not have resulted in timely action to prevent the damage to the plaintiff's barn. I should have said that the plaintiff's barn, or that part of it on which the tree fell, was not inhabited, there was no danger to life and limb, but it was used for storing wood, machinery and wool. H

The evidence of experts on both sides was that this tree which fell was a beech tree some 120 or 130 years old and that beech trees (it was not clear whether they usually do or only can exceptionally) live for 200 years. There was undisputed evidence that it was leaning over the defendant's boundary in the direction of the barn and that its crown was weighted in the direction of the barn, that the prevailing wind from the south-west would further improve its chances of falling on the barn if it did come down, but that it did present a healthy appearance. After it came down it was discovered that it was suffering from butt rot, but there was no visible sign, the judge having rejected the suggestion that J
there was any significance in the fact that the roots had lifted, of its being diseased or more likely to fall than any old tree leaning in that way and subjected to the storms of winter. However, it was the evidence of the plaintiff's expert, at any rate, that an expert looking at it would have recommended lopping the crown, and I think would have recommended felling the tree after some preliminary lopping of the top, and that it would have required a highly-skilled tree-feller to do what was not a simple job, to bring down the tree without bringing it down on to the barn.

Mr Moger has submitted on behalf of the defendant that it could not have been negligent for Mrs Hext not to have K
inspected the tree within 11 days of that conversation with the plaintiff, Mr Knight. I am bound to say that I accept that submission. If the conversation with Mr Knight had contained, what Mr Knight said it contained, then there would have been urgency in the matter and the judge's finding that Mrs Hext ought to have inspected or had a look at the spinney, including this tree, that day or the next, that is to say, December 16 or 17, would not, I think, have been open to serious complaint. But on the version of the conversation which the learned judge accepted, I cannot myself see that there was any urgency, any reason for Mrs Hext to think that the matter was urgent or that what she had been told L
required her to go before Christmas, or before the New Year, or before December 27 at any rate, to look at this spinney.

As it seems to me, even without the fact that a good deal of her time was no doubt taken up by worry about her husband's illness, and we are told that there was also a 21st birthday party for her daughter, Christmas was coming, and, although as a countrywoman she no doubt realised, with old trees and one having come down, that winter storms might bring trees down, unless she was not merely put upon inquiry that there might be old trees which could come down, including old trees in this clump, but put upon such inquiry as to suggest that some immediate action was called for, it was M
perfectly reasonable for her to take the view "I will get round to having a look at these trees after Christmas or in the New Year." If the duty imposed on her was to inspect this clump on one of the five remaining working days (and that is all there were) before the tree fell down, it seems to me to impose an unreasonably high duty upon this landowner. The conversation having taken place on the Saturday, she could have, Mr Toomer submitted, easily taken half an hour in going round to look at the trees. But, in all the circumstances, I cannot see that there was any duty on her to go and look at them at that weekend or on any weekday the

A following week, or to spend any part of her Christmas in taking half an hour to go and look at them. On that ground I would allow this appeal.

In case I am wrong about that, I think I ought to go on to consider both Mr Moger's other two points. One of them I think is formidable, the other not. He says, secondly, that even if there was such a duty and Mrs Hext had gone and looked at this tree well before December 27, she would not have called in an expert and it would not have been reasonable to call in an expert. She would have seen a 60-ft tall apparently healthy tree, it is true leaning in this way, it is true weighted in this way, and liable to fall, if it did fall, on the
B barn. But why should she say to herself "I must do something at once. I must call in an expert?" It is said by the learned judge that she had only to telephone, but it would not have been reasonable to expect her to do that on the Saturday or Sunday. So there were only five days on which she could do that—December 18, 19, 20, 21 and 22. Then we are at the weekend, the 23rd would be Saturday, then Christmas Eve, Christmas Day and Boxing Day, with the tree down on the 27th. It does not leave much time to take action. For my part, if I had any doubt on the first point, I would have considerable doubt whether it could be found that her inspection of the tree in time would have resulted or ought to have
C resulted in her telephoning an expert before it fell.

There remains the third point submitted by Mr Moger. If she had looked at the tree and if she had called in an expert as a matter of urgency—I mean telling the expert that she regarded the matter as urgent or, as Mr Toomer suggests, saying: "Well, I am not sure whether it is urgent or not as a result of my inspection, but I would like you to come and look right away and say what you think about it"—would she have been able to get a tree-feller in time? Mr Moger has submitted that that, too, is very doubtful. On that point I am not prepared to put my opinion against the opinion of the learned judge, based as it was on evidence that there was a
D tree-feller with a winch, and so on, working some 400 yards away from this property at the material time and that the judge had judicial knowledge that there were skilled tree-fellers quite capable of doing this job in that neighbourhood.

But, for the reasons which I have given, I have come to the conclusion, differing, as I say, reluctantly from a decision of this learned judge, that he gave the wrong answer to the question and this appeal ought to be allowed and judgment entered for the defendants.

Agreeing, WALLER LJ said: I realise that, in allowing this appeal we are differing from the view of a very experienced judge, but I have come to the conclusion that it is not possible
E to reconcile his finding that there was no question of urgency in the conversation with the finding of negligence in failing to take action at once. The timetable shows that Mrs Hext moved into the house on Tuesday, December 12, and that the conversation or conversations—she speaks of two conversations—took place on Saturday, December 16. There was then the following week, with Christmas Day on Monday week. So that there were only five working days, or 5½ if the Saturday before Christmas on Monday could be counted, in which action could be taken.

The test of the standard of care is that of a prudent non-expert landowner. The first question which has to be con-
F sidered is: was she under an obligation to take action at once to inspect trees, as the judge found, or was the information given to her, with no question of urgency, such as it would be proper for her to make an inspection of the trees at some time, namely, within a reasonable time? In my judgment, there was no question of urgency raised by the conversation and therefore, although the nature of the trees had been drawn to Mrs Hext's attention, all that was required of her was that, within a reasonable time, she should go and make an inspection. I find it impossible to say that a failure to do so before Christmas was a failure to take action within a reasonable time.

But even if "reasonable time" is to be construed as within G
two or three days, it seems to me that it is quite impossible to say that, at any time during that period, one can import such a sense of urgency as would make it necessary to obtain the services of a tree expert and a tree-feller within the two, three or four days which were then left.

So I agree with my Lord that, at the first stage, there was no breach of duty on the part of Mrs Hext, and even at some stage during the next two or three days there would be no failure to take action within a reasonable time if nothing was done before Christmas. I agree that this appeal should be allowed.

Also agreeing, CUMMING-BRUCE LJ said: It is easy to make H
carping criticisms of a judgment when one reads the note of the judgment and I hope that my criticisms are not carping. I have come to the conclusion that the judge did not really ask himself the right question when he considered the problem which he identified as the first problem. He asked this question: When a new landowner is told that there are old trees which could come down, in the context of a tree which had come down, should that put him on inquiry? The judge answered that question in the affirmative, adding that there was no question of urgency.

Then the judge went on to postulate the hypothetical
answers that a reasonable landowner would give in the situation that had arisen. He put it this way: "After the conversation with the plaintiff a reasonably prudent landowner would have said: 'Well, I have got old trees and they are somewhere near his building. I think I ought to take a look at them.'" That is an incomplete answer, because the problem posed is: would the prudent landowner, in all the objective circumstances, have said: "I think I ought to take a look at the trees in a reasonable time" or would the prudent landowner have said: "I think I ought to take a look at the trees immediately?" J

It seems to me clear that what moved the judge to decide
that the duty was to look at the trees immediately was founded on his appreciation of the fact that it was very easy for the defendant to go and look at the trees, just a matter of walking a few steps. But though it was very easy to do that, the question of the extent of the duty depends on whether, on the facts, the duty was to look at the trees as a matter of urgency. For the reasons stated by my Lords, on the judge's findings, the answer to that question is "No." K

I agree that the appeal should be allowed.

The appeal was allowed and the judge's order set aside, judgment for the defendants being substituted, with costs in the Court of Appeal and below.

L

COURT OF APPEAL

November 29 1979

(Before Lord Justice MEGAW, Lord Justice BROWNE and Lord Justice BRIGHTMAN)

FEDERATED HOMES LTD v MILL LODGE PROPERTIES LTD

Estates Gazette April 5 1980

(1979) 254 EG 39

Restrictive covenant—Enforcement of covenant restricting density of housing development—Effect of various transactions resulting in the vesting of separate parts of "retained land" in the respondents (plaintiffs) and the vesting of subject land in the appellants (defendants)—Whether benefit of covenant had vested in and was enforceable by respondents—In the case of one parcel the benefit had passed through an unbroken chain of assignments—In the case of the other the matter was put beyond doubt by section 78 of the Law of Property Act 1925—True construction of section 78 discussed—Not merely statutory shorthand to reduce size of legal documents —Section 78 has the effect that if a covenant "touches and M

concerns" the land of the covenantee it runs with the land for the benefit of his successors in title, persons deriving title under him, and other owners and occupiers—Narrower view taken by Lords Upjohn and Wilberforce in Tophams Ltd v Earl of Sefton of section 79 not applied by analogy to section 78—Covenant annexed to whole prima facie annexed to every part thereof—Judge's decision in favour of respondents upheld—Brightman LJ's judgment of interest for conveyancing and real property law

This was an appeal by the defendants, Mill Lodge Properties Ltd, from a decision of J W Mills QC, sitting as a deputy judge of the Chancery Division, granting the plaintiffs, Federated Homes Ltd, an injunction restraining the defendants from building 32 houses in breach of a restrictive covenant debarring the defendants from exceeding a density of 300 houses on a site near Newport Pagnell in Buckinghamshire.

A L Price QC and M E Mann (instructed by Blythe, Dutton, Holloway, agents for Coffin, Mew & Clover, of Havant) appeared on behalf of the appellants; M A F Lyndon-Stanford QC and R G B McCombe (instructed by Eatons) represented the respondents.

Giving the first judgment at the invitation of Megaw LJ, BRIGHTMAN LJ said: This is an appeal from a judgment of J W Mills QC, sitting as a deputy High Court judge of the Chancery Division. The dispute relates to a large development site near Newport Pagnell in Buckinghamshire. This site consists of four areas of land of roughly equal size which can, for convenience, be called the red, green, pink and blue land. There were also included in the development site certain additional bits of land which I shall ignore.

The plaintiff company is now the owner of the red and the green land. The defendant company, Mill Lodge Properties Ltd (which I shall call "Mill Lodge"), is the owner of the blue land. The plaintiff company claims to be entitled to the benefit of a restrictive covenant which is said to debar Mill Lodge from building more than 300 houses on the blue land. Mill Lodge is in the process of exceeding that density by building an additional 32 houses in conformity with a new planning permission which it has obtained. The judge decided against Mill Lodge and granted an injunction.

In September 1970 a company called Mackenzie Hill Ltd (which I shall refer to as " Mackenzie Hill ") was about to become the owner of the site. On September 18 the Buckinghamshire County Council, as planning authority acting through the Newport Pagnell Urban District Council as its agent, granted outline planning permission to Mackenzie Hill to develop the site by the provision of housing and associated amenities. There were a number of conditions attached to the permission, of which the important ones were these:

> 1. The approval of the County Council shall be obtained to the number, siting, design and external appearance of the buildings (except the schools), and the means of access thereto before the development is commenced. . . .
>
> 2. This permission shall be null and void if the approval of the County Council to all the matters referred to in the last preceding condition has not been applied for in writing within three years —ie by September 1973.
>
> 4. This permission shall enure only for the benefit of the applicants and their subsidiaries.
>
> 5. This permission shall relate to the erection of a church/community centre, a shopping cluster, a petrol filling station, a public house, and approximately 1,250 private residential dwellings. . . .

Condition 8 specified the rate of development, but this topic was covered in a revised form by the agreement to which I shall next refer.

On the same day as the planning permission Mackenzie Hill entered into an agreement with the urban district council which has been called the phasing agreement. By this agreement the council undertook to construct roads and sewers through the site and Mackenzie Hill agreed to contribute towards the cost. Clause 7, so far as material, reads as follows (in the clause the expression "Developers" means Mackenzie Hill and "H1" means the development site):

> (a) If the Developers shall desire to see the whole or any part of H1 they shall forthwith inform the Council and the terms and conditions of such sale or sales shall include the disclosure of this Agreement to the purchasers and provisions to safeguard the Council's position under this Agreement which shall be to the satisfaction of the Council . . . PROVIDED that this sub-clause shall not apply to the sale of any single completed dwelling to an individual purchaser.
>
> (b) The Developers shall have the right to assign the burden and benefit of this Agreement but shall not be released from their obligations hereunder on any such assignment unless the proposed assignees shall have first furnished the Council with a Bond. . . .

By clause 9(a) Mackenzie Hill agreed that the rate of development should not exceed 50 houses by the end of 1970 and a further 125 houses in each of the years 1971 and 1972. The rate of development was then to be reviewed with the possibility of an increase, but not a decrease, in the rate. On that basis the development would be completed in or before the year 1980 according to whether or not the rate of development ultimately exceeded a minimum of 125 houses a year. That rate of development differed slightly from the rate laid down in the planning permission and that, no doubt, was the reason for clause 9(e) of the phasing agreement, whereunder Mackenzie Hill covenanted with the council that they would enter into an agreement under seal with the council on or before December 14 1970 incorporating the terms of clause 9(a).

It seems to me, reading clauses 7 and 9(e), that the urban district council contemplated that Mackenzie Hill might not itself develop but might part with the development in favour of someone else. Furthermore, as it was the urban district council which had, on the same day as the phasing agreement, in its capacity as agent for the Buckinghamshire County Council, granted the planning permission to Mackenzie Hill, the planning permission ought fairly to be read in conjunction with the phasing agreement.

On Februry 26 1971 Mackenzie Hill and its mortgagee sold and conveyed the blue land to Mill Lodge. By clause 3 the conveyance was expressed to be subject to and with the benefit of the phasing agreement. Clause 5(iv) set out the covenant which is the subject-matter of this action. It reads as follows:

> The Purchaser hereby covenants with the Vendor that . . . (iv) in carrying out the development of the " blue " land the Purchaser shall not build at a greater density than a total of 300 dwellings so as not to reduce the number of units which the Vendor might eventually erect on the retained land under the existing Planning Consent.

There was a simultaneous conveyance of the pink land to a company called Gough Cooper (Midland) Ltd (which I shall abbreviate to " Gough Cooper "). Clause 6 of the Mill Lodge conveyance provided that the blue land was sold with the benefit, so far as the same related to the blue land, of the agreements and undertakings on the part of Gough Cooper contained in the Gough Cooper conveyance. Clause 7 of the Mill Lodge conveyance contained a covenant by Mackenzie Hill with Mill Lodge, expressed to be for the benefit of the blue land and every part thereof, that Mackenzie Hill would not build houses on the red and the green land, with an immaterial exception, before the date on which Mill Lodge had erected or had permission from the planning authority under any revised phasing agreement to erect 300 dwellings on the blue land or January 1 1975, whichever should happen first, but not in any event earlier than January 1 1974.

The Mill Lodge conveyance contains no express definition of the retained land. There is, however, a reference in clause

2 to "any adjoining or adjacent property retained by the Vendor." I read "the retained land" in clause 5(iv) as meaning just that. I do not accept the submission of the respondent's counsel that the retained land included the pink land; clause 6 makes it clear that the pink land was not retained, but was being simultaneously conveyed to Gough Cooper. Counsel invited us to look at the contract of sale to Mill Lodge for the purpose of resolving an ambiguity as to the meaning of the retained land, but I see no ambiguity. I, therefore, conclude that the retained land means the red and the green land and the small additional areas comprised in the site, other, of course, than the blue and the pink land. To avoid confusion, I think I ought to explain that the expression "the green land" is made use of in the Mill Lodge conveyance, but it means both the red and the green land as I use those expressions in this judgment.

The Gough Cooper conveyance is not, in my view, relevant to the construction of the Mill Lodge conveyance, but it is permissible to refer to it as part of the backcloth against which the Mill Lodge conveyance was made, since the Gough Cooper conveyance is referred to in the Mill Lodge conveyance as a document of simultaneous execution. Clause 4 of this conveyance was obviously intended to cover the same subject-matter as clause 5(iv) of the Mill Lodge conveyance and was in the following terms:

> The Purchaser for itself and its successors in title hereby covenants with the Vendor and its successors in title that the Purchaser shall not build on the land hereby conveyed at a greater density than a total of Three Hundred dwellings so as not to reduce the number of units which the Vendor might eventually erect on the land edged green on the said plan under the existing Planning Consent obtained by the Vendor in respect of the whole of the land edged red edged blue and edged green on the said plan.

Clause 6 of the Gough Cooper conveyance is the counterpart of and has much the same wording as clause 7 of the Mill Lodge conveyance, in effect binding Mackenzie Hill to give precedence to the purchasers' building programme in the operation of the phasing agreement.

A month later Mackenzie Hill sold and conveyed the red and the green land to William Brandt's Sons & Co Ltd (which I shall call "Brandt's"). Brandt's was in fact Mackenzie Hill's mortgagee of the blue and the pink land at the time of the earlier conveyance, but nothing turns on that. The conveyance to Brandt's was dated March 25 1971 and it contained an express assignment of the benefit of the covenant contained in the Mill Lodge and the Gough Cooper conveyances. Just under a year later Brandt's sold and conveyed the green land to the plaintiffs. This conveyance, which was dated February 25 1972, likewise contained an express assignment of the benefit of those covenants.

Shortly afterwards Brandt's and the plaintiff company conveyed the red land to BTA Trading Company Ltd. In that conveyance the plaintiff was the purchaser and BTA was the subpurchaser. This conveyance, which is dated March 1 1972, also contained an express assignment of the benefit of the covenants. Following that conveyance BTA's title became registered at the Land Registry. On March 18 1975 BTA, which by then had changed its name to UDT Properties Ltd, sold and conveyed the red land to the plaintiff company. This conveyance was in the form of a transfer applicable to registered land. It did not contain any express assignment of the benefit of the covenants in the Mill Lodge and the Gough Cooper conveyances.

The original outline planning permission granted in 1970 lapsed in 1973 because approval of the county council to all the matters referred to in Condition 1 had not been applied for in respect of the total site within three years. In fact, Mill Lodge proceeded with the separate development of the blue land under new planning permissions granted in 1971 and 1972 for a total of 300 dwellings.

Towards the end of 1977 the plaintiff company applied for planning permission in respect of the red and the green land. The balance of density left available for these areas, having regard to the Mill Lodge and Gough Cooper conveyances, was approximately 650 dwellings if the principle of the 1970 planning permission still applied. The application was, in fact, for a much greater density, but after a planning inquiry the 1970 density of about 1,250 dwellings for the entire site was reaffirmed. It was at about this time that the plaintiff company discovered that, on January 9 1975, Mill Lodge had obtained planning permission for the erection of an additional 32 dwellings on the blue land. It is not in dispute that the existence of such additional dwellings would or might prejudice the plaintiff company in relation to the development that might be permitted on the red and the green land; and, therefore, if the plaintiff company's rights would be infringed by the building of the further 32 dwellings, the plaintiff company would suffer damage.

In September 1978, after much prevarication on the part of Mill Lodge, the plaintiff company issued a writ to restrain Mill Lodge from building on the blue land at a greater density than a total of 300 dwellings in breach, it was alleged, of clause 5(iv) of the Mill Lodge conveyance. The defences raised by Mill Lodge so far as relied upon in this appeal were as follows: (1) the covenant in clause 5(iv) was said to be personal to Mackenzie Hill so that the benefit thereof was incapable of assignment to the plaintiff company; (2) alternatively, it was said that the covenant became spent when the 1970 planning permission became void at the end of the three-year period; (3) it was said that, if the covenant was assignable and was not spent, then the benefit did not become vested in the plaintiff company by assignment or otherwise.

That, in broad effect, was how the defence was pleaded so far as relevant for present purposes. In a reserved judgment the learned deputy High Court judge held that the covenant was not personal to Mackenzie Hill and was not spent when the original planning permission lapsed. As regards the transmission of the benefit of the covenant, he held that the benefit was not annexed to the red and the green land, so that it did not automatically pass upon conveyances of the red and the green land. However, he found, as was clearly the fact, that there was an unbroken chain of assignments between transferor and transferee of the green land, so that the benefit of the covenant was now vested, by reason of such assignments, in the plaintiff company as the present owner of the green land. There was no such unbroken chain of assignments in the case of the red land; but the judge considered that section 62 of the Law of Property Act 1925, which implies general words into a conveyance of land, was apt to carry the benefit of the covenant from UDT Properties Ltd, the previous assignee of such benefit, to the plaintiff company when the registered transfer in its favour was made. The defence, therefore, failed. The judge rejected a submission that damages would be the proper remedy. He granted an injunction against building in excess of the permitted density and gave liberty to apply for a mandatory injunction.

I deal first with the question of construction, upon which two issues arise: whether the covenant was personal to Mackenzie Hill and whether it is spent.

Mr Price, for the appellants, pointed out that the planning permission was expressed by condition 4 to enure for the benefit of Mackenzie Hill and its subsidiaries. That meant that a purchaser from Mackenzie Hill, not being a subsidiary company, had no legal right as between itself and the Buckinghamshire County Council to rely on the planning permission as authority to carry out development which would otherwise be contrary to planning legislation. Possibly the condition was inserted to enable the planning authority to object to development by somebody of whom it did not

A approve. Mr Price sought to argue from this that the covenant in the Mill Lodge conveyance was personal to Mackenzie Hill and not assignable. It was designed, he submitted, to protect a non-assignable planning permission and therefore should itself be treated as non-assignable.

I do not think that the appellants can gain much comfort from the form of the planning permission. The planning permission was certainly not exclusive to Mackenzie Hill because it was available to a subsidiary of Mackenzie Hill; so the benefit was clearly assignable to that extent. Theoretically, and not I think as a matter of reality, there was nothing in the planning permission to prevent the assignment B of the benefit. The only restriction was that the proprietor who ultimately relied upon the permission would have to be either Mackenzie Hill or a subsidiary. There could, in theory, be any number of intermediate transfers of the land, with the benefit of the permission, through persons who would not have been themselves qualified to rely upon the permission. I think this is a narrow point and I do not stress it. Mr Price also relied upon the fact that clause 5 is penned as a covenant with the vendor (no mention of assigns) and that it is linked with the number of dwellings which the vendor (again no mention of assigns) might erect on the retained land.

C But, apart from these considerations, I entirely agree with the learned deputy High Court judge when he said that it is neither necessary nor natural nor sensible to read the covenant as personal to Mackenzie Hill. Generally speaking, the benefit of a contract between businessmen is assignable without mention of assignability unless the contract is of a personal nature, which the restrictive covenant was not. Furthermore, clause 3 of the Mill Lodge conveyance stated that the property was sold subject to and with the benefit of the phasing agreement and, as I have already mentioned, clauses 7 and 9(e) of that agreement in terms contemplated D that Mackenzie Hill might sell the development site in whole or part. So it is hardly possible to argue that the parties *must* have contracted on the basis that Mackenzie Hill personally would develop the retained land. I conclude that the restrictive covenant was not personal, but was assignable, which seems to me to correspond with business realities.

If there were still any doubt, section 78 of the Law of Property Act 1925 in my view sets that doubt at rest. For it provides that a covenant relating to any land of the covenantee shall be deemed to have been made with the covenantee and his successors in title, which presupposes assignability. I shall have occasion in due course to examine E this section at greater length.

I turn to the defence that the covenant is now spent as the 1970 planning permission has lapsed. The concluding words of clause 5(iv) are: "so as not to reduce the number of units which the vendor might ultimately erect on the retained land under the existing planning consent." As no dwellings can now be erected in reliance upon the original planning consent, the covenant has, it was argued, lost its purpose. This resolves itself into the question whether the concluding words form an integral part of the restrictive covenant or are merely explanatory of the covenant without controlling it. I think that the answer is largely a matter F of impression and is not susceptible of prolonged argument. It does, however, seem to me a little unlikely that the parties intended to tie the restrictive covenant to the original planning permission so that the covenant and the permission should stand and fall together. There would seem no purpose in such rigidity. There must always have been a strong possibility, if not a likelihood, that a developer would have to apply for a new planning permission at some stage (as happened) because the three-year deadline for seeking planning approval for the siting, design and external appearance of the buildings and the means of access thereto does not fit easily into the 10-year phasing of the development.

G There is the additional consideration that, if the covenant were linked to the 1970 planning permission, it is difficult to see its value. The 1970 planning permission was to become void if detailed approval were not applied for in relation to the entire site within three years. As Mackenzie Hill was selling off and therefore not developing the blue and the pink land, the 1970 planning permission seems to me to have had a very doubtful future. It could hardly have been anticipated that Mackenzie Hill would be applying for detailed approval for the development of land which was going to be developed by others. There was no covenant by Mill Lodge or Gough Cooper to apply for detailed approval H within the three-year period; so the covenant, if linked to and dependent upon the 1970 planning permission, becomes somewhat of a nonsense as Mr Price was, I think, constrained to admit. I think that the more natural and businesslike construction of the subclause is to read the reference to the existing planning permission as explanatory and not as controlling.

Having reached the conclusion that the restrictive covenant was capable of assignment and is not spent, I turn to the question whether the benefit has safely reached the hands of the plaintiff company. The green land has no problem, owing to the unbroken chain of assignments. I am disposed J to think that that is sufficient to entitle the plaintiff company to relief, and that the plaintiff company's right to relief would be no greater at the present time if it were held that it also had the benefit of the covenant in its capacity as owner of the red land. However, the judge dealt with both areas of land and I propose to do the same.

An express assignment of the benefit of a covenant is not necessary if the benefit of the covenant is annexed to the land. In that event, the benefit will pass automatically on a conveyance of the land, without express mention, because it is annexed to the land and runs with it. So the issue of annexation is logically the next to be considered. K The judge said this:

> The next heading with which I must deal is "annexation," to which I will now come. It is a somewhat technical thing in the law of restrictive covenants. A good deal of argument was addressed to me on annexation by both sides. Submissions are made about express annexation, implied annexation, that is to say, annexation implied from surrounding circumstances, and annexation by assignment. In my judgment, there was in this case no "annexation" of the benefit of the covenant to the retained land or any part of it. Section 78, in particular, of the Law of Property Act does not have the effect of annexing the benefit of the covenant to anything. It is simply a statutory shorthand L for the shortening of conveyances, which it perhaps has done to some extent in this case. Annexation depends upon appropriate drafting, which is not here in this case, in spite of a recent process which can perhaps be called "a widening of the law" in these matters. The attendant circumstances moreover, positively militate against annexation because, as Mr Price rightly pointed out to me (though he did so in the course of his argument on construction), the restriction in this particular case is of limited duration and plainly not applicable to ultimate purchasers of plots of the land intended to be benefited. "Annexation," in my judgment, is for the parties to the covenant itself to achieve if they wish to, and (though those parties may no doubt provide for annexation at a later stage) I am not satisfied or prepared to hold that there M is any such thing as "delayed annexation by assignment" to which the covenantor is not party or privy.

The reference to "delayed annexation by assignment" is to a proposition that a covenant can, on a later assignment, thereby become annexed to the land by the act of the assignor and the assignee alone.

In my judgment the benefit of this covenant was annexed to the retained land, and I think that this is a consequence of section 78 of the Law of Property Act 1925, which reads:

> (1) A covenant relating to any land of the covenantee shall be deemed to be made with the covenantee and his successors in title

and the persons deriving title under him or them, and shall have effect as if such successors and other persons were expressed.

For the purposes of this subsection in connexion with covenants restrictive of the user of land "successors in title" shall be deemed to include the owners and occupiers for the time being of the land of the covenantee intended to be benefited.

(2) This section applies to covenants made after the commencement of this Act, but the repeal of section 58 of the Conveyancing Act 1881 does not affect the operation of covenants to which that section applied.

Mr Price submitted that there were three possible views about section 78. One view, which he described as "the orthodox view" hitherto held, is that it is merely a statutory shorthand for reducing the length of legal documents. A second view, which was the one that Mr Price was inclined to place in the forefront of his argument, is that the section only applies, or at any rate only achieves annexation, when the land intended to be benefited is signified in the document by express words or necessary implication as the intended beneficiary of the covenant. A third view is that the section applies if the covenant in fact touches and concerns the land of the covenantee, whether that be gleaned from the document itself or from evidence outside the document.

For myself, I reject the narrowest interpretation of section 78, the supposed orthodox view, which seems to me to fly in the face of the wording of the section. Before I express my reasons I will say that I do not find it necessary to choose between the second and the third views because, in my opinion, this covenant relates to land of the covenantee on either interpretation of section 78. Clause 5(iv) shows quite clearly that the covenant is for the protection of the retained land and that land is described in clause 2 as "any adjoining or adjacent property retained by the Vendor." This formulation is sufficient for annexation purposes: see *Rogers v Hosegood* [1900] 2 Ch 388.

There is in my judgment no doubt that this covenant "related to the land of the covenantee," or, to use the old-fashioned expression, that it touched and concerned the land, even if Mr Price is correct in his submission that the document must show an intention to benefit identified land. The result of such application is that one must read clause 5(iv) as if it were written: "The purchaser hereby covenants with the vendor and its successors in title and the persons deriving title under it or them, including the owners and occupiers for the time being of the retained land, that in carrying out the development of the blue land the purchaser shall not build at a greater density than a total of 300 dwellings so as not to reduce, etc." I leave out of consideration section 79 as unnecessary to be considered in this context, since Mill Lodge is the original covenantor.

The first point to notice about section 78(1) is that the wording is significantly different from the wording of its predecessor, section 58(1) of the Conveyancing and Law of Property Act 1881. The distinction is underlined by subsection (2) of section 78, which applies subsection (1) only to covenants made after the commencement of the Act. Section 58(1) of the earlier Act did not include the covenantee's successors in title or persons deriving title under him or them, nor the owners or occupiers for the time being of the land of the covenantee intended to be benefited. The section was confined, in relation to realty, to the covenantee, his heirs and assigns, words which suggest a more limited scope of operation than is found in section 78.

If, as the language of section 78 implies, a covenant relating to land which is restrictive of the user thereof is enforceable at the suit of (1) a successor in title of the covenantee, (2) a person deriving title under the covenantee or under his successors in title and (3) the owner or occupier of the land intended to be benefited by the covenant, it must, in my view, follow that the covenant runs with the land, because *ex hypothesi* every successor in title to the land, every derivative proprietor of the land and every other owner and occupier has a right by statute to the covenant. In other words, if the condition precedent of section 78 is satisfied —that is to say, there exists a covenant which touches and concerns the land of the covenantee—that covenant runs with the land for the benefit of his successors in title, persons deriving title under him or them and other owners and occupiers.

This approach to section 78 has been advocated by distinguished textbook writers; see Dr Radcliffe in 57 *Law Quarterly Review* 203, Professor Wade in the *Cambridge Law Journal* (1972) 157 under the apt cross-heading "What is wrong with section 78?", and *Megarry and Wade, Law of Real Property*, 4th Ed 764. Counsel pointed out to us that the 4th edition of Megarry and Wade's textbook indicates a change of mind on this topic since the 3rd edition.

Although the section does not seem to have been extensively used in the course of argument in this type of case, the construction of section 78 which appeals to me appears to be consistent with at least two cases decided in this court. The first is *Smith and Snipes Hall Farm Ltd v River Douglas Catchment Board* [1949] 2 KB 500. In that case an agreement was made in April 1938 between certain landowners and the Catchment Board under which the Catchment Board undertook to make good the banks of a certain brook and to maintain the same, and the landowners undertook to contribute towards the cost. In 1940 the first plaintiff took a conveyance from one of the landowners of a part of the land together with an express assignment of the benefit of the agreement. In 1944 the second plaintiff took a tenancy of that land without any express assignment of the benefit of the agreement. In 1946 the brook burst its banks and the land owned by the first plaintiff and tenanted by the second plaintiff was inundated. The two important points are that the agreement was not expressed to be for the benefit of the landowner's successors in title; and there was no assignment of the benefit of the agreement in favour of the second plaintiff, the tenant. In reliance, as I understand the case, upon section 78 of the Law of Property Act, it was held that the second plaintiff was entitled to sue the Catchment Board for damages for breach of the agreement. It seems to me that that conclusion can only have been reached on the basis that section 78 had the effect of causing the benefit of the agreement to run with the land so as to be capable of being sued upon by the tenant.

The other case, *Williams v Unit Construction Ltd* (unreported in the usual series of law reports but fully set out in the 19th volume of *The Conveyancer* 262), was decided by this court in 1951. There a company had acquired a building estate and had underleased four plots to Cubbin for 999 years. The underlessors arranged for the defendant company to build houses on the four plots. The defendant company covenanted with Cubbin to keep the adjacent road in repair until adopted. Cubbin granted a weekly tenancy of one house to the plaintiff without any express assignment of the benefit of the covenant. The plaintiff was injured owing to the disrepair of the road. She was held entitled to recover damages from the defendant for breach of the covenant.

We were referred to observations in the speeches of Lord Upjohn and Lord Wilberforce in *Tophams Ltd v Earl of Sefton* [1967] 1 AC 50, which appear at pp 73 and 81, to the effect that section 79 of the Law of Property Act (relating to the burden of covenants) achieved no more than the introduction of statutory shorthand into the drafting of covenants. Section 79, in my view, involves quite different considerations and I do not think that it provides a helpful analogy.

It was suggested by Mr Price that, if this covenant ought to be read as enuring for the benefit of the retained land, it should be read as enuring only for the benefit of the retained land as a whole and not for the benefit of every part of it; with the apparent result that there is no annexation of the benefit to a part of the retained land when any severance takes place. He referred us to a passage in *Re Union of London and Smith's Bank Ltd's Conveyance, Miles v Easter* [1933] Ch 611 at p 628, which I do not think it is necessary for me to read.

The problem is alluded to in *Megarry and Wade Law of Real Property*, current edition, at p 763:

> In drafting restrictive covenants it is therefore desirable to annex them to the covenantee's land "or any part or parts thereof". An additional reason for using this form of words is that, if there is no indication to the contrary, the benefit may be held to be annexed only to the whole of the covenantee's land, so that it will not pass with portions of it disposed of separately. But even without such words the court may find that the covenant is intended to benefit any part of the retained land; and small indications may suffice, since the rule that presumes annexation to the whole only is arbitrary and inconvenient. In principle it conflicts with the rule for assignments, which allows a benefit annexed to the whole to be assigned with part, and it also conflicts with the corresponding rule for easements.

I find the idea of the annexation of a covenant to the whole of the land but not to a part of it a difficult conception fully to grasp. I can understand that a covenantee may expressly or by necessary implication retain the benefit of a covenant wholly under his own control, so that the benefit will not pass unless the covenantee chooses to assign; but I would have thought, if the benefit of a covenant is, on a proper construction of a document, annexed to the land, *prima facie* it is annexed to every part thereof, unless the contrary clearly appears. It is difficult to see how this court can have reached its decision in *Williams v Unit Construction Ltd* unless this is right. The covenant was, by inference, annexed to every part of the land and not merely to the whole, because it will be recalled that the plaintiff was a tenant of only one of the four houses which had the benefit of the covenant.

There is also this observation by Romer LJ in *Drake v Gray* [1936] Ch 451 at p 465. He was dealing with the enuring of the benefit of a restrictive covenant and he said this:

> . . . where one finds not "the land coloured yellow" or "the estate" or "the field named so and so" or anything of that kind, but "the lands retained by the vendor," it appears to me that there is a sufficient indication that the benefit of the covenant enures to every one of the lands retained by the vendor, and if a plaintiff in a subsequent action to enforce a covenant can say: "I am the owner of a piece of land that belonged to the vendor at the time of the conveyance," he is entitled to enforce the covenant.

In the instant case the learned judge in the course of his judgment appears to have dismissed the notion that any individual plot-holder would be entitled, even by assignment, to have the benefit of the covenant that I have been considering. I express no view about that. I only say this, that I am not convinced that his conclusion on that point is correct. I say no more about it.

In the end, I come to the conclusion that section 78 of the Law of Property Act caused the benefit of the restrictive covenant in question to run with the red land and therefore to be annexed to it, with the result that the plaintiff company is able to enforce the covenant against Mill Lodge, not only in its capacity as owner of the green land, but also in its capacity as owner of the red land. For these reasons I think that the judge reached the correct view on the right of the plaintiff company to enforce the covenant, although in part he arrived there by a different route.

There remains only the question whether we ought to interfere with the remedy granted by the judge of an injunction against the building of the 32 extra dwellings. The case of *Shelfer v City of London Electric Lighting Co* [1895] 1 Ch 287 is authority for the proposition that a person who has the benefit of a restrictive covenant is, as a general rule, entitled to an injunction on the trial of the action as distinct from an award of damages unless (1) the injury to the plaintiff's legal rights is small, (2) it is capable of being estimated in terms of money, (3) it can adequately be compensated for by a small payment and (4) it would be oppressive to the defendant to grant an injunction. In my view, the first, third and fourth of these conditions have not been shown to be satisfied.

I would, therefore, uphold the injunction and I would dismiss this appeal.

Browne LJ agreed.

Also agreeing, Megaw LJ said: There is only one matter on which I would desire to add a few words. That is in respect to the passage quoted by Brightman LJ towards the end of his judgment from the judgment of Romer LJ in *Drake v Gray*. It is right to observe that the passage which my Lord read has been the subject of criticism by Buckley J, as he then was, in *Russell v Archdale* [1963] Ch 38. The criticism appears at p 47. Buckley J suggests that the passage in question in Romer LJ's judgment is *obiter dictum* and that there is difficulty in accepting the distinction which Romer LJ there drew between, on the one hand, land described as the land marked yellow on the plan or the estate of the vendor known as such and such a name and, on the other hand, reference to the remaining land or retained land. It is, however, right to point out that Buckley J at p 47 says:

> No doubt every case of this kind, being one of construction, must be determined on the facts and the actual language used

and he goes on to say that, with the utmost respect to Romer LJ, he cannot see that the distinction is a valid one.

For myself, I would regard the observations made in the passage which Brightman LJ read from *Megarry and Wade* at p 763 as being powerful reasons, and I find great difficulty in understanding how, either as a matter of principle, or as a matter of practical good sense in relation to a legal relationship of this sort, it can be said that a covenant which *ex hypothesi* has been annexed to the land as a whole is somehow or other not annexed to the individual parts of that land.

I agree that the appeal should be dismissed.

The appeal was dismissed with costs. Leave to appeal to the House of Lords was refused.

TOWN AND COUNTRY PLANNING

COURT OF APPEAL

October 23 1979

(Before Lord Justice BUCKLEY, Lord Justice BRIDGE and Lord Justice TEMPLEMAN)

L A H AMES LTD v NORTH BEDFORDSHIRE BOROUGH COUNCIL

Estates Gazette January 5 1980

(1979) 253 EG 55

Town and Country Planning Act 1971—Section 52 agreement between development company and planning authority—Agreement that development should not be commenced until such time (not to exceed 21 years from date of agreement) as necessary for works of improvement to sewage disposal system "to be placed in contract"—Planning permission given for development to be begun not later than three years from date of permission—Sewage contract not let within the three years —Literal result that there was never any time when, in accordance with agreement and the planning permission, the development could be commenced—Various submissions by developers to avoid this frustrating conclusion rejected in view of the clear condition of the planning permission—Judge's decision upheld

This was an appeal from a decision of Oliver J in an action by the plaintiffs (the present appellants), L A H Ames Ltd, against the North Bedfordshire Borough Council, the defendants and present respondents, for a declaration as to the true construction of an agreement under section 52 of the Town and Country Planning Act 1971 and a planning permission dated the day after the agreement. Oliver J held that in the events which had happened the planning permission was spent and was no longer of any use to the plaintiffs.

K R Bagnall QC and P de la Piquerie (instructed by Wedlake Bell, agents for C C Bell & Son, of Bedford) appeared on behalf of the appellants; K Schiemann (instructed by Sharpe, Pritchard & Co) represented the respondents.

Giving judgment, BUCKLEY LJ said: This is an appeal from a decision of Oliver J in a question which has arisen on the construction and effect of an agreement entered into between the plaintiff company, L A H Ames Ltd, and the Bedford Rural District Council, who were the predecessors of the defendants, the North Bedfordshire Borough Council, on March 28 1974, and of a planning permission which was granted to the plaintiff company by the Bedford Rural District Council, acting as the agents of the planning authority, on the following day, March 29 1974.

The plaintiff company had acquired a considerable area of land during the period from early March 1972 until March 1974, and they were anxious to develop that land by the erection upon it of 81 houses with ancillary works. At that time the sewage installation in the development area was inadequate to take any further sewage and consequently the authority was unwilling to grant planning permission for further developments there until additional sewerage had been provided. For some time the authority had been refusing to grant planning permissions for that reason.

Some correspondence passed between the parties shortly before the date when the agreement was entered into. On March 25 1974 the rural district council wrote to the plaintiff company's solicitors a letter which included the following statement:

> . . . I now write to confirm that the council are only prepared to grant planning permission for the erection of 81 dwellings and garages off Rotten Row, Riseley if the applicants, L A H Ames Ltd, enter into an agreement under section 52 of the Town and Country Planning Act 1971, to the effect that the proposed development will not be commenced until improvement works to the overloaded main drainage system at Riseley are in contract.
>
> In this connection I would point out that the government's recent restrictions on local authority expenditure have delayed the implementation of this improvement scheme, and it now seems unlikely that it will be commenced during the 1974/75 financial year. As from April 1 1974 this matter will be the responsibility of the Anglian Water Authority, and I am endeavouring to find out the latest position, but for the purposes of the agreement I have estimated the date of commencement as September 1 1975.

The writer then included two copies of the draft agreement for consideration by the plaintiff company.

So I think it is clear from the terms of that letter that the agreement was prepared upon the footing and in anticipation that the date for the commencement of the sewage improvement works would be about September 1 1975, by which date, of course, the contract for those works would have been entered into.

The agreement in question, which is contained in an exhibit to Mr Ames' affidavit in this matter, is an agreement between the Bedford Rural District Council of the one part and the plaintiff company of the other part. It recites that the council are the agents of the Bedfordshire County Council for planning purposes, that the company is the owner of the relevant land and that the owner has by written application dated December 4 1973 applied to the council for permission to develop the land in accordance with certain plans, specifications and particulars deposited with the council. It then contains this recital, which is numbered 4:

> The council is satisfied that the development disclosed by the said plans specifications and particulars is such as may be approved by the council under the said Act but that the proposal must be regarded as premature in the light of the Council's decision on June 2 1973 that no new development should be permitted within the area served by the overloaded sewage disposal system at Riseley.

The operative part of the agreement is in the following terms:

> 1. This Agreement is made in pursuance of Section 52 of the Town and Country Planning Act 1971 and in consideration of the covenant by the Owner hereinafter contained. 2. SUBJECT to the completion of this Agreement the Council proposes to issue its written approval of the said development for the purposes of the said Act subject to the imposition of certain conditions therein contained.

I pause to interpolate that at the date of this agreement the plaintiff company did not know what those conditions were to be.

Then:

> 3. The owner hereby covenants with the Council that the said property shall be subject to the restrictions and provisions regulating the development as specified in the Third Schedule hereto—

I will return to that in a moment.

Then:

4. The Owner hereby agrees to carry out the said development in strict conformity with the said plans specifications and particulars and to use the said property and all erections to be made thereon in strict accordance with this Agreement and not otherwise.

Then there is a clause which says that the expressions "the Council" and "the Owner" include their successors in title.

In the Third Schedule there is set out this restriction regulating the development:

The said development as set out in the Second Schedule hereto shall not be commenced until such time (not exceeding twenty-one years from the date of this Agreement) as is necessary for the works of improvement to the sewage disposal system at Riseley to be placed in contract which said time shall be notified by the Council to the Owner and is estimated at the date of this Agreement as September 1 1975.

That agreement was executed under seal by both parties, and on the following day the council granted planning permission in respect of the land for the erection of 81 dwellings, garages, roads and ancillary works subject to conditions, and condition no 1 is in these terms: "The proposed development to which this permission relates must be begun not later than the expiration of three years from the date of this permission"—that is to say, three years from March 29 1974. The reasons for that condition are stated as being: "In accordance with section 41 of the Town and Country Planning Act 1971, to prevent the accumulation of unimplemented planning permissions and to expedite housing development," and secondly to preserve the amenities of the area. There are other reasons given, which I do not think I need read.

In fact the sewage contract was not let within the three years; consequently there was never any time at which, in accordance with both the agreement and the conditions on the planning permission, it was possible for the plaintiff company to commence its development in accordance with the terms of the permission. That would not, of course, have precluded the plaintiff company from applying for a new planning permission as soon as the contract for the sewage works had been entered into, or when it was in prospect. At the time the matter was before the learned judge they had not made any such application; they have since made an application, but it has been refused.

The learned judge came to the conclusion that on the true construction of the planning permission the permission was in these circumstances spent and was no longer of any use to the plaintiff company, and the plaintiff company, by the terms of its originating summons, sought a declaration that upon the true construction of the agreement and upon the true construction of the planning permission the period of three years referred to in condition 1 of the planning permission did not commence to run until such time as the works of improvement to the sewage disposal system at Riseley had been placed on contract. In this court an alternative argument has been presented, to the effect that upon the true construction of the planning permission the period of three years specified in condition 1 ought in the circumstances to be construed as though it referred to a period of 21 years from the date of the planning permission.

Mr Bagnall, appearing for the plaintiff company appealing in this court, has submitted that the agreement and the planning permission should be read together as one document, being virtually contemporaneous and relating to one and the same transaction. He has referred us to *Smith v Chadwick* (1882) 20 ChD 27, where Sir George Jessel MR did treat two documents as representing a single transaction and construed them accordingly. Mr Bagnall has submitted that it cannot be right to construe the two documents together as having the consequence that the power of the plaintiff company to develop the land should be restricted to development commenced within three years from the date of the planning permission. He has invoked the principle that when construing a deed the earlier of two provisions in a deed which are inconsistent with one another should prevail over the later; and he has contended that the terms of the agreement, being earlier in date than the planning permission, must be regarded as the dominant provision and that accordingly we ought to treat the 21-year period as being the significant period and ignore the limitation in the planning permission to three years from the date of the granting of the permission.

In my judgment that principle, that the earlier of two provisions in a single deed must prevail over the later if the two provisions are mutually inconsistent, is really of no assistance to us here. In any event, the principle only applies where the provisions are incapable of being read literally and consistently with one another. The section 52 agreement of March 28 1974 is, by the terms of the section, binding not only on the plaintiff company but also on the company's successors in title, and it clearly prohibits the company and its successors in title from commencing the proposed development—that is, the development set out in the plans, specifications and particulars deposited with the council by the plaintiff company—at any time before the contract for the new sewage installation had been entered into. It is also clear on the face of that agreement, as well as from the earlier correspondence which I have already mentioned in my judgment, that the agreement was entered into in the expectation that that contract would be let before September 1 1975. In my judgment the agreement does not impose any contractual obligation upon the council to grant any planning permission; it merely states what the council proposes to do, and what it proposed to do was to issue a planning permission which contained certain conditions but, as I have said, the plaintiff company was ignorant of what those conditions would be. If this clause had been intended to have contractual effect, it would have been giving the council a blank cheque to meet its contractual obligation by granting a permission subject to any kind of condition that it sought fit to impose, no doubt for good planning reasons. It is very unlikely that such a clause would be intended to have contractual effect, and in any case the language of the clause seems to me to indicate that it was not intended to have contractual effect, but merely to state what the intention of the authority at that time was.

In due course the council granted the permission which it did grant, on March 29, and it made that permission subject to the condition numbered 1 which I have read. With deference to Mr Bagnall's argument, that condition does not seem to me to be in any respect inconsistent with the terms of the agreement. If, as seems to have been the case, the parties were at that time under the impression that the sewage contract would be let in or about September 1975, condition 1 of the planning permission is a perfectly reasonable and sensible condition, and if that anticipation of the parties was disappointed in the event, there would have been nothing to have prevented the plaintiff company applying for a further planning permission. Whether they would have got it or not would no doubt have depended upon the exercise by the council of their proper discretions and functions as the agents for the planning authority for this area, but that was a hazard to which the plaintiff company was at all times subject; they could not develop without the approval of the planning authority.

With regard to the submission that condition 1 should be construed as though it provided that the proposed development must not be begun later than three years from the date of the letting of the sewage contract, which is one way in which Mr Bagnall has put his case, that seems to me to be an impossible construction when one considers the terms of section 41 of the Town and Country Planning Act 1971. Subsections (1) and (2) of that section provide as follows:

(1) Subject to the provisions of this section, every planning permission granted or deemed to be granted shall be granted or, as the case may be, be deemed to be granted, subject to the condition

A that the development to which it relates must be begun not later than the expiration of—(a) five years beginning with the date on which the permission is granted or, as the case may be, deemed to be granted; or (b) such other period (whether longer or shorter) beginning with the said date as the authority concerned with the terms of the planning permission may direct, being a period which the authority considers appropriate having regard to the provisions of the development plan and to any other material considerations.

(2) If planning permission is granted without the condition required by subsection (1) of this section, it shall be deemed to have been granted subject to the condition that the development to which it relates must be begun not later than the expiration of five years beginning with the date of the grant.

B So the permission must be granted subject to the condition either that the development must be begun within five years from the date of the permission, or that it must be begun within some other period, longer or shorter, beginning with the date of the permission. A condition which provided that the development should be begun not later than three years after some contingent event in the future, the date of which could not be foreseen at the date of the grant of the planning permission, would not comply with the provisions of section 41(1)(b), and if it did not comply with that provision it would I think follow that the permission would be treated as subject to the provision contained in section 41(1)(a); that is to say, C the development would have to be begun within the five-year period beginning with the date of the permission.

The condition numbered 1 in the planning permission in the present case, if construed as Mr Bagnall asks us to construe it, would not comply with section 41(1)(b), but it is said to be imposed in accordance with section 41, and quite manifestly, in my judgment, that was the intention of the condition and it was intended to take effect in accordance with its terms, which do not seem to me to be in the least equivocal; it requires that the development must be begun not later than three years from the date of this permission.

For these reasons I think that the learned judge came to the D right conclusion, and that this appeal should be dismissed.

Agreeing, BRIDGE LJ said: The issue raised in the appeal is, in the end, a very short one: what meaning is to be placed on the phrase to be found in condition 1 of the planning permission issued to the appellants on March 29 1974, "three years from the date of this permission"?

The only contention raised in the originating summons is that those words should be read as meaning three years from the date when it becomes possible for the appellants, pursuant to the agreement made on March 28 1974, to implement the permission because the sewage improvement works were placed E in contract.

It seems to me that there are three clear reasons why that is not an acceptable contention: first, the language of condition 1 of the planning permission is perfectly clear and free from ambiguity, that "the date of this permission" means exactly what it says, the date when the permission was granted.

Secondly, in agreement with my Lord, I can see no conflict between the planning permission read in accordance with its plain terms and anything which is to be found in the agreement dated the previous day. As my Lord has pointed out, clause 2 of that agreement provides that, subject to the completion of the agreement the council proposes to issue its F written approval of "the said development"—that is by the building of 81 houses—for the purposes of the said Act, subject to the imposition of certain conditions therein contained. I cannot see that that clause in the operative part of the section 52 agreement imposed any obligation on the local planning authority to issue a planning permission for the carrying out of this particular development at any particular time or in any particular way.

But thirdly, and most importantly—this, as I understand it, was the main reason why the learned judge decided against the appellants in this case as my Lord has explained before—if the words in condition 1 of the planning permission were to G be read as Mr Bagnall submits they should be read, then the condition would not be in compliance with section 41 of the Town and Country Planning Act 1971, with which, as the reasons for the imposition of the condition make clear, it was clearly intended to comply.

The alternative argument that the condition should be read as referring not to three years but to 21 years is not in my judgment open to the appellants having regard to the terms of their originating summons; but if it were, it is in my view wholly unacceptable.

For these reasons, and in full agreement with the judgment that has been given by my Lord, I too would dismiss this appeal. H

Also agreeing, TEMPLEMAN LJ said: Some correspondence has been doubtfully admitted; some ingenious arguments have been forcefully submitted and some largely irrelevant authorities have been cited in order to contradict the plain meaning of a condition contained in a planning permission granted by a rural district council to a developer for housing development which required the council's sewage system to be improved.

At the material time, namely March 1974, it was envisaged that a contract for the improvement of the council's sewage system would be placed within three years or so. By an agreement dated March 28 1974, the developer covenanted not to carry out the proposed development "until such time (not J exceeding 21 years from the date of this Agreement) as is necessary for the works of improvement to the sewage disposal system . . . to be placed in contract which said time shall be notified by the council . . . and is estimated at the date of this Agreement as September 1 1975."

This agreement imposed a negative covenant on the developer, but it is conceded that it did not oblige the council to provide the sewage disposal system within 21 years or at all. By the planning permission dated March 29 1974 the council granted planning permission for the development to be carried out, but condition 1 provided that the development must be K begun "not later than the expiration of three years from the date of this permission." This condition contained in the planning permission enabled the council to allow or forbid the development to take place if for any reason it was not begun within three years. Whether or not the sewage system was enlarged, or was proposed to be enlarged, or was abandoned, or whether for any reason the development had not been commenced, nevertheless in the interests of the public the council might, at the end of the three-year period, come to the conclusion that the development should no longer be allowed to be commenced. The three-year period laid down by condition 1 of the planning permission expired; the sewage L contract had not been placed; the planning permission is dead and the appeal must be dismissed.

The appeal was dismissed with costs.

CHANCERY DIVISION

October 9 1979

(Before His Honour Judge John FINLAY QC, sitting as a judge of the High Court)

SOUTH GLAMORGAN COUNTY COUNCIL v HOBBS (QUARRIES) LTD M

Estates Gazette March 8 1980

(1979) 253 EG 1014

Town and Country Planning Act 1971 and Town and Country Planning General Development Order 1977—Erection by limestone quarrying company of a "ready-mixed concrete batching plant"—Whether permitted development under article 3 and class XIX, paragraph 2, of order—Was the plant required in connection with the treatment or disposal of the limestone?—The process carried out by the plant involved the loading of limestone into a hopper and its conveyance to

a silo which delivered it together with cement and water to a vehicle with a rotating drum—Operation neither a "treatment" nor a "disposal" of the limestone but a "utilisation" of it—Purpose to produce ready-mixed concrete—Not permitted development within class XIX—Declaration in favour of planning authority

In these proceedings the plaintiffs, South Glamorgan County Council, sought a determination as to whether development proposed by the defendant company was permitted development under the Town and Country Planning General Development Order 1977. The plaintiffs contended that the development in question, namely, the erection of a ready-mixed concrete batching plant, could not be carried out without planning permission being granted by the plaintiffs or by the Secretary of State for the Environment.

M G V Harrison (instructed by Sharpe, Pritchard & Co, agents for M D Boyce, County Solicitor, South Glamorgan County Council) appeared on behalf of the plaintiffs; I D L Glidewell QC and G R G Roots (instructed by Hallinan, Blackburn, Gittings & Co, of Cardiff) represented the defendants.

Giving judgment, JUDGE FINLAY said: The defendant company has for some years past won and worked limestone at a quarry known as Blaengwynlais and situated in part in the county of South Glamorgan and partly in the county of Mid-Glamorgan, these quarrying operations being carried out under and by virtue of a number of planning permissions granted from time to time, the area to the south-west of the quarry area being worked prior to the coming into force of the Town and Country Planning Act 1947 or pursuant to an application approved on March 15 1948, and other areas lying to the north and east of that being worked under planning approvals granted in 1961.

On July 21 1977 the defendant company applied for planning permission for the erection of a ready-mixed concrete batching plant to be sited in the quarry area, which was that which I have already indicated was earliest worked. The application was considered, and on October 21 1977 notification of the county council's refusal to grant the permission was communicated to the defendant company. The defendant company appealed against the refusal, and, after a determination by the Secretary of State for Wales to hold a public inquiry, subsequently on October 12 1978 notified the county solicitor that the defendant intended to withdraw his appeal and erect the ready-mixed concrete batching plant without permission, on the ground that the proposed development was permitted by virtue of article 3 and classes VIII and XIX of Schedule 1 to the Town and Country Planning General Development Order 1977.

The County Council of South Glamorgan, the plaintiff council, not accepting that view of the operation and effect of the General Development Order, issued this summons, which seeks the determination of the court on the question whether the proposed development is permitted under class XIX and class VIII, and, further, for a declaration that these provisions do not permit the proposed plant to be erected in the quarry without planning permission granted by the plaintiff or by the Secretary of State.

At the commencement of this hearing Mr Glidewell, who appears with Mr Roots for the defendant company, indicated that he did not propose to submit that the development was permitted under class VIII, taking the view that the same argument as was relevant to the submission that it was permitted under class XIX would be applicable to the question whether it was permissible under class VIII. In short, if the defendants are entitled to bring themselves under class XIX and thus to succeed, it is not necessary to do so under class VIII, whereas if they cannot succeed under class XIX, they would not succeed under class VIII. The argument accordingly has been directed before me to the construction and operation of class XIX of the order.

The Town and Country Planning General Development Order, 1977 provides in article 3 thereof that:

> Subject to the subsequent provisions of this order, development of any class specified in Schedule 1 to this order is permitted by this order and may be undertaken upon land to which this order applies, without the permission of the local planning authority or of the Secretary of State:
>
> Provided that the permission granted by this order in respect of any such class of development shall be defined by any limitation and be subject to any condition imposed in the said Schedule 1 in relation to that class.

Subparagraph (2) of the article provides that "Nothing in this article or in Schedule 1 to this order shall operate so as to permit any development contrary to a condition imposed in any permission granted or deemed to be granted under Part III of the Act otherwise than by this order." It is common ground that that limitation is not here applicable.

Class XIX, paragraph 2 provides:

> The erection, alteration or extension by mineral undertakers on land in or adjacent to and belonging to a quarry or mine comprised in their undertaking of any building, plant or machinery, or structure or erection of the nature of plant or machinery, which is required in connection with the winning or working of materials, including coal won or worked by virtue of section 36(1) of the Coal Industry Nationalisation Act 1946, but not any other coal, in pursuance of permission granted or deemed to be granted under Part III of the Act, or which is required in connection with the treatment or disposal of such minerals.

There is then a proviso, which is not material to the present question.

I have had the advantage of hearing the evidence of Mr Peter Norman Cope and Mr Alistair Mundy, both of whom, having sworn affidavits in the matter, gave further oral evidence and were also cross-examined.

The proposed development consists in the erection on land of what is described in the application for permission for development, which I earlier mentioned, as the erection of ready-mixed concrete batching plants, and the nature of the plant is well indicated by drawings which are part of the exhibit PMC4 to the affidavit of Mr Cope. The plant comprises a hopper, containing four compartments, with suitable arrangements for measuring the contents on discharge of the contents from those compartments on to a belt conveyor, which would convey the material to a silo, which has arrangements for the storage not only of water but also of cement. The operation of the silo is such that it would deliver measured quantities of the material coming to the hopper, together with measured quantities of cement and water, into a vehicle containing a rotating drum, the operation of which would result in the mixing of this composition and the resultant production of cement, the idea being that the cement is produced on the journey taken by the vehicle from this cement-mixing plant to the site where the cement is required.

The hopper is constructed, as I have said, with four compartments, and these are designed to contain the appropriate quantities of coarse aggregate and fine aggregate. Although there is not a total measure of agreement between Mr Cope (who does not claim to have any expertise in relation to the manufacture of cement) and Mr Mundy (who has a degree of experience connected with that operation), I have had little difficulty in coming to certain broad conclusions as to the nature of the cement-making process and how it would be proposed to be carried out with the use of the cement-making machine at Blaengwynlais quarry.

The proportions in which aggregate and cement are mixed depend upon the purpose for which the concrete which is to be produced by the mixture is required. The richest and, as I understand it, strongest mixture is one in which the proportions of coarse aggregate, fine aggregate and cement are as 3 is to $1\frac{1}{2}$ is to 1, and as the proportions of aggregate to cement are increased, the concrete loses structural strength.

A One may have, for example, a mixture consisting of 20 parts of stone aggregate and 1 part of cement, which is a mixture of a kind producing concrete suitable for road building and not for structural purposes. So far as structural cement is concerned, there is, according to Mr Mundy—and I accept his evidence on this—a rule under which you should not have a higher proportion of fine to coarse aggregate than 1 to 2. The fine constituents of the stone aggregate to be used in the concrete mixture can comprise sand or can comprise limestone fines, which can be produced at the quarry. It is a general rule, although one which admits of exceptions, that for structural purposes—that is, for the production of concrete to
B be used for structural purposes—there should be, so far as the fine content of the mixture is concerned, 50 per cent of sand and 50 per cent of limestone fines. The upshot is, however —and I so find—that the concrete could be mixed in this concrete mixer and could be produced from stone from the quarry either in the form of coarse aggregate or in the form of limestone fines, the latter being such as would pass through a quarter-inch mesh, and the proportion of material from the quarry would be 70 per cent of the total material used in the production of concrete. That would be so whatever the nature of the mix was, provided that at least 50 per cent of the fine content of the mixture was stone from the quarry
C in the form of limestone fines.

I have also heard evidence as to the number of quarries in South Wales where there are installed machines either similar to the ready-mixed concrete plant which I have already mentioned or of a different nature; that is, coating plants which are used for the production of coated road stones.

I am satisfied that there are numerous quarries in South Wales which have ready-mixed concrete plants. They are listed in the exhibit P1, which was produced by the plaintiff council, by the witness Mr Cope, but elicited in the course of cross-examination by Mr Glidewell for the defendant company, and on which Mr Mundy made a number of comments, the upshot
D of which is that rather more than the quarries listed as those having ready-mixed concrete plants have them. This is because, according to Mr Mundy, a number of the quarries listed as having neither ready-mixed concrete plants nor coating plants in the exhibit P1 *have* got such plants. There are at least two which had; and Mr Mundy's evidence also dealt with the reasons why some others would not require to have them. For example, two quarries supply stone directly to British Steel, and another three are quarries operated in connection with cement works owned by the quarry owners.

I do not consider, however, that that evidence assists me to resolve the question which is raised by the originating
E summons, although it throws a good deal of light upon the reason why the question arises. I am satisfied that it is no doubt commercially desirable to have a ready-mixed concrete plant associated with the operation of a quarry, and that emerges, if for no other reason, from the circumstance that the demand for raw limestone is a fluctuating demand, which at this present juncture appears to be not great. Consequently it is, as a matter of commercial prudence, desirable to have some mode of marketing the quarry production that does not depend upon a demand for raw limestone, and one such mode is to use the production of the quarry in the production of concrete. Mr Mundy told me that for some time past,
F because of the lack of demand for raw limestone and because no concrete-mixing plant has been in operation at Blaengwynlais quarry, the quarry production has simply been stockpiled.

I turn to consider the provisions of paragraph 2 of Class XIX of the General Development Order. That permits, as I have already noted, the erection, alteration or extension by mineral undertakers, and it is common ground that the defendant company are mineral undertakers as defined in article 2 of the order; and the erection is permitted on land in or adjacent to and belonging to a quarry or mine comprised in their undertaking. No question arises as to compliance with that part of the provision because the proposal
G is that the ready-mixed concrete plant should be erected on part of the quarry, which clearly is comprised in the defendant company's undertaking at Blaengwynlais. What is permitted to be erected is any building, plant or machinery, or structure or erection of the nature of plant or machinery. Again, there is no question between the parties but that the proposed erection is an erection of plant. Furthermore, it is not contended that the ready-mixed concrete plant is plant required in connection with the winning or working of the limestone. Moreover, it is the case that the quarry limestone is limestone which has been won or worked in pursuance of permission granted or deemed to be granted
H under Part III of the statute.

The question then is that which is raised by the final words of paragraph 2 before one comes to the proviso. Is the plant required "in connection with the treatment or disposal of such minerals"; that is to say, of the limestone won and worked at Blaengwynlais quarry?

I have been referred to the decision, both in the court of first instance—that is, the decision of Goulding J—and in the Court of Appeal, in the case *English Clays Lovering Pochin & Co Ltd v Plymouth Corporation*, reported at first instance in [1973] 1 WLR 1346 and also [1973] 2 All ER 730, and in the Court of Appeal at [1974] 1 WLR 742 and
J also in [1974] 2 All ER 239. That case, however, was not concerned with the question raised in these proceedings as to whether the plant in question is required in connection with the treatment or disposal of the minerals won or quarried, but with a question arising under the provisions corresponding to paragraph 2, class XIX, of the 1977 order, as to whether the situation on which the proposed works were to be carried out was part of the quarry or site on which the mining operations were carried on, or could be described as land belonging to the quarry and adjacent to it.

Mr Glidewell referred me to what was said by Goulding J at [1973] 2 All ER at p 737, letter E, in relation to the
K question of what constitutes treatment of the minerals:

> After considering all that authority and argument, I remain persuaded that I have to apply the ordinary interpretation of language and I remain of opinion that the essential idea in the term "winning or working of minerals" as it appears in item 2 of class XVIII is that of the extracting or separation of raw mineral from the solid earth in which it occurs. What is done afterwards may be called cleaning or dressing or treatment or refining, or some other appropriate term according to circumstances, but it is not, in my judgment, within the ordinary meaning of the words "winning or working of minerals." My interpretation of the words is strengthened by the distinction which is made in the Development Order itself between winning
> L or working on the one hand and treatment or disposal of the same minerals on the other.

I pause there to say that the reason for the distinction between winning or working, on the one hand, and treatment or disposal, on the other, appears to me to be that the reference to "winning or working" is qualified by the words "in pursuance of permission granted under Part III of the Act," because paragraph 2 of class XIX only applies where the winning or working has been lawfully carried out in accordance with the planning legislation, and it is only minerals that have been won or worked in that lawful manner that are dealt with when the paragraph goes
M on to refer to requirements in connection with the treatment or disposal of such minerals.

In the Court of Appeal Russell LJ, who gave the judgment of the court, at [1974] 2 All ER p 243 deals with the words "winning or working" and says this:

> It is perhaps not necessary to be dogmatic on the point in this case: but our present view is that to "win" a mineral is to make it available or accessible to be removed from the land, and to "work" a mineral is (at least initially) to remove it from its position in the land: in the present case the china clay is "won" when the overburden is taken away, and "worked" (at least initially) when the water jets remove the china clay together

with its mechanically associated other substances from their position in the earth or land to a situation of suspension in water. Thereafter it may be that the processes of separation out are more aptly described as treatment.

That case was, of course, concerned with the extraction and disposal of china clay, and the point there in question concerned the situation and not the nature and purpose of the proposed structure. As a result, I do not find great assistance from the decision in the *English Clays* case in resolving the question with which I have to deal. On that question I am told that there is no authority.

The words which I have to construe are "which is required in connection with the treatment or disposal of such minerals." In my judgment, "required" does not there mean necessarily required in the sense that the plant is a *sine qua non* for the treatment or for the disposal of the minerals in question.

It is submitted by Mr Glidewell for the defendant company that "required in connection with" means no more than "will be useful in," or "will assist in" the treatment or disposal. Mr Harrison, for the plaintiff council, submits that it means "reasonably required," and that it does not mean either needed as a *sine qua non* or the same as "will be useful in" or "will assist in" the treatment or disposal.

In my judgment, "required" means "reasonably required," and, furthermore, that in determining whether it is reasonably required, regard has to be had to the nature of the minerals which have to be treated and disposed of and the circumstances in which the treatment or disposal has to be carried out.

Mr Glidewell, although submitting that the ready-mixed concrete plant was required in connection with the treatment of the limestone, put his submission mainly, I think, on the basis that it was required in connection with the disposal of that mineral. He summarised his submission very helpfully in these four propositions:

(1) The process carried out by the proposed batching plant (which I have been referring to as the ready-mixed concrete plant) will properly be described as the disposal and possibly the treatment of the mineral won and worked in the quarry.

(2) The addition to that mineral of smaller quantities of sand and cement does not alter the nature of the process carried out by the plant.

(3) If the word "required" in class XIX, paragraph 2, imports an element of necessity, then in this case the evidence of demand for stone mixed with other materials to become concrete and of lack of demand for raw stone satisfies such requirement.

(4) As an alternative to (3)—and I think this was the main way he put it—he submitted that one should look at the whole phrase "required in connection with" and that that meant no more than will be useful in or will assist in, and he submitted that the plant in question would be useful in or would assist in the treatment of the stone.

I have already indicated my view as to the fourth proposition. So far as the third proposition is concerned, that "required" imports an element of necessity, I have also indicated my views on that. I would add only this, that "which is required in connection with the treatment or disposal of such minerals" does not, in my judgment, mean "which is required as a matter of commercial necessity" or "business prudence," nor indeed are these words to be equated with something to the effect of "which is desirable as a commercial matter in connection with the treatment or disposal of such minerals."

I turn then to the first two propositions, the first, that the process carried out by the proposed batching plant will properly be described as the disposal and possibly the treatment of the mineral won and worked in the quarry. In my judgment, the operation of this ready-mixed concrete plant cannot properly be regarded as an operation which effects treatment of the limestone which would be intended to be loaded into the hopper which is part of the plant. The loading of coarse and fine limestone into that hopper does not, in my judgment, as a matter of the ordinary use of the language, amount to treatment of the limestone except in the very general sense that to deal with any material in any way is to treat it in that manner. But "treatment" in paragraph 2 of class XIX means something, in my judgment, more than merely dealing with or handling, or anything of that kind. It conveys to my mind the suggestion that the mineral is to be dealt with in some manner which changes its quality, its nature or size, or something of that kind. The mere movement of the mineral is not, in my judgment, treatment of it; and all that happens, in my judgment, to the limestone, so far as the plant is concerned, is that it is moved by being put in the hopper, carried on the conveyor belt, dealt with by the silo and discharged into the lorry containing the revolving mixing drum.

The second proposition, that the addition to the mineral of smaller quantities of sand and cement does not alter the nature of the process carried out by the plant, I consider in relation to the first proposition, so far as it is that the process carried out by the plant can properly be described as disposal of the limestone. In my judgment, "disposal" in paragraph 2 means the physical removal of the limestone from the quarry or the land adjacent thereto where the minerals have been won, worked and treated. Now, in that sense, the ready-mixed concrete plant is involved in the process of disposing of the limestone, because at the beginning of the operation in which the plant is used, the limestone is situated in the quarry or the land occupied with the quarry, and when the operation has been completed, the limestone is in a lorry or truck, ready to be physically removed from that land. It would, however, in my judgment, be difficult to say that a plant carrying out the various functions, which this ready-mixed concrete plant was intended to effect, was reasonably required merely to effect the physical removal of the limestone from the quarry or the land in which the limestone had been treated. The plant clearly does something more than merely physically remove it. It also effects the mixing of the coarse and fine limestone with cement, or possibly with sand and cement.

It is submitted, as Mr Glidewell's second proposition, that the addition of smaller quantities of sand and cement—smaller, that is, than the quantity of limestone—does not alter the nature of the process carried out by the plant. I accept that it does not alter the fact that the process, so far as it affects the limestone, of disposing of limestone in the sense of removing it from the quarry is not affected by the fact that other materials are at the same time physically removed.

I consider that it is necessary to ask the question which Mr Harrison submitted was the question that I should ask: what is the purpose of the plant? Is it required in connection with the disposal or, for that matter, the treatment of the limestone, or for some other purpose? I am unable to avoid the conclusion that the purpose of the plant is to produce ready-mixed concrete, and what it is required for is neither the treatment of the limestone nor the disposal of the limestone but the production of a mixture of limestone and possibly sand, and certainly cement with water, which will produce ready-mixed concrete. The plant is, in my judgment, required neither in connection with the treatment of the limestone nor in connection with the disposal of it but in connection with the utilisation of the limestone. Paragraph 2 of class XIX does not, in my opinion, permit, without planning permission, the erection of a plant which is required for the utilisation of the minerals that have been won and worked in the quarry in question. The paragraph permits such development only where the plant is required in connection with the treatment or disposal of such minerals. That the utilisation of the limestone by being put into the hopper of the plant may result in its disposal in the sense of its removal from the quarry and the adjacent land does not mean, in my judgment, that the

A plant is required in connection with the disposal of the limestone.

Accordingly, in my judgment, the question raised by the originating summons and remaining for disposal—that is to say that the provisions of article 3 and class XIX(2) of Schedule I to the Town and Country Planning General Development Order do not permit the erection of the plant in question at Blaengwynlais quarry—is to be answered "Yes." These provisions do not, in my judgment, permit the erection of that plant without the permission of the planning authority or the Secretary of State.

The judge made a declaration that the provisions of article
B 3 and class XIX(2) did not permit the plant in question to be erected in the quarry without planning permission granted by the plaintiff planning authority or the Secretary of State. The parties having agreed on costs, no order was made.

CHANCERY DIVISION

October 18 1979

(Before Mr Justice BROWNE-WILKINSON)

C SUNBELL PROPERTIES LTD AND ANOTHER v DORSET COUNTY COUNCIL

Estates Gazette March 15 1980

(1979) 253 EG 1123

Town and country planning—Proposed development to be carried out by local authority—Procedure under Town and Country Planning General Regulations 1976—Permission deemed to be granted on the passing of a second resolution by the authority—Plaintiffs' motion for interim injunction to restrain authority from passing resolution which would result in "deemed" planning permission for authority's development
D —Such "deemed" permission, which included provision for office accommodation, would be likely to prejudice the plaintiffs' own proposals for office development—Various objections by plaintiffs based on alleged non-observance of procedural formalities rejected—Main argument that the procedure laid down in the relevant regulation implied that the development was to be carried out by authority itself, not by someone else—In present case it had been contemplated that a commercial developer would provide capital finance on a profit-sharing basis for authority's development—Plaintiffs had shown an arguable case that scheme did not comply with the regulation—Principles governing interim injunctions as explained in
E American Cyanamid case applied—Injunction granted as balance of convenience was in favour of plaintiffs

In these proceedings the plaintiffs, Sunbell Properties Ltd and Pinewood Homes, applied for an interim injunction to prevent the defendants, Dorset County Council, from passing a resolution the effect of which would be that permission would then be deemed to have been granted for the comprehensive development of a site in the centre of Bournemouth. This development was to include office accommodation. The plaintiffs were interested in obtaining permission for office development on that site and an adjoining site.

Sir Derek Walker-Smith QC and John Male (instructed by
F Wilde, Sapte & Co, agents for Baldocks, of Godalming and Rawlins, Davy & Wells, of Bournemouth) appeared on behalf of the plaintiffs; Miss Elizabeth Appleby QC (instructed by Sharpe, Pritchard & Co) represented the defendants.

Giving judgment BROWNE-WILKINSON J said: This motion concerns two sites in central Bournemouth. The first (which I will call "the red area") is a site in which there are a number of properties in which both the plaintiffs have substantial interests. The plaintiffs have applied to Bournemouth Borough Council for planning permission to develop the red area for office purposes. Their application not having been dealt with by the borough council within the statutory time-limit, it was deemed to have been rejected, and the plaintiffs have appealed G
to the Secretary of State and a public inquiry is to be held, probably starting in January 1980.

The second site is near the railway station and is known as the Bournemouth Central Station area (I will call it "the grey area"); it adjoins the red area. The defendants, Dorset County Council, are concerned that the grey area should be comprehensively developed as a travel interchange area with incidental services and with commercial offices. The county council own certain land in the grey area. The first plaintiff also owns a substantial amount of land in the grey area for which it has very recently applied for planning permission. The rest of the grey area belongs to other persons. H

The office accommodation on the grey area proposed by the county council is for an area of some 218,000 sq ft. The south-east Dorset structure plan, which has been approved by the Secretary of State despite objections from the plaintiffs, provides for 1,020m sq ft of office space in the Lansdown Central Station area, which includes the grey area. The structure plan does not contemplate office development on the red area.

The county council are proceeding to obtain planning permission for the development of the grey area. Being themselves the local planning authority, they are seeking to use a statutory procedure for obtaining deemed planning permission, J
and I must first give an outline of that procedure.

The Town and Country Planning General Regulations 1976 modify the general procedure laid down by the Town and Country Planning Act 1971. Regulation 3 modifies the normal statutory provisions in relation (a) to development ". . . by a local authority being a local planning authority . . . of land within their area" and (e) "development of any land which is vested in a local planning authority and which is situated within their area . . . ." As part of the land in the grey area is not vested in the county council, the special procedure can only apply if the development is to be "by" the local authority. Cases where the land is to be developed by a local authority K
are dealt with by regulation 4, which is expressed to apply to a development which they (that is to say, the local authority) propose to carry out. Under regulation 4 the procedure is as follows. First, under paragraph (1) it is said:

> . . . where the authority require a permission for development which they propose to carry out . . . and where they resolve to seek permission for the carrying out of that development

the provisions of regulation 4 are to apply.

The resolution to seek permission for the carrying out of development has been called "the first resolution." Once the first resolution has been passed, by paragraph (2) the local L
authority has to give certain notices in writing, and it is common ground that those notices have been given in this case. Paragraph (3) requires notice of the resolution to be placed in Part I of the register of planning applications, and that has been done in this case. Then paragraph (5) provides that on the expiry of a period (which has, in fact, expired):

> . . . the authority may, unless the Secretary of State has required them to make an application to him for permission for the development described in the resolution referred to in paragraph (1) of this regulation, resolve, by a resolution which is expressed to be passed for the purposes of this paragraph, to carry out that development; and, subject to the provisions of regulations 8 M
> to 11, on the passing of such resolution permission shall be deemed to be granted by the Secretary of State for that development.

Accordingly, on passing the resolution under paragraph (5) of regulation 4, a deemed planning permission is at once obtained, unless the Secretary of State has called in the application. The resolution referred to in subparagraph (5) has been called "the second resolution."

Then regulation 6 provides:

> Where an authority proposing to seek permission for the carrying out of development under regulation 4 have made arrangements under section 101 of the Local Government Act 1972 for the

discharge of the relevant functions by an officer of the authority, the provisions of the regulation 4 shall apply as though references to the passing of a resolution by the authority to seek permission were references to the giving of written notice to the authority by that officer that he proposes to seek permission and references to a resolution to carry out the development were references to written notice given to the authority by that officer that the development is authorised pursuant to regulation 4(5).

During the course of this year numerous committees of the county council have considered the proposed scheme for the grey area and have resolved to apply for the necessary planning permission. On the basis that the function of applying for such permission under regulation 4 has been validly delegated to the county solicitor, on July 5 1979 notice of an application for planning permission by the county council was sent to Bournemouth Borough Council with copies to the planning officer of the county council. Notice of the application was given to both the plaintiffs and no point arises on this. Objections of a technical nature, similar to those urged before me, were made by the first plaintiffs. The second plaintiffs made no observations. The Secretary of State has not as yet called in the application and the defendant county council has called a meeting for Friday, October 19 (that is to say, tomorrow) to consider and, if thought fit, pass the second resolution, namely, a resolution to carry out the development on the grey area. If such resolution is validly passed, planning permission for development of the grey area by the county council will be deemed to have been granted.

The plaintiffs claim an interim injunction restraining the county council from passing such resolution, claiming that their prospects of success on their pending appeal relating to the red area will be prejudiced by the existence of a planning permission for 218,000 sq ft of office space on the grey area. The plaintiffs say that the passing of the resolution would be *ultra vires* on two main grounds: first, that there has been no strict compliance with the formalities prescribed in the first resolution—that is to say, the resolution to seek planning permission referred to in regulation 4(1); secondly, the procedure in regulation 4 does not apply to this case at all, since the county council is not proposing itself to carry out the development, since, they say, the land, or the commercial part thereof is to be developed by a commercial developer.

It must be borne in mind that this is an application for an interim injunction and must therefore be approached on the basis laid down by the House of Lords in the case of *American Cyanamid Co v Ethicon Ltd* [1975] AC 396. The plaintiffs, therefore, are required to show only an arguable case fit to go to trial. In the course of the argument it sometimes seemed to me that I had to decide the whole case on the material now before me. That is not so: if the plaintiffs have an arguable case and (as appears to be common ground) damages would not be an adequate remedy to either party, the decision whether or not to grant an interim injunction depends on the balance of convenience.

First then, have the plaintiffs shown an arguable case on their first ground, namely, that the formalities of the first resolution have not been complied with? On this point the case is fundamentally different from that which the plaintiffs originally came to make. The plaintiffs asked to be supplied with a copy of the first resolution and the county council replied in these terms:

> The County Council are proposing to carry out the development under regulation 4 of the General Regulations, and I set out below the Minute adopted by the Policy and Resources Committee at their meeting on February 8 1977 which is "the first resolution."

The letter then sets out the minute, which reads as follows:

> That the power to seek permission for development on behalf of the County Council under regulations 4 and 5 of the Town and Country Planning General Regulations 1976 be delegated to the County Solicitor.

Not surprisingly, the plaintiffs thought that such a resolution was wholly insufficient to satisfy the requirements of regulation 4(1). Indeed, Miss Appleby for the county council conceded as much as soon as the motion was opened. However, the county council now argue that the resolution I have just read constituted a valid delegation to the county solicitor for the purposes of regulation 6.

On July 5 of this year the county solicitor wrote to the borough planning officer as follows:

> I am authorised by the County Council to seek approval for development they are proposing to carry out, and I therefore enclose herewith two copies of the proposals relating to the above development. Please ensure that in accordance with regulation 4(3) the necessary information is placed in Part I of the Register of Planning Applications kept by you.

There then follows certain other matters, and it ends, "Yours faithfully"; then there are initials in manuscript "DSH" and underneath the initials in typewritten script "(signed) John Simmonds." I interpose to say that John Simmonds is the county solicitor.

The enclosure with that letter, which bears on its face the fact that a copy was also sent to the county planning officer, is a completed form which is headed "Proposal by Local Authority for Deemed Permission for Development." It describes the land and gives certain particulars, and then, in paragraph 14, reads as follows:

> I hereby seek deemed planning permission or approval in accordance with the details given in this proposal and described in the accompanying plan and associated documents

—and then a signature appears, "John Simmonds," and then in typescript "County Solicitor"—and then it goes on, "On behalf of Dorset County Council."

The county council contend that by virtue of regulation 6 the sending of this letter, plus the enclosure, constitutes compliance with the requirements of regulation 4(1). Sir Derek, for the plaintiffs, contends otherwise on a number of grounds. First, he says, that there was no valid delegation of the functions to the county solicitor, since only a limited function relating to the first resolution was delegated, and regulation 6 requires the delegation of the functions relating to both the first and the second resolutions. He points to the fact that regulation 6 modifies the wording of regulation 4 in relation to both resolutions. I do not accept this submission. Regulation 6 contemplates delegation of the relevant functions, and in my judgment the relevance referred to is related to the modifications thereinafter mentioned in regulation 6—that is to say, the machinery is modified to the extent that the functions are delegated.

Next, Sir Derek says, probably correctly, that the letter enclosing the planning application was not actually signed by the county solicitor but only has his name typed on it. Regulation 6 does not require the notice to be signed by the delegated officer and I see no reason to import any requirement of actual signature. Moreover, the planning application itself is signed by the county solicitor and to my mind is by itself sufficient notice. Finally on this point, Sir Derek says that regulation 6 requires the notice to be given to the acquiring authority, that is to say, to the county council, whereas the letter is directed to the Bournemouth Borough Council. Copies of the letter and its enclosure were sent to the county planning officer and that, in my judgment, was proper service on the county council if such service was required.

Therefore, I do not think that the plaintiffs have shown any arguable case that the formalities required by regulation 4 have not been duly complied with.

The plaintiffs' second main argument is much more formidable. Again, the case has changed somewhat during the course of the hearing. The basis of the argument is that the deemed planning permission procedure laid down in regulation 4 only applies where the local authority requires a planning permission for development which they propose to carry out. The procedure does not therefore apply if the development is proposed to be carried out by someone else. The plaintiffs' evidence-in-chief exhibited a minute of the county council's community land committee which referred to the fact that Taylor Woodrow had already been offered, and accepted, appointment as the county council's developer. This, said the plaintiffs, coupled with the facts that the office development is a commercial development and that Taylor Woodrow has acquired land in the grey area, showed that the proposed development was not to be carried out by the council but by a commercial developer.

The county council put in a large number of minutes of various of its committees. From these the following is to be deduced:

(1) The scheme in its present form has been under consideration by a number of different committees from at least January 1979.

(2) Initially the scheme, or at least the necessary land acquisitions, was to be financed under the Community Land Act.

(3) That in March 1979 it was envisaged that part of the financing was dependent upon financial contributions from developers. A confidential report from the chief executive made in March of this year reads as follows:

> Having regard to the multiplicity of ownerships of land involved in the redevelopment of the Central Station Area, the County Council have decided that Community Land Act powers should be used to effect the land assembly. A financial appraisal of the scheme indicates, that, dependent on adequate capital contributions by developers of the site and speed of implementation, it should be possible to secure an annual cash flow from the commercial development within the time required, sufficient to satisfy the criterion of the Department of the Environment for the issue of loan sanction.

If developers were to provide financial capital contributions to the development they must in return have been seeking to take some profit from the development.

(4) In May 1979 central government informed the county council that the Community Land Act powers could not be used except in exceptional circumstances and that the present case was unlikely to be considered exceptional.

(5) Following the removal of the likelihood of the land acquisition being financed under the Community Land Act, in June and July of this year the local authority committees decided to go ahead with the scheme, financing the land acquisitions by loans from the county funds.

(6) The minutes after May 1979 disclose no discussions of any kind relating to the financing by the county council of the construction works necessarily involved with the development of the land by the county council. Taylor Woodrow's position as county council developers is treated as still subsisting: see the minute of June 6 which I have read.

(7) The minutes do not disclose any information as to the basis on which Taylor Woodrow were offered the position as developers. Although negotiations with Taylor Woodrow have taken place, I was informed that no detailed agreement has been reached between Taylor Woodrow and the county council.

The county council evidence states that the county council consider it important that the whole of the grey area should be developed as one and that they intend to carry out this development themselves. They say they will acquire any land of Taylor Woodrow in the grey area, and they also say this:

> The Council do not employ sufficient staff to undertake the works involved in such a development and accordingly have to look outside their own staff resources for a suitable agent to act on their behalf.

Then, after referring to the appointment of Taylor Woodrow as the county council developers, they go on:

> At all times the Defendants will be in sole control of the development and the detailed plans and the carrying out of the development, and in no way will they delegate control to a developer.

Finally, Miss Appleby told me on instructions that following the demise of the Community Land Act powers the county council were no longer considering any scheme involving finance provided by developers in return for a share in the profits and, to my mind rather surprisingly, that no consideration had been given by the committees as to how the construction works were to be financed.

Regulation 4 obviously raises certain difficulties. In what circumstance can one say that the development is to be carried out by the local authority? Plainly, regulation 4 must cover a case where the development is carried out by the local authority by direct labour; equally it must cover a case where the local authority, as building owner, employs contractors to build according to the local authority's plans. At the other extreme, as at present advised it would seem to me not to cover a case where the local authority granted building leases of parts of the area to developers who, subject to overall guidelines, could develop in accordance with their own plans and would take the profits. But between these extremes there are an infinite number of possible arrangements which could be reached between the county council and Taylor Woodrow in this case.

Happily it is not necessary for me to grapple with these questions of degree, because the county council have accepted that a scheme under which a commercial developer took a share in the profits of the development (as opposed to a fixed remuneration) would not fall within regulation 4. On this basis, I think the plaintiffs have shown an arguable case that when the application for planning permission was made in July of this year, the scheme then in mind was not a development of the whole site by the county council alone. There is evidence that before the withdrawal of the Community Land Act power Taylor Woodrow had been appointed developers, and the capital contributions from developers were to provide part of the financing. After the demise of the Community Land Act financing, new arrangements were discussed for financing land acquisitions, but Taylor Woodrow's appointment as developer remained and the minutes disclosed no discussion of any alteration either in the financing of the building and construction works or the basis on which negotiations with Taylor Woodrow were to proceed. Accordingly, whatever may have been the intentions of the officers of the county council, on the basis of whose instructions Miss Appleby told me the current position, it seems to me at least possible that at the trial the plaintiffs will be able to establish that in July 1979 it was still contemplated that Taylor Woodrow would provide capital finance on a profit-sharing basis—in which case, it is common ground, regulation 4 would not have been applicable.

I think therefore on the facts before me, which I emphasise are only interim facts, the plaintiffs have shown an arguable case which it is possible will succeed at the trial. For *American Cyanamid* purposes, that is all that is necessary for them to show. I come then to the balance of convenience.

I will first consider the matter apart from the plaintiffs' interest in the red area, since it is submitted with some force that their interest in the planning status of the adjoining red area does not give them any legal *locus standi* in relation to the planning permission on the grey area. Looking at the first plaintiffs' interest in the grey area alone, if the second resolution is passed tomorrow there would be an apparent planning permission affecting the plaintiffs' land, which would debar consideration of the plaintiffs' own application for planning permission and effectively sterilise its use for other development until the validity of the county council's actions is determined. On the other hand, the county council are anxious to get ahead with their scheme over the whole grey area, including the public services traffic interchange and, for that purpose, need planning permission. An injunction will delay this. Moreover, the plaintiffs have been guilty of slight delay in bringing this motion, but so far as I can see the defendants have not been prejudiced by the delay, save in

A preparing their case to deal with the motion—which they have dealt with most efficiently.

On balance, I think the greater inconvenience will be suffered by the plaintiffs. It is true that the planning permission deemed to be granted under paragraph 4 is limited to a permission for the county council itself to develop, but the difficulties of disentangling who is developing at a later stage would be very considerable indeed. Since it is admitted that neither detailed plans nor finance have been prepared for the county council's scheme at this stage, obviously there must be some time-lag before effective work can be done. Moreover, for myself, I entertain considerable doubts whether the second resolution B for the purposes of regulation 4(5) would, or could, properly be passed at the present time. The regulation 4 procedure is a strange one in that the local authority is made a judge in its own cause, that is to say, it decides its own planning application. The final stage is a resolution by the local authority to carry out the development. As at present advised, I do not think Parliament intended such a resolution to be passed simply for the purpose of getting planning permission, but assumed that a responsible local authority would not pass a resolution to carry out the works unless and until satisfied that the development could be carried out. At present no consideration has been given to the question how the construction work C (as opposed to the land acquisition) is to be financed. From the minutes I am not sure to what extent the members of the various committees appreciate that financing the scheme from developers, based on profit sharing, is not possible if the regulation 4 procedure is adopted. I have doubts whether a second resolution would in any event be passed if the true position, namely, that no finance for the construction works has yet been worked out, were appreciated.

If, in considering the balance of convenience, it is legitimate for me to take account of the red area, the balance of convenience would tip overwhelmingly in favour of the plaintiffs. Their chance of success in an appeal relating to office develop-D ment must be prejudiced by the existence of an apparent planning permission on adjoining land absorbing all the proposed office space for the area. True, the plaintiffs have to overcome the difficulty of not having land within the structure plan designed for office use, but a planning permission on the grey area would present them with even greater difficulty.

It would, in my judgment, be better if the plaintiffs' and the county council's schemes could together be considered by the Secretary of State on an even footing, rather than one of the schemes having the advantage of an apparent planning permission.

For these reasons I propose to grant an injunction against E the passing of the resolution under regulation 4(5). I will consider the form of the injunction. I will also direct a speedy trial if either party desires it.

COURT OF APPEAL

October 22 1979

(Before Lord Justice STEPHENSON and Lord Justice BRANDON)

CO-OPERATIVE RETAIL SERVICES LTD v SECRETARY OF STATE FOR THE ENVIRONMENT AND OTHERS

F Estates Gazette May 10 1980
(1979) 254 EG 401

Town and Country Planning Act 1971—Secretary of State's refusal to postpone date of public inquiry—Whether this action was a "decision by the Secretary of State on an appeal under section 36" of the Act within the meaning of section 242(3)(b)—If so, an application under section 245 to question the validity of the decision was competent—Held that the Secretary of State's refusal to postpone inquiry was not a "decision" disposing of an appeal to him, which was the kind of decision contemplated by these provisions, but was merely a decision G allowing the appeal to proceed—As such it could not be challenged under section 245, but could, if there were any grounds, be subject to the supervisory jurisdiction of the Divisional Court by way of judicial review—In any case the action of the Secretary of State was not a breach of the rules of natural justice—His refusal to postpone did not deprive the inspector at the inquiry of his discretionary power to adjourn it—Decision of Phillips J, rejecting the application under section 245 on the grounds of want of jurisdiction, upheld by Court of Appeal

This was an appeal by Co-operative Retail Services Ltd from a decision of Phillips J rejecting an application under section 245 of the Town and Country Planning Act 1971 H which questioned the validity of the refusal of the Secretary of State to postpone a public inquiry. The inquiry related to a planning appeal by William Morrison (Supermarkets) Ltd, the third respondents to the present appeal to the Court of Appeal, who had been refused permission to carry out demolition work and erect a supermarket. The first respondent to the present appeal was the Secretary of State and the second respondents were the planning authority, City of Wakefield Metropolitan District Council.

Patrick Ground (instructed by Bower, Cotton & Bower, agents for Bury & Walkers of Leeds) appeared on behalf of the appellants; David Latham (instructed by the Treasury J Solicitor) represented the first respondent; and Duncan B W Ouseley (instructed by Last Suddards & Co, of Bradford) represented the third respondents. The second respondents were not represented and took no part in the proceedings.

Giving judgment, STEPHENSON LJ said: This is an important appeal and I regret that in view of the lateness of the hour and the urgency of the matter I must deal with it shortly and, therefore, run the risk of doing a considerable injustice to the careful submissions made by both counsel, particularly those made by Mr Ground on behalf of the appellants.

What we have before us are orders of Phillips J made at the end of last week, in effect refusing to interfere with a K decision of the Secreary of State for the Environment, refusing to grant an adjournment of a public inquiry fixed for tomorrow, at the suit of the appellants, the Co-operative Retail Services Ltd, and the City of Wakefield Metropolitan District Council, the planning authority for the area. The public inquiry is concerned with an appeal by the third respondents, William Morrison (Supermarkets) Ltd, against a decision refusing them planning permission to carry out some demolition work and erect a large supermarket, which is obviously going to have a considerable effect, if it is permitted, on the nearby premises of the Co-operative Retail Services Ltd and it could have a considerable effect, one would have thought, on the L planning of the whole area.

In the correspondence which we have seen, namely in the months of August and September of this year, the Co-operative, as I will call them, have been repeatedly attempting to get the Secretary of State to adjourn the hearing and holding of this public inquiry on the ground that they cannot be ready in time. I would not like to be thought insensitive to the difficulties of getting specialist counsel and specialist experts to support opposition to an appeal which, of course, may be unsuccessful even if unopposed, because the Co-operative and the planning authority are not appellants; they are resisting this appeal of Morrison, as I will call them. M

The Secretary of State has, in letters which I shall not read, taken, if I may say so, a hard line, not always choosing his words particularly well and being perhaps rash enough to indicate an unbending policy which might be thought not to take adequate account of the objections that were being put: but I bear in mind the authority which Mr Latham has cited to us, the *Ostreicher* case, reported in [1978] 1 WLR 810, and what in particular the Master of the Rolls has said at p 816:

> . . . there is a distinction between an administrative inquiry and judicial proceedings before a court. An administrative inquiry has to be arranged long beforehand. There are many objectors to

A consider as well as the proponents of the plan. It is a serious matter to put all the arrangements aside on the application of one objector out of many. The proper way to deal with it, if called upon to do so, is to continue with the inquiry and hear all the representatives present: and then, if one objector is unavoidably absent, to hear his objections on a later day when he can be there. There is ample power in the rules for the inspector to allow adjournments as and when reasonably required.

That, of course, was said in reference to a very different case to this, but nevertheless it states matters which have to be borne in mind in considering the second ground on which we are asked to allow this appeal.

B Phillips J dismissed the motions before him on the ground that he had no jurisdiction to hear them. He was sitting as a single judge of the High Court to hear an appeal under section 245 of the Town and Country Planning Act 1971, and by order 94, rule 1(1) a single judge is the person to hear a matter of that kind.

What was said by Mr Ground to the learned judge, and has been said to us, is that what is being questioned is a decision of the Secretary of State on an appeal under section 36 of this Act, as provided by section 242(3)(b). Section 36 of the Act deals with appeals against planning decisions. It sets out a number of matters and by subsection (6) states: "The decision of the Secretary of State on any appeal under this section C shall be final." That, like all the other subsections of the Act, has to be read with section 242, which provides by subsection (1):

Except as provided by the following provisions of this Part of this Act the validity of . . . (e) any such action on the part of the Secretary of State as is mentioned in subsection (3) of this section, shall not be questioned in any legal proceedings whatsoever.

By section 242(3) the action referred to in subsection (1)(e) includes:

(b) any decision of the Secretary of State on an appeal under section 36 of this Act.

D The opening words of section 242(1) let in the provisions of section 245 for questioning the validity of, *inter alia*, such a decision on an appeal: Section 245(3).

The learned judge was referred to sections 36, 242 and 245, and the first question he had to decide was, "Are these decisions decisions of the Secretary of State on an appeal under section 36?" "It is a short but not necessarily easy question," he said, in a note of his judgment which has been agreed but has not been submitted to the learned judge for approval because it had not been typed, "namely whether in section 242(3)(b) they constitute a decision of the Secretary of State on an appeal. In my judgment they do not."

E He was referred to *Chalgray Ltd v Secretary of State for the Environment*, which is a decision of Slynn J, reported in (1977) 33 P&CR 10, and the case of *Button v Jenkins*, which was applied in that case, reported in [1975] 3 All ER, 585: but he took the view that what was in contemplation when Parliament referred to a decision on an appeal, both in section 36(6) and in section 242(3)(b), was a decision which disposes of an appeal, not a decision in the course of an appeal but one dealing with its final outcome. Then Phillips J considered that this decision, and this cannot be doubted, was not a decision disposing of the appeal: it was a decision letting the appeal go on but refusing to alter the date on which it was to F be started; and so he came to the conclusion that it was not for him under the section which empowered him to deal with the matter to grant the relief sought, but it was possible to apply for judicial review to a Divisional Court.

It is now conceded that he was right about that possibility and if this were a decision on an appeal under section 36 there would also be that possibility because by section 242 subsection (4) it is provided that:

Nothing in this section shall affect the exercise of any jurisdiction of any court in respect of any refusal or failure on the part of the Secretary of State to take any such action as is mentioned in subsection (3) of this section.

G It seems to me that if the Secretary of State refuses to do his statutory duty, or fails to do his statutory duty by making a decision which he ought to make, he is then subject to the supervisory jurisdiction, the judicial reviewing jurisdiction, of the Divisional Court.

For my part, after hearing the interesting arguments addressed to us, I am in complete agreement with the view of Phillips J on this point. It may be in a sense a procedural wrangle. It is a matter, as the judge said, of some importance, but it seems to me that looking at section 242 and section 245, this decision of the Secretary of State is not a decision on an appeal under section 36 of the Act. I think some support for the view which the learned judge took is to be derived from H section 246 of the Act, which starts off in subsecton (1) with these words: "Where the Secretary of State gives a decision in proceedings on an appeal. . . ." The case of *Button v Jenkins* on which Mr Ground relied, both here and before Phillips J, was a decision on that section and on those words, and it may well be that the decision in this case was a decision in proceedings on an appeal, but it was not in my view a decision on appeal.

If there is anything in Slynn J's decision which indicates a different view from that taken by Phillips J, or can be taken to define the words used in a different way to that in which they are defined by Phillips J, I would not be disposed to follow J it. I see no reason to suppose the decision by Slynn J was wrong and have every reason to suppose that Phillips J was right in saying that he was not departing from that decision in reaching the decision which he did on the main point. I would, however, like to add, that if I were wrong on this point, and Phillips J was wrong on this point, we have listened to full argument on the question whether the Secretary of State's decision does result in a breach of the rules of natural justice or, in other words, does deny justice to the appellants or to the planning authority, the district council. In my view, it does not. It is, of course, true, that the inspector will know that the Secretary of State has repeatedly refused to adjourn K the date of this inquiry, and he will not be able to put that out of his mind. Nevertheless, he has, as indicated in the passage quoted from the Master of the Rolls' decision in the *Ostreicher* case, a complete discretion of his own under the rules to adjourn if he thinks it is necessary in the interests of justice to any objector or any party to do so. We were referred to the terms of the Town and Country Planning (Inquiries Procedure) Rules 1974, rule 10(8), which gives him a wide and unfettered discretion from time to time to adjourn. One knows that adjournments are ordered by inspectors handling public inquiries in order to enable further objections to be taken or pursued. L

I am not indicating in any way whether the inspector should yield to arguments addressed to us in the course of this appeal, if and when he is asked to do what the Secretary of State refuses; but it is very apparent that the Secretary of State by refusing to delay the start of this inquiry has in no way decided that the inquiry is to be continued and to proceed to a determination by the inspector without adjournments if an adjournment is asked for.

The notice given was adequate. There are, no doubt, great administrative pressures in these cases and I have carefully considered the possible injustice to the parties who complain that they have not been given more time. If I thought that M they would have no opportunity of pressing those objections but would be bound to do the best they can without any possibility of a further adjournment being granted, I might feel able to take the view that natural justice was being breached by the hard line taken by the Secretary of State. But I am not satisfied on the material we have that it would not be possible for all the parties to put adequately before the inspector, at any rate before the end of the inquiry, the objections which they wish to put before him, and I am by no means satisfied that if they were to try to persuade the inspector to grant an adjournment, their attempts would

A necessarily be defeated. There seems to be a real possibility that they may be successful in obtaining an adjournment, but that is a matter for the inspector and he has that duty to consider any applications that may be made to him, and to decide them independently of what the Secretary of State has already decided. There is a difference between a decision to postpone the start of a widely advertised inquiry which may cause great inconvenience, and a decision *de die in diem* to adjourn in order to enable particular points or material to be put before the inspector as the inquiry goes on. For those reasons I would dismiss the appeal.

Agreeing, BRANDON LJ said: I agree that the appeal should B be dismissed on the two grounds stated by my Lord. The first ground is that the decision of the Secretary of State to refuse a postponement of the inquiry was not a decision on an appeal under section 36 of the Town and Country Planning Act 1971 within the meaning of section 242(3)(b) of that Act. There was accordingly no jurisdiction in the judge to deal with the application. The second ground, which is alternative, is that, if that is wrong, then the decision to refuse postponement of the inquiry has not been shown on the evidence before us to have been contrary to justice. Therefore the court has no power to interfere with that decision.

I should like to make some observations about the case of C *Chalgray Limited v Secretary of State for the Environment* (1977) 33 P&CR 10, which was relied on by counsel for the appellants. In the headnote of that case, the second part of the decision is stated in this way: " . . . That the words in section 242(3)(b) 'any decision of the Secretary of State on an appeal under section 36 of the Act' were not necessarily limited to the decision or orders or final result specified in section 36(3); and that the Secretary of State's declining to consider the appeal was a decision on an appeal under section 36." Even if that decision made by Slynn J is correct, it does not assist the appellants on the facts of this case because there has not been any refusal to consider the appeal. I am bound D to say, however, that I have doubts about the correctness of that decision on the law by Slynn J. It seems to me very arguable that the expression "any decision of the Secretary of State on an appeal under section 36," as used in section 242(3)(b), is limited to decisions or orders or final results arrived at under section 36(3). I further think that, where there is a refusal to consider an appeal, the case might well come within section 242(4) of the Act. It is not, however, necessary to decide that question in this case. I only wish to express my doubts about this because the case has been relied on and I would not like it thought that I regard it as necessarily correct.

E The appeal was dismissed with costs. Leave to appeal to the House of Lords was refused.

COURT OF APPEAL

February 1 1980

(Before Lord Justice MEGAW, Lord Justice DONALDSON and Sir Patrick BROWNE)

BILBOE AND ANOTHER v SECRETARY OF STATE FOR THE ENVIRONMENT AND ANOTHER

Estates Gazette May 24 1980

F (1980) 254 EG 607

Town and Country Planning Act 1971—Enforcement notice—Appeal from decision of Divisional Court—Validity of notice upheld by Secretary of State and Divisional Court—Question as to whether materials which could be tipped into rubbish dump in old quarry were restricted to sand, soil, rock and clay or whether other inert material, such as builders' rubble, could be tipped—Condition in 1950 planning permission that only approved materials could be tipped—From about 1950 other materials, including builders' rubble, were tipped in breach of condition—Confusion throughout proceedings between development consisting of carrying out operations and development consisting of material change of use—In present case development which was in breach of planning control took place about 1950—Further tippings did not constitute further development or further breaches—By section 22(3)(b) the tipping did not constitute an operation, but a change of use—Section 88(1)(d), not 88(1)(c), applied and enforcement action was time-barred because breach occurred before January 1 1964—Appeal allowed and matter remitted to the Secretary of State for re-hearing and determination under RSC Order 94, rule 12(5) G

This was an appeal by John Francis Bilboe and John Francis Bilboe (Junior) from a decision of the Divisional Court (1978) 248 EG 229 dismissing an appeal from a decision of the H Secretary of State, who had upheld an enforcement notice served by the second respondents, West Lancashire District Council, on the appellants in respect of the tipping of materials, in particular builders' rubble, into a worked-out quarry, Sandfield Quarry, Aughton, Lancashire.

Iain Glidewell QC and David Gilliland (instructed by Compton Carr, agents for Russell & Sutton, of Southport) appeared on behalf of the appellants; Simon Brown (instructed by the Treasury Solicitor) represented the first respondent, the Secretary of State for the Environment; Bruce Martin QC and Miss Cherie Booth (instructed by the Solicitor, West Lancashire District Council) represented the second respondents. J

Giving the judgment of the court at the invitation of Megaw LJ, SIR PATRICK BROWNE said: This is an appeal from a decision of the Divisional Court given on July 17 1978, dismissing an appeal from a decision of the Secretary of State dated November 25 1976 by which he dismissed an appeal against an enforcement notice dated April 29 1976 served by the West Lancashire District Council on the appellants.

The land concerned is a worked-out quarry called Sandfield Quarry at Aughton, in Lancashire, which is owned and occupied by the appellants and was used by them as a rubbish tip from August 1975 until about July 1976. The practical issue is whether the materials which the appellants are permitted to tip K are restricted to sand, soil, rock and clay, as the council contend and the Secretary of State has held, or whether they can also tip other inert material, in particular what may be generally described as builders' rubble.

In 1938 an agreement was made between the then owner of the quarry and the West Lancashire Rural District Council under which the quarry was to be regarded as a private open space in consideration of the owner being allowed to tip approved materials in two parts of the quarry. Before these two areas were filled in, the quarry was sold to the Wellington Haulage Co, and in June 1948 it was agreed that tipping should cease on March 1 1950, by which time it was estimated that the L areas described in the original agreement would be used up. On February 14 1950 a Mrs Hartley applied for planning permission for development, the purpose for which the land was to be used being described as "Tip." It appears from the plan attached to the application that it was proposed to fill most of the quarry in addition to the areas already filled. On August 9 1950 planning permission was granted by the district council to the Wellington Haulage Co. The planning permission recited the particulars of the application, "Development forming the subject of the application" being described as "Proposed Tip." The decision was:

> M The West Lancashire Rural District Council, as agents for the Lancashire County Council, . . . hereby give notice . . . that permission for the carrying out of the development referred to in Part I hereof in accordance with the application and plans submitted has been granted subject to the following conditions: (1) That the materials to be deposited shall be approved by the Surveyor to the West Lancashire Rural District Council. (2) That no materials of a noxious nature or materials likely to give rise to overheating or noxious fumes or smell shall be deposited.

Then there are two other conditions which we do not think we need read. Then the reason given for the conditions was: "In order to protect the amenities of the area." Mr Glidewell,

A for the appellants, accepted that this permission was needed for the proposed tipping.

Tipping began in 1950-51. In 1952 about 40 tons of mussels and barnacles were tipped on the site, and the Wellington Haulage Co were warned not to continue tipping organic materials.

The inspector who conducted the public local inquiry into the appellants' appeal against the enforcement notice (held in August 1976) found the following facts:

> (m) The district planning authority have been unable to find any record of an approval or disapproval by "the Surveyor" to the tipping of any particular materials and none was produced by the appellants. (n) There is evidence that in 1952 some 40 tons of
> B mussels and barnacles were tipped; the then Medical Officer of Health called for them to be covered as soon as possible; apart from a subsequent letter on the same subject from the county council to the rural district council. there is no record of any further action being taken by the authorities. (o) Excavations carried out for the appellants in July 1975 show that various materials other than sand, soil, rock and clay were tipped down the bank on the north side of the bottom level in the past and it was found that the slope bore vegetation including trees up to 10 years old. (p) At the inquiry a local resident stated that there had been no tipping for 12 to 14 years before July 1975; that about 12 years ago. at the request of the then owner, he put a gate across the entrance to the site and secured it and that it so
> C remained until 1975; his evidence was not challenged.

We were referred without objection to some of the evidence on which these findings were based, as set out or referred to in the inspector's report. On finding (n) we were referred to letters dated September 15 1952 from the medical officer of health to the Wellington Haulage Co and October 7 1952 from the divisional planning officer of the county council to the surveyor of the district council. Mr Glidewell relied on the latter letter as showing that the council knew that builders' rubble was then being tipped on the site. As to finding (o) we were referred to the inspector's summary of the evidence of Mr Boardman, the engineer who carried out the investigation, whose findings
D were accepted by the council. Mr Boardman summarised his conclusions as follows:

> The investigation shows that this site has been used as a tip for builders' rubble and similar non-noxious materials for a period of more than 20 years. There is no evidence of any change in the type of materials tipped during this period, and these materials do not appear to be different from the materials which Bilboe Bros propose to tip.

Finding of fact (p) shows that these materials must have been tipped—at latest—before the end of 1963.

In August 1975 the site was bought by the appellants. After correspondence and a site meeting, the district council wrote to
E the appellants' solicitors a letter dated July 8 1975 (ie before the appellants bought, which included the following statements:

> I would point out that it is the responsibility of your client to submit full details of his proposed scheme and programme of tipping to this council. . . . A meeting was held on the site on July 3 1975, when the following matters were raised and agreed with Messrs Bilboe Bros. . . (d) that all materials deposited shall accord with conditions numbered 1 and 2 in planning permission 8/6/781. The responsibility falls wholly on Bilboe Bros to ensure that only such materials are tipped. As a guide such material shall only include inert builders' rubble, such as bricks, mortar, glass, porcelain, undiseased wood and metal.

F

The inspector found as facts that "In a reply dated July 17 1975 the appellants stated among other things that they would conform to the list in July 8 letter and asked for confirmation that tipping could proceed. The council did not reply, but on November 19 1975 they wrote to the appellants' solicitors asking for a written request for approval of materials."

Meanwhile, in October 1975 the council had "resolved to take action under section 51 of the Town and Country Planning Act 1971 to discontinue completely the use of the site for tipping and meanwhile to adopt a restrictive interpretation of the planning permission conditions; in February 1976 they G reaffirmed these resolutions."

On February 2 1976 the appellants' solicitors wrote to the district council asking for formal approval under condition 1 of the tipping of "inert builders' rubble and other materials of a non-noxious nature", including a list of items they set out—eg bricks, mortar, putty, glass—and also sand, soil, rock and clay; the solicitors said: "We would also make it clear that the making of this application is not to be taken to be any admission that approval under condition 1 of the planning permission is required for these materials or that it has not been given in the past." The council replied on February 27 1976:

> After careful consideration I am writing now to inform you H that the only materials taken from the list included in your letter of February 2 which I am prepared to approve for tipping at Sandfield Quarry are sand, soil, rock and clay. I consider that the deposit of other materials will constitute a breach of planning control and as such will render your clients liable to enforcement action.

The appellants continued to tip other materials, in particular builders' rubble, until about July 1976, a stop notice having been served on May 13.

The enforcement notice to which these proceedings relate was dated April 29 1976. Recital (iii) is as follows: "On the ninth day of August 1950, the West Lancashire Rural District Council J as agents for the Local Planning Authority upon an application in that behalf made under Part III of the Town and Country Planning Act 1947 granted permission for the following development of the said land namely the carrying out of operations consisting of Proposed Tip." Then recital (iv) recited condition 1 of the permission. Recital (v) was:

> It appears to the council that within the period of four years before the date of service of this notice there has been a breach of planning control in that the said development has been carried out but the said condition has not been complied with in the following respects namely: Materials have been deposited other than sand, soil, rock and clay which in a letter dated February K 27 1976 the Director of Planning of the West Lancashire District Council as the Officer who by virtue of Article 16 of the Local Authorities (England) (Property etc) Order 1973 has succeeded the Surveyor to the former West Lancashire Rural District Council approved for the purposes of condition 1 of the consent dated August 9 1950.

Then the notice required the appellants to cease depositing upon the land any materials other than those specified in the last paragraph of the letter of February 27 1976.

The appellants appealed against the enforcement notice to the Secretary of State under section 88 of the Town and Country Planning Act 1971. The Secretary of State dismissed the appeal, subject to a variation of the plan attached to the notice. The L appellants appealed to the Divisional Court under section 246 of the Act and RSC Order 94, rule 12, and their appeal was dismissed. They now appeal to this court; the Divisional Court gave leave to appeal, if it was required.

We think it is convenient to refer at this stage to the relevant provisions of the 1971 Act. We do not propose to read them but we must be taken to have incorporated in this judgment the following sections: section 22(1) and (3)(b); section 23(1); section 87(1) (as amended by the Local Government Act 1974), (2), (3) and (6); section 88(1)(a), (b), (c) and (d); section 246(1); and the definition of "use" in section 290.

We feel bound to say that it seems to us that this case has been M complicated and confused throughout by a failure on all sides to keep clear the distinction between the two mutually exclusive forms of development to which the Town and Country Planning Acts relate—development by carrying out "operations" and development by making a "material change of use." Recital (iii) of the enforcement notice alleges that the 1950 permission "granted permission for the following development of the said land, namely, the carrying out of operations consisting of Proposed Tip", and recital (v) alleges that "within the period of four years before the date of service of this notice there has been a breach of planning control in that the said

A development has been carried out but the said condition has not been complied with . . . ." (The reference to four years implies that this was an operations case—see sections 87(3) and 88(1)(c).) Mr Glidewell takes no technical point on the wording of the notice, but he submits that the "development" with which we are concerned was not the carrying out of operations but the making of a material change of use.

The appellants' appeal against the enforcement notice was based on section 88(1)(a), (b) and (c) of the Act. We are not concerned with section 88(1)(a) (that planning permission ought to be granted), but if Mr Glidewell's submission is right the relevant provision is section 88(1)(d), not section 88(1)(c).
B Mr Brown, for the Secretary of State, did not object to our considering section 88(1)(d), because the point was raised in the Divisional Court, but he said that the Secretary of State was giving no indication that he would not object in future to a challenge to an enforcement notice on a ground not advanced to him. This confusion persisted as late as the notice of appeal to this court, ground (2) of which refers to "a period of more than four years prior to 1964."

The inspector concluded: (a) that whether or not the council's letter of July 8 1975 could be regarded as an approval of the tipping of the materials to which it referred (builders' rubble), the council was entitled later to limit its approval to
C other materials, and had done so by its letter of February 27 (sand, soil, rock and clay)—the appeal under section 88(1)(b) should therefore fail; (b) as the tipping of materials other than those specified in the letter of February 27 1976 took place within the period of four years before the service of the notice, the appeal under section 88(1)(c) should also fail.

The Secretary of State concluded: (a) that the council's letter of July 8 1975 was not an approval under condition 1 of the planning permission, and that, although such approval need not be in writing, the appellants had failed to establish that any such approval had been given before the council's letter of February 27 1976; (b) that "whether or not an earlier breach
D of planning control occurred, the present breach is unconnected with it, the tip having been unused for some 20 years before 1975, and it is therefore capable of being made the subject of enforcement action", and that the inspector's conclusion on section 88(1)(c) was accepted.

In the Divisional Court Mr Glidewell, for the appellants, contended: (1) that the Secretary of State was wrong in holding that no approval had been given by the council under condition 1 to the tipping of materials other than those specified in the letter of February 27 1976—in particular builders' rubble, which had in fact been tipped between 1950 and 1963—and that the only reasonable inference was that
E such approval had been given, probably between 1950 and 1952; (2) alternatively, if no approval was given, that the breach of planning control on which the enforcement notice was based took place before January 1 1964 and that enforcement was therefore time-barred; (3) that the letter of July 8 1975 gave rise to an estoppel against the council.

The Divisional Court rejected all three contentions. In this court, Mr Glidewell renewed contentions (1) and (2) but did not pursue contention (3) in face of the decision of this court in *Western Fish Products v Penwith District Council* (1978) 38 P&CR 7.

As to contention (1), an appeal from the Secretary of State
F lies only on a point of law (section 246(1)). Mr Glidewell accepts that he cannot succeed on this point unless the only reasonable conclusion on the facts found is that approval was given for the range of materials found by Mr Boardman on his site investigation. He submits that this is the only reasonable and proper inference. He does not rely on the letter of July 8 1975 as an approval, but he relies on the finding of the Secretary of State that approval need not be in writing. He says that the facts are—(a) that the materials found on the site investigation (in particular builders' rubble) were tipped between about 1950 and 1963; (b) that in 1952, at the time of the mussels and barnacles incident, the district council knew that such materials were being tipped, but made no G objection to anything except the mussels and barnacles. He submits that the inescapable inference from the combination of these facts is that the council had before 1952 approved the tipping of builders' rubble etc. We are not satisfied that we should necessarily have come to the same conclusion as the Secretary of State, but this does not matter. In our judgment it is impossible to say that the only reasonable conclusion is that approval had been given and that the Secretary of State was therefore not entitled to make the finding he did make. In our judgment, therefore, this contention must fail.

As to Mr Glidewell's second contention, Mr Brown for the Secretary of State and Mr Martin for the council submit that H the Secretary of State and the Divisional Court were right in holding that although no enforcement action could now be taken in respect of the tipping before 1964 there were further breaches of planning control when tipping was resumed from August 1975 onwards of materials which had not been approved under condition 1, and enforcement action can now be taken in respect of these failures to comply with the condition. They rely by analogy on the decision of this court in *Thomas David (Porthcawl) Ltd v Penybont RDC* [1972] 1 WLR 1526 and submit that Lord Widgery CJ was right in the present case when he said:

> The view that I take is that in that situation the local authority J can proceed in respect of the infringements which occurred after January 1 1964, notwithstanding the fact that they are of course time-barred in respect of anything happening before that day. I do not recognise a principle which says that, if you have a certain number of incidents amounting to a breach of planning control occurring before 1964, that necessarily prevents the local authority from proceeding in respect of those which occur afterwards. Indeed, I take exactly the opposite view and think there is nothing in what one might call the four-year principle, or now the 1964 principle, to prevent that from happening.

Mr Glidewell submits that in this case (assuming that no approval was given) the breach of planning control took place K when tipping of builders' rubble etc began about 1950 without approval. This was "development" by making a material change of use, not by carrying out "operations," and *Thomas David v Penybont RDC* is therefore distinguishable. If no permission at all had been given, the breach would have taken place once and for all when the change of use was made about 1950 and no enforcement action could now be taken; the same must apply to non-compliance with condition 1, subject to which permission was granted. He submits that it would be absurd that (as Mr Martin accepts) the appellants are now worse off than if they had made the change of use without any permission at all. L

Once the distinction between development by operations and development by making a material change of use is fully appreciated, it is in our judgment plain that Mr Glidewell's contention is right. The distinction between "development" by carrying out "operations" and "development" by making "a material change of use" is fundamental to the scheme of the Town and Country Planning Act 1971 (and of the earlier Town and Country Planning Acts). It might be difficult in the abstract to say into which category the tipping of refuse falls, but section 22(3)(b) makes it clear that it is to be regarded as a "change of use" and not as "operations." The Divisional Court so held in *Roberts v The Vale RDC* (unreported February 8 1977).* M *Thomas David v Penybont RDC* was concerned with continuing mining operations, and it was held that "the moving of every shovelful was a separate act of development and could be the subject of an enforcement notice if carried out without permission" (Lord Denning MR at p 1531-B). It followed that although no enforcement notice could be served in respect of operations carried out more than four years before the service of the notice, it could be served in respect of continuing operations after that date. But in that case

* Reported as *Roberts v Vale Royal District Council* (1977) 242 EG 811 and (1977) JPL 369.

A Lord Denning MR said at p 1531: "What I have said relates only to mining operations. They are *sui generis.*" This recognised the distinction between development by operations and development by change of use and distinguished *Britt v Buckinghamshire County Council* (1962) 60 LGR 430 which related to the latter type of development. Edmund Davies LJ in *Thomas David v Penybont RDC* (at pp 1531-H to 1532-A) said:

> Decisions relating to change of use do not necessarily provide a sure guide in relation to questions regarding the other activities which come within the definition of "development" contained in section 12 of the Act of 1962. The former can usually be
> B pinpointed in time, though the process may be difficult; but the "carrying out of operations" is a far more expansive phrase and may well have more elastic implications in point of time.

The situation in this case in 1950 was exactly within section 22(3)(b). Two parts of the quarry had already been used for tipping, but they had been filled up and the occupiers wished to extend the area of the deposit.

In our judgment, the "development" in this case took place when tipping restarted in or about 1950. That development would have been a breach of planning control if no planning permission had been granted and was a breach of planning control because the occupier had not complied with
C condition 1, subject to which permission was granted.

The foundation of the power to serve an enforcement notice is "a breach of planning control after the end of 1963" (section 87(1)), and "there is a breach of planning control if development has been carried out... without the grant of planning permission... or if any conditions or limitations subject to which planning permission was granted have not been complied with" (section 87(2)). In this case, the relevant "development" was the change of use by the tipping of builders' rubble, and so on, outside the two areas previously used which took place in or about 1950, and the relevant "breach of planning control" was the non-compliance with
D condition 1, which also took place then. It is not necessary to identify exactly the point of time at which the change of use took place, but plainly it had taken place long before the end of 1963. In our view, the further tippings which took place after the change of use and before 1963 did not constitute further development or further breaches of planning control. Sections 87(3) and 88(1)(c) (except section 87(3)(c), which is irrelevant) apply only to development by "operations" and so do not apply to this case, but by virtue of sections 87(1) and 88(1)(d) no enforcement action can now be taken in respect of the breaches before the beginning of 1964.

In the course of the argument, my Lord, Megaw LJ,
E suggested that the tipping which was resumed in 1975 might constitute a fresh "change of use" within the meaning of section 22(3)(b) although the "land" on which this new tipping took place was "comprised in a site already used for that purpose," if the new tipping was on a part of the "site" on which there had not previously been tipping. This point was not investigated at the inquiry, and there is no evidence one way or the other and no finding about it. Mr Brown and Mr Martin asked us not to express any view on this point. We shall therefore assume that the tipping in 1975-1976 was on the same "land" as the tipping between 1950 and 1963. On this assumption, the tipping in 1975-1976 was not a change of
F use and was not the carrying out of development, and did not constitute a breach of planning control. The "development" (ie the change of use) and the breach of planning control (ie the non-compliance with condition 1) had taken place once and for all in or about 1950. In our judgment, therefore, the council was not entitled to serve the enforcement notice dated April 29 1976, and the appeal should succeed under section 88(1)(d).

It might be argued that because of the gap in tipping between the time when tipping stopped about 1963 and when it restarted in 1975 there was a change of use when it started again. But this would involve a finding that there was an abandonment of the tipping use in or about 1963. This is a G
question of fact, which does not seem to have been investigated at the inquiry and on which there is no finding of fact. Further, if such development were to be relied on as the foundation of an enforcement notice, the notice would have to be framed differently from that on which the council rely in this case.

The appeal will, therefore, be allowed and the matter remitted to the Secretary of State under Order 94, rule 12(5) of the Rules of the Supreme Court for rehearing and determination by him.

Leave to appeal to the House of Lords was refused.

H

QUEEN'S BENCH DIVISION

November 1 1979

(Before Mr Justice FORBES)

MILLARD v SECRETARY OF STATE FOR THE ENVIRONMENT AND ANOTHER

Estates Gazette May 31 1980

(1979) 254 EG 733

**Town and Country Planning Act 1971—Application to quash decision of Secretary of State dismissing a planning appeal relating to development of a property in Bournemouth as a J
management training centre—Appeal actually determined by an inspector as an "appointed person" under Schedule 9 to the Act—Different grounds on which decisions can be challenged explained by judge—Seven criticisms of inspector's decision put forward by applicant, of which six are rejected by judge—Remaining one, however, proves fatal—Inspector fell into error in making a comparison which was ill-founded on a material point—There was in existence a planning consent, not so far implemented, for demolishing the subject property (a large house) and putting in its place seven town houses—Inspector compared the traffic generation and movement of people involved in the proposed management K
training centre scheme with the present position on the footing that the existing planning consent was not implemented—True comparison would have been between the proposal under consideration and the position if the existing consent were implemented—Although some of the attacks on the inspector's decision reminded the judge of the disputations of the medieval schoolmen ("angels dancing on pinpoints"), this last was a matter of substance—Decision quashed**

In these proceedings the applicant applied in accordance with section 245 of the Town and Country Planning Act 1971 for an order quashing a decision made by an "appointed person" which, under Schedule 9 to the Act, was to be treated L
as made by the Secretary of State. The decision which was challenged dismissed a planning appeal relating to the proposed development as a management training centre of a large house in half an acre of ground in Bournemouth. The applicant was a tenant of part of the subject property.

Miss M Viner (instructed by Cartwrights, of Bristol) appeared on behalf of the applicant; Simon Brown (instructed by the Treasury Solicitor) represented the Secretary of State. The local planning authority were not represented and took no part in these proceedings.

Giving judgment, FORBES J said: In this case Miss Viner moves for an order to quash a decision of the Secretary of M
State in a matter in which he dismissed a planning appeal in relation to some land in Bournemouth. This was one of those cases in which the appointed person, the inspector, as one usually prefers to call him, decided the matter himself under those provisions of the Act and not one in which the inspector made a report to the Secretary of State and the Secretary of State then made a decision. So that the report and the decision letter are one document, and that is a document dated December 21 1978.

Taking the matter quite shortly or as shortly as I can, the premises about which the dispute arose consisted of a large

A house built in 1911 in half an acre of ground in a residential area in Bournemouth. I have seen photographs of the house and also plans. It is a large house with either 18 rooms or 18 bedrooms and at present it is divided into two, 5 and 5A Dunbar Road. It is on the corner of that road and another road called Berwick Road, and it gives the appearance of the sort of house one is familiar with in what used no doubt to be very well-to-do portions of towns of the character of Bournemouth. It has got an attractive garden and it is described as an attractive house. There have been two earlier appeals relating to this site. They were concerned, as I understand it, the first one with the possible establishment of a B language school, and the second with the establishment of what was described as a management centre. Those two appeals have been dismissed.

There were two applications in this case to the local planning authority both of which have been turned down and therefore gave rise to the two appeals with which the inspector was concerned. They were slightly different ways of dealing with another and different proposal for a management centre. What was proposed, and I will take it shortly, was that there should be established at this house a course, a 12-week course I think it was, in management expertise based on what is said to be a highly successful course under the aegis of the C University of Aston in Birmingham. It was intended that there should be another centre running a course of exactly the same kind for the benefit of the south and west of England in giving managerial expertise of that kind. It is said that the courses were designed to equip specialists in large firms for senior management positions in small firms and to fit specialists in small firms with greater managerial responsibility. A great point was made, and not as I understand challenged, that the students, if one may mentally place that term in inverted commas, at this establishment would not be youngsters of the kind one might expect at other educational establishments. They would be mature businessmen aged 35 and over, because D one of the requirements was to be that the student had to have 10 to 15 years' experience in business.

The appellant, as I understand it, was the tenant of one part of this site, but the application was made in relation to the whole site. It has not been made clear to me what the appellant's standing was in relation to the whole site, but I do not think that that is relevant. The necessary notices and so on have clearly from the papers been served on the owner. But what the appellant was really saying was that this was a house of which his part was far too big and too expensive for him to keep up on his own; there was only one person inhabiting the other half of the house; and that the proposal E would mean that the whole house would be properly utilised, that the grounds would be properly kept up and that there were advantages in retaining this building rather than pulling it down and putting up something else. So far as pulling it down and putting up something else was concerned, there was a current planning consent for demolishing the house and putting in its place seven town houses. And the applicant apparently gave evidence, because it is so recorded, that if he was not successful with this planning appeal he intended to implement that planning consent.

There were, as one may imagine, many points which were raised by the appellant at the inquiry countered by the local F authority, and raised by the local authority and countered by the appellant. It is the extent to which those points are correctly or adequately dealt with in the report and decision of the inspector which forms the argument in this case. I do not think I need say anything in very great detail about the principles of law involved. It seems to me that very broadly there are three ways in which a decision of the Secretary of State, for such it is in form, may be challenged in these courts. One is if the Secretary of State misdirects himself in law, if he takes a view of the law which is wholly wrong, and that seems to me to be well indicated in a case called *Hope v Secretary of State for the Environment*, a decision of Phillips J reported at (1975) 31 P&CR 120. It is G quite plain when looking at that report that the ratio of Phillips J's decision was that the inspector had taken a wholly erroneous view of the legal effect of agreements entered into under section 52 of the Town and Country Planning Act 1971, and having done so of course he had erred in law and therefore the courts interfered. That is one way.

The second way is that the Secretary of State may err by going outside the powers of the Act. And the normal way in which this is argued is on the well-known principle in the *Wednesbury* case, that is *Associated Provincial Picture Houses Ltd v Wednesbury Corporation* [1948] 1 KB 223. I need not traverse the ground traversed by Lord Greene MR in his judgment H in that case. The passages are well known. I only pause to say that I do not consider that the *Wednesbury* principle, as it is frequently referred to, is confined to cases of extreme perversity. It is not merely the dismissal of the red-headed school teacher or that kind of case that is covered by or enshrined in the *Wednesbury* principle. The matters are all set out *in extenso* on pp 228 and 229 of the Master of the Rolls' judgment and they cover all the matters which are commonly canvassed in appeals of this kind before this court and can conveniently be described as cases where it is alleged that the Secretary of State has failed to take into account a material consideration or taken into account an immaterial considera- J tion.

And thirdly there are the cases where there has been an error of procedure in this sense, that the Secretary of State has not followed the rules or the provisions of the Act in some way. And again the most usual source of argument on that head in these courts is the giving of reasons in the case of inspectors who are charged with the duty not only of holding an inquiry but also of making a decision thereon. Those are enshrined in rule 16 of the Town and Country Planning Appeals (Determination by Appointed Persons) Inquiries Procedure Rules 1974 (SI 1974 no 420), and rule 16 provides so far as is material: "the appointed person shall notify his K decision and his reasons therefor in writing." There is a considerable body of learning on what that means, starting with the well-known decision in *Re Poyser and Mills' Arbitration* [1964] 2 QB 467 and going on to the *Givaudan* case (*Givaudan & Co v Minister of Housing and Local Government* [1967] 1 WLR 250) in both of which Megaw J, as he then was, indicated that when the rules say that reasons must be given, this means proper, adequate reasons which grapple with the major points which were exercising the inquiry.

Against that background one must I think look in a little more detail at what were the material points. And one of course should add that it is clear from the authorities that it L is not every point which either party thought they wanted to argue that must feature in the decision of the tribunal, be it the inspector or the Secretary of State. Nor is it necessary to look at the language in which the decision is couched with, to borrow Willis J's words, the analytical attention of the medieval schoolman. There were moments during the argument in this case when I could see the shadow of angels dancing on pinpoints. But it is plain that that type of analytical examination of the Secretary of State's letter is alien to the procedure in the special paper on cases of this kind.

Miss Viner for the appellants puts her submissions under three main heads. First what one might call the taking into M account of an immaterial point or failure to take into account a material point. Then the question of reasons, and finally what she called the *Wednesbury* principle point. But as I have already indicated, that was directed to an argument on perversity, and I do not think that it is necessary, when arguing the *Wednesbury* principle, to go as far as to say that the Secretary of State has been perverse. It is sufficient to say that he has erred in not taking into account material considerations and so on. So I prefer to deal with this really only under the two heads, and out of deference to Miss Viner's argument I should add that, despite the way she invited me to look at the letter as a whole and come to the conclusion that the inspector's conclusions were perverse, that is not a conclusion to which I can find myself coming on reading the decision letter in that way.

Looking at the notice of motion the points of criticism are set out in a number of paragraphs, 1 to 7. It would be sensible, I think, if I took those one by one and explained the situation in relation to each of them and the arguments that arise.

The first argument is that the inspector failed to take into account the importance and effect of the existing detailed planning permission for seven town houses at the appeal site. It is only necessary, I think, to go to the conclusion of the inspector's report straightaway to show what the situation was. Before doing so may I just say this. The format of the inspector's report in this case is not perhaps the usual one, and I am not suggesting it is any the worse for that. But the traditional way in which inspectors' reports are normally laid out is that having considered the site and its surroundings, the inspector normally goes on to detail the case for the appellant, then the case for the local authority, followed by the cases put forward by any other interested party, and then usually to make a series of findings of fact and finally the conclusions and recommendations which he or she makes. But that is not the way this report is laid out. Rather the inspector has taken each issue separately and set out the arguments for the appellant and the arguments for the local authority in respect of each issue in consecutive paragraphs. As I say, the report is none the worse for that, though there are one or two portions of it in which it is not quite clear with what voice the inspector is speaking; in other words whether he is setting out the contention of one or other of the two sides or interjecting a view of his own. At paragraph 25 of the report he starts his conclusions. I should read, I think, paragraphs 25 and 26:

> 25. Having heard the evidence given during the inquiry and having seen the appeal site and its environs, it seems to me that although certain other factors must be borne in mind, the central issue is whether or not the proposed use would harm either the character or the appearance of this residential area allocated for primarily residential use in the approved development plan.
>
> 26. In my opinion much of the evidence given at the inquiry tended to draw attention away from this central issue. And this evidence concentrated on two main factors neither of which in my view are determining factors. The first factor concerns the planning consent for the seven town houses. It is obviously right that the existence of this consent should be known, but I consider that it would be quite wrong to arrive at a conclusion drawn from a direct comparison between that development and the form of development proposed here. It may well be that your client may wish to build those houses should his appeals fail, but it is not incumbent upon him to do so, either in the form at present approved or in any other form. In the circumstances of this particular case the present applications should be considered on their merits and not evaluated against some other form of development which might or might not be an alternative.

Now Miss Viner says that that is to fail to take into account the importance and effect of the existing detailed planning permission. I do not think I can accept that argument. It seems plain to me that the inspector had taken into account the existing planning permission. He had adverted to it and he said it is obviously right that the existence of this consent should be known. He goes on to give reasons for his evaluation of that matter as a planning concern. And if there is one thing that is certain in this jurisdiction, it seems to me to be this: that while this court will interfere if something is not taken into account that should have been, it will never interfere if the question is simply, having taken something into account, what weight has been given to it by the Secretary of State. The weight to be given to any material consideration is entirely a matter for the Secretary of State. To fail to take into account a material consideration entirely is an error of law. It may be sometimes difficult to sort this distinction out, particularly if the result in the end is that the inspector or the Secretary of State having taken something into account evaluates it as of no value. But it seems to me that if it is right that the amount of weight that should be attributed to a material fact is entirely a matter for the Secretary of State, the Secretary of State's decision cannot be validly challenged if what he said is: "In my scale this matter weighs nothing." The thing is he has got to weigh it. That is, he has got to take into account the material consideration, but he cannot be faulted for saying, having done that, "in the way I look at it, making a planning judgment, the value of this point is nil" or "I consider that against that background this is irrelevant," or whatever it may be. That seems to me what the inspector has been doing in the passages I have read, paragraphs 25 and 26. He has evaluated the matter, he has said, "it is there, I have to consider it. But in the view I have taken of this matter its value is nil and I disregard it."

So it seems to me that Miss Viner's first argument fails. It is quite impossible in my judgment to say that he has failed to take into account the existing detailed planning permission. The notice of motion of course in a sense gives the game away, because the words of it are not those I have just mentioned. The words are: "Failed to take into account the importance and effect of the existing detailed planning permission." Well, whether the existing detailed planning permission was important or whether it had any considerable effect were entirely matters of value judgment for the Secretary of State. As I say, I am satisfied that the inspector took into account the existing detailed planning permission and Miss Viner cannot succeed on this point.

I pass over item 2, because I want to return to it later, and go to no 3, and that is: "Failed to take into account the fact that the proposed uses of Northover," that is the appeal site, "were mainly or predominantly residential uses or were residential educational uses, or alternatively, wrongly held that the said uses were business uses." I do not think I need spend very long on that argument. The question of what label in planning terms you attach to a particular use is surely entirely a planning matter. It matters not (this is where visions of angels dancing on pins came to my mind) whether you call it residential, educational or business. The point about this case was it was said that the user proposed was a user alien to a residential area. It was not residential. And in fact if one looks through the inspector's report it is plain, particularly perhaps in one passage where he records the views of local residents, that you can find the proposal described as a business use and an educational use in the same paragraph. I do not think it is proper to consider that the inspector erred as a matter of law in describing this use as a business use or even as an educational use. The main point was it was not in his planning judgment a residential use, and he was, it seems to me, perfectly entitled to come to that conclusion without interference by the courts.

Then in no 4 it is said that, "he failed to take into account the differences between the two proposed uses of Northover." Pausing awhile just to indicate what those two proposed uses were; there were two alternative plans put forward. One was for the management centre to be operated in this house with two flats above it, one for the person running the centre and the other a perfectly ordinary residential flat for letting, confining entirely the management centre use to the ground floor and providing in effect that the students would have to live elsewhere in boarding houses or hotels. The second proposal was for a residential centre, that is that the students would be accommodated in bedrooms on the property, and in this case there would be a smaller flat, as I understand it looking at the plans, for the person who was running the management centre. Those were the two alternatives. But the criticism is that he failed to take into account the differences between those two proposed users. In paragraph 27 the inspector said: "I accept that in both the present applications the first and second floors of Northover would remain residential in one form or another." That of course is right. One form would have been two flats; in the other form it would have been one flat and a number of dormitory bedrooms. "But the use of

A the ground floor as a whole," he goes on, "would be a nonconforming use, and it is the effect of this use that will determine the outcome of these appeals." Then he goes on to say in paragraph 29: "Similarly, there would be up to 22 people living on the site depending on whether the centre were residential or not," and then, dealing with the movement of those people: "...they could not fail to generate greater visible activity," he says, "either coming or going," which seems to me to be an indication of what would happen if the students lived elsewhere, "or in their off-duty periods," which no doubt is a reference to what would happen if they lived on the premises. And finally in paragraph 33 he
B says: "Although there are two applications for slightly different forms of development, they both in my view suffer the same disadvantages, and I therefore conclude that the proposals are unacceptable if judged against the test suggested on your client's behalf."

One cannot read those passages without coming to the inevitable conclusion that the inspector did indeed take into account differences between the two proposals, but in his view those differences were insufficient to make in the end any differentiation between those two forms in the view he had taken of the proposal. And it seems to me one only really has to read the passages I have read for it to be apparent that
C it cannot with any real force be said that he failed to take into account the differences. It may be very disappointing of course, if there was an argument that there was a considerable difference between the two, to find the inspector dismissing that argument in a slightly cavalier fashion. But that he was entitled to say that the difference between the two was so minimal that it did not affect his mind seems to me to be clear. Once again it is a matter of value judgment on a planning matter and therefore fully within the competence of the inspector, and not, may I say, within the competence of this court to interfere as if he had gone wrong on a point of law.

D Then 5: "failed to take into account the existence of a national and regional need and/or wrongly held that there was no special need for the proposed development." I confess with all respect to Miss Viner that when she first developed this argument I could not really understand what it was, and she very gallantly attempted to dispel my misunderstanding or failure to understand it when she returned to it in her reply. I had better read the paragraph in the conclusion. The inspector said this in paragraph 31:

> There remains the question of need. From the evidence given I accept that the course run at the Aston Small Business Centre is of value to small businesses and that indirectly it may contribute
> E to the country's economy. I accept, too, that benefit might be gained if another such centre were to open, as present courses at Aston are oversubscribed. Thirdly, I accept the south west might be an acceptable area for such a second centre. I do not accept this centre must be in Bournemouth, and even less do I accept that it must be in Dunbar Road, however convenient this may be. It follows that there is no special need in this case which might justify my overriding relevant planning considerations and approved development control policies in the borough.

Miss Viner seeks to make a distinction between what she calls "development need" and "site need," development need being a need shown for a development of a particular char-
F acter and site need being the need for having it on the site in question. She goes on to say that the local authority volunteered that there were various other sites available or at least methods of dealing with this need within Bournemouth and that a very considerable time was spent during the inquiry in examining the good and bad features of those sites. And she says there is not a word in the inspector's conclusion about the suitability or unsuitability of those alternative sites. She put it in this way, that the planning authority chose to put forward alternative sites, and there was therefore a duty on the inspector to consider those sites. He did not consider them, and having not done so, it was his duty then to consider whether need overrode the planning considerations. As I say, I do not G follow that argument at all. In the first place I do not follow that there is a legitimate dichotomy between what Miss Viner calls development need and site need. The question that has to be asked in all these planning cases is whether or not a particular development should be permitted on a particular site. If there is an argument about need it may of course be that that argument indicates that such is the need for a particular development and so difficult is it to find any other suitable site for it that one is driven to the conclusion that there is a special need which can only be satisfied on the site in question. That seems to me to have been the proper way of looking at the need, and it is idle to suggest that just because H one can show that there is a need for a particular development one must then go off and ask the planning authority for alternative sites, or if the planning authority produces alternative sites for those to be canvassed at great length and so on. The point is here that it is quite apparent that, if one accepts the inspector's view, there may well be a national need for a centre of this kind, and there may well even be a need for it in the south west. He is still fully entitled to take the view that there was no evidence before him that it was necessary to fulfil that need in Bournemouth, and that is what he said. Equally he is entitled to say that he could not see why it should be fulfilled in Dunbar Road, and if he had J come to that conclusion it would follow, as he said, that there was no special need in this case which would justify overriding relevant planning considerations. For the life of me and giving all due weight to the matters urged by Miss Viner and trying to understand them as best I can, it seems to me that nothing can possibly be said about the manner in which the inspector treated need in this case to indicate that he failed to take into account any material considerations or took into account an immaterial one. It seems to me he applied his mind properly to the question of need and its impact on a particular site. And all he is saying is there is nothing in the evidence which indicated to him that there was a special need to have this K development on this site. And in those circumstances I cannot see that need overrides the relevant planning considerations. That is an intelligible way of dealing with it and one which it seems to me deals with it as a planning question and on a planning basis, and I do not think the courts can possibly say in those circumstances that there is anything which requires interference by the court.

Then 6: "failed to take into account the fact that the imposition of appropriate conditions in the grant of planning permission could regulate the proposed development and use at Northover and/or safeguard the character and appearance of the Northover neighbourhood." This seems to me to be L quite a short point. In paragraphs 20 and 21 under a heading "Conditions" the inspector said this:

> 20. It was suggested on behalf of your client that many of the council's fears about the proposed use could be allayed by the imposition of suitable conditions on any planning consent which might be granted. Such conditions could limit the use to a management centre, and could limit the age and number of students and would thus accord with the advice given in Circular 5 of 1968. 21. The council's view is that some conditions are extremely difficult to enforce, and quoted from experience in the borough. It would, for example, be difficult to enforce a condition limiting the age of students. It is also sometimes hard to resist an application to vary conditions, particularly if the M alterations were minor and gradual. In any event, it is not the purpose of conditions or section 52 agreements to manoeuvre a nonconforming use into a residential area.

Now it is true that in his conclusions the inspector nowhere dealt with conditions at all. He never mentions them. But in paragraph 30, as Miss Viner points out, he said this:

> Finally, although I accept that it is your client's present intention to have the house well-curtained and discreetly lit, this need not always be so if, for example, he were to sell the property, and any change in this sense could make activities within the building a great deal more obvious to the public.

I would like to deal with this in two different compartments. First of all the argument or criticism that he never made any reference to conditions in his conclusions. It is apparent from the paragraphs I have read, 20 and 21, that the conditions there being spoken of were conditions suggested by the appellant to meet a criticism by the local planning authority that the management centre plan for mature students, as it was being put forward, could change, either with a change of ownership or with the present owner, and that the age of students might alter and the management centre might become a language school and various other things of that kind. And conditions restricting the permission to a management centre or restricting the age of students and so on are of course only relevant once one gets to the situation that the planning permission is granted. If you have conditions designed to ensure that a planning permission for a particular use is not abused by increase in the use or subtly altering that use within the Use Classes Order and so on, then conditions designed to prevent that are, of course, relevant. But before they become relevant you have to reach the point where the planning permission is granted. If in fact you come to the conclusion that the planning permission ought not to be granted, the value of attaching conditions to make certain that the proposed development does not go off and become something else is, of course, absolutely nil. And this inspector came to the conclusion that he did not wish to give planning permission for the proposal itself. Once he had reached that situation the question of whether or not he should attach conditions to a non-existing planning permission simply does not arise. Any suggestion therefore that he failed to take into account the appropriate conditions seems to me to be very wide of the mark. It is difficult, I think, or may often be difficult in cases where the inspector himself makes the decision to take the view that he failed to take into account a material consideration if he detailed that consideration as part of the argument in the case. It is clear that it must have impinged upon his mind; otherwise he would not have recorded it. But perhaps that must not be taken too far. But it seems to me in any event that in this case, as I have indicated, conditions about restricting the user as a management centre and the age of the students would be quite idle once one had come to the conclusion that a management centre should not be allowed.

So far as paragraph 30 is concerned it is apparently true, and of course I accept from Miss Viner, that there was considerable argument about whether or not one ought not to have conditions about drawing the curtains in this house and so on. Whether or not a condition referring to something as mundane as curtain drawing is in any event a matter which affects the interior of the house and not its exterior, like whether or not you have fluorescent lighting inside, I leave. I do not know, it has not been fully canvassed before me. But all I can say is this. I am certainly not disposed to say that this planning appeal should be quashed because the inspector has not seen fit to set out his rejection of a question about conditions on curtaining. It seems to me to be so small a proportion of the whole matter that even if he had erred in not referring to conditions in that context, it is certainly a matter which should be regarded as *de minimis.*

Then paragraph 7: "failed to take into account that the grant of permission would result in the preservation of Northover in accordance with development control policy no 20, whereas the refusal of permission would result or would be likely to result in the demolition of Northover." As I indicated in the course of the argument, there are two quite distinct points wrapped up in that paragraph. The first is whether or not the situation fell within the ambit of paragraph 20 of the development control policy. The second is whether or not the inspector took into account the effect, the different effect if one likes, of refusing permission and of granting it in this regard, that granting permission would avoid the demolition of the house whereas refusing it, if the appellant's intention was to be taken at face value, would result in its demolition. On that point the inspector recorded this. Paragraphs 6 and 7:

> Your client contends that Northover is an ideal venue for such a course, since it offers all the facilities that are needed.... Its use for this purpose would ensure the rentention of the house and grounds in good condition. The first and second floors would also continue to be in residential use.
>
> 7. Development Control Policy Note Number 2 sets out government policy for development in residential areas, and although paragraph 20 refers to the desirability of keeping houses in residential use, the main and only test in this case is covered by paragraph 19.

Let us just have a look at paragraphs 19 and 20 of the policy note. They read in this way:

> 19. The allocation of land for primarily residential use does not exclude all other uses. Some are appropriate to a residential area, for example a doctor's surgery or the occasional corner shop. Others may be acceptable if the local conditions are right. Some business uses serve residential areas and need to be sited in or near them. But some uses, including most industrial uses, are almost always inappropriate. The test in most cases is whether the development would, because of its appearance or the noise or traffic it would generate, harm the character of the area and make it a less pleasant place to live in.

That is the paragraph which the inspector says lays down the main and only test. Then paragraph 20:

> Where it is a question of converting a house to some other use, the need for houses in the area and the desirability of keeping the building in residential use will be taken into account. In some areas this may be a decisive objection to the development. But there are sometimes cases in which the conversion of an old house of architectural or historic interest to some other use is to be welcomed as the best way of preserving it.

Now apparently there was some evidence given at the inquiry by the planning consultant called on behalf of the appellant that this was a house which fell within the ambit of that last sentence, ie an old house of architectural or historic interest. It seems plain to me that that sentence was meant to cover the rare case where there is a building, an old building of architectural or historic interest, something that is part, as it were, of the nation's heritage and which it would be better to preserve as it is even at the expense of converting it from residential use to some other use. But this is a house built in 1911 and, from the photographs, of the mock Tudor type with half-timbered elevations, and how it could conceivably be argued that that was either an old house or one of architectural or historic interest within the meaning of that paragraph I do not know. And I am not in the least surprised that the inspector brushed the suggestion aside in the way he did. It seems to me he was perfectly entitled to do so. It is not the function of this court to say he ought to have paid more attention to it or dealt with it at greater length in his report. What he said about it clearly is that it is not paragraph 20 at which he wants to look, it is paragraph 19. And it seems to me to be a value judgment once again which he was fully entitled to make. It may be that the planning consultant for the appellant took a different view, but that again is a value judgment and of course matters of architectural appearance are notoriously subjective. It does not seem to me that one can make very much out of that point.

As to the other question which is bound up in this paragraph, the question of whether it was not a better thing from the point of view of the appearance of the neighbourhood to preserve the house rather than to demolish it and erect seven town houses in its stead, that it is a matter, it seems to me, which stands or falls with the other argument which I have already dealt with, namely taking into account the detailed planning permission for seven town houses. If I am right, the inspector was entitled to say: "I note the existence of that planning permission which of course entails the demolition of the existing house. I do not think that, so far as I am concerned, is a planning matter which carries any weight at all." Having come to that conclusion, it seems to me he comes to that conclusion on both heads simultaneously.

First of all he is dealing, as it were, logically in the right order, though historically in the wrong one, with the erection or the use of seven new town houses, Secondly, he is obviously dealing with the prior demolition of this present house. He cannot have taken into account the planning permission for the erection of seven town houses without also taking into account the demolition of the existing one. Having done that I do not really think that it was incumbent upon him to spell out any more in his decision letter.

Finally we come back to paragraph 2 of the particulars in the notice of motion, and it is this one that has given me the greatest difficulty in this case. I have already read the inspector's paragraphs 25 and 26 so far as is relevant. I should I think read paragraphs 27 and one or two of the other paragraphs in the conclusion. Paragraph 27:

> Whatever the precise term used, the area around the appeal site is a predominantly residential area in the accepted sense of the term. It is true that there are some business or quasi-business uses, but these are in the main wholly or predominantly residential uses, even if some of them are run for gain. I accept that in both the present applications the first and second floors of Northover would remain residential in one form or another, but the use of the ground floor as a whole would be a nonconforming use, and it is the effect of this use that will determine the outcome of these appeals.

It seems to me that if he had stopped there, if he had dealt with paragraphs 25, 26 and 27 as I have read, nobody could really, with any show of validity, have challenged these conclusions. But he did not; he went on in this way:

> 28. It may be difficult accurately to quantify the effect that a small group of people may have upon a neighbourhood, but in this instance any effect would be related to the presence and movement of cars, students and staff, to modifications to the grounds, principally for car parking, or to modifications to the house. It has been suggested that up to 22 parking spaces would be provided within the curtilage, some on the two parking areas in front of the house and with access from Dunbar Road and some in a larger park off Berwick Road. Any cars parked in front of the house, including staff cars, would be obvious from the street and would be obtrusive, whilst even the screening provided by the wall and by trees could not wholly conceal from passers-by the park off Berwick Road. I do not suggest that traffic generated by the centre could not safely be absorbed on the neighbouring streets, but there would clearly be materially more vehicle movement than there is now. 29. Similarly, there would be up to 22 people living on the site depending upon whether the centre were residential or not. However quiet and well-behaved these people might be, there would be many more than there are now and they could not fail to generate greater visible activity, either coming or going or in their off-duty periods.

Then he goes on in paragraph 32:

> If taken separately, the various factors I have covered might not individually preclude the introduction of a business use into this area. When they are taken together it is a different matter. I do not consider that the use of Northover as a business centre is acceptable, since it could not fail to be evident to passers-by and other residents of the area and since it would in my opinion materially harm the character and appearance of a neighbourhood that is at present quiet, secluded and of high quality despite the changes that may have taken place during recent years.

It seems plain to me that what the inspector has done is this. He has said to himself: "I do not regard the development for which planning permission exists," that is the seven town houses, "as from the planning point of view a relevant consideration in this case." If he had gone on to say something like: "The reason I say this is that however similar the vehicle movement and the movement of persons may be in the two situations, the character is wholly different," or something of that kind, then it seems to me that would have been a perfectly reasonable and legitimate thing to do. But he did not do it. What he went on to do was in fact to compare the vehicular movement and the movement of people, not with what would occur if the development for which planning permission existed were carried out but with the situation now, as he put it. Mr Brown suggested that "now" was a sort of shorthand, meaning "with the present building as it is and without the benefit of any further planning permission." And Miss Viner says it cannot mean that. Putting it at its highest in Mr Brown's favour and taking it that what the inspector was meaning was: "I propose to compare traffic movement and the movement of people with the premises divided into two residential units," which is really what Mr Brown's argument amounts to, it seems to me he is still making the wrong comparison; unless he were to say, which he does not, "comparisons of that kind cannot properly be made in this case because the planning permission that has been granted is of a wholly different character from the proposal which is now before me." There is a certain amount of overlapping always when one considers the question of giving reasons for a decision and the question of whether or not an immaterial point has been taken into consideration or a material point not taken into consideration. If on a reading of the decision it is not plain what an inspector or the Secretary of State has done with a particular point, it may be impossible so say, as long as the point is material, whether he has taken it into account or not. Similarly if the language of the decision letter is very obscure it may be impossible to say whether he has taken into account an immaterial consideration. There is more than one case in the books illustrating that sort of point. It seems to me here that try as one may, however one may approach this decision letter in an endeavour to preserve its integrity, one cannot read anywhere in it any reason for making the comparison which the inspector has made. It seems to me to be plainly illogical to make a comparison between the proposal and the existing state of affairs. If, as the inspector set himself in this case, the exercise is to determine what is the effect of the use on the surrounding area, and that is what he says at the end of paragraph 27, and if he quantifies that effect as far as he may by dealing with it in terms of the movement of cars, students and staff, it must, it seems to me, be right to make that comparison with what would otherwise be the situation if this planning appeal is dismissed. It is true that in paragraph 26 he indicates that his reason for that is that it is not incumbent upon the appellant to carry out that form of development. But in the face of the clear record which he himself has put down (this is in paragraph 4): "If the present proposals are not permitted and planning permission for change of use withheld, your client intends to implement a current planning consent for the demolition of Northover" and so on, in the face of that record it seems to me to be unfortunate that the inspector should consider it as unlikely, which he must have done, that the appellant would carry out his intention, and in any event it seems to me inevitable that the only proper comparison was between the traffic generation and movement and the movement of people in the proposal and in the development for which planning permission existed.

If one wanted any authority for that proposition it is to be found in the case of *Wells v Minister of Housing and Local Government* [1967] 1 WLR 1000. I do not think I need deal with the facts of that case. They are well known. A passage from the judgment of Lord Denning MR at p 1008 G reads:

> In considering the appeal, the Minister ought to have taken into consideration the fact that the builders' merchants had in their favour an irrevocable determination that, in respect of the 27 ft 6 in plant, no planning permission was necessary. He seems to have ignored it. It seems that the traffic would be just as great a nuisance with the 27 ft 6 in plant as with the 48 ft plant. In those circumstances he might have allowed the 48 ft plant to remain. In the circumstances I think that the case should be remitted to him to take it into account.

Now I do not resile at all from my earlier view that the way in which the inspector took into account the planning consideration and then came to the conclusion that from the planning point of view its value was nothing was a value judgment which he was entitled to make. What I am saying is that when he condescends from that and in particular when

A he points out as he does that it is the effect of traffic movement and the movement of people on the locality that is the overriding question in determining whether or not a nonconforming user of this kind should be allowed in a residential area, it seems to me inevitable that in making that comparison he fell into error in not comparing it with the effect of implementing the existing planning permission but comparing it instead with the situation which then existed as if no such planning permission would be implemented.

In the result it seems to me that on that ground and on that ground alone it can be said that the inspector in this case unfortunately fell into error. I do not think that that error
B is saved by the perfectly unexceptional way he dealt with it in paragraph 32, a paragraph which I have already read. It seems to me unfortunate perhaps that he should have set out as he did in paragraphs 28 and 29 the comparisons he therein made. No doubt he was goaded into it by the fact that he is bound to give reasons for his decision and that those reasons must embrace at any rate the major points at issue between the parties, and there is no doubt of course that traffic generation and movement of people was a major issue between the parties. But unfortunately it seems to me he erred in this case in taking the wrong comparable, and in those circumstances I think that the decision must be quashed.

C The order made by the Secretary of State dismissing the appeal was quashed. The applicant was awarded costs against the Secretary of State.

QUEEN'S BENCH DIVISION

November 19 1979

(Before Sir Douglas FRANK QC, sitting as a deputy judge of the division)

RANN AND ANOTHER v SECRETARY OF STATE FOR THE ENVIRONMENT AND ANOTHER

Estates Gazette June 21 1980
D (1979) 254 EG 1095

Town and Country Planning Act 1971—Application to quash decision of Secretary of State—Whether use of a house as holiday accommodation for mentally-handicapped persons came within Class XIV of the Town and Country Planning (Use Classes) Order 1972—Whether a material change from previous permitted use as an hotel/guest-house—Secretary of State held to be wrong in deciding that the use as holiday accommodation for the mentally-handicapped had the feature of semi-permanent institutional care which characterised Class XIV—Purpose of Use Classes Order not to define what is
E **development, but, on the contrary, what is not—Secretary of State's decision quashed.**

The applicants, Jack Rann and Ellen Barbara Rann, were the owners of a three-storey building in St Mildred's Road, Westgate-on-Sea, Kent, for which planning permission had been given in 1975 for use as an hotel/guest-house. This was before the purchase of the property by the applicants. They started to use it in 1976 to provide holiday accommodation for mentally-handicapped persons. The local authority requested the Ranns to apply for planning permission to change the use to a use within Class XIV of the 1972 Use Classes Order. Their application was made but was refused. An enforcement notice was
F served requiring the discontinuance of the holiday accommodation use. On appeal to the Secretary of State against both the planning refusal and the discontinuance order, the Secretary of State granted permission for the use of the house as a holiday home for mentally-handicapped persons, but only subject to a condition that the use should cease on or before December 30 1981. The applicants applied to the High Court for the Secretary of State's decision to be quashed on the ground that he was in error in deciding that the use came within Class XIV.

D Robins (instructed by Parrys, of Herne Bay) appeared on behalf of the applicants; Simon Brown (instructed by the Treasury Solicitor) represented the respondants.

Giving judgment, SIR DOUGLAS FRANK QC said: This is an G
appeal from a decision of the Secretary of State in which he granted planning permission for the use of a former large dwelling-house at 63 St Mildred's Road, Westgate-on-Sea, Kent, as a holiday home for mentally-handicapped persons but subject to a condition that that use shall cease on or before December 30 1981. The matter came before the Secretary of State on an appeal against an enforcement notice, and the local planning authority's main contention, which was accepted by the inspector, was that the present use falls within Class XIV of the Town and Country Planning (Use Classes) Order 1972. The facts so far as material to the issue before me were found by the inspector in these terms: H

2. The appeal property is a three-storey building, containing a lounge, dining-room and kitchen, bedrooms for 23 mentally-handicapped persons and three supervisors, and the owners' flat.

5. In mid-1975 63 St Mildred's Road was in use as holiday flatlets. In November 1975 planning permission was requested, and granted, for change of use to an hotel/guest-house and for no other purpose.

6. In March 1976 the appellants completed the purchase of 63 St Mildred's Road, and after redecoration and installation of a fire-alarm system, started to provide holiday accommodation for mentally-handicapped persons.

7. In April 1976 the local authority requested that application should be made for planning permission for a change of use of the appeal premises to a use within Class XIV of the Use J
Classes Order 1972. This was done under protest in July 1976, and refused (one subject of the present appeals) in October 1976.

8. Mentally-handicapped persons on holiday at the appeal premises come in groups from hospitals for periods of one or two weeks, accompanied by their own qualified supervisors. The holiday season lasts for 21 weeks from May to September. There are no mental after-care patients.

9. Whilst on holiday, apart from medication which is administered by their own supervisors, mentally-handicapped persons requiring treatment are returned to their own hospitals. During 1976, 1977 and 1978, out of 1,089 persons holidaying at the appeal site, only five had made any call on local medical services; these cases were unconnected with mental conditions. K

In the course of his decision letter the Secretary of State said:

On the question of the aptness of the allegation—ie of the Class XIV use—the view is taken that a more accurate description of the use would be "a holiday home for mentally-handicapped persons."

However, in the following paragraph he said:

The view is taken that the use of the premises for the accommodation of parties of mental patients on holiday from hospitals where they are normally cared for as in-patients involves an institutional use, with care and supervision akin to that provided in hospital, and falls within Class XIV of the Town and Country Planning (Use Classes) Order 1972. Such use is considered to be materially different from both the permitted use of the premises as L
an hotel/guest-house, whose essential function is to provide accommodation and meals to members of the general public as paying guests, and from the previous use as holiday flatlets. It has been concluded, therefore, that the introduction of the holiday home use involved a material change of use and constituted development requiring planning permission. As such permission was not obtained there was a breach of planning control and the appeals accordingly fail on ground (b).

It is common ground that the sole question which I have to decide is whether the present use is one defined in Class XIV of the Town and Country Planning (Use Classes) Order 1972 and that I am not concerned with any other question. M
Class XIV is in these terms:

Use as a home or institution providing for the boarding, care and maintenance of children, old people or persons under disability, a convalescent home, a nursing home, a sanatorium or a hospital.

Mr Robins, for the applicants, first drew attention to the apparent inconsistency in the Secretary of State's letter: with one breath saying that the Class XIV definition was not an apt description of the use and with the next stating that the use falls within that class. Mr Brown conceded the apparent inconsistency but submitted, in my view correctly, that the question which I have to decide is whether, on the facts as found, the use falls within Class XIV. Turning to the Class XIV definition, Mr Robins said that the expression "holiday home" is a home and not a place where one has a holiday, nor is the word "home" or "institution" apt to describe the guest-house operated by the applicants: it is a guest-house specialising in mentally-handicapped persons. Further, it is inherent in the definition that the home or institution must provide for the care and maintenance of the residents, which this guest-house does not.

Mr Brown described the premises as "a home from home" and as "an arm of a more permanent institution." It was an inevitable corollary to the provision of accommodation for the mentally-handicapped that they would come from a small preselected section of the community as a composite group accompanied by institutional staff. He argued that "providing for" does not mean that the appellants must be the provider of the care and maintenance but only that they have to offer the facilities. He also said that it is permissible to take into account the effect on the neighbourhood, but I am bound to say that I do not understand how that can be.

I think it important to bear in mind the origin and purpose of the Use Classes Order, as from the foregoing arguments it might be thought that its purpose is to define certain kinds of development, whereas the converse is true. The order derives from section 22 of the Town and Country Planning Act 1971, which in subsection (2) provides that:

> The following operations or uses of land shall not be taken for the purposes of this Act to involve development of the land, that is to say:
>
> (f) in the case of buildings or other land which are used for a purpose of any class specified in an order made by the Secretary of State under this section, the use thereof for any other purpose of the same class.

It follows that the Use Classes Order is being borrowed in this case for a purpose for which it was not intended. Its intended purpose is to put outside the ambit of the Act a change of use which has taken place within the same use class. It is not uncommon to find the definitions in the order used as a shorthand to describe what is not being permitted, but it is important to bear in mind the effect of the order. Thus, if the Secretary of State is right, then a guest-house providing holidays solely for children could be changed to a hospital or a home for the mentally sick without the grant of planning permission. I can think of a number of examples of such guest-houses—residential sailing and riding schools, mountaineering and physical education centres. The inquisitive bystander would be surprised to hear such places described as homes or institutions and if he were a neighbour he might be dismayed to know that planning permission would not be required to change the use to, say, a home for alcoholics.

I do not think that Class XIV bears the extensive meaning ascribed to it by the Secretary of State. I think it inherent in the words used that the care and maintenance referred to is of a special nature to be provided by those running the home or institution. For example, in all the cases described in the order I would expect to find a matron employed. Moreover, the word "home," and by application of the *ejusdem generis* rule "institution," connotes permanence. I think I can best relate what I here have in mind by adapting words used by Lord Widgery CJ in *Mayflower Cambridge Ltd v Secretary of State for the Environment* (1975) 30 P & CR 28 at p 32: I have no doubt that it is well understood and is acceptable as an ordinary use of English, namely, that the real difference between use as a home and use for the purpose of a guest-house turns on the instability of the population in the premises and the extent to which they are making the guest-houses their homes. The essence of a guest-house is that it takes transient passengers. Of course there may be an individual here and there who stays for a long time if it suits him, and there are buildings which are wrongly described as guest-houses or as residential guest-houses. But the basic feature of a guest-house as the word is used in the English language is that it contains a transient population because it is there to serve people travelling who require short stays only. By contrast, the home in the way in which the phrase is used in English is somebody's home. It is where somebody lives. It is where somebody remains for a substantial time. Accordingly, one has in a home or institutional use a far more stable population than one has in a guest-house use. For the removal of doubt I have used the phrase "guest-house" as it is used in Class XI as being the same kind of use as an hotel, that is, for providing holiday accommodation and not as it is sometimes used, say in Earl's Court, as a place providing permanent accommodation. It does not necessarily follow that I hold that the use in this case falls within Class XI, for I have heard no argument on that point nor whether the use is *sui generis*, as was held in the case of a students' hostel: see *Mornford Investments Ltd v Ministry of Housing and Local Government* (1970) 21 P & CR 609. Accordingly, I do no more than hold that the Secretary of State wrongly construed the use in this case as falling within Class XIV of the Use Classes Order.

The Secretary of State's decision was quashed with costs.

QUEEN'S BENCH DIVISION

February 15 1980

(Before Mr Justice PHILLIPS)

BULLOCK v SECRETARY OF STATE FOR THE ENVIRONMENT AND ANOTHER

Estates Gazette June 21 1980

(1980) 254 EG 1097

Town and Country Planning Act 1971, section 245—Application to quash tree preservation order in respect of a small coppice—Applicant, a farmer, had made an unsuccessful attempt to obtain planning permission for a housing development on the land and he now challenged the tree preservation order—He submitted that as a matter of law it was not possible to make a tree preservation order in respect of a coppice and, alternatively, that it was in practice impossible to reconcile such an order with the normal good management of a coppice, which required regular felling and regeneration from the stools—Meaning of "trees" and "coppice" discussed—Held that trees in a coppice were subject to the power to make preservation orders and that, so far as good management was concerned, it was not impossible to devise a practical scheme—Application dismissed

This was an application by John Bullock under section 245 of the Town and Country Planning Act 1971 to quash the Brook Farm Drive Malvern Tree Preservation Order 1978 no 36, made by the Malvern Hills District Council and confirmed by the Secretary of State for the Environment.

M A Hutt (instructed by Hewitt, Woollacott & Chown) appeared on behalf of the applicant; Simon Brown (instructed by the Treasury Solicitor) represented the first respondent, the Secretary of State. The second respondents, Malvern Hills District Council, were not represented and took no part in the proceedings.

Giving judgment, PHILLIPS J said: This is an application under section 245 of the Town and Country Planning Act 1971 and order 94 of the Rules of the Supreme Court, by which the applicant, John Bullock, seeks to have quashed or suspended the Brook Farm Drive Malvern Tree Preservation Order 1978 no 36.

The order relates to a wood of about half an acre. It is the remains of a much larger wood and on one side or one and a half sides, particularly to the north-west and the north-east, it is now surrounded or approached by a fairly new estate, and, as is often the case, it is much cherished by the people who live in the estate as an amenity. The applicant has a farm which is off to the north-east and this wood now, if not divorced from the farm, is only tenuously connected to it; but he still uses it.

A Among the facts found at the inquiry is that the land "comprises a coppice, primarily of stool grown ash over an area of about ½ hectare. The north and south-west boundaries adjoin recent residential development while open farmland surrounds the remainder of the wood. Public footpaths run along the north and south-east boundaries. There are few other woods in the area." The land is throughout described as a coppice, and it is found as a fact that: "The coppice was originally used for hop poles and had been clear felled on a rotational basis over 20-25 years. It is now used for farm fencing purposes and would be cut in the same manner."

The applicant, thinking that the land in question was suitable B for housing, made an application for planning permission. The application was for eight houses and garages and vehicular access. That was refused on January 9 1978, the grounds of refusal being, broadly speaking, that the land was not needed for residential purposes because there was available other unused land. The grounds also contained this: "It is considered that this coppice should remain in its present form in order to provide both aesthetically and biologically a useful oasis of natural vegetation on the periphery of an existing residential area."

It being discovered that the applicant had felled a number of trees in the centre of the wood, the second respondents to C this application, the Malvern Hills District Council, made an interim tree preservation order. In due course that order was confirmed by the Secretary of State.

The Secretary of State accepts the inspector's conclusions regarding the amenity value of this coppice, and says that its removal would have a significant adverse impact on the environment and its enjoyment by the public. The Secretary of State then turns to consider matters relating to the continued management of the wood. Various matters had been canvassed before the inspector, and the Secretary of State goes on:

> A condition permitting the removal of ash poles more than 6 inches in diameter is not however thought either appropriate or D practical. Whether felling should be selective by diameter of pole or on some rotational basis is a matter for future decision under the procedures contained in the order and explained in paragraph 58 of the department's circular 36/78.

The Secretary of State goes on to make some general observations as to the kind of basis upon which the future management of the wood might be arranged. That is the background of the application before me.

Mr Hutt, who has argued the application persuasively and interestingly, makes two major submissions. First of all, he submits that it is not possible as a matter of powers to make what he calls a coppice preservation order. That is not a E term of art, and what is meant is that there is no power under the Act to make a tree preservation order in respect of a coppice. Secondly, and alternatively, he says that if there is power in theory there is not in practice, because there is a total inconsistency between the purpose of such a tree preservation order and what is necessary in the interests of continuing good management, and without which, indeed, the objects of a tree preservation order cannot be secured.

I should have said that in the schedule to the tree preservation order the land in question is described under the fourth possible category as "Woodlands," and after identification is described as "Mixed hardwoods consisting mainly of Ash."

F I turn then to the first submission. This submission is based on the fact, which seems to be accepted in the various documents, that this wood or woodland is a coppice. I wondered sometimes when listening to the case whether the applicant really wants it as a coppice, because it is quite clear that he was trying to have it cut down in the hope that houses could be built upon it. But I ignore that feature of the matter for present purposes, and consider the submissions in a more abstract form.

A coppice is defined in the dictionary as "a small wood of underwood and small trees grown for periodical cutting," and in *Dashwood v Magniac* [1891] 3 Ch 306 Kay LJ says this: G "That definition"—he is talking about what is meant by timber wood—"would include coppice"—and then he goes on—"which, properly speaking, is oak, ash, or other wood cut at intervals of less than 20 years, so that it springs again from the same stool or stub; and the cutting of this at the proper periods by a limited owner is not waste." So the essential of a coppice is that it is in the form of a crop and that when the trees are ripe, and are cut at the proper height in the proper way, there will be regeneration from the stools, and so it will go on from generation to generation.

Mr Hutt's argument is that the nature of a coppice and the form of silviculture involved are inconsistent with the powers in relation to tree preservation orders. Those powers derive H from the re-enactment in the Town and Country Planning Act 1971 of earlier provisions, those relevant for present purposes being section 59 and the following sections.

Section 60 provides:

> If it appears to a local planning authority that it is expedient in the interests of amenity to make provision for the preservation of trees or woodlands in their area, they may for that purpose make an order (in this Act referred to as a "tree preservation order") with respect to such trees, groups of trees or woodlands as may be specified in the order; and, in particular, provision may be made by any such order—(a) for prohibiting . . . the cutting down, topping, lopping or wilful destruction of trees. . . . (b) for securing the replanting. . . . J

Section 59 provides:

> It shall be the duty of the local planning authority—(a) to ensure, whenever it is appropriate, that in granting planning permission for any development adequate provision is made, by the imposition of conditions, for the preservation or planting of trees. . . .

and there is provision in section 62 for the replacement of trees.

There is no definition of what is meant by "trees." According to the dictionary a tree is "a perennial plant having a self-supporting woody main stem, usually developing woody branches at some distance from the ground and growing to a K considerable height and size," and the description in the book I have been shown is in similar terms.

There seems to me nothing to prevent those things which grow in this woodland from being "trees" as ordinarily understood; and indeed Mr Hutt does not, I think, contend that this woodland is not a woodland within the meaning of section 60, and he does not contend that, ordinarily speaking, what is in the woodland might not be regarded as trees. But he says that they are not trees for the purpose of this part of the Act when one looks more closely at the purpose of it and the nature of a coppice. He relies on some observations of Lord Denning MR in *Kent County Council v Batchelor* (1976) L 33 P & CR 185. There a tree preservation order had been made, an injunction had been granted restraining the appellant from cutting down or uprooting any trees on the land which was the subject of the tree preservation order, and Lord Denning says at p 189:

> Furthermore, I must say that there is an ambiguity in this Act and in the order. We are not told what is a "tree." Many bushes and saplings are certainly not "trees." In woodland like this, it is often, from the agricultural point of view (especially in a derelict area such as this) very important to get out the bushes, scrub and saplings and to replant—as, indeed, Mr Batchelor was doing. There is no definition of "tree." I should have thought that in woodland it ought to be something over seven or eight inches in diameter. I M do not know that any of the trees referred to here would be of that diameter.

That way of looking at the matter does not appear in the judgments of Stephenson LJ or Shaw LJ, nor is it the only ground upon which Lord Denning proceeded to quash the committal order in respect of which the appeal was to the Court of Appeal.

Obviously those observations, though *obiter*, are entitled to great respect, but I have come to the conclusion that I ought not to follow them. In the first place, it is plain upon the terms of the order that there is room and proper provision for applying for consent; so that the fact that the order goes to small trees does not prevent, provided consent has been given, getting out bushes, scrub or saplings. Bushes and scrub nobody I suppose would call "trees," nor indeed shrubs, but it seems to me that anything which ordinarily one would call a tree is a "tree" within this group of sections in the 1971 Act. It seems to me that if it were not so it would be difficult to apply section 59, which relates to the imposition of conditions for the planting of trees, which in the nature of things are quite likely to be saplings; or section 62 which makes provision for the replacement of trees, and which again in the nature of things are likely to be replaced by saplings.

Furthermore, contrary to Mr Hutt's submission, I think that way of looking at it is supported by another Act which is not directly in point, but is in *pari materia;* that is the Forestry Act 1967. Section 9(1) provides:

> A felling licence granted by the Commissioners shall be required for the felling of growing trees, except in a case where by or under the following provisions of this Part of this Act this subsection is expressed not to apply. (2) Subsection (1) above does not apply—(a) to the felling of trees with a diameter not exceeding 3 inches or, in the case of coppice or underwood, with a diameter not exceeding 6 inches. . . .

So subsection (1) contains a general prohibition. It follows from subsection (2) that it is assumed that but for subsection (2) that prohibition would extend to those things growing in a coppice, in other words it is assumed—it must be—that what grows in a coppice generally speaking would be trees, because otherwise there would be no need to have an exception in relation to those trees in a coppice with a diameter not exceeding six inches.

So I reach the conclusion that "trees," for the purpose of the relevant sections, has the ordinary meaning that the word does have, and I see no reason, as a matter of principle, subject to remaining points, why the trees in a coppice should not be within the powers under which tree preservation orders can be made.

Next it is said—and this overlaps, I think, to some extent, with the second submission—that the proper management of a coppice is impossible if there is a tree preservation order. And that what would otherwise pass as trees cannot pass if they are to be found in a coppice, because the Act will not work, and it is not the policy of the law to try to compel the impossible. I do not accept that. No doubt there are some practical problems, but there is ample provision for consents and for schemes of management, and I see no reason why a practical scheme of management could not be produced in the case of a coppice as in the case of any other group of trees or of a woodland. It may be inconvenient, and I can see that it perhaps is, but that does not mean that it is impossible. Nor am I satisfied that it leads to an absurdity in that every time consent is refused to cut a bean pole the applicant would have to apply (separately on each occasion) for compensation. I would have thought that consents could be dealt with in categories and that what is generally refused could be the subject of an application for compensation.

As to the second way of putting it (and, as I say, to some extent it overlaps with the first), it has two separate limbs. First, that there was no material here on which the order could have been made in the interests of amenity because no reasonable Secretary of State could possibly have taken the view that the interests of amenity would be served by the order. With respect, I regard that as a virtually unarguable proposition. It seems to me that there was plenty of evidence, and that the facts are far short of those which would entitle me to interfere with what is the exercise of the Secretary of State's discretion.

Secondly, the matter is put like this. It is said, in effect, that the Secretary of State in making no provision for management, abdicated all responsibility in the matter, and made the order regardless of the possibility of future management, with the result that nobody knows whether it would be possible to manage the coppice, and that the interests of amenity cannot be served because the coppice will decay and deteriorate. There is a contradiction in terms, it is said, between making a tree preservation order in relation to this woodland and its maintenance and management in proper heart.

I do not accept that. Again I can see it may be inconvenient, but it does not seem to me that it is more than that. I do not see why the Secretary of State should not do what he did, which was to postpone this, hoping that the parties could get together and make a sensible arrangement for the future management of the coppice on the lines of paragraph 58 of the circular 36 of 1978.

Then it is said in reply to that: it appears from paragraph 41 of the same circular that while a tree preservation order may be appropriate to secure the protection of a tree or groups of trees or perhaps a small woodland, in the case of anything larger it is dangerous because the consent procedures and the management under particular consents may be so difficult and cumbersome that owners may be induced by an order to allow their woodland to run derelict. "It is better therefore," paragraph 41 says, "for authorities to seek agreements with landowners for the proper management of their woodlands using grants available from the Countryside Commission or the Forestry Commission's small woods schemes and to use orders as a last resort."

That is a useful point, but the answer to it seems to me to be twofold. First, that this was an emergency, because the applicant appears to have been bent on cutting the coppice down. And, secondly, to some extent which procedure to adopt must be a question of degree and discretion and of judgment, and that is a matter for the Secretary of State and not for the court, unless, following the *Wednesbury* principle, in *Associated Provincial Picture Houses Ltd v Wednesbury Corporation* [1948] 1 KB 223, I can say that the result which he has adopted is so absurd that it cannot be one that a reasonable Secretary of State might have arrived at. In other words, put less politely, that it is perverse. I cannot say that.

In those circumstances I reject the second, as the first, submission and it follows accordingly that the application must be refused.

The application was dismissed with costs in favour of the first respondent.

QUEEN'S BENCH DIVISION

February 25 1980

(Before Mr Justice LLOYD)

SEARS BLOK v SECRETARY OF STATE FOR THE ENVIRONMENT AND ANOTHER

Estates Gazette June 28 1980

(1980) 254 EG 1195

Town and country planning—Applications to quash decision of Secretary of State dismissing appeals from local planning authority's refusal to grant permission for use of first floor of premises at Southwark as solicitors' offices—Decision challenged both as perverse and as disclosing errors of law on its face—Permission had been given in 1973 for the ground floor, previously used as a bakery, to be used for solicitors' offices—Submission that Secretary of State's decision was perverse rejected by judge—It was not possible to hold that decision was so unreasonable that no reasonable person could have reached it—Allegations of errors of law on the face of the decision also rejected—No ground for saying that the wrong questions had been asked—Decision letters should not be subjected to hypercritical analysis nor construed as if they were statutes—Applications dismissed

The applicants in these proceedings, Sears Blok, a firm of solicitors, sought to have quashed the decision of the Secretary of State dismissing their appeals against the refusal of the

London Borough of Southwark to permit the use of the first floor at 18 Camberwell Church Street, London SE5, as solicitors' offices. The firm was already occupying the ground floor for this purpose, planning permission having been given in 1973 to change the use from a bakery to solicitors' offices.

Matthew Horton (instructed by Sears Blok) appeared on behalf of the applicants; Jonathan Ruffer (instructed by the Treasury Solicitor) represented the first respondent, the Secretary of State. The second respondents, the London Borough of Southwark, were not represented and took no part in the proceedings.

Giving judgment, LLOYD J said: In these proceedings Sears Blok, a firm of solicitors, move to quash a decision of the Secretary of State for the Environment relating to premises at 18 Camberwell Church Street in the London Borough of Southwark. There were two appeals before the Secretary of State against the refusal by the borough to grant planning permission for the use of the first floor of the premises as solicitors' offices. The Secretary of State dismissed both appeals. They were dealt with together by the inspector in a decision letter dated November 1 1977. The applicants attacked the decision on the ground that it was perverse. Alternatively they say that there are errors of law on the face of the decision letter and in particular that the inspector misconstrued the provisions of the relevant development plans and in consequence asked himself the wrong question.

The facts can be stated very briefly. Camberwell Church Street is a busy shopping street in South-East London. No 18 is on the south side and is four floors high. Between 1968 and 1974 the ground floor was used as a bakery. The first floor was residential but vacant. The second and third floors were also residential. On September 4 1973 the borough granted planning permission to change the use of the ground floor from a bakery to solicitors' offices. In November 1974 the applicants acquired the property, no doubt for the purpose of carrying on their practice. In March 1975 they applied for permission to change the first floor from residential to office use. In July 1975 they moved in and they used both the ground and first floor for their offices. On October 28 1975 planning permission for use of the first floor as offices was refused. In 1976 they made a further application for planning permission. This time they sought permission to continue the use of the first-floor premises as offices and to convert the second and third floors to a single self-contained maisonette. On December 10 1976 the borough once again refused permission. On March 14 1977 the applicants appealed to the Secretary of State. There was a hearing before the inspector, J Marsden, on October 12 1977 and he issued his decision letter, as I have already mentioned, on behalf of the Secretary of State on November 1 1977.

In the decision letter the inspector states the "determining issues" as follows:

> In the case of both schemes whether the proposals would be so significantly contrary to the provisions of the Greater London Development Plan concerning restriction of offices as to be unacceptable; and whether the suggested change of use and alterations would materially depart from the land use allocation of the Initial Development Plan for Greater London and its intention that residential accommodation should be retained, with or without adaptation, except in very special circumstances.

In paragraph 6 the inspector identifies the two primary strategic policies of the Greater London Development Plan as those relating to residential and office accommodation. Paragraph 6 continues:

> Paragraph 3.37 of the written statement states, inter alia, that so long as an overall shortage of housing persists in London, planning permission will not normally be given for a change from residential use of any residential building that can still be used, with or without adaptation, for residential purposes. As to offices the plan seeks to restrict such development to preferred locations. The evidence showed that the appeal site does not fall within such an area. Even small office suites, not requiring an office development permit, should only be allowed in an appropriate zoning; and except in very special circumstances no dwelling, or part of a dwelling, should be allowed for office use.

In pararaphs 7 to 12 the inspector considers a number of arguments put forward by the applicants to justify a departure from the policies set out in the development plans. Then in paragraph 13 he states his conclusions:

> I have considered all the other matters raised at the inquiry and in written representations but, in my opinion, they are outweighed by the factors set out above which have led to my decision. While I believe that, in some respects, the issues are finely balanced I am not convinced that any overriding special need has been proved for the specific use of the appeal property to allow an exception to the current policies of the Greater London Development Plan and the Initial Development Plan regarding office location and the restrictions to prevent the loss of existing or potential housing accommodation.

Mr Horton on behalf of the applicants argues that the decision of the inspector was perverse, that is to say it was a decision to which no reasonable inspector could have come. He relied on the admitted need for solicitors' offices in the area, having regard in particular to the proximity of the recently-opened Camberwell Green Magistrates' Court. He relied also on the fact that Camberwell Church Street is a busy street, not particularly suitable for residential accommodation and that in the initial development plan it is zoned for commercial purposes with a shopping frontage. Since the borough itself granted permission for office use in the selfsame building in September 1973 no possible objection could be taken to the proposal on the ground of its location. As for the loss of residential accommodation, Mr Horton stressed that the first floor had been vacant for a number of years, that the loss of a single unit of accommodation would be so insignificant as to be almost *de minimis*, and that the reinstatement for residential use while not wholly impracticable would be expensive.

I have given these and the other matters relied on by Mr Horton careful consideration. They are strong reasons why the decision might well have gone the other way. The inspector himself regarded the matter as being finely balanced. But the question for me, of course, is not whether the inspector reached the right decision but whether his decision was so unreasonable that no reasonable inspector could have reached it. I am myself quite unable to take that view.

I was referred to the decision of Sir Douglas Frank, sitting as a deputy judge of the High Court, in the case of *Niarchos (London) Ltd v Secretary of State for the Environment* (1977) 35 P&CR 259. In that case the Secretary of State had, as in the present case, dismissed appeals against refusal of planning permission for the continued use of premises in Mayfair as offices. The decision of the Secretary of State was quashed on the ground that it was perverse and unreasonable. In that case the inspector was treated as having found as a fact that the costs of adapting the premises for residential use were such that the proposed schemes were not economically viable. The Secretary of State took the view that the financial considerations prevailing at the time were not of such importance as to persuade him to permit an exception to the policy under which further extensions of office use were not normally permitted. The court held that the "true and only reasonable conclusion" on the facts was that the premises could not reasonably have been adapted for residential accommodation. Accordingly the decision of the Secretary of State was quashed.

I do not regard the decision in *Niarchos v Secretary of State for the Environment* as providing any support for the applicants in the present case. In the first place the decision rested on the finding of fact that the proposals were not economically viable. In the present case there is no equivalent finding. On the contrary, in paragraph 9 of the decision letter the inspector states: "My inspection confirmed that reinstatement of the first floor to residential use, albeit expensive, was not impracticable." Mr Horton argued that "not impracticable" in that context meant not *physically* impracticable, and that the expense might still be such as to make the scheme *economically* impracticable. I do not agree with that construction.

Secondly, the decision in *Niarchos v Secretary of State for the Environment* turned, as I read it, on the particular provisions of paragraph 2 of the initial development plan whereas in the present case it turns on the much broader question whether there should be an exception to the approved policy. The cost of reinstating the first floor to residential use would, as it seems to me, be only one factor in considering that question and not the sole factor as it was in the *Niarchos* case: see report *(supra)* p 264.

I now turn to the other main argument advanced by Mr Horton, that there are errors of law on the face of the decision letter. It is covered by an amendment to his notice of motion. I take first the change from residential to office use. Paragraph 3.37 of the Greater London Plan provides: "So long as an overall shortage of housing persists in London, planning permission will not normally be given for a change from residential use of any residential building that can still be used, with or without adaptation, for residential purposes." In paragraph 13 of the decision letter the inspector says that he had not been convinced of any "overriding special need" for an exception to the policy set out in the Greater London Development Plan. Mr Horton argued that in asking whether there was an overriding special need for an exception the inspector has asked himself the wrong question. For it suggests a much stricter test than anything to be found in paragraph 3.37 of the written statement. The argument is even stronger, says Mr Horton, in relation to the location of offices, which is covered by paragraph 4.14 of the written statement as follows: "Within the framework of the overall policy for office development (and subject to general considerations of labour supply and distribution) the general locations in London where offices can be sited with benefit are listed in Table 2." Mr Horton argues that there is nothing in that paragraph to suggest that an exception should only be allowed in the case of overriding special need. If the inspector had asked himself the right question he might have reached, says Mr Horton, a different result.

I would agree that if the inspector has asked himself the wrong question his decision cannot stand. But I can see no reason for taking that view. It is clear from paragraph 13 of the decision letter, and indeed from the decision letter read as a whole, that the inspector thought of the initial development plan and the Greater London Development Plan as being complementary to each other. The two documents have to be read together. He cannot in my view be faulted in that respect. It is true that in the event of any conflict the Greater London Development Plan must prevail. But there is no question of any conflict here. Paragraph 17 of the initial development plan provides: "Throughout the county, permission (temporary or permanent) will not, except in very special circumstances, be given for a change from residential use of any residential building which can still be used, with or without adaptation, for residential purposes of any kind." That language is reflected both in paragraph 2 and paragraph 6 of the decision letter. Mr Horton argued that there is a conflict between paragraph 17 of the initial development plan and paragraph 3.37 of the Greater London Development Plan which I have already quoted. Again I cannot agree. It may be that the initial development plan provides a more stringent test in relation to conversion from residential to office use than is to be found in the Greater London Development Plan. But that does not mean that the two are in conflict or that the very special circumstances required by the initial development plan should therefore be ignored.

The inspector was in my judgment entitled to ask himself the question: are there any very special circumstances here to justify the change from residential to office use? He was entitled to conclude that if not the policy set out in paragraph 17 of the initial development plan should prevail. It seems to me that on a fair reading of the decision letter that is precisely the question which the inspector did ask himself. It is true that in paragraph 13 he refers to overriding special need rather than very special circumstances, and that need is only one of the factors to be taken into account. But paragraph 13 is only a summary of the inspector's conclusions. The decision letter as a whole is clear enough. It would be quite wrong to require the inspector, as was suggested by Mr Horton, to deal explicitly with every argument advanced by the applicants, to dot every "i" and to cross every "t." Like Forbes J in *George Wimpey & Co Ltd v Secretary of State for the Environment and Another* (1978) 250 EG 241 I would deplore any attempt to subject decision letters to that sort of hypercritical analysis and to construe them as if they were a statute.

There is in my judgment no sound basis for the argument that the inspector misconstrued either the Greater London Development Plan or the initial development plan or that he asked himself the wrong question. For myself I can see no error of law on the face of the decision letter. Nor is this a case where it can be said that the inspector has failed to give any intelligible reasons at all. It follows that neither of Mr Horton's arguments in my view succeed and the applications must be dismissed.

The applications were dismissed with costs.

VENDOR AND PURCHASER

COURT OF APPEAL

February 5 1980

(Before Lord Justice MEGAW, Lord Justice LAWTON and Lord Justice ORMROD)

OYSTON ESTATE AGENCY LTD v G C T CONSTRUCTION LTD

Estates Gazette May 24 1980

(1980) 254 EG 611

Action by estate agents dismissed for want of prosecution—Appeal from decision of Vice-Chancellor of the County Palatine of Lancaster—Agents sought declaration that a contract made in 1972 for the sale of land by the vendors, their principals, contained a stipulation that the agents were to have sole selling rights in respect of any houses to be erected on that land—Plaintiffs' writ issued in 1972—Defence pleaded that prospective purchasers had withdrawn the offer on which plaintiffs' claim was founded—Various steps in action taken at long intervals, including a gap of nearly 2½ years between January 1976 and June 1978—Plaintiffs now stated that, whether or not original action was dismissed for want of prosecution, they intended to issue a fresh writ—Original claim was for declaration only—They now wished to add a claim for damages as houses had been sold in breach of alleged stipulation giving them sole selling rights—Court of Appeal refer to disadvantages of two separate actions arising out of substantially the same subject-matter proceeding simultaneously—Delay in present action was "grave, inordinate and inexcusable"—Vice-Chancellor's decision dismissing it for want of prosecution upheld—Birkett v James distinguished

This was an appeal from an order made by Judge Blackett-Ord, Vice-Chancellor of the County Palatine of Lancaster, dismissing for want of prosecution an action started by the plaintiffs, Oyston Estate Agency Ltd, in the Blackpool District Registry against the defendants, GCT Construction Ltd. The action arose out of certain negotiations in 1972 concerning land in Kirkham in the County of Lancaster.

L J Porter (instructed by Baden, Barnes, Nabb & Co, of Blackpool) appeared on behalf of the appellants; P V St J Chambers (instructed by Stephenson, Harwood & Tatham) represented the respondents.

Giving judgment, MEGAW LJ said: This is an appeal from an order made by Vice-Chancellor Blackett-Ord on November 2 1979 in an action in the Chancery Division, started in the Blackpool District Registry. The learned Vice-Chancellor ordered that the action should be dismissed for want of prosecution. The plaintiffs appeal, asking that this court should set aside that order and allow the action to proceed.

The plaintiffs in the action are Oyston Estate Agency Ltd and the defendants are GCT Construction Ltd. The action arose out of events which took place in the month of March 1972; that is now nearly eight years ago. The statement of claim was delivered following a writ issued by the plaintiffs on September 18 1972. In it the plaintiffs say that they are estate agents carrying on business in north-west Lancashire. In the month of March 1972 a gentleman called Owen Oyston, who is a director of the plaintiff company, was in negotiation with another gentleman, called Derek Barnes, who was acting on behalf of the defendants. The negotiations concerned the sale of certain lands, in Kirkham in the County of Lancaster, by principals of the plaintiffs, not named in the statement of claim, to the defendants. It is alleged that, during the course of the negotiations, Mr Oyston told Mr Barnes orally that contracts for the sale of the lands would not be entered into by the plaintiffs' principals, the prospective vendors of the lands, unless the defendant company agreed to grant, not to the prospective vendors but to the plaintiffs, sole selling rights in respect of any houses to be erected upon the land, after it had been sold. It is alleged that that stipulation was made by Mr Oyston with the prior knowledge and approval of the plaintiffs' principals, whoever they were. If it were not with the prior approval of the plaintiffs' principals it would, of course, be subject to criticism. It is then alleged that, on March 9 1972, Mr Barnes, on behalf of the defendants, executed an agreement to buy the lands and also an agreement which conferred on the plaintiffs sole selling rights in respect of any dwelling-house to be erected on the lands. Then it is said that, on March 10, the following day, an agreement for the sale of the lands themselves was executed on behalf of the plaintiffs' principals, whoever they were, and the part of the agreement so executed was posted that day by registered post to the defendants. It is alleged that with the posting of that part of the agreement the agreement came into full force and effect and that, because of the stipulations made by Mr Oyston during the negotiations, the sole selling rights for any buildings erected on those lands thus sold would, by contract, belong to the plaintiffs. Paragraph 6 of the statement of claim asserts "The defendants now deny that the agreement conferring sole selling rights on the plaintiffs in respect of any dwelling-house to be erected on the said lands had come into effect." The prayer is as follows: "(1) A declaration that the said sole selling rights agreement is now a valid and binding contract. (2) Costs. (3) Further or other relief." So the substance of that claim was a claim for a declaration that it was a valid and binding contract, which included provision for the plaintiffs, who in other respects were acting as agents in making the contract, to have rights to sell any buildings erected on the lands sold.

By the defence, it is asserted—I summarise the defence as it was pleaded—that at about 10.30 am on March 10 1972 Mr Blackburn, a solicitor who was acting on behalf of the defendants, telephoned Mr Oyston and orally withdrew the defendants' offer to enter into the proposed agreements. It is said that, at about 10.45 on the same day, Mr Blackburn also telephoned the solicitors acting for the plaintiffs and the proposed vendors and informed them of the withdrawal. It is alleged that the agreement for the sale of the land was not posted by or on behalf of the plaintiffs until after the defendants' offer to enter into the contract had been withdrawn and that, therefore, there was no contract. In other words, it is asserted that there was an offer by the defendants but that that offer was withdrawn before it was accepted and, on the ordinary principles of the law of contract, there was, therefore, no contract.

What happened thereafter is conveniently summarised in the judgment of the Vice-Chancellor. The writ was served on

September 19 1972. The statement of claim accompanied it. The defence was served on November 1 1972. There was a reply on August 23 1973—that is, nearly a year later. Then there was a summons to dismiss for want of prosecution, which, in the circumstances, I do not think matters. It was dismissed. Then the defendants sent an amended defence to the plaintiffs in August 1974. In June 1975 a receiver was appointed in respect of the defendants' business and the plaintiffs were duly notified a few days later. A summons for directions was heard and an order for directions given on June 13 1975. That order provided that the plaintiffs should deliver their list of documents within 14 days, and that the defendants should do the same; that there should be an inspection within seven days; and that the action should be tried at Preston, its estimated length being five hours, and that it should be set down within 56 days.

The learned Vice-Chancellor says, "The action proceeded with reasonable expedition." Some might regard that as a somewhat generous view of events up to the middle of 1975, but that is the view he took. "By the middle of 1975 an order had been made for discovery. One side had served their list"—that was the plaintiffs—"the other"—the defendants—"had not, when the defendant company went into receivership." Then the judge refers to a letter from one of the joint receivers, who had been appointed for the defendant company, to the solicitors acting for the plaintiffs, telling them of the receivership and indicating (to put it broadly) that there was not very much chance of the plaintiffs getting very much out of their action, even supposing their action succeeded.

The plaintiffs set down the action for trial on August 7 1975. After that, nothing happened until January 30 1976, when a firm of solicitors, Stephenson Harwood & Tatham, acting for the receiver, wrote to the plaintiffs' solicitors the letter which is before us, asking for further documents. The letter indicated that the plaintiffs' solicitors would be aware that their clients were in the position of unsecured creditors for the whole amount of their costs. The letter contained this sentence: "The receiver instructs us that he is prepared to agree for the action to be discontinued on the basis that each party pays its own costs incurred to date." The solicitors, Stephenson Harwood & Tatham, were not on the record for the defendant company. The receiver had not formally adopted the action, but he was, as the Vice-Chancellor says, the man in control.

The next letter referring to the action was a long time later, nearly 2½ years. On June 29 1978 the plaintiffs' solicitors swung into action with a letter in which they said that they had heard no more since January 1976 and so they understood that there was presumably no division of assets in which their client company could join. They said they were under pressure from the High Court Registry, because the action had been set down in the middle of 1975 and the court were anxious to know what was happening about it. So the plaintiffs' solicitors asked for confirmation of the position with regard to any possibility of recovery.

That letter apparently was not received or was somehow lost. It was not answered by Stephenson Harwood & Tatham. A reminder was sent by the plaintiffs' solicitors on November 17 of the same year, setting out the effect of their previous letter of June. That drew a reply from Stephenson Harwood saying that they had not received the letter of June 29 and answering what they were now told was in that letter. They said, "The situation outlined in our letter of January 30 1976 still applies, and the receiver would still be agreeable to dealing with this matter on the basis that the action be discontinued with each side paying its own costs. We trust this assists you in relation to the listing of the action."

That was November 1978. There was silence until May 31 1979. The plaintiffs' solicitors then purported to give notice that they were resuming the carrying on of the action and that they were going to take out an application for further documents to be submitted to them. In fact that was procedurally in error, because they had to give one month's notice of their intention to proceed further with the action, as nothing had been done in it for more than a year. That error was subsequently corrected.

So the matter went on, and on July 23 1979 the receiver of the defendant company decided to adopt the case and Stephenson Harwood went on the record as solicitors for the defendant company. On the same day the defendants notified the plaintiffs of their intention to apply to dismiss the action for want of prosecution. That was the application which came before the learned Vice-Chancellor, in which he gave his decision from which this appeal is brought. The learned Vice-Chancellor decided that the action should be dismissed and made the order accordingly.

That decision of his is attacked in this court by Mr Porter, on behalf of the plaintiffs, on two grounds. The first ground is that the learned judge erred in taking the view, as he must, by implication, have done in order to arrive at the conclusion that he did, that the delay which had occurred was not merely (to use the well-known words) inordinate but was also inexcusable. Mr Porter contends that the judge was in error in holding that the delay here was inexcusable. Having regard to the guidance given by Lord Diplock in the second paragraph of his speech in *Birkett v James* [1978] AC 297 at p 317, I take the view that that point cannot succeed. I see no reason to suppose that the learned judge exercised his discretion in any way wrongly or failed to take into account any material consideration in respect of the question whether the delay was inexcusable. That the defendants would be prejudiced was held by the learned Vice-Chancellor and I do not think that that finding is challenged in this court, and in any event it seems to me it was plainly justified. So the first ground of attack fails.

The second ground of attack was that, having regard to the House of Lords decision in *Birkett v James*, the learned Vice-Chancellor had arrived at a wrong decision in law. I will read the relevant paragraph from the Vice-Chancellor's judgment. He says:

> It is said that if the action is dismissed, nevertheless it will be open to the plaintiffs to issue a fresh writ because the limitation period has not run. I do not accept that the limitation period has not run. If it has not run, as was pointed out, I think it would be because there was no relevant prejudice, and it would seem that the action could not be struck out until the plaintiffs in effect thought fit to put forward a claim. But I have to bear in mind that the action is one for a declaration only, which is equitable relief subject to the doctrine of laches. It is very relevant that this dispute goes right back to 1972 and that in 1978 the plaintiffs regarded the action as dead. I think that it is reasonable for the defendants to take the view that it would be impossible to obtain satisfactory oral evidence of events so long ago, particularly in view of the history of GCT Construction and Mr Barnes since that date. I hold that the defendants are likely to suffer serious prejudice by the delay and I think it is right that the action should be struck out for want of prosecution.

What is said here, on behalf of the plaintiffs, to put it in summary form, is that, if this present action were indeed dismissed for want of prosecution, they would be entitled to issue a fresh writ and they intend to issue a fresh writ. Indeed, it was intimated that that would be done, whether or not the present appeal succeeds. The purpose of the fresh writ is that the present statement of claim does not contain any claim for damages. It is a claim merely for a declaration that there is a contract. The defendants deny that there is a contract. It emerges from what we are told in this court that they would also be likely to put forward the defence, as an alternative, that, if there were, indeed, a contract that was binding when it was made, they (the defendants), within a day or two, unequivocally repudiated that contract. If there were a contract, that would be a wrongful repudiation, but if that repudiation had been accepted by the plaintiffs—and the defendants say that, in those circumstances, it had been accepted—then the cause of action, and the only cause of action, which the plaintiffs would have would be a cause of action for a breach of

A contract which occurred, once and for all, on a date in March 1972; and any fresh writ claiming damages for that breach of contract would be subject to an unanswerable defence based on the Statute of Limitations as being in respect of a cause of action which had accrued more than six years before the issue of the hypothetical fresh writ.

The further writ which the plaintiffs desire to issue, whether or not the present action is dismissed, is one in which they would be claiming damages. They say that there was a contract and they say that they had not suffered any damage at the date when they issued the original writ and, therefore, they could not put in a claim for damages in that writ or B statement of claim. The damage occurred, they say, only when, in breach of contract, houses having been erected, they were sold other than by a commission coming to the plaintiff company. They say that has occurred since the date when they issued the writ in the action now in question and that they cannot amend that writ or that statement of claim to include this after-accrued loss. Therefore, they would be issuing a fresh writ in which they would claim the making of the contract, breach of contract and damages.

If the present action is not dismissed, that means that there will be running contemporaneously two legal proceedings between the same parties. In both of them the first issue will C be: was there a contract made or not? That is precisely the same issue in each of the two intended actions. It may be that the defendants could, as Mr Porter, for the plaintiffs, suggests, have amended or could now amend their defence in the current action so as to plead, in the alternative, that, if there were a contract, the plaintiffs had accepted a wrongful repudiation of it by the defendants. But whether that is so or not, that is something they would be entitled to raise in the new action which the plaintiffs propose to put forward. Also there would arise in the new action the question whether or not there had been a breach of contract, if contract there were and it remained unrepudiated, and what the damages would be.

D In giving this judgment, of course, I am expressing no view whatever as to what is likely to be the result, ultimately, of any of the issues here involved. That is not the question. I am only looking to see what the issues are or would be. The broad result is that, if this present action is not dismissed, as the learned Vice-Chancellor held that it should be, there would be these two proceedings running contemporaneously. To my mind, this would be something which would, at any rate, be close to an abuse of the process of the court. If it is something that can be remedied without injustice, the remedy without injustice seems to be perfectly clear. It is that which the learned Vice-Chancellor arrived at, namely, to dismiss the E present action. The result, apart from the question of costs, would be in no way, as far as I can see, prejudicial to the plaintiffs. There has here been this grave delay, inordinate and inexcusable. In my judgment, the facts which I have disclosed take this case outside any rule or principle laid down in *Birkett v James*.

Birkett v James also is not applicable here for another reason. The claim in the present action is a claim solely for a declaration. That is a discretionary remedy and the court exercises very great care before it grants such a remedy. One basis on which it would refuse such a remedy is laches—delay. Here the remedy is a discretionary one. In *Birkett v James*, F if the action there in question had been struck out for want of prosecution, the statutory limitation of six years, which applied in that case, not having run, there would have been nothing to prevent the plaintiffs from issuing a new writ on the next day. That would have been a new writ claiming precisely the same relief as had been claimed in the first action. Here the relief which is sought to be claimed in the proposed new action is one which is materially different from the discretionary remedy which is being sought in the present action. G

In my judgment, there can be no doubt, in all the circumstances, that the learned Vice-Chancellor was right in the order which he made. I am glad to think that the order which he made corresponds with commonsense as well as with the law. I would dismiss the appeal.

Agreeing, Lawton LJ said: When the plaintiffs issued their writ on September 18 1972 they thought they had a contract with the defendants which gave them a selling agency in respect of houses which the defendants intended to build on some land. The defendants had not started to build any houses at the date of the issue of the writ. So it follows that the only possible remedy at that time which the plaintiffs could have was a H declaration. But if the plaintiffs are right in their submission that they did have a contract with the defendants, by December 5 1972 the defendants had already started to break that contract; in other words, they had made themselves liable for damages to the plaintiffs. Unless the plaintiffs are an unusual body, it was damages which they wanted. From December 5 1972 onwards, from time to time, the defendants sold houses, some 50 or so by 1975. All those sales, if the plaintiffs are right in their submissions, were in breach of contract and each sale would have attracted some damages. It follows, therefore, that by 1975 the claim for a declaration was useless; it was not going to do the plaintiffs any good at all. But instead J of getting on with the claim for a declaration, there was the inordinate and inexcusable delay found by the learned Vice-Chancellor.

So we have this situation: a claim for discretionary relief, which had already been started; a claim for damages arising out of, substantially, the same subject-matter, which had not been started but which we were told by the plaintiffs they now intend to start. The result would be two actions dealing with, broadly, the same subject-matter.

It is in this connection that I invite attention to a note in the *Annual Practice*, under Order 15, rule 16, note 2A: K "Where specific relief, other than a declaration, is not claimed, the jurisdiction to make a binding declaration of right should be exercised with great caution; and a declaration ought not to be made upon a preliminary point in an action brought for that purpose, where the substantive relief must be claimed in another action." That is this case.

In those circumstances, it seems to me that the Vice-Chancellor was right in dismissing the claim for a declaration. I, too, would dismiss the appeal.

Also agreeing, Ormrod LJ said: The original action for a declaration, for reasons which my Lords have given, was a perfectly sensible claim at the time the writ was issued. It has L ceased to be a sensible form of relief, because the plaintiffs now, if they are right, have a substantive claim for damages. *Birkett v James* has no application to this case. In that case, immediately after the court had dismissed his first action for want of prosecution, the plaintiff was in a position to issue a new writ claiming precisely the same relief; in this case, he is not in that position; the relief claimed in the second action would be substantive.

One only has to look at the case and test it in this way: if the equitable relief claimed, instead of being a claim for a declaration, had been a claim for an injunction (assuming it would be possible to formulate one), it is inconceivable that M any court, eight years later, would listen to such a stale claim for an injunction. Similarly, in my view, the court should not entertain an action for a declaration after such a long time.

For these reasons, and for those given by my Lords, I too would dismiss the appeal.

The appeal was dismissed with costs.

LANDS TRIBUNAL

COMPENSATION

A CORRIN AND OTHERS (TRUSTEES OF NORTHAMPTON (MUNICIPAL) CHURCH CHARITIES) v NORTHAMPTON BOROUGH COUNCIL

(REF/45/1979)

(V G WELLINGS QC)

January 14 1980

Estates Gazette February 2 1980

(1980) 253 EG 489

Part of agricultural land at Northampton acquired for road extension—S 17 certificate granted for open space/playing fields or institutional development—Quantification of hope B value for residential purposes—Subject-to-contract agreement to sell part of claimants' land not taken to consortium of adjoining house-owners should "be treated with considerable caution"

C McAra (solicitor, Browne & Wells, of Northampton) appeared for the claimants; the borough solicitor for the acquiring authority.

Giving his decision, MR WELLINGS said: This is a reference for determination of the amount of compensation to which the claimants are entitled for the compulsory acquisition of undeveloped land, 1.1 acres in extent, at Hollylodge Drive, Northampton. The land lies just outside the northern boundary C of the designated area of the new town of Northampton. It immediately adjoins that boundary which runs along the west, south and east boundaries of the land. As Timothy John Hadland BSc ARICS FRVA, called on behalf of the claimants, said, the land slots into the boundary.

The land was included in the County Borough of Northampton (Hollylodge Drive Extension) Compulsory Purchase Order 1973, made on May 3 1973 and duly confirmed on November 2 1973. The purpose of the acquisition was for the construction of a road across the reference land as an extension of a road which previously had been brought up to the southern boundary of the land. Notice to treat was D served on November 19 1975, notice of entry was served on November 19 1975 and possession of the land was taken by the acquiring authority on January 5 1976.

The reference land was part of a field of 11.94 acres, the fee simple in which was vested in the claimants, who are the trustees of a registered charity known as the Northampton (Municipal) Church Charities. The taking of the reference land severed the 11.94-acre field, leaving 3.71 acres lying to the north-west of the reference land and 7.13 acres to the south-east thereof.

Adjoining uses

The area immediately to the west and south-west of the E reference land has been developed for residential purposes. Indeed, houses immediately adjoin the southern boundary of the 3.71 acres. Immediately to the east of the reference land is a school and immediately to the south of the school are the grounds of Nene College, a college of further education. The area to the east of the school is the Moulton Park employment area, which is at present undergoing development with industrial and warehousing units and with campus office buildings. The area immediately to the north of the reference land is in agricultural use.

F On May 15 1973 the claimants applied to Northamptonshire County Council for a certificate of appropriate alternative development under section 17 of the Land Compensation Act 1961. The application specified residential development as being appropriate for the reference land. The county council on July 24 1973 granted a certificate under section 17 of the Act of 1961. The certificate stated that in the opinion of the local planning authority planning permission could not reasonably have been expected to be granted for residential development of the reference land because:

> the site is beyond the limits of development shown in the approved village plan for Boughton and is outside the boundary of the Northampton designated area, and the development of the site G for residential purposes would constitute an undesirable and unwarranted extension of development into open countryside.

However, the certificate stated that in the opinion of the local planning authority, if the reference land were not required by the acquiring authority, planning permission might reasonably have been expected to be granted for the following development of the land:

> (a) Public or private open space/playing fields, subject to conditions that the means of access to the site and off-street parking provision shall be as may be approved by the local planning authority.
>
> (b) An establishment with large grounds attached for the purposes H specified in Class XIV of the Town and Country Planning (Use Classes) Order 1972, ie a house or institution providing for the boarding, care and maintenance of children, old people or persons under a disability, convalescent home, a nursing home, a sanatorium or a hospital, subject to conditions that the siting, design and external appearance of the building, the means of access thereto and the associated off-street parking arrangements shall be as may be approved by the local planning authority.

The 11.94-acre field was at all material times let to W Johnson for agricultural purposes and it was an agricultural holding within the meaning of the Agricultural Holdings Act 1948. The rent paid by Mr Johnson was £120 per annum. J He was also the freeholder of some 12 acres of adjacent land fronting Boughton Green Road. That field and the claimants' 11.94-acre field constituted his farm. His freehold land was also severed by the road scheme, the part acquired being purchased with vacant possession with compensation devaluing at £1,000 per acre.

The following local values are agreed:

(1) The value of land for leisure or recreational use in January 1976 was in the region of £1,500 per acre.

(2) The value of land for residential use in January 1976 was in the region of £16,000 per acre.

The claimants claim compensation for the interest acquired. K They make no claim for severance or injurious affection.

Parties' respective valuations

Mr Hadland expressed the opinion that there was hope value attaching to the reference land for development for residential purposes and that a vendor would be unwilling to sell the land at a price which did not reflect this factor. In Mr Hadland's view, a willing vendor of the reference land would require a price calculated by reference to half the value per acre of residential building land in the area in order to induce him to sell. An institutional developer would expect to pay the same price. Mr Hadland's valuation was as follows:

1.1 acres at £8,000 per acre	£8,800
Deferred one year at 10% and tax at 35% PV of £1 =	0.93897
present value =	£8,262
Less compensation to the tenant annual rent for 1.1 acres at £120 per annum for 11.94 acres = £11 compensation, 5 years' rent	55
	£8,207
Say	£8,200

Mr Hadland assumed that there would be some demand for institutional development of the reference land by a local authority or a local health authority. A new institutional home, he said, is being built very close by the reference land. In 1976 the local authority might have chosen the reference land for that home rather than acquire the existing site. Institutional developers in purchasing land would have to compete with other uses.

Philip Neville White BSc ARICS, an estate officer employed by the acquiring authority, was called on its behalf. He expressed the view that there was no demand locally for land for development by institutional users of the kind contemplated in paragraph (b) of the certificate. He thought that there was a modicum of hope value for development for residential purposes but that it was already accounted for in the price of £1,000 per acre paid by the acquiring authority to Mr Johnson for his freehold agricultural land. But for that hope value his land would have been worth £750 per acre, so that the hope value was worth £250 per acre. The value of the land based on hope value for residential development was therefore exceeded by its value as land for leisure or recreational use, namely £1,500 per acre. There being no demand for institutional development, he said, that was the correct value to apply. He said that deferment should be for 15 months because as at January 26 1976 the earliest date on which the agricultural tenancy could have been determined was March 25 1977. His valuation was as follows:

Before-valuation	
11.94 acres @ £1,500 per acre (Vacant possession value for use as described in section 17 certificate)	£17,910
After-valuation	
7.13 acres } @ £1,500 per acre	£16,260
3.71 acres }	£1,650
PV £1—15 months @ 10% (tenancy determinable in March)	0.89
	£1,468.50
Less: Tenant's compensation — Agricultural Holdings Act 1948 — Agriculture (Miscellaneous Provisions) Act 1968 Apportioned on area basis: 5 × £120 × 1.1/11.94	55.20
Compensation Payable	£1,413.30

Plus surveyor's fees RICS Scale 5A and proper legal costs

Mr White thought that the wording of the section 17 certificate ("an establishment with large grounds attached") was important. Both he and Mr Hadland regarded that wording as referring to a larger area of land than was constituted by the reference land itself. On the question of lack of demand for institutional development, Mr White said that there was no evidence of sales to show that there was any such demand. There had been nine planning applications received by the acquiring authority since the beginning of 1976 relating to Class XIV of the Town and Country Planning (Use Classes) Order 1972. Six were for the conversion of existing buildings. The remaining three were for the erection of new buildings, all applications by Northamptonshire County Council for building on new housing estates. Daventry District Council (the planning authority for the reference land) received one application in Class XIV for a change of use of an existing building.

Agreement with consortium of house-owners

Although there is no evidence of comparable properties or transactions, there is one matter to which both parties refer and which they interpret differently. The claimant has agreed to sell the 3.71 acres to a consortium of adjoining house-owners for the sum of £10,000 (approximately £2,700 per acre), subject to Mr Johnson's agricultural tenancy and subject to contract. As part of the agreement the house-owners are required to enter into a restrictive covenant for the benefit of the claimants' remaining 7.31 acres that the land sold will not be developed.

The agreement took place in 1978 in correspondence between Mr Hadland and a representative of the consortium. Delay in putting the agreement into effect was attributable to the need to obtain the sanction of the Charity Commissioners. That has now been obtained and both Mr Hadland and Mr White treat the agreement as firm in the sense that they expect it to be carried into effect.

They both accept that the reason why the consortium of owners wish to purchase the land in question is in order to protect the rear of their properties from residential development. They both rely upon the "transaction," but in different ways. Mr White regards it as evidence that the hope value for residential development is very much less than that spoken for by Mr Hadland. Mr Hadland did not see the "transaction" in that light; he regarded it as consistent with his valuation because the price was a sitting-tenant price, there being no immediate prospect of obtaining possession from the tenant because the land in question is unaffected by any planning permission, deemed or otherwise, for institutional development. He thought that 1978 prices represented a 10 per cent increase over 1976.

Reference was also made to a piece of land of 1.29 acres sold by the claimants to the acquiring authority in July 1973 for the construction of a road. The land was situate in the south-eastern portion of Northampton within the flood plain of the Riven Nene. The land was subject to an agricultural tenancy and the section 17 certificate was in favour of sports, recreation and horticultural use. The land was sold at a vacant possession value of £1,350 per acre. This transaction appears to have no usefulness in the present case except in so far as it may support the agreed value of £1,500 per acre for the reference land for leisure or recreational use.

Tribunal's findings

I have inspected the reference land and the immediate locality. I do not have to decide whether hope value for residential purposes exists because both valuers agree that it does. My task is to quantify it. I think that the so-called agreement between the consortium and the claimants with respect to the 3.71 acres should be treated with considerable caution. It is subject to contract which means, of course, that it is in no way binding on either party and that they are still in negotiation. It appears to me that there is very little evidence of demand for institutional development of the reference land. It is implicit, however, in Mr White's valuation that there is demand for use of the reference land as public or private open space/playing fields. Any purchaser who seeks to use the land for such purposes must expect to pay the market value of the land.

Having regard to all the circumstances, including more particularly:

(1) The nearness of the existing residential development to the southern boundary of the 3.71 acres and, therefore, to the southern part of the reference land;

(2) That boundaries of designated areas are not necessarily inflexible and that planning policies from time to time do change,

a reasonably minded vendor or purchaser would, in my judgment, regard the hope of planning permission for residential development of the reference land as being worth more than £250 per acre. Mr Johnson's freehold land is rather further away from the existing residential development than is the reference land and, if developed for residential purposes at all, is likely to be developed at a rather later date. I think that the value of the reference land for use as public or private open space/playing fields or for institutional development plus hope value for residential purposes exceeds £1,500 per acre. On the other hand I think that Mr Hadland's valuation is much too high.

I shall take as the starting figure £2,500 per acre. 1.1 acres at that rate becomes £2,750. It seems to me that the period of deferment should be 15 months and I shall apply Mr White's multiplier of 0.89, which gives a product of £2,447.50. Deducting the tenant's compensation of £55.27, the balance is £2,392.23, say £2,400.

I award to the claimants the sum of £2,400 plus surveyor's fees on scale 5(a) of the scales of the Royal Institution of Chartered Surveyors. The acquiring authority will pay to the claimants their costs of these proceedings, such costs, if not agreed, to be taxed by the Registrar of the Lands Tribunal on the High Court Scale.

PORTSMOUTH ROMAN CATHOLIC DIOCESAN TRUSTEES REGISTERED v HAMPSHIRE COUNTY COUNCIL

(REF/231/1976 and REF/79/1978)
(V G WELLINGS QC)
November 15 1979

Estates Gazette March 22 1980 and March 29 1980
(1979) 253 EG 1236 and 1347

Land at Winchester acquired for distributor road—Question of betterment under s 222(6)(a) of Highways Act 1959—Whether claimants' contiguous lands had benefited—S 17 certificate for residential development—No doubt that planning permission for contiguous land would have been refused until road actually built—Grant of planning permission not kind of benefit to which tribunal required to have regard—Too remote—Benefit is one directly referable to purpose for which land acquired, such as providing access—Award of amount of agreed valuation assuming no betterment—Alternative finding that notwithstanding s 222(6)(a) claimants entitled to same amount—Retained land not benefited—In no-scheme world residential development could reasonably have been expected—Tribunal's view that s 17 certificate, being part of machinery for assessment of compensation for land taken, has no part to play in any assessment in applying s 222 (6)(a)—Words "shall have regard to" give tribunal a discretion —If retained land had benefited and value of benefit had equalled or exceeded award figure, tribunal would have felt justified in awarding more than nil compensation

John Drinkwater QC and Daniel Robins (instructed by Larcome, Winter & Linington, of Portsmouth) appeared for the claimants; Jeremiah Harman QC and Miss Judith Jackson (instructed by the solicitor to Hampshire County Council) for the acquiring authority.

Giving his decision, MR WELLINGS QC said: These two references, which were heard together, are for the assessment of the amount of compensation to which the claimants are entitled for the acquisition of land to the east of the Romsey Road (A31), Winchester, Hampshire. The land lies in the vicinity of the village of Pitt and Oliver Cromwell's Battery.

REF/231/1976 relates to 5.04 acres of land, compulsorily acquired by the acquiring authority under section 214(2) of the Highways Act 1959 for the purpose of constructing a new distributor road in part thereon. I will call the land so acquired "the pink land."

REF/79/1978, which is a reference by consent, relates to six small parcels of land, separate from one another but each adjoining the pink land, which were the subject of a contract for sale between the claimants and the acquiring authority. Their aggregate area is 0.65 acre. The contract provides that the purchase price for the six parcels is to be such price as may be determined and contained in an award of the Lands Tribunal on a reference thereto under section 1(5) of the Lands Tribunal Act 1949 as if the purchaser had made a compulsory purchase order under the provisions of the Highways Act 1959 and had served a valid notice to treat thereunder on May 15 1972 and had taken possession on April 3 1974. I will call the six parcels together "the contract land" and the pink land and the contract land together "the reference land."

Question to be determined

The central question for the Lands Tribunal to determine is whether the amount of compensation for the reference land which would otherwise be payable (and which in effect is agreed) ought, as the acquiring authority contend, to be reduced to nil by reason of betterment to land retained by the claimants resulting from the construction of the road. The land retained by the claimants adjoins the pink land on the north, I will call it "the green land." It is 6.64 acres in extent. The question arises under section 222(6) of the Highways Act 1959, which provides as follows:

> In assessing the compensation payable in respect of the compulsory acquisition of land by a highway authority under powers conferred by the said section 214 (except subsection (8) thereof), or by the said section 215, the Lands Tribunal—
>
> (a) shall have regard to the extent to which the remaining contiguous lands belonging to the same person may be benefited by the purpose for which the land is authorised to be acquired;
>
> (b) without prejudice to the generality of the foregoing paragraph, shall, in the case of land authorised to be acquired for widening a highway, set off against the value of the land to be acquired any increase in the value of other land belonging to the same person which will accrue to him by reason of the creation of a frontage to the highway as widened; and
>
> (c) shall take into account, and embody in its award, any undertaking given by the highway authority as to the use to which the land, or any part of it, will be put;
>
> and the Acquisition of Land (Assessment of Compensation) Act 1919 shall, in its application to a compulsory acquisition by a highway authority under either of the said sections, have effect subject to the provisions of this subsection.

The question in the present case arises under paragraph (a) of the subsection, not paragraph (b) thereof.

Agreed facts

In an agreed statement of facts, the following facts or matters were agreed:

(1) The land, the subject of both references, lies on the south-western outskirts of Winchester and between the former line of the Romsey Road and the side of the property known as The Croft, which has a frontage to Oliver's Battery Road and was, in April 1974, also fronting the private road known as Seldon Close.

(2) The Hampshire Western Distributor Road Winchester (Compulsory Purchase) Order 1970 was made by the Hamp-

A G

B H

C J

D K

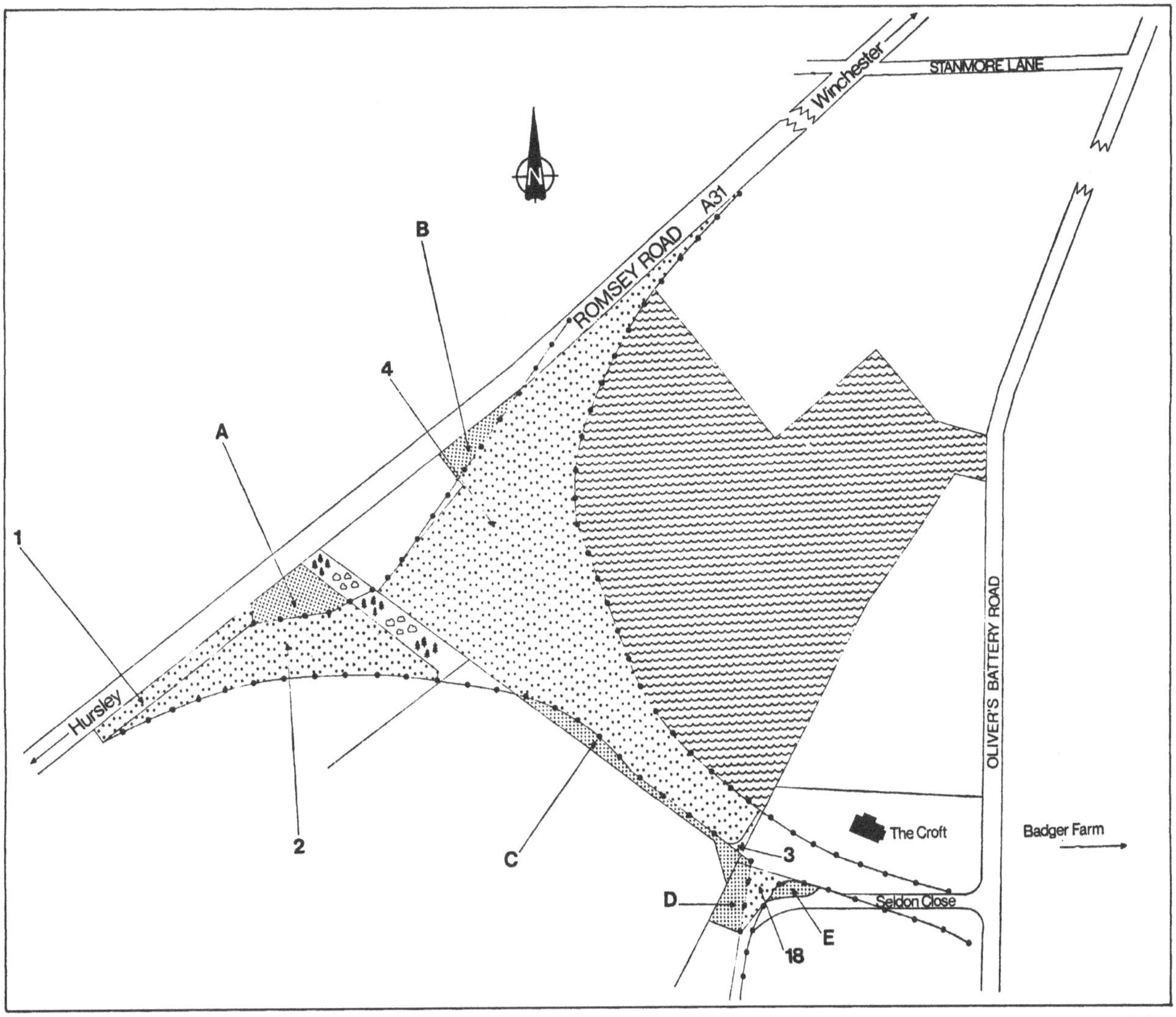

A sketch map of the reference area. The linked dots mark the line of the proposed distributor road. The map does not form part of the tribunal's decision.

E L

shire County Council on December 23 1970 under section 214 of the Highways Act 1959 as extended and supplemented by section 44 of the Highways Act 1971. The order was confirmed (without modification) by the Secretary of State for the Environment on March 30 1972 and notice to treat and notice of entry were served on May 15 1972. Possession of the land (the pink land) was taken on April 3 1974, which is the relevant date for valuation.

(3) The pink land is as follows:

F M

Number on compulsory purchase order schedule and plan on reference 231/76	*Number and colour on plan on ref 79/78*	*Area in acres*	*Description*
1	1 Pink	0.19	Tree-covered area of scrubland lying between the Romsey Road and an agricultural field in the ownership of the claimants.
2	2 Pink	0.88	Part of agricultural field above referred to, used for grazing purposes.
3(a)	3 Pink	0.02	Small strip of land to south-east of plot 4. The schedule to the compulsory purchase order shows this land as being in an ownership other than the claimants. Subsequently it was accepted that this land was that of the claimants and it was included in the notice to treat.

A

Number on compulsory purchase order schedule and plan on reference 231/76	*Number and colour on plan on ref. 79/78*	*Area in acres*	*Description*
4	4 Pink	3.96	Part of school playing-field and the tree-covered area adjacent to former line of the Romsey Road and part of a field adjoining the playing-field and lying between the former line of the Romsey Road and plot 3(a) on compulsory purchase order schedule.

B

(4) The contract land is as follows:

Number and colour on plan for reference 79/78	*Area in acres*	*Description*
A Blue	0.17	Lying adjacent to former line of the Romsey Road and plots 1 and 2 comprising part of the tree-covered area of scrubland and part of the agricultural field.
B Blue	0.10	Lying adjacent to the former line of the Romsey Road and plot 4 and comprising part of a thin tree-belt to the field adjoining the playing field.
C Blue	0.12	Strip of land adjoining the south-west boundary of plot 4.
18 Pink	0.08	The schedule to the compulsory purchase order shows this land as being in an ownership other than the claimants. It is now accepted that this land is that of the claimants. Scrubland.
D Blue	0.14	Scrubland fronting Seldon Close, adjoining plot 18.
E Blue	0.04	Scrubland fronting Seldon Close adjoining plot 18.

C D

(5) A small part of the green land has been retained by the claimants for use as an extension to the local authority school which has been built on land formerly belonging to the claimants north of the green land. The rest of the green
E land has been sold and is being developed with 27 houses. A further area of land remains in the claimants' ownership to the south of the pink land but is used and will continue to be used for agricultural purposes only.

(6) Accommodation works in the form of a fence along the boundary between the pink land and the green land were carried out by the acquiring authority.

(7) On March 27 1975 the claimants applied to the acquiring authority for a certificate of appropriate alternative development under section 17 of the Land Compensation Act 1961. Such a certificate was issued by the acquiring authority to the claimants on October 10 1975.

F The section 17 certificate, which related to approximately 4.2 acres consisting of that part (the larger part) of the pink land north of a belt or thicket of trees together with two of the parcels of contract land, stated that in the opinion of the Hampshire County Council, as the local planning authority, planning permission for development of that land might in the relevant circumstances reasonably have been expected to have been granted for residential development thereof at a density of four dwellings per acre subject to the conditions referred to in the certificate.

As to the part of the pink land lying south of the belt or thicket of trees together with another of the parcels of the contract land the certificate stated that no development G might in the relevant circumstances have been expected to have been granted. The reasons given for that opinion were:

(1) The proposed development (residential development) would be contrary to the provisions of the county development plan in that it would represent an undesirable intrusion of development into an area of open countryside that is within an area of great landscape value and that lies beyond the limits for development set by the Winchester town map.

(2) The proposal constitutes undesirable ribbon development along the Romsey Road and would create danger to and interfere with the free flow of traffic.

H The conditions relating to the 4.2 acres mentioned above were *inter alia* as follows:

(1) The usual conditions relating to layout, landscaping, siting and design of buildings, the layout of sewers and surface water drains and so on.

(2) "The landscaping scheme shall include the provision of a 50-ft wide belt of trees along the boundary with Romsey Road and along the south-west boundary of the site."

(3) "The land shall be developed comprehensively with adjoining land to the north east," that is to say, the green land.

J (4) No access, vehicular or pedestrian, is to be made into the Romsey Road.

Agreed valuations

The following valuations were agreed in writing by the parties. I have made minor changes in the statement of the valuations (in the numbering of paragraphs, the description of plots of land and the section 17 certificate and in order to avoid duplication and repetition), but the statement which follows is substantially verbatim. The valuations and contentions which accompany them are:

Valuation 1: K
It has been agreed that based upon the section 17 certificate the valuation at April 3 1974 of the 4.2 acres (plots 3, 4, B and C) to which the certificate related should be full residential value of £30,000 per acre. Therefore, the capital value of this land is agreed at £126,000. The value of the part of the pink land and one parcel of the contract land (plots 1, 2 and A) in respect of which the certificate stated that no permission could be expected (a total of 1.24 acres or thereabouts) should be based upon agricultural value of £1,000 per acre, the capital value therefore being £1,240. The remaining area, comprising three parcels of the contract land (plots 18, D and E) totalling 0.26 acre, has been agreed L as constituting one building plot with frontage to Seldon Close, the value being agreed as £7,500.

Valuation 2:
Valuation assuming no betterment of the green land:
It is contended by the claimants that if there had been no scheme for the construction of a distributor road planning permission would still have been granted for comprehensive development of the 4.2 acres and the above-mentioned building plot for residential development in conjunction with the green land. This development would have been at four units to the acre and such development would have commenced M without deferment or other restraint. If this assessment of the planning situation is correct, it is agreed that the valuation of the land to be acquired would be an aggregation of all the values agreed, ie £134,740, say £135,000.

Valuation 3:
Valuation assuming betterment to the green land:
A planning application was made on behalf of the claimants on October 30 1972 in respect of the green land. The planning authority stated that it would not consider the planning application on this area of land until the distributor road

A construction was complete and the road opened. The acquiring authority's valuation is based upon the information contained on the statutory town map together with the resolution of the city planning committee of July 1971 that no residential development on any part of the land owned by the claimants would be permitted in advance of the construction of the distributor road. In these circumstances the green land, it is agreed, would have been worth "hope value" of about £2,500 per acre, giving a total capital value of £16,600. Following construction of the road, as planning permission for low-density residential development could be assumed to be forthcoming immediately on the green land (and in fact B such an outline planning permission was issued in respect of this land dated October 24 1975), it is agreed that the value of the green land after the scheme would be—on the basis of £30,000 per acre—£199,200. The increment in value, it is therefore contended, as a result of the scheme is £199,200 less £16,600, ie £182,600. When this increment in value has been deducted from the value of the land to be acquired as above, the compensation properly payable is nil.

Some other facts

It is common ground that the distributor road was opened for traffic on January 20 1975. Construction of the road C appears to have been begun in or about September 1973. The western extremity of the road is situate on the pink land at a point where it fronts upon Romsey Road. The road runs in generally a south-easterly direction for about 1½ miles until it links up with St Cross Road. It has had the effect of converting the southern part of Oliver's Battery Road, which was formerly one of the thoroughfares into Winchester, into a cul-de-sac.

Planning history

(1) In the Hampshire development plan, officially approved on June 29 1955, the reference land and Badger Farm (here- D inafter mentioned) were shown as white land.

(2) In the report of survey (to accompany the reviewed county development plan, which was never approved) which was prepared in or about 1959, five major areas were selected for residential development in the period 1959 to 1981. Three of the areas amounted to 94 acres in total. A fourth area, the Oliver's Battery Road area, was stated to have been already substantially developed. The fifth area, an area of 160 acres, less than 1 mile south-east of Oliver's Battery Road, the reference land and the green land, was stated to be for development in the period after 1971 at an average overall E density of seven dwellings per acre (1,120 dwellings). The total number of new dwellings contemplated in the Oliver's Battery Road area and at Badger Farm in the period 1959-81 was 1,257.

(3) In the Winchester town review map, officially approved in 1966, the reference land and the green land were shown as being within an area of great landscape value and the Oliver's Battery Road area and Badger Farm area were allocated primarily for residential development. In the written statement accompanying the review map, instead of the five major areas previously selected for residential development, four such major areas were mentioned. Two of them totalled F 72 acres. The Oliver's Battery Road area was again stated to have been already substantially developed. The Badger Farm area was referred to, this time as an area of 135 acres, and was described as being intended for development to an overall density of no fewer than 10 dwellings per acre.

(4) On March 21 1967, the Minister of Housing and Local Government dismissed appeals from a decision of the Hampshire County Council refusing planning permission for residential development of a substantial part of Badger Farm and did so on the ground that no suitable access for the proposed development would exist until a new link road had been built. Badger Farm is adjacent to the line of the distributor road, which, of course, in 1967 was no more than in G contemplation.

(5) On August 5 1970 the claimants made two applications for planning permission in respect of an area of land including the green land and the 4.2 acres (about 12.5 acres altogether), the first seeking permission for the erection of a private secondary school with playing fields, and the second permission for residential development at a density to be agreed. Both applications were refused on December 21 1970. The reasons for refusal were:

(a) It is not intended that the land shall be developed for residential purposes and the county development plan provides that the existing use of land in this area, which H is within an area of great landscape value, shall remain for the most part undisturbed.

(b) The site is affected in part by the Western Distributor Road.

(6) On July 16 1971 the Hampshire county development plans subcommittee resolved "that the informal views of the local planning authority be as follows:

(i) Until and unless the distributor road is constructed no development would be permitted;

(ii) Since access can only be from Oliver's Battery Road an extension of the school or other institutional use would J be appropriate, subject to the extra traffic generated being minimal;

(iii) Alternatively, residential use at not more than four dwellings per acre would be acceptable, subject to a scheme of tree-planting between the development proposed and the new road boundary being carried out by the developer to the satisfaction of the local planning authority."

(7) On July 30 1971 the county planning committee, acting under delegated powers, approved that resolution.

(8) On October 30 1972 the claimants applied for planning permission to erect 27 dwellings on the green land. On Feb- K ruary 8 1973 chartered surveyors acting for the claimants wrote to the director of technical services, City of Winchester, proposing that that application be held in abeyance pending a final decision as to the distributor road. They added: "At this stage it is accepted without prejudice that the granting of planning consent on this site is contingent upon the distributor road being constructed. We suggest that the application be revived for a decision of the committee as soon as construction of the new road is commenced." On September 10 1973 the same surveyors wrote to the director of technical services asking that the application might proceed because construction of the road had commenced. On Sep- L tember 12 1973 the deputy director of technical services replied: "I am afraid that your interpretation of the planning situation regarding this land is not quite correct. It was a decision of the county planning committee in July 1971 that no development of this land would be permitted until the distributor road is constructed not 'until construction is commenced'." On September 14 1973 the surveyors again wrote to the director suggesting that outline planning permission be granted subject to a condition that no house be occupied until the new road is completed and in use. The reply, on September 18 1973, was as follows: "As you know, although this land is not allocated for residential development M in the development plan, the local planning authority reached agreement with the owner, in connection with the acquisition of land required for the distributor road, that an application for residential development would be supported when the road is constructed. The owner was fully aware of the agreed timing for the development when the agreement referred to was reached and I cannot foresee that this will be altered now." Although this correspondence treats the application dated October 30 1972 as being held in abeyance it is shown in the register of planning applications as having been withdrawn on December 12 1972.

(9) In January 1974 the stage two report of the Winchester and District Study joint steering committee was published. That committee was set up jointly by the former Winchester City Council, Winchester Rural District Council and the former Hampshire County Council. The report was intended for public discussion and as a consultative document but it disclaimed any intention to represent the policy of the councils concerned. It may nevertheless have been treated as a policy control document. It was treated as such by the Secretary of State for the Environment in a planning decision relating to land at Teg Down Farm dated August 1 1974. The report spoke of the supply of housing in the study area not being sufficient to meet all the demands made upon it and spoke of ways in which the short-fall might be made up. In paragraph 4.15 the report suggested a draft policy as follows: That the local planning authority would release no further land for housing before 1981 except for certain exceptions which it mentioned (not relevant for present purposes). At the same time a plan, which is part of the report, showed the reference land and the green land as being "existing committed" for residential develoment in the period up to 1981.

(10) On June 17 1974 the acquiring authority themselves applied for planning permission to develop the green land for local authority housing with approximately 70 dwellings. In the application they described themselves as "prospective purchaser." No action was taken upon that application although it does not appear to have been expressly withdrawn.

(11) On March 27 1975 the claimants applied for planning permission for residential development of the green land. On October 24 1975 the acquiring authority, as local planning authority, granted outline planning permission for residential development of the green land.

(12) No structure plan affecting this area has as yet been approved or, so far as I know, prepared. The Winchester area plan was adopted in June 1976 and represents the third stage of the study referred to above and is intended to be a local plan when a structure plan for mid Hampshire has been prepared and approved. This plan does not appear to have belied the views expressed in the stage 2 report. Looking ahead to the late 1980s one of its express policies is "to encourage the development of new housing and rehabilitation of existing dwellings especially within the centre of Winchester."

The case for the claimants

The claimants' case is succinctly expressed in the contention which is at the beginning of valuation 2. The claimants seek to apply section 222(6) of the Highways Act 1959 by comparing what might reasonably have been expected as to the grant of planning permission as at April 3 1974 in the light of the coming of the road, which at that date was a virtual certainty, with what might reasonably have been expected with respect to the grant of planning permission as at that date if no such road had been in contemplation.

In examining the no-scheme world the claimants seek to rely on the section 17 certificate. They say (quite rightly) that it is, in relation to the 4.20 acres, the equivalent of an actual planning permission (see Act of 1961, section 15(5)), that it envisages no deferment or other restraints and that one of its conditions expressly envisages contemporaneous development of the green land. They contend that the certificate represents the opinion of the local planning authority as to what would have been the planning situation of the 4.2 acres as at the relevant date in the no-scheme world. For these reasons they rely on the certificate as indicating in plain terms, more especially when considered together with the evidence of Mr Trott, that in the no-scheme world planning permission for development of the green land at a density of 4 units to the acre could have been expected.

Further on that evidence, they say that, without regard to the certificate (if as a matter of law the certificate did not have the effect contended for), then, nevertheless in the no-scheme world as at April 3 1974 planning permission for residential development of the 4.2 acres and the green land at such a density could be expected.

Leonard John Trott JP FRICS is a chartered surveyor of much experience in private practice. He said that the section 17 certificate reflected the no-scheme world; that if an application to develop the green land at a density of four units to the acre had been made in 1974 he would have expected approval to have been given in the no-scheme world. The decision of the minister in 1967 to refuse planning permission for the Badger Farm development was, he said, disastrous. In the no-scheme world, that is to say, in the absence of the road, development of Badger Farm (other than a small part of it) could not be achieved because there would be no suitable access to it. In these circumstances, the demand and need for housing land in the City of Winchester would have been such as to ensure the success of such an application for planning permission. The demand and need would have overridden all objections including the GLV notation (as the section 17 certificate demonstrated), and amenity and traffic objections.

Mr Drinkwater said that if the acquiring authority failed to satisfy the tribunal that no residential development on any part of the land owned by the claimants would be permitted in advance of the construction of the distributor road, they would lose. No intervening position could arise. The acquiring authority had taken upon themselves the burden of proving the negative position: that planning permission would not have been granted for the green land. The onus of proof on this issue was on them.

As to section 222(6) of the Highways Act 1959, Mr Drinkwater submitted that the words "shall have regard to" in paragraph (a), in contradistinction to the words "shall . . . set off against the value of the land to be acquired . . ." in paragraph (b) and the words "shall take into account" in paragraph (c) and subsection (7), gave a discretion to the tribunal as to the manner in which the benefit or its value should be treated.

The acquiring authority's case is represented by valuation 3. As I understand the contention there expressed, it is that the comparison as at April 3 1974 is, under section 222(6) of the Act of 1959, required to be entirely in the real world, or with-scheme world, between:

(a) A situation in which planning policy takes account of the scheme for the building of the road and forbids development of the claimants' land until the road is actually built; and

(b) A situation in which the road has been or is certainly about to be built, enabling that planning policy to be fulfilled by the grant of planning permission (such as that which was in fact granted), thus benefiting the green land by giving to it a development value which otherwise it would not have had.

Mr Trott accepted that in the with-scheme world, in which the compulsory purchase order had been made, planning permission for the green land could not have been expected until the road had been actually constructed. The acquiring authority relied upon that admission and on the failure of the claimants to appeal any of the planning refusals relating to the green land.

As to the no-scheme world, the acquiring authority's witnesses expressed the opinion that, for reasons which I shall mention, planning permission could not reasonably have been expected for the development of any of the claimants' land even if the development of Badger Farm were not to proceed. Those reasons were:

(a) That residential development would be detrimental to amenity;

A (b) That, while there was plenty of demand in Winchester and its environs for land for residential development there was an absence of need for such land;

(c) Traffic grounds.

In support of these reasons Patrick Ernest Pym MRTPI, assistant director of planning in the development control department of the county council, and Malcolm Roger Noyce BSc, chartered engineer, a section engineer employed by the county council, gave evidence.

Mr Pym's evidence, summarised, was as follows: the GLV notation was of great importance; the green and pink land were prominent in the landscape. They sloped gently to the
B south-west and were conspicuous from high ground at points on the Romsey Road in the direction of Hurley, further to the south-west. In 1974, any residential development on the claimants' land would have been conspicuous from the south-west notwithstanding such tree screen along the southern boundary of the 4.2 acres as then existed. The width of the thicket of trees (some part of which had been cut down in the process of road-marking) had been 40 ft, but to the east of the thicket there was no significant hedge or other screen.

If an application for development of the claimants' land had been made in 1974 it would have been refused on grounds of detriment to amenity, absence of need and traffic
C considerations. In the no-scheme world, with no scheme for Badger Farm in existence or contemplation, the construction of housing on the claimants' land would not have filled the need created by the absence of that scheme. The need in Winchester was land for high density development. The claimants' land was not suitable therefor. Development of the claimants' land would have made the traffic situation in Oliver's Battery Road and Stanmore Lane worse. An appeal would have been refused on the same grounds.

The coming of the road had reduced the traffic flow along Oliver's Battery Road by 80 or 90 per cent. In the result the green land had become more acceptable as a site for
D development. The urban fence had been moved out slightly. If the road had not been constructed planning permission for such development would not have been granted; if constructed, such planning permission would have been granted. There was no similarity between the claimants' land and Badger Farm. The former was part of a large rural area stretching away to the south-west. Badger Farm lay within the urban bowl of Winchester.

Mr Noyce said that if in 1974 he had been asked to tender advice on highway considerations relating to an application for permission to develop the green land, he would have
E made an objection on highway grounds, that is to say, that residential development of the green land would add significantly to pre-existing traffic congestion in peak hours in Oliver's Battery Road and Stanmore Lane. He produced a "prediction" of traffic flows as at 1974 hurriedly calculated by him from a document which he had discovered shortly before giving evidence. That document purported to be a record of a census of traffic flows at the junction of Stanmore Lane and Oliver's Battery Road and bore the date March 12 1965. He had found it among the county council's highway records. The author of the census was unknown to him. By applying appropriate percentages to figures taken
F from the census his findings of traffic flows as in 1974 were:

(a) 8 am to 9 am.

(i) In Stanmore Lane: 180 vehicles proceeding in an easterly direction, and 185 in a westerly direction;

(ii) In Oliver's Battery Road: 180 vehicles entering Stanmore Lane;

(b) 16-hour flow:

(i) In Stanmore Lane 2,475 vehicles proceeding in an easterly direction and 1,840 vehicles in a westerly direction;

(ii) In Oliver's Battery Road 1,260 vehicles entering Stanmore Lane.

The distance from the give-way line on Romsey Road to the left-hand edge of Oliver's Battery Road at the junction was 41m. From the centre line of the one junction to the centre line of the other the distance was 50m. He thought that the traffic generation at peak hour would be 1.5 vehicles per household, ie $27 \times 1.5 = 40$ vehicles. That compared with Mr Trott's view that traffic generation at peak hour would be 0.7 per household. In cross-examination Mr Noyce was willing to reduce his figure to 1.2 vehicles per household. Mr Noyce thought that the traffic generation would be significant and would in the existing conditions preclude the grant of planning permission for development of the green land. Mr Trott had expressed the view that traffic generation, even of 40 vehicles, would be insignificant. Both Mr Pym and Mr Noyce had observed traffic jams in Stanmore Lane and Oliver's Battery Road although Mr Noyce had seen them rarely. G H

Mr Harman said that Mr Drinkwater was effectively directing attention to the wrong question. In order to determine whether the green land had been benefited by the scheme for construction of the road, regard should be had to the position in the real world. On that basis, in April 1974, no planning permission for residential development of the green land would have been granted until the road was constructed, because that was the declared policy of the local planning authority. If regard were required to be had to the no-scheme world position as at April 1974, the evidence of Mr Pym and Mr Noyce was clear and convincing as to the planning and other reasons why in the no-scheme world planning permission would not have been granted. Mr Harman urged the tribunal to accept that evidence. The section 17 certificate could not, he said, be prayed in aid in order to strengthen the no-scheme world position. The statute required the acquisition to be ignored but it did not require there to be an assumption that there was to be no distributor road. (As I understand this line of argument, it is that section 17 does not preclude an assumption that the road is to be built but on a varied line avoiding the claimants' land.) J K

Decision

There is no doubt that in the conditions actually subsisting on April 3 1974 any application to the local planning authority for permission to develop the green land would have been refused until the road was actually built. That that was the declared policy of the local planning authority is clear. Why the local planning authority was so insistent upon that policy is not so clear and has not been explained; it may have been because of a desire to avoid further congestion in Oliver's Battery Road and Stanmore Lane; it may not. Why the local planning authority was unwilling to impose the condition suggested in the correspondence by the claimants' former surveyors in their letter dated September 14 1973 to the director of technical services (that no house be occupied until the new road be completed and in use) I do not know. I do not know what the attitude of the Secretary of State on an appeal might have been. L

While I accept that the actual completion of the road on January 20 1975, as it were, fulfilled the local planning authority's policy and enabled it, in accordance with that policy, to grant planning permission for residential development of the green land, I do not believe that the grant of a planning permission is the kind of benefit to which the tribunal is required to have regard under section 222(6)(a) of the Highways Act 1959. M

The kind of benefit to which the tribunal is required to have regard is, in my opinion, one which is directly referable to the purpose for which the land is authorised to be acquired, such as where the coming of the road will provide access to the retained land of a new or improved kind (including the creation of a frontage to a widened highway), which benefit increases the value of that land. It appears to me that the grant of planning permission by the local planning authority in respect of the green land was an indirect effect of the purpose for which the land taken was acquired; it was too remote. *A fortiori* is this so where, as in the present case, the planning permission was granted in pursuance of a policy which was itself referable to the purpose for which the land taken was acquired.

These reasons would justify me in holding that valuation 2 applies and that the claimants are entitled to an award of £135,000. However, in case I should have come to a wrong decision in law, I will proceed to consider what would be the position if the grant of a planning permission were a benefit of the kind referred to in section 222(6)(a) of the Act of 1959.

In order to determine whether the retained land is benefited by the purpose for which the land taken is authorised to be acquired it is necessary, in my judgment, to compare the situation existing as at April 3 1974, first, on the assumption that no purpose of building a distributor road exists and, secondly, on the assumption that that is the relevant purpose. In other words, the comparison is to be between the no-scheme world and the real or with-scheme world, both as at April 3 1974.

I appreciate that the wording of the paragraph may not entirely rule out reference to the possibility of the road being built but over a route avoiding the taking of land from the claimants. I can rule out consideration of that possibility in the present case because there is no evidence whatsoever as to whether another or varied route for the road would have been possible and as to where it might have run: see *Margate Corporation v Devotwill Investments* [1970] 3 All ER 864. I shall therefore proceed to compare the no-scheme world with the with-scheme world as at April 3 1974.

I may add that a comparison of that kind would seem to be required by "corresponding enactments": see section 7 of the Land Compensation Act 1961; section 14 of the Light Railways Act 1896; paragraph (2)(c) of the Schedule to the Development and Road Improvement Funds Act 1909 and paragraph 4 of Schedule 4 to the Housing Act 1957.

The section 17 certificate is part of the machinery for the assessment of compensation for the land taken. It appears to me that it properly has no part to play in any assessment which may be necessary in order to apply section 222(6)(a) of the Act of 1959. Accordingly the claimants' reliance on the certificate in relation to the green land is, in my view, misconceived.

I nevertheless hold that the green land was not benefited by the purpose for which the land taken was authorised to be acquired because I am satisfied on the evidence that in the no-scheme world, even without regard to the certificate, planning permission for development of the green land at four residential units to the acre could reasonably have been expected. I agree with Mr Trott that if no road were in contemplation the Badger Farm development would not be on stream and that, in the result, there would be pressure for development of the green land, if not the 4.2 acres and the pink land. In those circumstances there would be a need for housing land which could not be met elsewhere in Winchester. I agree with Mr Trott that the pressure would overcome objections on amenity and traffic grounds. I may add:

(1) That I have made an inspection of the areas in question, including the relevant high points on the road to Hurley and I have asked myself the question why, bearing in mind Mr Pym's observations with respect to detriment from amenity from a proposed development of the green land, the actual development now taking place on the green land has been permitted, in circumstances where the screen (however inadequate it may have been in the no-scheme world) is certainly less adequate now than it was before the coming of the road. I think that the answer is that the development, both proposed and actual, though regrettable, no doubt, is and would be not particularly conspicuous or offensive because it lies and would lie against the backdrop of the tree screen which at all times has subsisted along the Romsey Road and Oliver's Battery Road frontages of the green land.

(2) I reject the evidence of the traffic census. The maker was not called and indeed is not known. There is no evidence that he or she was properly qualified or experienced. The census was in part in writing and in part in pencil and contained 28 alterations. Altogether I regard the document as unreliable. In rejecting it, the consequence is that I reject also Mr Noyce's prediction. I however accept that as at April 3 1974 the junctions of Romsey Road/Stanmore Lane/Oliver's Battery Road were at peak hour subject to some congestion and that the local planning authority would not willingly have added substantially to it. Nevertheless I prefer Mr Trott's figure of 0.7 per household for traffic generation from the 27 houses to be built on the green land to Mr Noyce's ultimate figure of 1.2. I note that no highway ground appeared in any actual refusal of planning permission for the green land and in all the circumstances I believe that in the no-scheme world a highway objection of the kind described by Mr Noyce would not have been pursued.

For these reasons I conclude that notwithstanding section 222(6)(a) of the Act of 1959, the claimants are entitled to an award under valuation 2 of £135,000. I award that sum to them together with surveyor's fee under scale 5(a) of the scales of the Royal Institution of Chartered Surveyors.

I ought to add that had I been of the opinion that the land retained by the claimants had been benefited by the purpose for which the land taken was authorised to be acquired and that the value of the benefit was equal to or in excess of £135,000 I would have wished to award and would have felt justified in awarding to the claimants more than nil compensation. I agree with Mr Drinkwater that the words "shall have regard to" in section 222(6)(a) give to the tribunal a discretion. They are a guide and not a fetter; see *per* Farwell LJ cited in *Perry v Wright* [1908] 1 KB 441 at 458. I think it would be wrong if the compensation for the land acquired were less than existing use value (£1,000 per acre) or indeed less than existing use value plus hope value (£2,500 per acre). So far as the contract land is concerned the "price" can hardly be nil.

The acquiring authority shall pay to the claimants their costs of the proceedings, such costs if not agreed, to be taxed by the Registrar of the Lands Tribunal on the High Court Scale.

LORBRIGHT LTD v STAFFORDSHIRE COUNTY COUNCIL

(REF/66/1978)

(W H REES FRICS)

January 15 1980

Estates Gazette April 5 1980

(1980) 254 EG 53

Land at Stoke-on-Trent acquired for road-widening and improvement purposes—Cost of works and land acquisition required for suitable access road to subject land held to exceed value of land and nominal £1 awarded—No finding therefore necessary on amount to be deducted for increase in value of claimants' adjoining land—Dictum in Portsmouth Roman Catholic Diocesan Trustees case referred to

John C Taylor (instructed by Hancock & Willis) appeared for the claimants; Christopher Cochrane (instructed by the Solicitor to Stoke-on-Trent City Council) for Staffordshire County Council.

Giving his decision, Mr Rees said: This is a reference by consent of the parties to determine the compensation to be paid in respect of two plots of freehold land, in all about an

A acre ("the 1 acre"), off Etruria Road and Forge Lane, Etruria, Stoke-on-Trent. By the time the matter came before me only two questions remained:

(1) The amount to be deducted from an agreed figure of £24,000 for the cost of works required for access to the 1 acre; the claimants sought a reduction of £10,000 to give a net figure of £14,000 and the acquiring authority £30,000 or £31,000 (which included the cost of necessary works and land to be purchased) to give a nominal figure of compensation, £1.

(2) The amount to be deducted in respect of the increase in value of the claimants' adjoining land, also freehold, nearly 4 acres ("the 4 acres"); the claimants put this at nil but the
B acquiring authority contended for a minimum of £30,000.

Counsel for the claimants called F E Noble, a civil engineer, and S M Garside FRICS; counsel for the acquiring authority called P Thorley BSc (Eng), a municipal engineer, and M Astbury ARICS, a principal valuer employed by the authority.

Etruria Road runs east-west from Stoke-on-Trent to Hanley. About a mile from the town centre of Hanley, a short distance west of the bridge carrying Etruria Road over the Trent and Mersey Canal, Forge Lane runs north. The main access to the claimants' land was off Forge Lane but there was a secondary access to Etruria Road alongside the Trent and Mersey Canal.
C The agreed amount of £24,000 includes the value of the 1 acre taken, the rental value for two years of a small piece of land used as a working area and depreciation to about 2½ acres of the 4 acres. The 2½ acres were not built on and the depreciation was caused by the loss of the secondary access to them. The other 1½ acres had buildings on them; these were let. The whole of the 5 acres was allocated primarily for industrial purposes in the City of Stoke Development Plan.

The 1 acre of land was acquired by agreement under section 214 of the Highways Act 1959 and sections 44 and 47 of the Highways Act 1971. The relevant order was the City of Stoke-on-Trent (Etruria Road A53 Scheme) Highways Compulsory
D Purchase Order 1972 made on December 20 1972 and confirmed on October 17 1974. Notices to treat and entry were served on January 29 1975. The parties agreed that the valuation date was December 10 1975 and that "compensation shall be assessed on a compulsory purchase basis." The 1 acre was taken to widen Etruria Road and to improve the junction at Forge Lane.

In his valuation, from the agreed figure of £24,000, Mr Garside deducted £10,000, being the cost of a traffic lights scheme at the junction of Forge Lane and Etruria Road. Mr Noble described such a scheme and gave it as his opinion that this would suffice to satisfy the requirements of the authorities
E as to access; he did not consider that it would be necessary to purchase any land on either side of Forge Lane at its junction with Etruria Road to meet these requirements. He put the cost of a suitable traffic lights installation at £10,000 and gave evidence as to the traffic using Etruria Road.

Mr Thorley's view was that such a control as traffic lights at the junction was insufficient on its own and that it would be necessary to acquire land to widen Forge Lane. He spoke to two schemes which in his opinion would be suitable and he put the costs of these at £30,000 and £31,000 respectively. Neither included traffic lights although he said that they would be a desirable addition.

F I prefer Mr Thorley's evidence, mainly having regard to the width of Forge Lane, which varies from 16 to 20 ft, and to the probable effect on the traffic along Etruria Road of the installation of traffic lights at the junction of Forge Lane with Etruria Road. Visibility at that junction was substandard and would have remained so had no land been acquired, although Mr Noble said that there would be no need to improve sight-lines at the junction if lights were installed. If no land were acquired it would be difficult for long vehicles to turn into and to come out of Forge Lane. I find that the cost of a suitable access road would entail the acquisition of land and the cost of the whole scheme would exceed £24,000; traffic lights on their own would not be sufficient to meet the requirements of G
the authorities. Thus the compensation to be paid is a nominal amount, £1.

Both counsel put arguments to me as to the effect of sections 6 and 7, and Part I of the 1st Schedule to, the Land Compensation Act 1961 and section 222 (6) of the Highways Act 1959 on the valuation of the 2½ acres. In view of my finding above I do not need to consider these. I have since had an opportunity of reading the decision of this tribunal (V G Wellings QC) in *The Portsmouth Roman Catholic Diocesan Trustees Registered v Hampshire County Council* (1980) 253 EG 1236, heard in March 1979 and dated November 15 1979 in which similar arguments are considered in detail; the H
following sentence taken from a paragraph towards the end of that decision is relevant to the subject reference:

> The kind of benefit to which the tribunal is required to have regard is, in my opinion, one which is directly referable to the purpose for which the land is authorised to be acquired, such as where the coming of the road will provide access to the retained land of a new or improved kind (including the creation of a frontage to a widened highway), which benefit increases the value of that land.

It seems to me that the circumstances of this reference accord at least to some extent with what the learned member there J
had in mind.

I heard submissions as to costs. The claimants will pay to the acquiring authority their costs of this reference. Such costs unless agreed to be taxed by the Registrar of the Lands Tribunal on the High Court Scale.

AFZAL v ROCHDALE METROPOLITAN BOROUGH

(REF/108/1979)

(R C WALMSLEY FRICS)

January 21 1980 K

Estates Gazette May 17 1980

(1980) 254 EG 512

Corner shop and living accommodation in predominantly Asian community in Rochdale acquired as "added lands"—Claimant's business as general provision merchant and grocer well established when authority's redevelopment scheme intervened—Claim for value of interest in property and extinguishment of business—Evidence of falling sales due to scheme—Deduction from net profit of interest on loan not accepted by tribunal—Mortgage advance not a matter that L touches on profitability of business—Tribunal's reminder that for business extinguishment on compulsory acquisition the measure is "value to claimant"—Claimant required to re-establish business "in locality where there is a community of similar specialist customers"—Loss sustained, in terms of YP of net profit, more than double what a purchaser would pay

C Shelton-Agar (instructed by Thomas Jones, Son & Carden, of Manchester) appeared for the claimant; J Gleeson (solicitor to the Metropolitan Borough of Rochdale) for the acquiring authority.

Giving his decision, MR WALMSLEY said: This is a reference as to the compensation payable to Mohammed Afzal by the M
Rochdale Metropolitan Borough Council on the acquisition of the claimant's interest in a property 11 Trafalgar Street (including 25a Bradshaw Street) in the Wardleworth district of Rochdale.

The reference property was included as "added lands" in the Borough of Rochdale (Bradshaw Street, Rochdale) Compulsory Purchase Order 1977 made under Part III of the Housing Act 1957. A resolution to make this order was passed by the Housing Services Committee of the council on May 24 1977. The order itself was confirmed by the Secretary of State for the Environment on June 13 1978; notice to treat

A was served on October 2 1978; and the property became vested in the council on October 12 1978 following a general vesting declaration made on September 8 1978. The date that is relevant for my purposes is October 12 1978.

The premises, which were originally built in 1872 as two houses, now comprise a corner shop with linked living accommodation. The interest being acquired is the leasehold for the residue of a term of 999 years from May 1 1869 at an annual ground rent of £5.

At the hearing C Shelton-Agar appeared for the claimant and called valuation evidence from Derwent Archer. J Gleeson appeared for the authority and called valuation evidence B from Alan Pickles ARICS. Mr Archer is an estate agent and describes himself as a " member of the Royal Estates Association." Mr Pickles is the principal valuation officer of the Rochdale Metropolitan Borough Council.

In the notice of reference, drawn up by Mr Archer and dated June 25 1979, the claim was stated to be for £34,863.35 made up of:

The property	£10,000.00
Business goodwill	21,119.00
Stock-in-hand	1,744.35
Fixtures and fittings	2,000.00
C	£34,863.35

At the hearing this claim became amended to £37,645 made up of:

Interest in property	£16,000
Business goodwill	19,432
Stock-in-hand	1,744
Fixtures and fittings	469
	£37,645

The background circumstances, as I understand them, are these. Mr Afzal's parents are of Asian origin and came to D Rochdale some 10 years ago, when the claimant was aged about 16. Some two years later the claimant started to buy the reference property from an uncle of his. According to an entry in the Land Registry the property became registered in his own name on July 20 1973, and this entry shows " Price paid £2,665 "; albeit Mr Afzal, in evidence, was unshaken that he had paid £3,200. The purchase had been financed partly with moneys borrowed from Lloyds Bank Ltd and the United Dominions Trust Ltd and partly by loans totalling £2,200 from friends. There was already a grocery business established in the premises and this the claimant took over without any payment for goodwill. He built up a trade as general provision E merchant and grocer, mainly by sales to immigrant Asians, who comprised a predominant part of the community in the Wardleworth district.

Early in 1974 the claimant extended the former small corner shop into the other house at a cost which he stated to have been £2,300. To finance this alteration a first mortgage of £4,000 had been taken out in January 1974 with Cedar Holdings Ltd, and at the same time the former mortgages had been redeemed. Following the extension of the shop, profits increased, and by 1977 Mr Afzal had been able to repay the £2,200 borrowed from his friends. But regrettably he had fallen into arrears with Cedar Holdings Ltd, and by F the relevant date in October 1978 the amount owing on this mortgage had risen to £6,107.95.

The claimant continued trading after the vesting date and indeed is still living in the reference property with his wife and three young children. This being so, it is not possible at this stage to determine the total amount of compensation payable pursuant to the notice of reference. In particular, I was given no details as to fixtures and fittings; nor is it yet practicable to assess any claim in respect of loss on forced sale of stock, or removal expenses, or *Harvey v Crawley* costs. For this reason I am making this an interim decision, and there will be leave to apply. The present decision deals only with the first two heads of claim, namely interest in property, G and business goodwill. The reason why it is possible to deal now with business goodwill is because the parties agree that the appropriate basis for this head of claim is to assume the extinguishment of the business.

Interest in the property

The figures respectively contended for in respect of the claimant's leasehold interest are £16,000 by the claimant and £6,500 by the authority.

Mr Archer told me that the reason why he had increased the claim figure from £10,000 to £16,000 was because, on investigation, he had come to the conclusion that if the claimant had to reinstate he would not now be able to buy H a suitable property for under £16,000. As evidence Mr Archer referred to two properties now, or recently, on the market, *viz* 43 Dale Street, Milnrow, for which the vendor was asking £16,000, and 891 Manchester Road, Castleton, for which £15,000 was being asked. I have seen both these properties but can derive no assistance therefrom: they are both two miles from Wardleworth (in opposite directions) and both are in more prominent trading positions; and in any event evidence as to an asking price is helpful only to the extent of indicating that the property in question is likely to be worth *not more than* the price asked.

J Mr Pickles relied for his figure of £6,500 on two comparables, one of which he regarded as primary. This is 9 West Street (including 2 North Street), a corner grocery shop with linked living accommodation, only two streets away from the reference property and of a similar type and character. The council had negotiated the purchase by agreement of this comparable (with Mr Archer) in February 1978 at a price of £7,000. Mr Pickles provided comparative particulars as shown at the head of the next column.

Mr Archer rejected 9 West Street as a dependable comparable, on the following grounds: the vendor had just come out of hospital in February 1978 after suffering a heart attack; K

	Gross value	*Shop size*	*Floor area*	*Value*
Subject property	£220	13 ft × 26 ft	1,224 sq ft	Valuation as at Oct 1978 £6,500 = £5.2 per sq ft
9 West Street	£293	15 ft × 13 ft	1,323 sq ft	Price paid as at Feb 1978 £7,000 = £5.2 per sq ft

his wife suffered from asthma; his aunt, who lived with them, was dying, the owner was anxious to sell because the neighbourhood had become Asian in character, with corresponding L loss of trading income from Europeans, and the price had been accepted against his advice. It is to be added also that this transaction was a purchase in advance of the exercise of compulsory powers. These circumstances, combined, are sufficient to preclude my placing reliance on this comparable as evidence of open market value.

I am left, in the outcome, simply with a second comparable that was relied on by Mr Pickles, but, fortunately, one which was accepted by Mr Archer as helpful. This is 237-239 Entwistle Road, within a quarter-mile of Mr Afzal's shop, in sounder condition, sold on the open market in March 1978 at £12,000. The comparative particulars provided by Mr Pickles M are:

	Gross value	*Shop size*	*Floor area*	*Value*
Subject property	£220	13 ft × 26 ft	1,224 sq ft	Valuation as at Oct 1978 £6,500 = £5.2 per sq ft
237/239 Entwistle Road	£425	15 ft × 14 ft plus 15 ft × 15 ft	2,496 sq ft	Price paid in March 1978 £12,000 = £4.8 per sq ft

A In considering the above particulars it is, I think, common ground that property values rose over the seven-month period March to October 1978. I bear in mind also that the price paid for this comparable was a figure realised in the real world, whereas what I have to assess is a value in a no-scheme world. (I shall be referring later to the scheme and its effect.)

Having considered such evidence as was made available, I find the value of the claimant's interest in property to be £8,000.

Extinguishment of the business

The figures respectively contended for in respect of B extinguishment of business are £19,432 by the claimant and £2,526 by the authority. The calculations resulting in these figures may be expressed thus:

Mr Archer, for the claimant

Net profit for year to April 30 1978	£2,776
YP	7
	£19,432

Mr Pickles, for the authority

Net profit for year to 30 April 1976		£2,590
do 1977		2,039
do 1978		2,776
Unadjusted average net profit		£2,468
Deduct for interest of capital employed:-		
Stock-in-trade	£1,870	
Fixtures and fittings	300	
Cash	200	
	£2,370 @ 8% £189	
Rental value of shop premises	416	605
Adjusted average net profit		£1,863
Less, interest on loan obtained	£4,000 @ 15%	600
Compensatable net profit		£1,263
	Y.P.	2
		£2,526

C D

It will be convenient to deal with this head of claim by taking each of the elements of the authority's calculation in turn, *viz:* (a) Net profit prior to adjustment; (b) Interest on capital employed; (c) Stock-in-trade; (d) Fixtures and fittings; (e) Cash; (f) Rental value of shop premises; (g) Interest on loan; (h) Compensatable net profit; (i) Capitalisation.

(a) *Net profit prior to adjustment*

Accounts, prepared by a firm of chartered accountants, were E produced. In a condensed form the trading and profit and loss accounts show the following results for the three years to April 30 1978:

	1975/76		1976/77		1977/78	
Sales		£19,420		£17,123		£16,296
Stock brought forward	847		1,470		1,744	
Add purchases	15,358		13,373		12,076	
	16,205		14,843		13,820	
Less stock carried forward	1,470	14,735	1,744	13,099	2,398	11,422
Gross profit	(*sic*)	£4,485		£4,024		£4,874
F Outgoings		1,895		1,985		2,098
Net profit		£2,590		£2,039		£2,776

As can be seen, the sales fell steadily over the above period. It was contended for the claimant, and I accept, that a sufficient reason for the falling sales was the authority's scheme. In the initial years of Mr Afzal's trading, the area surrounding his shop was populous with Asians. The locality now is semi-deserted. Demolition has not been extensive, but the subject property is within a housing action area where the council are pursuing a policy of housing improvement linked with selective clearance of unfit property, and there are many houses empty. G On the face of things, in order to assess profitability in a no-scheme world it would accordingly seem proper to adopt the net profit figure for the *first* of the three years above. But, unexpectedly, the accountants' figures show the highest profit to have been earned in the *last* of the three years. There seem to be two reasons for this:

(i) The figures adopted for stock. Over a three-year period of falling sales and of diminishing purchases the value of stock-in-hand at the year-end is shown as having steadily increased, from £1,470 to £1,744 to £2,398.

(ii) *A casting error* in the calculation of profit for the first H year. The corrected gross profit, on the figures, for the year 1975-76 was £4,685 and the corrected net profit £2,790.

I propose to adopt a figure of £2,750 as the unadjusted net profit in a no-scheme world.

(b) *Interest on capital employed*

I agree with Mr Pickles that for compensation purposes an adjustment falls to be made for interest on capital employed in the business, and I adopt his interest rate of 8 per cent.

(c) *Stock-in-trade*

J There is capital locked up in stock. In the notice of reference the value of stock-in-hand is shown as £1,744. Looking at the trading account for the year 1977-78, the value of stock at the relevant date in October 1978 could be taken to be well over £2,000. The figure of £1,870 as adopted by Mr Pickles is an average of the figures shown in the accounts for the three years. A more realistic figure would, I think, be £1,500.

(d) *Fixtures and fittings*

Further capital is locked up in fixtures and fittings. In the notice of reference this item is shown at £2,000, but at the hearing Mr Archer amended this to £469, being a figure taken from K the balance sheet as at April 30 1976. In adopting, as I do, Mr Pickles' figure of £300 I make no finding other than that, for the purpose of the present calculation, such a figure seems not unfair to the claimant.

(e) *Cash*

For a business of this type and size I think it appropriate to include no more than a token figure of £100 for cash that, notionally, might have been earning interest elsewhere.

(f) *Rental value of shop premises*

I agree with Mr Pickles that for compensation purposes an L adjustment falls to be made also for the rental value of the shop premises. Having seen certain let shops on which Mr Pickles relies for his assumed rental value of £416 I adopt a rental value of £400.

(g) *Interest on loan*

In Mr Pickles' calculation there appears a further deduction, of £600, as "interest on loan obtained £4,000 @ 15 per cent". This had been done because, although the balance sheet as at April 30 1978 includes as a fixed asset "building" at a value of £4,650 (representing "£2,350 cost" plus "£2,300 capital intro- M duced"), no mention is made of the £4,000 advance from Cedar Holdings Ltd. Apparently the accounts prior to 1975-76 were prepared by another accountant who was unqualified.

Be this as it may, I do not regard a mortgage advance as a matter that touches on the profitability of the business and I delete the £600 deduction.

(h) *Compensatable net profit*

At this stage it is possible to quantify the figure of compensatable net profit to be capitalised in order to arrive at a sum payable for extinguishment of business. Mr Archer's figure for this purpose is £2,776. Mr Pickles' figure is £1,263.

My own figure, resulting from the foregoing findings, is £2,200 thus:

Net profit in a no-scheme world				£2,750
Deduct for interest on capital employed				
viz Stock-in-trade	£1,500			
Fixtures and fittings	300			
Cash	100			
	£1,900	@ 8%	£152	
Rental value of shop premises			400	552
				£2,198
			say	£2,200

(i) *Capitalisation*

Mr Archer capitalises his net profit @ 7 YP, on the basis that in the absence of the authority's scheme the business would have continued to be carried on from the subject premises for not less than a further seven years. Mr Pickles uses 2 YP, on the basis that various settlements for extinction of business, resulting from a large council redevelopment scheme in the town centre, had been negotiated by him within the range of 2 YP to 3 YP of an agreed average net profit.

I enquired of Mr Pickles what figure he was seeking to arrive at, and his reply was: "What a purchaser would pay for the business". But, with respect, this is not right. The market value of the claimant's goodwill is not the measure of what the council are required to pay. In respect of a business extinguished on compulsory acquisition the measure is always that of "value to the claimant", not that of "value in the market". The loss suffered by an expropriated trader is the ability to derive a future profit out of the premises from which he has been dispossessed. The tribunal has considered it desirable to comment on this point before, as for instance in *Remnant v London County Council* (1952) 160 EG 209 and in *Shulman (Tailors) Ltd v Greater London Council* (1965) 196 EG 453, and more recently in *Watson and Another v Warrington Borough Council* (1975) 250 EG 977.

I dare say that the value in the market of Mr Afzal's goodwill would be not more than 2 YP of net profit, if indeed it would be as much. Here, however, is a young man of Asian origin who in 1971 took over a retail business in a predominantly Asian community: the business did well and in 1974 the sales area was extended: then in 1977, when the businesss had become well established, the scheme intervened. In order to re-establish his business elsewhere Mr Afzal will require to find premises that are not only suitable in themselves but are also situated in a locality where there is a community of similar specialist customers. In terms of years' purchase of net profit the loss he has sustained is, in my opinion, more than double what a purchaser would pay for the business. I propose to allow £9,000.

Award

The figure of £9,000 for extinguishment of business becomes added to the figure of £8,000 for interest in property.

The acquiring authority will pay the claimant the sum of £17,000 to which will be added a surveyor's fee based on scale 5 of the Scale of Professional Charges of the Royal Institution of Chartered Surveyors, also legal costs or accountant's fees properly incurred from the date of the notice to treat up to the date of the notice of reference.

Mr Shelton-Agar has asked for costs on scale 3, and Mr Gleeson does not demur. The acquiring authority will pay the claimant his costs of this reference, such costs unless agreed to be taxed by the Registrar of the Lands Tribunal on scale 3 of the County Court Scales of Costs.

GATES v SOUTHWARK LONDON BOROUGH

(REF/145/1979)

(E C STRATHON FRICS)

April 15 1980

Estates Gazette May 24 1980

(1980) 254 EG 617

Freehold corner shop with living accommodation in Bermondsey, London, subject to 21-year lease with 7-year rent reviews—Procedure for first review in 1973 not operated—Claimants' arguments that lessor had right to review up to relevant compensation date in 1978—Tribunal's view that purchaser would feel unsure about operating rent review clause and would not take risk of adding to purchase price unless court had ordered in vendor's favour—Effect on value of residential subtenancy in breach of covenant considered "marginal"—Comparables of sales with vacant possession of no assistance—Rental value at second (1980) revision assessed at level current at relevant date in 1978

David Neuberger (instructed by Slaters) appeared for the claimants; Brian Leech (instructed by the Town Clerk of the London Borough of Southwark) for the acquiring authority.

Giving his decision, MR STRATHON said: This is a reference as to the compensation payable to Clive Robert Gates and Roland John Fellerman by the London Borough of Southwark Council on the acquisition of the claimants' interests in a property, 55 Bramcote Grove, Bermondsey, London SE16.

The reference property was within the Bramcote Grove Development Area as declared by the council in November 1971. The council, under the provisions of the Housing Act 1957 Part III, resolved to make a compulsory purchase order on December 4 1974 and the reference property was included and classified as "unfit" in the London Borough of Southwark (Bramcote Grove) (no 1) Compulsory Purchase Order 1976 made on April 27 1976. The reference property was subsequently reclassified as "added land" in the order as confirmed by the Secretary of State for the Environment on May 24 1977. A notice to treat was served on October 7 1977 and possession of the reference property was taken by the council on June 30 1978, which the parties agree is the relevant date for the assessment of compensation.

The premises, which were built in 1876-77, comprise an end-of-terrace, two-storey corner shop with living accommodation on the first floor. The ground floor was used for the lessee's grocery and provision business. At the rear there is a garage building which the claimants say was used as a secondary shop and the council describes as a store. The interest being acquired is the freehold subject to a lease dated November 25 1966 between the claimants and the lessee, A J Wilkins, for a term of 21 years from November 25 1966 at £450 per annum with rent reviews at the end of the 7th and 14th years. The property is zoned residential in the Greater London Development Plan and it is proposed to redevelop the Bramcote Development Area for housing to the west, separated by a linear public open space running from north to south from redevelopment of the eastern part for industrial use.

D Neuberger appeared for the claimants and called G H Moss, a solicitor who advises the claimants; C R Gates, chartered accountant, one of the two claimants, was also called; valuation evidence was called from D H Case FSVA, whose practice is in London SE15. B Leech appeared for the authority and called valuation evidence from A H B Watkins ARICS, borough valuer and property surveyor of the London Borough of Southwark since mid-1979. Mr Watkins' previous local government appointments had been with the councils of the Borough of Ealing and the London Borough of Enfield.

Before dealing with the valuation evidence, I will refer to two issues of law which, Mr Neuberger submitted, affected the determination of compensation. They concern clauses in the lease relating to rent reviews and subletting. The claimants purchased the reference property in November 1966 from

A A J Wilkins, who forthwith took the property as lessee by the lease of November 25 1966. The lease was assigned on November 3 1967 to R Strudwick, who died on November 21 1972, and the lease devolved upon the Midland Bank Trust Co Ltd, who sought licence to assign to the beneficiary, A C J Smithson, to whom the grocery business had passed. The terms as to rent reviews and subletting are reproduced as follows:

Rent Reviews

To hold the same unto the tenant for the term of twenty one years from the twenty fifth day of November 1966 paying therefor yearly during the first seven years of the said term and so in pro-
B portion for any less time than a year the yearly rent of four hundred and fifty pounds (£450) or such sum whichever be the higher as shall be assessed as a reasonable rent for the demised premises for the appropriate period such assessment to be made in the following manner, that is to say, either (a) such assessment as shall be agreed between the parties hereto in writing before the twenty fifth day of August one thousand nine hundred and seventy three or (b) in the event of the parties hereto failing to reach such agreement as aforesaid on or before the twenty fifth of November one thousand nine hundred and seventy three (in respect of which time is to be deemed to be of the essence of the contract) then the rent for the second seven years shall be fixed or assessed by an independent surveyor appointed for that purpose by the parties hereto or failing agreement as to such appointment by the
C twenty fifth day of December one thousand nine hundred and seventy three (time in this respect to be deemed to be of the essence of the contract) then by an independent surveyor appointed for that purpose by the President for the time being of the Royal Institute *(sic)* of Chartered Surveyors.

Provision is made for a further review to be assessed in the same manner save that the dates referred to shall be 1980 in lieu of 1973.

Subletting

Not to assign transfer underlet or part with the possession of part of the premises.

There is a further covenant by the tenant in the usual terms
D not to assign transfer underlet or part with the possession of the whole of the premises without the previous consent in writing of the landlord.

As to rent review, the procedure in the lease for review at 1973 has not been operated. The rent payable continued at £450 per annum. As to subletting, in correspondence in 1973 Mr Moss was informed by the surveyor then acting for the claimants that the lessee appeared to have sublet the first floor; reference to a subletting was also made in correspondence in 1974 and 1975.

On the issue of rent review, Mr Neuberger submitted that on the face of the lease the lessor, in the absence of expressly
E giving up the 1973 review, had a right to review at any time between 1973 and 1980; in the early steps to be taken time is to be deemed to be of the essence until application for appointment by the president of the Royal Institution of Chartered Surveyors, for which application no time is barred. Mr Neuberger relied on *United Scientific Holdings Ltd v Burnley Borough Council* [1978] AC 904. Mr Leech submitted that the rent review clause is not operable in the circumstances of the instant case. Time is of the essence for the first two events in the procedure but it is ambiguous as to the third event. The clause is to operate step by step, so that the third step, an application to the president of the Royal Institution
F of Chartered Surveyors, cannot be taken without going through the earlier two stages. Mr Leech contended that the lease on its proper construction, and the speeches in the House of Lords in *United Scientific Holdings Ltd*, supported his submission that the review clause in respect of 1973 must be deemed inoperable by 1978.

The second issue, subletting, refers to the residential use of the first floor of the reference property. Mr Leech argues that the claimants were made aware of a subletting in a surveyor's report in 1973 and in subsequent correspondence in 1974 and 1975, but, having taken no action, the claimants must be deemed to have waived the breach of covenant, and consequently at June 1978 regard must be had to the existence G of a subtenancy of the first floor. Mr Neuberger submitted that there was no evidence that a Mr Sawyer, who was rehoused by the council before the date of possession, was a subtenant, but if I found to the contrary, Mr Neuberger conceded that the acts and letters during the years 1973, 1974 and 1975 do amount to waiver. But by that earliest time the claimants were concerned solely with fighting to prevent the premises from being classified "unfit," and if I were to find that Mr Sawyer was a subtenant, it was argued that the *Pointe Gourde* principle should apply to exclude from the assessment of compensation any diminution in value of the premises due to the subtenancy. H

The respective valuations of Mr Case and Mr Watkins, slightly rearranged for ease of comparison, are as follows:

Valuation of Mr Case (for the claimants)

Estimated lease rent (fixed on review November 25 1973) for 2½ years from June 1978	£850	
Years' purchase for 2½ years at 11% and 2½%	1.977	£1,680
Estimated rental value at rent review on November 25 1980	1,458	
Years' purchase in perpetuity deferred 2½ years at 12% J	6.287	9,166
		10,846
Value of freehold interest, subject to existing lease		say £10,850

Valuation of Mr Watkins (for the acquiring authority)

Lease rent received at June 1978	£450	
Years' purchase for 2½ years to November 1980 at 11%	2.078	£935
Revision in 1980 to net rental value as on June 20 1978 K	1,225	
Years' purchase in perpetuity deferred 2½ years at 17½%	3.834	4,696
		5,631
		say £5,750

It can be seen that the surveyors adopt the same method of valuation, namely, capitalising rents or estimated rents, but they are apart at all other stages save one, the principal L interest rate of 11 per cent adopted for capitalising a rent estimated or received for the first 2½ years from June 1978. Taking these differences *seriatim*:

(1) *Rental for 2½ years from 1978* Mr Case at £850 a year estimated that this would be the rent fixed on a rent review with effect from November 1973. Mr Watkins took the view that as the lease rent review clause had not been operated, the original rent, £450, continues to be paid until the next rent review in November 1980. But Mr Watkins accepted £850 as a fair estimate of rental value in 1973.

(2) *Reversionary rent from November 1980* Mr Case estimated that rental values of £450 in 1966 and £850 in 1973 M would increase to £1,458 at the second rent review in 1980. Mr Watkins accepted a rising trend from 1973, but as a principle of valuation he considered that the rental value to capitalise was the value at the relevant date, not a rental value estimated to be attained at a future date.

(3) *Reversionary value percentage rate* Mr Case applied 12 per cent, a percentage based on his valuation experience; Mr Watkins' factor of 17½ per cent came from an analysis of auction sales of investment properties.

But there is a further, more fundamental, difference than those to which I have referred. It is the basis of the approach

A to value. It is revealed by the comparables on which Mr Case relies for support of his valuation; they are sales of properties with vacant possession. They were selected by Mr Case as transactions with which he had personally dealt. He considered that Mr Smithson (the lessee of the reference property), with an interim schedule of dilapidations to meet, a 1973 rent review yet to be operated and a further review in 1980, ought to do a horsedeal so that, broadly, the value of the reference property could also be looked at on a vacant possession basis. The comparables and sale prices on which Mr Case relied were:

B 62 Ilderton Road, London SE16, within the same compulsory purchase order as the reference property	£11,750
149 Southampton Way, Camberwell, SE5	£13,500
160 Southampton Way, SE5, £6,000 repair expenditure required	£9,000
97 Southampton Way, SE5	£11,750

Mr Case was satisfied that by the comparative process of plus and minus for locality, physical condition and so on, the transactions supported his valuation of the reference
C property at £10,850.

Mr Watkins sought support for his estimated reversionary rental of £1,225 from three rented properties and for his adopted 17½ per cent capitalising rate from four reported auction sales which were analysed to produce percentages of 17.2, 17.51, 20 and 20.83.

Before coming to my findings on details of valuation I will consider the submissions of counsel on the two issues of law. In my view the evidence confirms that the reference property was subject to a subtenancy of the first floor, but in terms of effect on value I accept Mr Case's opinion that it would be marginal rather than Mr Watkins' view that its
D existence would warrant an increase in the capitalising rate from 17½ per cent to 19 per cent.

As to the 1973 rent review clause, I found the arguments of counsel most helpful, but it is not a problem which stands on a legal solution alone. In terms of value, Mr Case considered that of £1,800 rent, if recoverable over the five years 1973 to 1978, a purchaser would add £900 to the purchase price; or if, by reason of the anti-inflationary legislation, only 2½ years could rank for payment, the amount recoverable would be limited to £1,000 and a purchaser would add £500 to the purchase price. Mr Case himself says that the tenant may be awkward and so cause the landlord to be involved
E in legal action. Mr Watkins considers that a purchaser would not take the risk of adding to purchase price unless the court has already ordered in the vendor's favour. I have come to the conclusion that a purchaser would feel very unsure about operating the clause and would take the view held by Mr Watkins.

I inspected the site of the reference property and eventually all the properties to which I was referred by the witnesses.

My conclusion as to recoverable rents applies also to the rents for the first 2½ years in the first stage of the valuations. The proper rent to take for this period is £450 and, capitalised at a rate of 11 per cent, the years' purchase is 2.078. In my view, Mr Watkins correctly adopts a rever-
F sionary rental value effective from November 1980 but assessed at a value level current at June 1978; it is a matter of adopting for the valuation of the reference property the tone that results from the analysis of the market. I am satisfied that at June 20 1978 the rent to take for this second stage of the valuation is £1,225.

The final question is the rate per cent at which the rent should be capitalised. Mr Case's opinion that 12 per cent is the proper rate is unsupported; it seems to me that this is the point at which Mr Case broadly takes into account an element of vacant possession. On the other hand, Mr Watkins is right to put the rate at less than the two transactions which

G showed returns of 20 per cent and 20.83 per cent, but on the evidence, confirmed by my inspections, I have come to the conclusion that the rate should be less than the 17½ per cent which Mr Watkins has adopted. There was evidence that one of the authority's rental comparables, 64 Ilderton Road, London SE16, a property within the relevant compulsory purchase order, supported a rate of 14.5 per cent on an agreed compulsory purchase price. In my judgment the proper rate to apply is 15 per cent. I find Mr Case's comparables of sales with vacant possession of no assistance. It is true that Mr Case considered that the difference between value subject to a lease and value with vacant possession could be taken
H to be about 10 per cent, but there was no evidence in support and Mr Watkins' view was that the difference is much wider.

Calculations concerning the "marriage value" of the interests in the leasehold and in the freehold-in-possession were put to Mr Watkins in cross-examination and to Mr Case, but it seems to me that the matter is carried no further.

I summarise my findings below.

Lease rent received at June 1978	£450	
Years' purchase for 2½ years to November 1980 at 11%	2.078	£935
J Reversion in 1980 to net rental value as at 20 June 1978	1,225	
Years' purchase in perpetuity deferred 2½ years at 15%	4.71	5,769
		£6,704

The acquiring authority will pay the claimants the sum of £6,704 compensation upon the acquisition of the freehold interest, subject to the lease, in 55 Bramcote Grove, London SE16. The acquiring authority will also pay a surveyor's fee based on scale 5 of the scale of professional charges of the
K Royal Institution of Chartered Surveyors and legal costs properly incurred from the date of the notice to treat up to the date of the notice of reference.

Having opened the sealed offer of compensation lodged by the acquiring authority, I find the amount thereof is less than the amount of my award. The acquiring authority will pay the costs of the claimants of this reference, such costs if not agreed, to be taxed by the Registrar of the Lands Tribunal on the High Court Scale.

L

WILSON v MINISTER OF TRANSPORT

(REF/214/1978)

(W HALL FRICS)

January 11 1980

Estates Gazette June 7 1980

(1980) 254 EG 875

Farm holding in Cumbria divided by acquisition of part of land for bypass—Value of land taken—Claim for injurious affection and severance—Tribunal holds that purchaser in reducing his
M bid would make overall assessment rather than calculate under various heads including obligation to maintain walls and fences, nuisance from laybys and insufficient height of walls—Claim for disturbance, including cost of land acquired for new sheep pen, loss of barley crop and time spent by farmer and man on problems arising during road construction works—£13,631 awarded against DV's figure of £7,097 and claim of £29,911

J K J Fisher (solicitor of Kendal Fisher & Co, of Ulverston) appeared for the claimants; J A R Grove (instructed by the Treasury Solicitor) for the acquiring authority.

Giving his decision, MR HALL said: This reference is to deter-

[A] mine the compensation payable to Edward I Wilson and Edward B Wilson by the Minister of Transport following the acquisition of land at Wilson Hills, Lindale [near Grange over Sands, Cumbria], together with an easement and licences over certain smaller areas in connection with the Levens Bridge to Carlisle Trunk Road A590 Lindale Bypass. The compulsory purchase order was made under the Highways Acts 1959 to 1971. Entry was taken on July 2 1975 and this is agreed as the valuation date.

At the hearing K J Fisher, solicitor, appeared for the claimants and J A R Grove of counsel appeared for the authority. Valuation evidence for the claimants was given by [B] G P Wilson FRICS FCIA, senior partner in Wilson & Co. Campbell Smith FRICS, of the Inland Revenue Valuation Office (South Lakeland), gave valuation evidence for the authority.

The claim as presented to me after adjustments was £29,911.45. The authority proposed a figure of £7,097. Both these figures were exclusive of surveyors' fees.

The land acquired was part of a holding extending to 89 acres. Edward I Wilson farmed, with his father, not only this land but also the main farm at Sandgate, Flookburgh. The total area farmed was 630 acres, including this land and 400 acres of tidal land. The farm enterprise is mainly livestock [C] fattening of beef and lamb with the growing of barley, wheat and potatoes. The new bypass road cuts through the 89 acres. The land to be acquired was originally calculated at 10.904 acres, but the final area taken is agreed as 8.482 acres.

The holding has been divided into two parts, of which 61.764 acres is to the north and 18.720 acres to the south of the new road. An underpass of a height and width sufficient only for the passage of cattle has been constructed under the new road. A new stock pen has been built by the department on the northern area; the existing stock pen serves the southern area. Stone walls have been erected as boundaries between the new road and the fields. The agreement on the accommodation [D] works was that the wall should be 4 ft 6 in in height. The height of the wall above the ground level on the field side is, in parts, less than 4 ft 6 in. On the lengths where no wall has been built, post and rail fences have been erected.

The claim was presented in a different form from the valuation lodged by Mr Smith. For comparison purposes, I adopt Mr Smith's presentation, which follows the conventional form, and alongside show the claimants' figures (see opposite).

It will be noted that the claim figure is less than as stated in the presented claim. This is due to a casting error.

Land taken

[E] Edward I Wilson's evidence was that this land was "hard" land in a district where land tended to be wet. It was therefore useful in that cattle could be put on to the land early and late when other land was unsuitable.

G P Wilson's valuation evidence was that the whole 89 acres would have been of considerable interest to adjoining farmers if placed on the market. In his opinion, the value of the land taken was £4,250. He produced no comparable sales to support his opinion but said he was familiar with the ones listed by Mr Smith. He did not consider them to be of much assistance in determining the value. He used, however, the sale price of approximately 80 acres at Witherslack, adjacent [F] to the reference land, which was sold by public auction in October 1973 for £35,000, which is £431 per acre. Mr Wilson considered that this price, for what he described as "very wet land, badly drained and low lying," justified his £500 per acre for "hard" land.

Mr Smith, for his part, said that the majority of the 89 acres was rough grazing and hilly with rock outcrop and scrub tree growth. About 23 acres was more level moss land and capable of cultivation—part was used for growing barley. In valuing the level land at £600 per acre and the rough grazing at £300 per acre, which produced a value of £2,962, he had regard to public auction sales within 5 or 6 miles of Wilson Hills. The £2,962 for the 8.482 acres represents an overall rate [G] of £349 per acre. The quality of land involved in these sales varied substantially. In 1973 four sales produced the following rates per acre: £716 per acre for approximately 8 acres of hard arable land at Fern Hill; £431 per acre for 80 acres at Witherslack; £393 per acre for 51 acres at Back o'th Fell Road, which was hard land with some outcrop and in his view considerably better grazing land than at Wilson Hills; and £211 per acre for about 7.50 acres again at Back o'th Fell Road, which was wet land. The only sale price in 1975 was for land extending to 9.29 acres at Allithwaite, analysing at £861 per acre, which was hard land with good grass and in Mr Smith's view much better than the moss land at Wilson Hills. He said [H] that the price of land had dropped after 1973 and did not recover till 1976. In 1976 a price of £1,243 per acre was obtained for just over 14 acres at Broughton. In support of his opinion, he also referred to three agreements with surveyors of the value of the land taken in compensation settlements with the owners of adjacent land on the west. The date of valuation was July 1975 and the rates per acre agreed were £812 and £800 for good-quality land, superior to the moss land of the reference, and £525 for quite good grazing land and £325 for rough grazing which was better than the reference land.

Description	*Claimants*		*Acquiring Authority*		
Land taken			1.390 acres @ £600	£834	
8.5 acres @ £500		£4,250	7.092 acres @ £300	£2,128	£2,962 [J]
Easement and licences		205			205
Injurious affection and severance—depreciation to land due to:					
(a) obligation to maintain walls	£6,748			500	
(b) obligation to maintain fences	in (a)			199	
(c) nuisance from lay-bys	in (f)			100	
(d) wall height being under 4ft 6in	1,285			844	
(e) sterilised area at underpass	310			48	[K]
(f) overall effect of bypass	13,520	£21,863		750	£2,441
Disturbance					
(a) cost of land acquired for new sheep pen		650.95			530
(b) loss of barley crop		140			94
(c) reinstatement of land adjacent to new wall		250			70
(d) loss of profits on 2.422 acres		150			150
(e) harvesting costs		in (f)			37
(f) general disturbance		1,703.50			608
(g) area south west of underpass		150			— [L]
		£29,362.45			£7,097

Mr Smith agreed that at the time of the sale a gas pipe was about to be installed the whole length of the 80 acres at Witherslack, but he did not consider that the price would be reduced because of this.

I have inspected the holding at Wilson Hills and the comparables referred to.

The sale closest in date was of the 9.29 acres at Allithwaite [M] in May 1975 at £861 per acre. This is a fairly level field in grass and appears to be the best land of the comparables. It compares with the 7.817 acres at Fern Hill, Witherslack, sold in October 1973 at £716 per acre, although that land is higher and more remote than Allithwaite. The 81.168 acres, which in location is closest to the reference land and sold in October 1973 for £431 per acre, is level and mossy. At the inspection, the land was soft. This is the land where, at the time of the sale, there was a proposal to put a gas pipe throughout the length of the field. The 50.948 acres at Back o'th Fell Road, Witherslack, is fairly steep land with rock outcrop and used for grazing. The price per acre in October 1973 was £393 per acre. The remaining sale before the valuation date is the 7.596 acres immediately opposite the 50.948 acres. It is mostly low-lying wet land and sold for £211 per acre in October 1973.

The land which Mr Smith settled the compensation for at £800 per acre appeared to be quite good hard pasture land. The land settled at £525 per acre and £325 per acre was not of high quality, but Mr Smith said it had deteriorated since the valuation date.

Wilson Hills has varying qualities of land within it and I consider a purchaser would, in his mind, assess these variations, as I now do. I determine a rate of £700 per acre for the well-drained arable part on the west. For the rough grazing, where there is a substantial rock outcrop, I determine £360 per acre. This leaves the central part which is not as good quality as the arable part but which is quite good grazing. I decide on £550 per acre.

Applying these rates to the land taken, I determine a value of £3,800 calculated as follows:

1.390 acres @ £700 per acre	£973	
1.377 acres @ £550 per acre	£758	
5.715 acres @ £360 per acre	£2,057	£3,788, say £3,800

which is equivalent to £448 per acre overall for the 8.482 acres

Easement and licences

These have been agreed at £205.

Injurious affection and severance

The major difference between the claim of £29,362.45 and the authority's proposal of £7,097 is due to items (a) and (b), involving the future maintenance of the walls and fences, and item (f) involving the overall effect of the bypass which, it is claimed, is the disadvantage due to the severing of the holding into two parts. I deal, however, with the various headings of claim as follows:

(a) and (b) Obligation to maintain walls and fences

G P Wilson said that as the wall is of inferior construction, firstly, it will have to be put in reasonable order and, secondly, it will have to be maintained at a cost of £280 a year. The £280 he calculated as follows: one week per year with a five-day week; on an eight-hour day and at a rate of £5 per hour for the farmer and £2 per hour for a man, this gives £200 for the farmer's time and £80 for the man's time. The £280 multiplied by 20 years' purchase equals £5,600 and to that was added the cost of £1,148, being an estimate to repair the wall, producing a total of £6,748 as his claim. He had measured the wall and fence at 1,220 yds and 383 yds respectively.

Mr Smith disagreed with the approach of the claimants' valuer. He said that he had considered the depreciation in value of the land held with the acquired land.

Evidence about the wall was given by N W Berry ARICS MIHE. He is a member of the North Western Road Construction Unit of the department and was resident engineer on the bypass. He said that the wall was completed on September 7 1976. Mr Wilson had complained about the condition of the wall and work had been done. In his view, the wall was sound in all respects. He agreed that there were some loose stones but said that if they were not pulled out the wall would remain sound. His measurement for the wall and fence was 1,126 yds and 440 yds respectively.

In Mr Smith's view, the maintenance of a new wall would not normally depreciate the value, but as this wall was a relatively long length, and doubts had been expressed about the condition, he had estimated that a purchaser would make a reduction of £500 in his bid to cover the possible liability. He had allowed £199 for the reduction in value due to the obligation to maintain the fence. Although he did not consider Mr Wilson's approach to be relevant, he submitted that the rates per hour were too high and that a multiplier of eight (representing 12½ per cent) would be more appropriate than 20.

My inspection confirmed the claimants' assertion that the wall is damaged in part by impact and that parts of the wall are not of a high standard. I would expect a purchaser to take into account the cost of improving it and thus the reduced burden of maintenance in the future. On the assumption that there is about 400 yds of fence to maintain or renew, this again is a liability which a purchaser would take into account. I consider that the work would be done as part of the ongoing liability he would have for the maintenance of the other boundaries of the farm.

(c) Laybys

No specific amount was claimed under this item. Mr Smith did not consider that the presence of the two laybys would be a disadvantage to a purchaser but allowed £100 for any possible nuisance. Having inspected, I consider there could be a nuisance by the deposit of litter, including plastic bags, which could harm cattle.

(d) Wall height

The fact that the wall is, in parts, less than 4 ft 6 in in height means that sheep may get over the wall on to the bypass road. Edward I Wilson said that this has happened and he has been informed by the police and others. He has then had to go and collect the sheep. An estimate dated January 1979 of £1,285 was submitted by the claimants for building up the walls to a height of 4 ft 6 in where they are under that height and that is the amount claimed.

Mr Smith said that the practical way of dealing with the problem was to erect a post and wire fence above the level of the wall, as was commonly done in the district. His information was that in 1975 it would have cost £844 to erect such a fence along the whole length. He considered that a purchaser would allow for that in bidding for the remaining land. The £844 in 1975 more or less equated with the claimants' £1,285 in 1979. He said that interest at the statutory rate will have accrued on the £844 to allow the claimants either to fence or raise the height of the wall.

I saw at the inspection that a temporary post and wire fence had been erected at the most vulnerable parts. A purchaser would consider this of importance for immediate action. He would either instruct that the wall be raised or have a post and wire fence erected. If he erected a fence, which would appear to be the quickest and most practical way of dealing with the problem, he would have further maintenance liability.

(e) Sterilised area at underpass

G P Wilson claimed that, as the approaches to the underpass had been bottomed to avoid poaching, an area of about 0.50 acre was sterilised. At his rate of £500 per acre for the land, he claimed £250. In addition, he claimed that with cattle tracking to the underpass an area of 0.25 acre will be reduced in value by 50 per cent. The claim is £60, making a total under this heading of £310.

Mr Smith accepted this depreciation in principle but had calculated the area at 0.159 acre, which at his rate of £300 per acre represented £48.

I have roughly checked the area involved and find it to be less than 0.50 acre. I consider about 0.25 acre, including the tracking, is involved and as the land was better than the rough grazing, land worth about £140 has been lost.

(f) Overall effect of bypass

G P Wilson said that the severance will cause additional time in operating the land, particularly due to the underpass. He had arranged that his client kept a diary of the time involved in dealing with the road works. From the diary, he had extracted the time spent by his client between June 19 1976 and September 19 1976 on work which he claimed was due to severance. For the three months he listed 19 hours for the farmer and 33.5 hours for the man. At the £5 and £2 per

A hour, this gave a figure of £162. He therefore multiplied that by four for the year, which is £648. At 20 years' purchase this produced a capital sum of £12,960. In addition to that G P Wilson estimated that the additional loading and gathering time due to the location of the new pen would be four hours per year minimum for the farmer and his man. At a combined rate of £7 per hour this is £28 per year, which multiplied by 20 is £560. These two items added together total £13,520.

Mr Smith, consistent with his assertion that the proper way to assess compensation was to consider the depreciation in value of the remaining land, said that in his opinion the severance only affected the smaller area of 18.72 acres to B the south of the new road. The value of that part before the road was built was, in his view, £7,500. In his opinion, the reduction in value due to the overall effect of the bypass was 10 per cent of that, namely £750.

My inspection gave me the impression of a compact unit, containing some arable and hard rough grazing with a pen near the main road and a pen with a more difficult access, being quite seriously affected by the bypass going through it. I do not consider a purchaser of the 80 acres would reduce his bid by only £750. My view supported the evidence that the underpass is not as convenient as an open area; the pen with an access from Nab Lane is inconvenient; the arable part is C now unevenly split; although equipment may have been difficult to take through the fields there is now no option and it may be necessary on occasions to cross the new road. I saw the gates at the road which are used by people on the public footpath which runs through the farm and which, if left open, will allow stock on to the road.

Although the approaches by the valuers were different in calculating the severance and injurious affection claim, I do not think, by the end of the hearing, the principle involved in the application of section 7 of the Compulsory Purchase Act 1965 was in dispute.

Section 7 states:

D

> In assessing the compensation to be paid by the acquiring authority... regard shall be had not only to the value of the land to be purchased by the acquiring authority, but also to the damage, if any, to be sustained by the owner of the land by reason of the severing of the land purchased from the other land of the owner, or otherwise injuriously affecting that other land by the exercise of the powers conferred by this or the special Act.

Accommodation works are usually agreed to be carried out by the acquiring authority to reduce the compensation payable for severance and injurious affection. Such agreements, however, have no statutory basis and the tribunal has no E power to direct the authority to do work on the owners' land nor to consider whether an agreement between the authority and the claimants, whereby the authority undertook to build a good and substantial wall, has been broken. *(Cripps on Compulsory Acquisition of Land,* 11 ed at para 4-291).

In my view, and I think this was accepted by Mr Fisher for the claimants, it is necessary to assess what reduction a purchaser would make in his bid for the remaining 80.484 acres having regard to:

(a) the obligation to maintain the additional walls and fences,
(b) the wall height being less than 4 ft 6 in in parts and therefore not stock proof,
F (c) the possible nuisance from the laybys and the sterilised area at the underpass, and
(d) the overall effect of the construction of the bypass on the working of the land.

Mr Fisher quoted from the recent decision in *Cuthbert v Secretary of State for the Environment* (1979) 252 EG 921, 1023 1115, 1176 and correction 253 EG 593 in which the tribunal (R C Walmsley FRICS) found, at p 1117, that "a purchaser, in deciding his bid... and looking at the responsibility for maintaining a 3-mile length of... fencing, might well *express* the value-effect of this in terms of £s per acre... but he would (if well advised) have first made as shrewd an estimate as possible G of the financial burden that he would be taking on." He asked me to have regard to the financial burden as set out by G P Wilson and to take into account the fact that the wall and fence obligation was related to an area of only 80 acres.

Although a purchaser might well make some calculations under the various heads, in the end he would make an overall assessment in deciding by how much he would reduce his bid for the land not taken. If there were competition for the land, as Mr Wilson contends, the intending purchaser would have to consider what other purchasers would be likely to bid.

As has been seen, Mr Wilson's total claim for injurious affection and severance is £21,863 and Mr Smith's total figure is H £2,441. These, in effect, provide a range of the possible reduction a purchaser would make in his bid for the remaining 80 acres. Neither valuer gave his opinion of the market value of the 80 acres after the roadworks, but Mr Wilson expressed the view that the value of the 89 acres before the acquisition for the road was £45,000 and Mr Smith said his view was £30,000.

Having regard to the remaining holding with its liabilities and disadvantages and to the accommodation works carried out, I determine the sum of £7,000 as the reduction in value due to injurious affection and severance.

Disturbance J

(a) Cost of land acquired for sheep pen

As part of the accommodation works it was agreed that a new pen be built on the north area. It was found, however, that an area of land would have to be acquired from an adjacent owner. Discussions about this took place between the claimants and the representatives of the authority. The authority had no powers to acquire and so the claimants had to negotiate this themselves. The owner insisted on their acquiring 0.6 acre and Mr Wilson, the claimant, said a price of £1,000 was asked. Eventually, a price of £600 was agreed. The sheep pen was then built. The claim is for the £600 plus costs, a total of £650.95.

Mr Smith's response to that claim is £530. As I understand K his calculation, he reduced the £650 claimed by £120 to give effect to his view that a part of the land acquired could be used for grazing and was surplus to the land required for the pen.

G P Wilson said he had understood that the department would reimburse the cost of this land. Frank Lawton ARICS, group engineer (land) of the Construction Unit, gave evidence about discussions, but it transpired that his recollection was about 39 sq yds for access purposes only. Whatever was arranged, my inspection revealed that the surplus area was a small triangular part of about 0.20 of an acre beyond the pen and used for grazing. The farm entrance to the pen is L restricted and the additional area appears to me to be no more than a small holding area. I have come to the conclusion that there is little or no advantage to the claimant. I therefore allow the £650.95

(b) Loss of barley crop

Mr Wilson, for the claimants, said that 1 acre had been destroyed. The crop was six weeks before harvesting and a price of £80 a ton, in his view, would have been realised. The yield would have been 2 tons. Allowing for the cost of harvesting, his claim was £140.

Mr Smith proposed £94. He calculated the area at 1.39 acres. He said that the crop was not mature and it would be two M months before harvesting. The yield which he would anticipate was 1.5 tons to the acre and the price he had ascertained at that time was £60 per ton. He considered that a purchaser of the growing crop would only pay three quarters of the ultimate price of £125 which was £94. I allow £120.

(c) Reinstatement of land adjacent to new wall

Mr Wilson claimed £250 under this heading. He calculated that on his length of 1,220 yds and a width of 2 yds, the area involved was 0.60 acre. He maintained that a grazing value of £25 per annum would be lost for 10 years.

A Mr Smith agreed the item in principle. His view was that if the land were properly seeded the loss would be restricted to two years. He proposed £70 for that loss. The criticism of Mr Wilson's figure was that it represented, in his value terms, a total loss.

I have inspected the land and confirm there is little growth. Cattle appear also to be tracking along the wall line. The 0.60 acre is virtually sterilised and I allow £160 for this item.

(d) Loss of profits on 2.422 acres

Although 10.904 acres was the original area of acquisition and was within the temporary fences, the final land taken was 8.482 acres. There was, therefore, an area of 2.422 acres which B could not be used for three years. The loss on this area is agreed at £150.

(e) Harvesting costs

This is an item which is included in the claimants' general disturbance claim. It is for additional work at the harvest in 1975 which was due to the road works. I adopt the authority's proposed figure of £37.

(f) General disturbance

This claim is substantiated by a detailed diary kept by Edward I Wilson of the time he spent, in his view, in C connection with the road scheme. It has been priced at £5 per hour for the principal and £2 per hour for a man. Mr Wilson was described by Mr Fisher as a working farmer who not only has to do the physical work but also the running of the farm (as Mr Fisher put it—" the thinking "). The gross income of the farming enterprise for 1975-76 was about £50,000. The livestock purchases were £10,747.

The diary reveals substantial difficulties over the period from July 1975 to July 1979. Mr Wilson said that no adequate fence was erected at the start and as the weather was dry the contractors " pushed on." This caused problems and he and his men had to leave attending to matters on the main farm seven miles away to come over to Wilson Hills. At times he was deal- D ing with the contractor direct at the suggestion of the authority's representative. He agreed that he would have to go to Wilson Hills in any event, particularly in the spring for lambing, but in July and August he would not be there as much. It was not always convenient for him to go there when he was called out. The claim is £1,703.50.

Mr Smith has gone through the diary, accepting only the entries which appeared to involve additional expense on time over the normal visit. He has challenged the £5 per hour for the principal and the £2 for the man. His proposal is £2 for the principal and £1 for the man. The £2, which was in effect £3 E less tax, can, however, be amended, he said, to £3 if the tribunal found itself satisfied that such a figure had been established by proof. Mr Smith has added in travelling costs of £2 for each accepted visit, which the claimants had not claimed. There are items which he considered should be directed to the contractors, but as they fall to be met by the department if the claimant cannot recover them from the contractor, he has included them in his proposal.

He has excluded any items which occurred after September 7 1976 when the walls were completed, as these were not a matter of disturbance. They should, in his view, properly be reflected in the severance claim. The claimant had to mitigate his loss F and that included taking steps to make the boundaries stock proof. He has also excluded an item of damage for £50 caused, it was alleged, by the contractor depositing soil outside the limits of acquisition. This was a matter of trespass and not the responsibility of the department.

Mr Smith's proposal was as follows:

Items accepted by him as the department's responsibility	£414.75
Items caused by contractors but accepted by the department	93
General allowance	100
	£607.75, say £608

If £3 per hour is accepted, he said that a sum of £145 can be G added to to his figure.

The cause of the difference between the figures of £1,703 and £608 may be summarised thus. Firstly, the rates per hour being taken at £2 and £1 as against the £5 and £2. Secondly, the entries after September 7 1976, when Mr Smith says the walls were complete, are not accepted by him. Thirdly, there are items which Mr Smith considers were in connection with the farming enterprise and not with the road works.

I therefore deal with these aspects:

(i) The rate per hour of £5 is arrived at by dividing the net profit by the hours which would normally be worked in a year H by the farmer. The figures were taken from the accounts for the year ended March 31 1976. The accounts for the year ended March 31 1977 were also produced and showed slightly higher figures for income and net profit. Although the rate was based on net profit, it was argued for the claimants that the time spent on dealing with the roadworks really affected the gross margin. By this, I understand it meant that the principal's efforts were to run the farm and to maximise the output and minimise the variable costs. The fixed costs had to be met even if the gross margin declined.

Mr Smith said that he would require three years' accounts, including the balance sheet, to assess the true net profit. The true net profit would take into account interest on capital. The J assets in the balance sheet would require to be adjusted to the market value at the time. From the limited information derived from one balance sheet he considered that the proper rate derived from the net profit was £3 per hour. As the farmer would have to pay tax on that he reduced the rate to £2 per hour. He conceded, however, that it was perhaps inappropriate to deduct tax.

The claim for the man's time was £2 per hour, although G P Wilson conceded this could be on the high side. He agreed that in 1975 the minimum wage for a grade 1 agricultural worker was £39.65 or £1 per hour but said that it was usually necessary to pay over £50 per week for a skilled worker. Mr K Smith had adopted £1 per hour and in support said that the total wages and national insurance figure in the accounts for the year ended March 31 1976 was £3,590. In the diary, I note that the time is often for two men. As we are dealing with a relatively short period in 1975 and 1976 I do not consider it necessary to see three years' accounts. The figures in the profit and loss account for 1975-76 are supported by those in 1976-77. These cover the period of the disturbance. The extent of Mr Wilson senior's involvement was not explained, but in any event I take the claimants' point that the farmer's loss is more related to the gross margin than the net profit. I therefore allow the £5. So far as the man's rate is concerned, the accounts do not indicate a L rate of more than £1 per hour and I accept that as the rate.

(ii) Mr Smith said that once the walls were completed on September 7 1976, any problems which arose after that date could not be accepted as disturbance. The claimants argue that the walls are not yet completed because of the condition and the height. The items in the diary for the period after September 7 1976 are mainly concerned with sheep being on the road, apparently having jumped the wall and on one occasion because a gate at the footpath had been left open. Mr Smith said that the claimants, on realising the wall was too low, should have mitigated their loss by taking steps to make the boundary M stockproof. I note from the diary that discussions about the height of the wall were still taking place with the engineer in November 1977. I find it difficult to accept the proposition that the claimants should have incurred quite considerable expenditure to remedy a deficiency in the agreed accommodation works, namely the wall being less than 4 ft 6 in, while negotiations were still going on about it. I allow £250 under this heading.

(iii) I have studied the items which Mr Smith says are normal to the farm enterprise and agree with his views. He has allowed £100 for general disturbance which could cover some points

A which are arguable. Included in that category is the £50 for the rutting when soil was alleged to be deposited outside the limits of the fence, but about which no detailed evidence was given.

The result of these conclusions is that an addition of, say, £500 will be allowed for the increase to £5 per hour and an additional £250 for the disturbance after September 7 1976. These added to the £608 proposed by Mr Smith total £1,358.

(g) Area south west of underpass

The claim for the £150 is for soiling, seeding, cleaning and generally making good an area of 250 sq yds south west of the
B **underpass. The evidence about this was vague, but my inspection showed that this was an untidy area where projecting rock had been cut. I consider the claimants are entitled to the £150 claimed to enable them to improve this part.**

I determine the total compensation at £13,631 summarised opposite.

The acquiring authority will pay the claimants the sum of £13,631 to which will be added a surveyor's fee based on scale 5 of the scales of professional charges of the Royal Institution of Chartered Surveyors, also any legal costs properly incurred from the date of the notice to treat up to the date of the notice C
of reference.

Having read the decision in this matter and having then opened the sealed offer of compensation lodged by the acquiring authority, I find that the amount thereof is less than the amount of my award; accordingly the acquiring authority will pay to the claimants their costs of this reference, such costs unless agreed to be taxed by the Registrar of the Lands Tribunal on the High Court Scale.

Land taken		£3,800
Easement and licences		205
Severance and injurious affection		£7,000
Disturbance		
(a) cost of sheep pen land	£650.95	
(b) loss of barley crop	120	
(c) reinstatement of land	160	
(d) loss of profit on 2.422 acres	150	
(e) harvesting costs	37	
(f) general disturbance	£1,358	
(g) area south west of underpass	150	£2,625.95
		£13,630.95
		say £13,631

D

LANDS TRIBUNAL

RATING

A **BRITISH AIRWAYS BOARD v RIDGEON (VO)**

(LVC/541-546/1976)

(J H EMLYN JONES FRICS)

May 24 1979

Estates Gazette January 12 1980; January 19 1980 and January 26 1980

(1979) 253 EG 154, 265 and 388

Maintenance base at Heathrow Airport comprising, in the main, hangars, workshops and ancillary accommodation—Ratepayers' contention that it is wrong to have regard to rents and assessments of other properties within airport perimeter, such as warehouses, not upheld—High plateau of values B within perimeter—Supporting contractors' test valuation by ratepayers' valuer described by tribunal as "very much second-best approach" and criticised in certain details—Would involve examination of rental evidence—General level of values applied by VO accepted but reductions made for certain units —Point of law on date of valuation for new valuation list—VO must have regard to all evidence available at last possible date and take account of "anticipated future trends"—"Ladies' Hosiery" dictum on correcting inaccuracies and securing uniformity not overruled by s 20 of 1967 Act—Tribunal's comment that it would be helpful if Parliament could con^ider providing precise date—No marked recession of values C at airport at end of 1971—In any event level of values adopted by VO accorded broadly with those of mid-1972—Ratepayers awarded only three-quarters costs

David Widdicombe QC and Christopher Cochrane (instructed by Lovell, White & King) appeared for the appellants; Sir Derek Walker-Smith QC and Konrad Schiemann (instructed by the Solicitor of Inland Revenue) for the respondent valuation officer.

Giving his decision, MR EMLYN JONES said: These appeals arise from decisions of the Greater London (North-Western) Valuation Court, given on May 13 1976, and relate to an D engineering base at no 1 maintenance area at Heathrow Airport. There are six appeals in all, owing to the fact that proposals were made on three occasions during the course of the first year of the new valuation list, at various dates between April 1 1973 and March 31 1974. Proposals dated August 10 1973 and March 20 1974 were made on behalf of the ratepayers and proposals dated March 28 1974 were made by the valuation officer. The base lies partly within the London Borough of Hillingdon and partly within the London Borough of Hounslow, so that at each stage there were two hereditaments, one in each of the two London boroughs. It is not suggested that the difference in the dates of the various proposals has any material E effect on the values to be determined. I propose therefore, in the course of my decision, to deal with the case as if it were a single appeal in respect of the whole of the maintenance base relating to its value during the 1973-74 rating year. The question of apportionment between the two boroughs can be dealt with at the end of my decision.

The maintenance base was formerly in the occupation of British European Airways (BEA) and has been constructed in phases over a period of some 25 years, starting in or about 1950. Since the dates of the proposals in the present case, BEA and the British Overseas Airways Corporation (BOAC) have become merged into the British Airways Board, who are the F appellants. But at the relevant date the adjoining maintenance base in the occupation of the BOAC was a separate occupation and was separately assessed. The assessment of that part of what is now the combined British Airways Board maintenance base is not before the tribunal in this appeal.

London Heathrow Airport is the principal airport for the United Kingdom and lies some 15 miles west of London. The main approach to the airport is from the M4 motorway or the A4 trunk road, from which access is gained through a tunnel to the central terminal area with its three passenger terminals and their associated buildings. All passenger traffic is handled there and the principal control tower complex and the Queens G Building, used for operation purposes, are also situated there.

Apart from the central terminal area there are three other main locations within the airport perimeter. Firstly, there is the north side, which contains a variety of buildings such as a police station, customs house, hotel, administrative buildings of the British Airports Authority, motor transport depots, cabin service facilities for many of the airlines and so on. On the south-west side of the airport is an area known as the cargo terminal area, connected with all other areas by the perimeter road and also connected directly with the central terminal area by a vehicular road tunnel, controlled by HM Customs, which enables cargo to be transported between the warehouses in the H cargo terminal and passenger aircraft loading and unloading in the central terminal area.

The third area lies on the south-east side of the airport and is known as the maintenance area, connected to the perimeter road but having no direct road link with the central terminal area. Since the relevant date the Piccadilly underground line has been extended through the station known as Hatton Cross, adjoining the maintenance area, to the central terminal area, but this is a factor which does not affect the value for the purpose of these appeals. The area known as no 1 maintenance area contains the appeal hereditament and the premises forming part of the sister maintenance base, formerly BOAC. The J buildings in this part of the maintenance area lie beyond the eastern extremity and to the north of runway no 5 which is one of the two principal east-west runways on the airport. To the south of runway no 5 is an area known as no 2 maintenance area, which is occupied and used by other airlines mainly for the servicing of aircraft. Parts of this area remain undeveloped.

The following passage taken from an agreed narrative sets out a description of the maintenance base and of its function:

> The function of a maintenance base is to provide short-term servicing, routine overhaul and checking, as well as major maintenance and overhaul of aircraft. It therefore requires not only hangarage, that is the covered space equipped with the appropriate K services into which an aircraft can be brought for the necessary stripping, overhaul, maintenance and repair to be carried out, but also the associated workshops, to which components can be taken for detailed attention. Inevitably there must be associated with this accommodation extra accommodation for the workforce, by way of toilets, canteen, medical accommodation and the office facility to administer the staff and the engineering functions. Certain aircraft, for example, because of their dimensions, will fit only into certain hangars. The dimensions refer not only to the space within the hangar, but also the dimensions of the doorways into the hangar.

Areas are also required for the storage of spare parts and part assemblies. With a very large maintenance base, the appropriate control is nowadays done by computer, which in turn brings in further building facilities, further grades of staff, as well as extra numbers of staff.

Inevitably, airlines that only "visit" a particular country or airport, will have very limited repair and overhaul facilities at that airport or country. Thus, a foreign airline flying into London may well have no facilities of its own at Heathrow at all. If any form of check or overhaul is required, this will be carried out on a contract basis by an airline which does have the appropriate facilities at that airport. Some airlines which have a heavy schedule of flights into a particular airport may have limited facilities there. Thus, at London Heathrow, Pan American has a small maintenance base there for its own use, as does Air India. The "home base" of any airline, however, that is the airport which is the major accommodation in its own country, will have major overhaul facilities. Thus, for example, Lufthansa has its primary maintenance base in Frankfurt. It is no surprise that BEA has its home base at Heathrow, and therefore it is at Heathrow that we have the major BEA maintenance base.

Apart from the BOAC base, the BEA base was the largest maintenance base at Heathrow. Indeed, it is by far the largest maintenance base occupied by any airline in any airport in the United Kingdom.

Attached is document No 1, which sets out the size and diversity of the BEA and BOAC fleets at the appeal date.

It will be seen that BEA has more aircraft than BOAC, but BOAC have larger aircraft. The facility requirements, therefore, are not identical.

Heathrow London Airport is operated by the British Airports Authority, and it is the policy of that authority not to sell off freehold land. Facilities therefore such as the appeal hereditament were held on ground leases or a series of ground leases. The BEA maintenance base is, in fact, held on a ground lease.

The BEA maintenance base is now some 84 acres and is held under a consolidated ground lease for a term of 84 years from June 1 1971, at a ground rent of £64,500. There is thus no rack rent payable on the property. The buildings were erected by the airline.

The details set out on document no 1 are as follows:

BEA and BOAC Fleet Numbers
(Maintained at Heathrow)

Fleet	31/3/69	31/3/70	31/3/71	31/3/72	31/3/73
BEA					
Comet 4B	12	Operation moved to Gatwick (but component overhaul remained at Heathrow)			
Viscount	35	31	24	20	—
BAC 1-11	10	Operation moved to Manchester (but major maintenance and component overhaul remained at Heathrow)			
Vanguard	19	19	19	18	15
Trident 1 & 2	33	38	40	42	41
Trident 3	—	—	3	17	25
Tristar	—	—	—	—	—
	109	88	86	97	81
BOAC					
Standard VC10	12	11	11	11	11
Super VC10	16	17	16	16	16
Boeing 707.336	5	8	9	11	11
Boeing 707.436	18	18	18	16	11
Boeing 747	—	—	6	11	12
	51	54	60	65	61

The appeal hereditament comprises three principal hangar blocks each with ancillary buildings, a main office block on ground and six upper floors, a computer and training building on ground and two upper floors and two cafeteria blocks. There are about 10 acres of land used for car parking and most of the rest of the site consists of concreted areas used as aprons giving access to the hangars.

The west block was the first to be built and was completed in 1954. It comprises 10 hangar bays with a clear height of 43 ft, together with some workshops of the same height and ancillary workshops having a height varying between 12 and 15 ft. The east block came next and was completed in 1960. It comprises 10 hangar bays in two banks of five with a clear height of 50 ft and between them workshops with a clear height of 30 ft, running their full length and across the southern end of the block. Both these blocks have small ancillary areas used for storage, cloakrooms and so on. Finally, the most modern hangar block, the servicing hangar, was completed in 1971. This has a clear height over most of its area of 70 ft, a small area with headroom of 43 ft and one or two ancillary plant rooms. The servicing hangar was built immediately adjoining the north end of the west block and it has had the effect of preventing access by aircraft to hangar bay 1 of the west block and to some extent impedes access to hangar bay 2 because of the need to share a common area of apron.

The total floor area comprised in the hereditament amounts to some 119,833 m². Of this total the 10 hangar bays in the west block provide about 20,000 m², the 10 hangar bays in the east block about 22,000 m² and the servicing hangar about 8,000 m². The total area used for car parks is rather more than 4 ha (about 10 acres) and the apron area comes to 9.239 ha (some 23 acres).

On April 1 1973 the assessments in the two valuation lists (Hounslow and Hillingdon) in respect of the appeal hereditament totalled net annual value £2,120,900. The valuation officer's proposals of March 28 1974 proposed an increase in both rating areas with a total net annual value of £2,189,750. The local valuation court, sitting on March 15 and 17 1976 and giving their decision, as I have said, on May 13 1976, determined an assessment in respect of all six proposals, producing a total of net annual value £1,872,411.

The British Airways Board appeal against this decision and the valuation of the whole put forward on their behalf by their expert witness is net annual value £1,250,000. The respondent valuation officer defends the decision of the local valuation court.

The issue between the parties is primarily one of valuation but certain matters were put to me by counsel on either side as matters of law relating to the correct approach to be adopted by the valuer. I can best sum up the arguments in this way. For the appellants Mr Widdicombe called Hector Wilks BSc (Est Man) FRICS, a distinguished rating surveyor who has acted for BEA and BOAC since 1953 and has been concerned with their rating assessments at Heathrow since that date. He also acts for many of the big companies operating in the aircraft industry.

It is urged on me that the best evidence to support Mr Wilks' valuation is to be found, firstly, in the rent of three hangar buildings at no 2 maintenance area occupied by a firm known as Field Aircraft Services (FAS) and used for the maintenance of executive jets and other smaller types of aircraft; and secondly, the rents of three hangar buildings—referred to as hangars 2, 3 and 4—occupied by the former BOAC as an extension of their maintenance base at no 1 maintenance area and used primarily for the maintenance of road vehicles.

The appellants say that it is quite wrong in seeking to arrive at the net annual value of the appeal hereditament, which is a maintenance base comprising, in the main, hangars and workshops with ancillary accommodation, to have regard to either the rents or the assessments of other properties within the airport perimeter used for other purposes—for example, the warehouses in the cargo terminal, offices in the central terminal area and small parcels of land on the north side of the airport used for car parking. It is wrong, firstly, they say, because warehouse values provide no true indication of industrial values and, secondly, because the values attributed to comparatively small properties used, for example, as offices do not give any indication of the value of offices forming an integral part of a large composite hereditament such as the appeal hereditament.

Appellants' primary valuation

In his primary valuation Mr Wilks values the different buildings of the appeal hereditament on the traditional basis at so much per m² at prices which, in his opinion, are supported by the evidence of his two comparables.

In arriving at his basic values Mr Wilks takes a number of factors into account. He starts with a devaluation of the rents of his two comparables. In the case of the FAS buildings the lease was dated March 8 1971 but rent was payable from September 1 1970. The tenant was responsible for inside repairs, and in order to convert this into terms of net annual value, that is to say with the tenant responsible for all repairs, Mr Wilks makes a deduction of 10 per cent. On this basis the rent devalues at £7.50 per m² in respect of the main hangar accommodation. Making the more normal adjustment of 5 per cent for internal repairs this figure becomes £8.70 per m². In the case of hangars 2, 3 and 4, where the leases run from March 9 1970, Mr Wilks arrives at basic values for the hangar space varying between £7.10 and £7.41 per m², making what he considers to be a realistic adjustment for the repairs liability. Making the traditional 5 per cent adjustment, the values vary between £7.90 for hangar no 2 and £8.24 for hangar no 4. The two lettings, says Mr Wilks, satisfy the four criteria which he puts forward as the basis for comparability:

1. The property should be in similar economic circumstances.

2. The function should be similar.

3. Its physical qualities should be broadly similar.

4. The dates when the rents are agreed should be fairly close to the relevant date of valuation.

In the light of these comparisons Mr Wilks attaches a value of £9.50 per m² to the hangars in the west block of the appeal hereditament, £11.50 per m² to the hangars in the east block and £14.50 per m² to the servicing hangar. The value of these three main blocks, including their ancillary workshops, accounts for 72.1 per cent of Mr Wilks' valuation of the whole. The corresponding proportion extracted from the valuation of the respondent valuation officer is 72.6 per cent, so that there is no significant difference between the two valuers on the relative values of these parts of the hereditament to the whole.

In arriving at his values Mr Wilks is influenced by what he considers to be the general circumstances affecting aircraft maintenance at the end of 1971. It is at this date, said Mr Wilks, that the valuation officer was in a position to analyse the rental evidence available in the course of preparing the new valuation list which came into force some 15 months later on April 1 1973. At the end of 1971, says Mr Wilks, there was a period of uncertainty affecting civil aviation. BEA had been required by the Government to hand over certain profitable routes. The aircraft were engaged in fewer flying hours. That led to a lessening in the demand for aircraft maintenance and this coincided with a time when there was a reduction in income.

The evidence of Mr Doggett MRAeS, who is the planning manager to the engineering department of the board (formerly BEA), indicated that with the increase in the size of aircraft it was becoming increasingly difficult to work conveniently with the hangars in the east and west blocks. The airline, in 1969, had ordered more Trident 3s and the decision was then taken to build the new servicing hangar and, in arranging for a clear height of 70 ft, BEA were looking ahead to the TriStar, the Douglas DC10 and other similar types of aircraft. It was also true that with the improvements in aircraft design there was a reduction in the frequency in major servicing of aircraft and between 1969 and 1971 there was a 35 per cent improvement in the length of time during which aircraft were kept on the ground. For example, in 1969 the Trident 1 and the Trident 3 required major overhaul after 3,000 flying hours whereas in 1971 this had been extended to 5,000 hours.

There was thus, said Mr Wilks, an increase in the amount of available hangar space at a time when the demand for hangar space was falling. As a result five of the hangars in the west block were, in fact, used as workshops and consequently their height was if anything a disability. Mr Wilks also considered that the decision of the Government, taken in April 1971, to adopt the minority recommendation of the Roskill inquiry, namely, to develop a fourth London airport at Maplin, led to a reduction in the value of the various installations at Heathrow. In support of this decision, Mr Wilks drew attention to a number of rents entered into at the end of 1971 which, in his opinion, indicated a fall in values at that time.

In support of his rental valuation Mr Wilks put forward a valuation on the contractors' basis. He started with an estimated capital cost of £24,715,582. This was agreed by the valuation officer to be the appropriate figure based on costs in the fourth quarter of 1971. He then converted this to effective capital value by making deductions of 20 per cent in respect of the east block and about 40 per cent overall (in varying amounts on the different parts) in the west block. He then added figures for plant and machinery, roads, fencing and surfacing of car parks and aprons and arrived at a total effective capital value of £19,581,809. This he took at 5 per cent to produce a net annual value of £979,090. He made no specific addition in respect of the fees on the building costs because, in his opinion, such an amount was offset by an allowance for the disabilities which arose from the aggregation of the different blocks into one composite whole. Finally, it was necessary to add for the land. Mr Wilks had no information which led him to an estimate of effective capital value of the land forming part of the hereditament. He took instead an annual value of £3,250 per acre, making a total of £273,000 and producing a total net annual value of £1,252,090, say £1,250,000. The annual value of £3,250 per acre was derived by Mr Wilks from the actual rent paid by the ratepayers in respect of certain parcels of land taken by them on lease from the British Airports Authority. The land consisted of six or eight small parcels around the perimeter of the hereditament, incorporated within a new subsidiary lease which came into force on April 1 1971 as a result of negotiations in 1970. At the same time there were certain small areas of taxiway leased to the ratepayers by the British Airways Authority at £200 per acre, it being understood that no development was possible on that land. In his valuation, Mr Wilks took the whole of the 84 acres at £3,250 per acre.

VO's rental valuation

The respondent valuation officer is D W Ridgeon ARICS. His valuation is made on the rental basis in which he values the west block hangars at £13.50 per m², the east block hangars at £15 per m² and the servicing hangar at £19 per m². He derives these values from a schedule giving devaluations of agreed assessments of eight hereditaments, five within the cargo terminal area, including the main British Airways cargo terminal, where the agreed assessment is gross value £1,031,000; two small properties adjoining the appeal hereditament; and the British Airways offices in the central terminal area.

In addition the valuation officer puts in a schedule analysing the rents of some 35 other hereditaments, including 11 in the cargo terminal area, seven on the north side and 15 in the central terminal area. He also devalues the rents of the FAS hangars payable under three consecutive leases showing £8.70 per m² for the hangars under the September 1970 agreement, £10.40 under an agreement operating from January 1974 and £22.60 under an agreement running from February 1977. Lastly, he devalues the rents of hangars 2, 3 and 4, in the occupation of BOAC, producing the same figures as those produced by Mr Wilks after making a 5 per cent adjustment for repairs.

Mr Ridgeon considers that in the light of all this evidence the values to be attributed to the appeal hereditament can be clearly established both by relation to the rental evidence and to the established tone of the valuation list. In these circumstances he does not consider that any useful purpose can be served by looking at a contractors' basis valuation. If Mr Wilks' valuation on that basis is to be looked at he criticises it on three principal grounds. Firstly, he does not accept the deductions from cost to value put forward by Mr Wilks, although, as I understand him, he does not suggest that no adjustment is called for. Secondly, he says that in a valuation on this basis the value of the land must be included in the effective capital value of the whole as the value of land freehold with vacant possession, and that it is wrong to introduce land value by way of a ground rent. Finally, Mr Ridgeon says that in any case the land values applied by Mr Wilks are quite inappropriate. In Mr Ridgeon's opinion, the land forming part of the appeal hereditament has a capital value of £225,000 per acre which, at 5 per cent, would produce overall a net annual value of £940,000 instead of £273,000 as suggested by Mr Wilks.

Both valuers produced a large number of supporting documents, about 50 in all, from which they sought to support their opinion of value and reveal what they considered to be flaws in the valuation put forward on the other side. For example, there was an attempted analysis of the rents and agreed assessments of some of the valuation officer's comparables related to agreed construction costs. From these the valuation officer attempted to show that the actual rate per cent to be applied to the adjusted construction costs was substantially more than 5 per cent to net annual value or 6 per cent to gross value as suggested by Mr Wilks. In the alternative, if the rate per cent was calculated on that basis then the land values were considerably higher than those suggested by Mr Wilks. For his part Mr Wilks set out at some length criticisms of the valuation officer's devaluation of rents and agreed assessments on his main supporting schedule and in particular challenged the valuation officer's estimate of land values.

I realise that this summary of the valuation evidence does less than justice to the meticulous care and skill shown by the two expert witnesses in the presentation of their evidence to me during the course of a hearing which lasted for some 16 days. I have found the case a most difficult one partly because of the wealth of evidence, often conflicting, and also because there is such a wide divergence between the valuations put forward by two experienced valuers each of whom clearly holds an honest opinion on the matters under review. The difference between the two sides is wide, the valuation of Mr Wilks being net annual value £1,250,000 and that of Mr Ridgeon being net annual value £1,872,411. A difference of this magnitude can sometimes be attributed to a different approach in law but in the present case I do not think that explanation can apply.

Matters of law

I will deal first with the matters of law which were put to me by counsel. Mr Wilks, in giving his evidence, said that he derived no assistance from the valuation officer's schedule of comparables. So far as the rented properties were concerned, this was because they related to small, separate occupations and in arriving at the value of, for example, the offices forming part of the whole hereditament, the rents paid for other offices at London Airport separately occupied as offices ought not to be taken into account. Mr Wilks referred to a passage in the decision of the Lands Tribunal in *British Home Stores Ltd v County Borough of Brighton and Burton (VO)* (1958) 172 EG 223 at p 224:

> What is being ascertained is the value of the hereditament as a whole, and the values of basements and upper floors should be fixed bearing in mind the values of the ground floor. The rents which they would fetch if separately occupied is a matter of no moment; it is not the value as offices or whatever may be the alternative type of occupation that is being sought, it is the value as part of the whole.

In that passage it seems to me that the tribunal is enunciating a principle of valuation rather than a proposition of law. But in any case the valuation officer is not here seeking to value the parts of the whole as if they were separate hereditaments. I agree with Sir Derek Walker-Smith. He argues that the whole hereditament has to be valued as one but it is unrealistic to assume that the hypothetical landlord and the hypothetical tenant would exclude from their minds what they know about the levels of values in the locality for different classes of hereditaments. I find that the evidence of the comparables put forward by the valuation officer is admissible evidence and the question to be decided is the weight to be attached to it.

Mr Wilks challenged also the relevance of warehouse rents when considering the value of industrial properties, more specifically the agreed values of the warehouses in the cargo terminal. During the course of the argument there was a small excursion into the field of town planning law. Here again I consider the question to be one of fact, namely, to what extent, if at all, would the parties to the negotiations, that is the hypothetical landlord and the hypothetical tenant of the appeal hereditament, be influenced by the levels of rent deemed to be established, by reference to the agreed assessments in the cargo terminal. Here again I think Mr Wilks is wrong to exclude that evidence from his mind.

Date of valuation

The remaining matter which was put before me as a question of law is the date of valuation. As I have already indicated, Mr Wilks detected what he considered to be a reduction in values in London Airport as a whole and particularly in the no 1 maintenance area at the end of 1971. It was at that date that the valuation officer had to "bring down the shutters" on the collection of evidence needed for the preparation of the valuation list which came into force on April 1 1973. That evidence established the tone of the valuation list and any valuations supported by higher rents which came to hand at a later date were above the established tone of the valuation list and therefore incorrect. This principle was also applied by Mr Wilks in his valuation on the contractors, basis, where he worked from a starting point of building costs in the fourth quarter of 1971.

Mr Widdicombe submitted that the approach adopted by Mr Wilks in his valuation was the correct one. In his submission it was certainly proper for the valuation officer, in preparing his valuation list, to project forward to April 1 1973, that is to the date on which the valuation list came into force, but such projection must be related to the evidence available to him at the end of 1971, since information collected after that date could not have been taken into account. The valuation officer would have no knowledge of it in preparing his valuations. There was, thus, an established level of value, in other words the tone of the valuation list. In Mr Widdicombe's submission the law required that there should be fairness in the assessments of all hereditaments for rating purposes.

"Ladies' Hosiery" Case

Sir Derek Walker-Smith referred to the well-known case of *Ladies' Hosiery & Underwear Ltd v West Middlesex Assessment Committee* [1932] 2 KB 679, but Mr Widdicombe said that this case was to be considered as an aberration in that it led inevitably to unfairness in the assessment of different hereditaments within the same valuation list. Thus, following that case uniformity of assessment was sacrificed at a time of rising values where, if an assessment could be shown to be correct in the light of rental evidence available at the date of valuation, it was not possible to hold that it was incorrect merely because it was on a higher level than all other similar hereditaments in the valuation list. This was the mischief, said Mr Widdicombe, which section 17 of the Local Government Act 1966 (now section 20 of the General Rate Act 1967) was designed to remedy. So section 20, which applied in the present case, should be construed in such a way as to ensure uniformity and that meant valuations were to be made by reference to the level of values at the date when the valuation officer closed his books. The evidence was that values were rising until 1971 but that they levelled off at the end of that year and did not start rising again until some time in 1972. At the key date, therefore, the valuation officer must have known that the rise in values had stopped. This was borne out, said Mr Widdicombe, by the level of agreed assessments.

For the valuation officer, Sir Derek Walker-Smith submitted that the end of 1971 was a date which had no relevance in law. April 1 1973 was the date when the list took effect and it is the value at that date which is the correct value to be entered in the list. If that value can be satisfactorily established by admissible evidence it cannot be overridden in the alleged interest of parity or the tone of the list. So, said Sir Derek, applying the correct principle of law the effect is that evidence of rents available before April 1 1973, albeit that it was not available at the end of 1971, is admissible and relevant without qualification. Similarly, said Sir Derek, evidence of rents after April 1 1973 can also be admitted in the same way or at least be adduced to confirm the projection of a rising trend commencing before 1971 and continuing up to the first quarter of 1973.

In my judgment the truth of the matter lies somewhere between these two positions. In the *Ladies' Hosiery* case the evidence was that the rent which might be expected for the appeal hereditament under the statutory terms was at least £325, which was the gross value determined by the assessment committee. At p 686 of the report Scrutton LJ said:

One would have thought that this evidence destroyed the ratepayer's objection, but he said: "No, I will show you that seven other hereditaments in the borough are assessed at sums lower than the rents which their hypothetical tenants would pay, and I require that I also should be assessed at a sum lower than the hypothetical tenant would pay. I do this because the essence of rating is fairness and uniformity, and I prefer my assessment to be inaccurate but uniform rather than it should be accurate but out of harmony with my neighbours. It is true that I can, if my facts are right, secure uniformity by correcting the inaccuracy of my neighbours' assessments, on objection to those assessments, and can get those assessments made uniform with my correct assessment. But I do not want this; I want to indulge my passion for uniformity by securing uniform inaccuracy, but the inaccuracy is to my pecuniary advantage." As I said during the argument, this looks all wrong, and I think it is a wrong view.

At p 688 Scrutton LJ continued:

The appellants here, however, say that besides the principle of independent valuation, there is another vital principle: that as between different classes of hereditaments, and as between different hereditaments in the same class, the valuation should be fair and equal. I agree, but in my view there is a third important qualification, that the assessing authority should not sacrifice correctness to ensure uniformity, but, if possible, obtain uniformity by correcting inaccuracies rather than by making an inaccurate assessment in order to secure uniform error.

In my opinion nothing contained in section 20 of the Act can be taken to overrule that dictum of Scrutton LJ, which is still good law. The 1967 Act provides as follows:

20(1) For the purposes of any alteration of a valuation list to be made under Part V of this Act in respect of a hereditament in pursuance of a proposal, the value or altered value to be ascribed to the hereditament under section 19 of this Act shall not exceed the value which would have been ascribed thereto in that list if the hereditament had been subsisting throughout the year before that in which the valuation list came into force, on the assumptions that at the time by reference to which that value would have been ascertained—

(a) the hereditament was in the same state as at the time of valuation and any relevant factors (as defined by subsection (2) of this section) were those subsisting at the last mentioned time; and

(b) the locality in which the hereditament is situated was in the same state, so far as concerns the other premises situated in that locality and the occupation and use of those premises, the transport services and other facilities available in the locality, and other matters affecting the amenities of the locality as at the time of valuation."

There is nothing in this subsection which can justify the insertion of an incorrect assessment in the valuation list merely on the grounds of uniformity or fairness. What it does do is to redefine the way in which correctness is to be determined so that the effects of inflation are to be disregarded. The level of values to be taken as correct throughout the life of a valuation list are those which can be shown to be correct at the date when the valuation list comes into force. These provisions therefore have the effect of eliminating unfairness during a period when values are changing from year to year and valuation lists remain in force for considerably longer than the five-year period which Parliament originally intended. As I said in *Boddington v Sture (VO)* (1970) 17 RRC 112 at p 116:

I take the view that in section 20 the words "the value which would have been ascribed" can only mean the value which would properly have been ascribed; and I do not see that the words of the statute can be invoked in order to give authority to an incorrect value merely because it was in fact the value ascribed.

The latest proposal in the instant case was made on March 28 1974. It is therefore a proposal which is covered by section 20. It is not suggested by anyone that the value at that date is lower than the value defined in section 20, so that what has to be determined is the value which would have been ascribed to the hereditament if it had been subsisting throughout the year from April 1 1972 to March 31 1973. It is common ground that the value to be ascertained is the value at April 1 1973 and that the valuation officer can quite properly project forward to that date in the light of the evidence available to him at the time when collection of evidence comes to an end.

"Peachey" Case

At this point it is clearly necessary to have regard to the realities of the situation. The whole question of the preparation of a valuation list was considered in great detail in *R v Paddington (Valuation Officer) ex parte Peachey Property Corporation Ltd* in the Divisional Court reported at [1964] 1 WLR 1186. Widgery J, as he then was, said at p 1205:

Both parties have assumed that the valuation list should reflect the values prevailing when it takes effect, and though there is no express statutory direction that he should do so, this is highly desirable if a spate of proposals to amend the list is to be avoided. It is, however, to be noted that there is no express duty upon a valuation officer to anticipate general changes in rental levels which may occur between the compilation of the list and its taking effect. There is, in any case, no need to anticipate changes in the value of money since these will not affect the relative value of the various hereditaments one to another.

In the opinion of the court a valuation officer should have regard to the desirability of his list being up to date when it takes effect. He should postpone the collection of rental evidence and the actual compilation of the list for as long as is reasonably possible having regard to the work to be done, but the actual timetable to be adopted is a matter for his own discretion if he acts in good faith. Having thus compiled his list on the material then available he may give effect to anticipated future trends if his skill and experience tell him that these trends are sufficiently established, but this question is one of professional opinion and his decision will rarely be open to challenge as being contrary to the law. He should hesitate to disturb the fairness and equality of his original list by giving effect to evidence of altered value which comes to him too late to be reflected in the list as a whole.

The finding of the Divisional Court was upheld in the Court of Appeal.

"All the evidence available"

There is thus the clearest indication that a valuation officer, in carrying out his statutory duty, is to have regard to all the evidence available to him at the last possible date before he settles the levels of value in the new valuation list. Inevitably

A this must be at least 12 months before the list can come into effect; he can nevertheless take account of anticipated future trends in so far as they appear to him to be well established; and having reached a conclusion on those matters, he then produces his valuation list in which the levels of value are correct for the purposes of that list, unless it can be shown that there is some error *ex facie* the list. Any alteration in levels of value is not to be undertaken by him unless it can be done without disturbing the correctness of the list as a whole.

This approach is supported by the provisions of section 68 of the Act (as amended by the General Rate Act 1975) which B reads as follows:

68(1) In the case of each rating area, new valuation lists shall be prepared and made by the valuation officer so as to come into force on April 1 in 1973, 1980 [now 1982] and each fifth year thereafter.

(2) The valuation officer by whom a new valuation list is prepared shall, not later than the end of the month of December preceding the date on which the list is to come into force (or if in any particular case the Minister, either before or after the end of that month, allows an extended period, then not later than the end of that period) sign the list and transmit it, together with a copy thereof, to the rating authority, who shall deposit the list at the C offices of the authority.

(3) Where, after the valuation officer has transmitted the list to the rating authority, but before the date on which the list is to come into force, it appears to him that, by reason of a material change of circumstances which has occurred since the time of valuation, the list needs to be altered in any respect, he shall cause the list to be altered accordingly before that date.

(4) In subsection (3) of this section, the expression "material change of circumstances" means a change of circumstances which consists of—

(a) the coming into occupation of a newly erected or newly constructed hereditament or of a hereditament which has been D out of occupation on account of structural alterations; or

(b) a change in the value of a hereditament caused by the making of structural alterations or by the total or partial destruction of any building or other erection by fire or any other physical cause; or

(c) the happening of any event whereby a hereditament or part of a hereditament becomes, or ceases to be, not liable to be rated; or

(d) a change in the extent to which any railway or canal premises within the meaning of section 32 of this Act are occupied for non-rateable purposes within the meaning of that section; or

(e) property previously rated as a single hereditament becoming E liable to be rated in parts; or

(f) property previously rated in parts becoming liable to be rated as a single hereditament; or

(g) a hereditament becoming or ceasing to be—

(i) a dwelling-house; or

(ii) a private garage or private storage premises within the meaning of Schedule 11 to this Act; or

(h) a hereditament being, in accordance with Schedule 13 to this Act, used to a greater or lesser extent for the purposes of a private dwelling or private dwellings,

and the expression "the time of valuation," in relation to a change of circumstances, means the time by reference to which the valu- F ation officer prepared so much of the list as is affected by that change of circumstances.

(5) The omission from a new valuation list of any matter required by law to be included therein shall not of itself render the list invalid.

(6) In respect of any new valuation list, it shall be the duty of the rating authority immediately upon receipt thereof to take such steps as the authority may consider most suitable for giving notice of the list, and of the rights of persons to inspect the list and to make proposals for altering it.

Subsections (3) and (4) clearly define the type of alteration which is to be made at a late stage. A change in the level of values is not a material change of circumstances for this purpose. G I think therefore that when it is said that the value to be ascribed in section 20 is the value properly to be ascribed this must be taken as meaning properly ascribed in the light of evidence available to the valuation officer and in the proper exercise of his expert opinion. The position is clearly one of some difficulty and it is significant that section 20 does not say that the ceiling value is to be the value of the hereditament on the day when the valuation list came into force. In using the expression "subsisting throughout the year" the draftsman does not entirely remove the uncertainty as to the correct level of value to be applied—particularly at a time when values are subject to rapid change. Indeed it would be H helpful if at some time Parliament could consider the possibility of substituting a precise date as to the date of valuation for a new valuation list, for example a date 12 months before the date on which the valuation list comes into force, when the evidence available to the valuation officer could be accurately assessed and could be generally understood.

The evidence in the present case is that the valuation list was despatched in accordance with the requirements of the statute to the proper rating authorities before the end of December 1972 and before that date the final preparation of the list and the typing took about 12 months. This was very much in line with the timetable referred to in the *Peachey* J case. It was therefore clearly physically impossible for the valuation officer to take account of changes in the level of value which came to light much after the end of 1971.

Level of values adopted

I am satisfied, nevertheless, that the valuation officer made some modest projection forward to April 1 1973 from the level of rents which were available to him for analysis. I do not accept that the level he adopted and applied represents the level of values at the end of 1971. Indeed, from an examination of all the rental evidence put before me it seems to me that the level of values finally adopted accords broadly K with the levels of mid-1972. There are, of course, variations and in no collection of individual rents is it possible to find a precise and uniform pattern. But I am not persuaded by the evidence of Mr Wilks that there was any marked recession in values at the end of 1971, although there are one or two rents which can be interpreted in this way.

The evidence regarding the possibility of a fourth London airport at Maplin does not, in my judgment, suggest any threat to the scale of operations at Heathrow but rather lends support to the contrary view that the time was coming when no further increase in traffic at Heathrow could be absorbed and that further provision was likely to be necessary else- L where. But in any case even if there had been such a "hiccough," as it was referred to, in the steady increase in rental values at that time I can see no sign that it was generally reflected in the actual assessments nor can I see any ground for incorporating it in the assessments in the valuation list coming into force some 15 months later, when values were clearly continuing to rise.

It is true that, owing to developments in the size and range of aircraft, some of the hangars forming part of the appeal hereditaments were becoming less suitable for aircraft maintenance. This problem is, of course, to be overcome in an M examination of the values of the constituent parts of the whole, but it does not justify Mr Wilks' opinion that in the negotiations between a hypothetical landlord and a hypothetical tenant the advantage lay with the tenant.

Buildings in other uses

Both valuers are agreed that there is a high plateau of values of land within the perimeter of Heathrow. This seems to me to be clearly established by an examination of the rents put before me. As I have indicated, I do not reject the evidence of the rents of the warehouses in the cargo terminal,

A and, indeed, in the rather unusual circumstances existing within the airport, I think that it is quite appropriate in arriving at the values on the hangars to look at the values settled and determined on the warehouses. This view is rather borne out by Mr Wilks' evidence. I find, for example, that one of the FAS hangars was described on the plan put in as being used as a warehouse and in his evidence relating to hangars 2, 3 and 4 he said words to this effect:

> The British Airports Authority say to BOAC—here is the accommodation. The rent is negotiated on the basis that the accommodation is to be used for the most valuable purpose which BAA would permit. If you use it for a less valuable purpose that is up B to you.

It is true in that case that BOAC wish to use the hangars for the maintenance of road vehicles, but Mr Wilks went on to say that he would be very surprised if the user clause in the lease had any bearing on the rent. Within very small limits he considered that the rents were the same as hangar rents and this certainly suggests to me some uniformity of values relating to the nature of the building rather than the actual use to which it is put. The hangars and workshops in the instant case are of a very high quality and materially better than the FAS hangars and the BOAC hangars 2, 3 and 4. Having inspected all the comparables except those in the C central terminal area I am not persuaded that there is anything wrong with the general level of values determined by the local valuation court, as now analysed and put forward by the valuation officer. I should record that the valuation officer's original figures which, as I have already said, totalled net annual value £2,189,750, were amended by him at the local valuation court, where he put forward revised figures of £2,080,800. The local valuation court made further reductions on the east and west block hangars and so arrived at their decision, £1,872,411. The detailed valuation now put forward by the valuation officer comes to that figure.

I can summarise my findings in this way. From the evidence D it is clear that, as the two valuers are agreed, the levels of value within the London airport are on a high plateau. I look at the evidence of the rents and in particular the rents of the FAS hangars and hangars 2, 3 and 4 and take into account the dates upon which they were agreed, the area of land occupied with them and reflected in the rents and the general quality of the buildings. The other rents quoted by the valuation officer are of much smaller units and are related only to parts of the appeal hereditament. In a sense this observation is true also of the FAS hangars and of hangars 2, 3 and 4 except that the parts to which they can be directly compared form a much bigger proportion of the whole. This E other rental evidence I regard as useful as background material.

I look also at the agreed assessments particularly those at the British Airways cargo terminal, which is another large hereditament with an agreed assessment of gross value £1,031,000. I do not think Mr Wilks is right to ignore these assessments, although I do not suggest that the two occupations are in any way interchangeable. At the same time I do not accept that aircraft maintenance was in such a poor state in the year before that in which the valuation list came into force that the hypothetical tenant would be able to persuade the landlord to depart from the general level of values attributable to modern purpose-built buildings of high quality.

F In short, the broad picture which emerges from a study of the rents and agreed assessments confirms the general level of values applied by the valuation officer and in particular his values per m² of £19 on the new servicing hangar, £15 on the hangar bays in the east block and £13.50 on the principal hangar bays in the west block.

Reductions for certain units

There are three areas, however, where I think some reduction is to be made. The valuation officer has applied a price of £16,000 per hectare for the area of car parks amounting to 4.11 ha and a figure of £22,500 per hectare for 5.03 ha of excess apron area. ("Excess apron area" is a term used by G both valuers to describe concreted areas of land, with thickness of concrete varying between 8 in and 20 in, in excess of the area which is equivalent to the total floor areas of the hangars—and which is deemed to be included in the values attributable to the hangars themselves.) These unit prices he seeks to justify in his analysis of car-parking and open-storage land in some of the areas at the cargo terminal and from two small areas of open car park on the north side of the airport. Mr Wilks challenges the valuation officer's analysis of the cargo terminal assessment and in particular claims that the values attributed to the car park are excessive. For his part, he refers to ground rents paid which show prices per hectare of about H £8,000 in the maintenance area in 1970, rather more than £13,500 for the cargo area in 1968 and about £16,000 for the north side in 1971. The north side figure is apparently derived from small car park lettings. The figure for the maintenance area comes from the price paid by BEA for the parcels of land taken around the perimeter of the hereditament.

Without accepting these prices as representing the correct differentials between the three areas, I nevertheless agree with Mr Wilks that the valuation officer's figures for the car parks and aprons are too high. At the same time I think that the ground rent of £8,000 per hectare derived from the 1970 rents for the parcels of land around the perimeter of the hereditament J does not adequately reflect the value of the land upon which the whole of the maintenance base stands.

I can see no justification for the values adopted by Mr Wilks, of £3,500 per hectare for the car parks and £5,000 per hectare for the excess apron area, values which are to be taken to reflect not only the value of the land but also the value of the surfacing. In my opinion, the values to be properly applied are £10,000 per hectare for the car parks and £14,000 per hectare for the excess apron areas. This leads to a reduction in the valuation officer's figures of net annual value £35,360 in Hillingdon and £32,055 in Hounslow.

Secondly, I think that the valuation officer has failed K adequately to recognise that, with the extension of the hereditament by the building of the servicing hangar in order to meet the requirements of the new generation of aircraft coming along in 1970, some parts of the existing hangars have lost a certain amount of value, partly because they are surplus to the requirements of the taxpayers and would, I think, be surplus to the requirements of any hypothetical tenant. The use of parts of the west block is also physically impeded by the new hangar and I think some further allowance is called for than that adopted by the valuation officer. In my opinion, hanger bays 6 to 10 and hangar bay 1 should be valued at £12 per m². This adjustment leads to a further reduction in L the Hillingdon net annual value of £16,985.

Finally, I consider some small reductions are appropriate in the values attributed to the engine overhaul offices at the north end of the east block, where the valuation officer has applied values of £20 and £22 per m² and Mr Wilks values of £13. Here I make a reduction of 10 per cent overall, which leads to a further reduction, again in Hillingdon, of £1,074.

Tribunal's views on contractors' basis valuation

Making these adjustments, I arrive at a net annual value in respect of the whole hereditament of £1,786,937. Before reaching a final conclusion it is appropriate that I should look at M the contractors' basis valuation put in by Mr Wilks. I should say straightaway that I regard this approach as being very much a second best, but I do not think it right for the valuer to close his mind to this evidence.

Briefly, I should say that the valuation of Mr Wilks on the contractors' basis does not persuade me that the valuation arrived at, having regard to the evidence of rents and assessments, is too high. I have already indicated that the value of the land on which the whole hereditament stands, taken by Mr Wilks at £3,250 per acre (approximately £8,000 per hectare), cannot, in my opinion, properly be equated with the rent

A payable by BEA for the parcels of land around the perimeter of the base taken in in 1970 or 1971. In the second place the adjustments from 1971 cost to effective capital value made by Mr Wilks in respect of the east block and the west block are, in my opinion, excessive. In any case any surplus hangar accommodation which there may be in the west block only occurs because of the aggregation of the various parts of the hereditament into a single whole. In so far as some reduction is called for in arriving at the effective capital value of the west block, then I think that Mr Wilks is duplicating his allowances when he makes an end-allowance for disabilities. It is, therefore, not correct to ignore an addition of 10 per B cent for fees in arriving at total cost.

At the end of the day the contractors' basis is merely a means of arriving at the rent which the tenant might reasonably be expected to pay. In reality it is not possible for a prospective tenant of a hereditament of this sort to say that in the absence of agreement with his landlord he will make his own arrangements to build the hereditament himself. Quite apart from any other consideration such an alternative hereditament would not be immediately available, and we are concerned with the value at the relevant date with the actual hereditament deemed to be standing there vacant and to let.

It is perhaps more helpful to envisage the hypothetical C tenant asking himself what he would have had to spend in order to provide the hereditament in the state in which it exists at the relevant date. If this is the right approach then the actual date on which the construction costs are calculated is of no more concern as evidence than the date of an actual rent. The hypothetical tenant in the instant case would ask himself D what would be the rent which he might reasonably be expected to pay on April 1 1973 in the light of information available during the preceding 12 months and in the knowledge of the actual costs of construction in the fourth quarter of 1971. It would be at this point that he would have regard to the fifth stage in valuation on the contractors' basis (see *Ryde on Rating*, 13th ed p 519). This would involve an examination of rental evidence available, which, as I have already indicated, in my judgment justifies the general level of values adopted by the valuation officer.

I am, therefore, not prepared to hold that that valuation should be overturned by the valuation on the contractors' E basis put forward by Mr Wilks.

The appeals are accordingly allowed to the extent which I have indicated. The assessment of the whole I determine at net annual value £1,786,937, which is to be apportioned in the following manner: in the London Borough of Hillingdon, net annual value £1,156,946 and in the London Borough of Hounslow, net annual value £629,991. Appropriate orders will be issued directing that the two valuation lists be altered accordingly.

I heard submissions as to costs. The appellants have been successful in their appeals, but a great part of the time at the hearing was spent on matters where I have found against F them. In the circumstances, my order is that the respondent valuation officer shall pay three-quarters of the costs of the appellants, such costs failing agreement to be taxed by the Registrar of the Lands Tribunal on the High Court Scale.

LANDS TRIBUNAL

TAXATION

A CHARLESWORTH AND ANOTHER (TRUSTEES OF THE WILL OF J H STRAKER (DEC'D)) v COMMISSIONERS OF INLAND REVENUE

CHARLESWORTH AND ANOTHER (EXECUTORS OF LADY E H STRAKER-SMITH (DEC'D)) v SAME

(DET/4 and 5/1978)

(W HALL FRICS)

January 11 1980

Estates Gazette March 1 1980 and March 8 1980

(1980) 253 EG 913 and 1020

Values of (i) Northumberland mansion house and grounds at August 1973 and (ii) adjoining tenanted farmland and in-hand
B **amenity woodlands at March 1971—" Dramatic gap " between opposing valuations of £15,000 and £80,000 for mansion—Difference of opinion due to effect of cost of repairs, adaptation and modernisation on price and effect of 1973 property boom on large country houses without agricultural land—Evidence of transactions in comparable houses between 1971 and 1978 —Considerable sum required to modernise subject house—Tribunal not convinced that addition of modernisation cost to price paid for such a house necessarily equates with value—Nor convinced by DV's opinion, not established by factual evidence, that prices of houses of subject type were the same in 1978 as in 1973—Determination of £30,000 for house.**

C **Rent level of farmland—Tribunal rejects DV's view that 1970 rent of £7 an acre was too low and that purchaser would expect higher figure, in 1970 money values, at rent review in 1973—Woodlands timber a shelter belt for arable fields the loss of which could affect rent—Possibility of competition for woodlands between owner of mansion and purchaser of farmland, realising more than nominal or " prairie " value for major part —Finding of £16,000 for land against appellants' £15,300 and DV's £28,000**

Malcolm Spence QC (instructed by Wilkinson, Marshall, Clayton & Gibson, of Newcastle) appeared for the appellants;
D Solicitor of Inland Revenue for the respondents.

Giving his decision, MR HALL said: These two references which I heard together arise because of appeals by the trustees of the will of Joseph H Straker (deceased) and the executors of Lady Edith Helen Straker-Smith against the determination for estate duty purposes of the Commissioners of Inland Revenue that (i) the principal value as at August 2 1973 of the freehold mansion house, gardens, lodge, two maisonettes and surrounding garden and grounds, Carham Hall, Carham, Northumberland, is £80,000 including timber valued at £3,000 and (ii) the principal value as at March 2 1971 of 120.694 acres of land adjoining Carham Hall is £30,000 including timber valued
E at £2,000.

The grounds of appeal are (i) that the principal values in the determination are in all cases excessive and ought to be reduced and (ii) that the value of the 120.694 acres has already been accepted for capital gains tax purposes at £15,300.

The appellants claim that the value of the said property should be:

(1) as to Carham Hall, £15,000 including timber;

(2) as to 120.694 acres, £15,300 including timber, and that the agricultural value for the purposes of section 23 of the Finance Act 1925 of the agricultural property should be £15,300.

The difference in valuation dates is due to Lady Straker-Smith F
disposing of her interest in the 120.694 acres on March 2 1971. The said interest therefore became liable for duty on her death on August 2 1973 within the statutory period.

The basis for the determination of value is contained in section 7(5) of the Finance Act 1894, which states:

> The principal value of any property shall be estimated to be the price which, in the opinion of the Commissioners, such property would fetch if sold in the open market at the time of the death of the deceased.

Where timber is involved this is dealt with in terms of section 61(5) of the Finance (1909-10) Act 1910 as amended by section 9 of the Finance Act 1912, which states: G

> Where an estate, in respect of which estate duty is payable on the death of a person dying on or after the thirtieth day of April nineteen hundred and nine, comprises land on which timber, trees, wood or underwood are growing, the value of such timber, trees, wood or underwood shall not be taken into account in estimating the principal value of the estate or the rate of estate duty, and estate duty shall not be payable thereon, but shall, at the rate due to the principal value of the estate, be payable on the net moneys (if any), after deducting all necessary outgoings since the death of the deceased, which may from time to time be received from the sale of timber, trees or wood when felled or cut during the period which may elapse until the land, on the death of some other person, again becomes liable or would, but H
> for this subsection, have become liable to estate duty, and the owners or trustees of such land shall account for and pay the same accordingly, as and when such moneys are received, with interest at the rate of three per cent per annum from the date when such moneys are received.
>
> Provided that if at any time the timber, trees, or wood are sold, either with or apart from the land on which they are growing, the amount of estate duty on the principal value thereof which, but for this subsection, would have been payable on the death of the deceased after deducting the amount (if any) of estate duty in respect of the timber, trees, or wood under this subsection since that date, shall become payable.

It is necessary in this case for the commercial value of the J
timber to be ascertained at the date of valuation.

As there are in effect two separate properties with two different valuation dates to be dealt with, it is appropriate to deal with them separately.

Carham Hall

Statement of facts

The following is agreed as a description of the house under appeal.

Carham Hall, which was built about 1880, is a substantial stone-built country house containing seven reception rooms, K
four bedroom suites and 11 other bedrooms. In addition there is a kitchen, servants' sitting room and other ancillary accommodation. An oil-fired central heating system is installed. There is a stable cottage which contains a sitting-room, kitchen, two bedrooms and bathroom. There is a lodge which contains a sitting-room, kitchen, three bedrooms and bathroom. Adjacent to the house there are extensive outbuildings containing stables, kennels and garages. The services are mains water and electricity; sewerage is a soakaway system. The main house has a reduced covered area of 18,000 sq ft.

A The property is generally in good order although there are considerable plaster cracks in the bedrooms in the west wing and subsidence indications at the steps in the back yard.

The ground attached to the house extends to 10.629 acres comprising amenity woodland, walled garden, shrubbery, lodge and drive and the house itself with outbuildings, lawns and grounds.

The nearest main railway station to Carham Hall is Berwick upon Tweed, which is about 16 miles away. Edinburgh (50 miles) and Newcastle upon Tyne (65 miles) are the nearest centres of population, each with airports.

The mansion house together with 8.59 acres of adjoining
B land was leased on a full repairing basis, partly furnished, from August 21 1974 for the term of 21 years at the rent of £1,000 for the first seven years, £2,000 for the next seven years and £4,000 for the final seven years.

The commercial valuation of the timber within the Carham Hall property is agreed at £3,000.

The following facts about the comparable houses are agreed.

Coupland Castle, Millfield, Wooler (appellants' comparable) was sold in November 1971 for a price in the region of £20,000. It is a substantial stone-built house, with a peel tower and keep dating back to 1584 forming the older part of the castle. The building is listed as being of special architectural
C and historic interest grade I. It contains four reception rooms and seven bedrooms, two of which have suites. There is a double garage, stables, stores, and two five-room modernised cottages. The ground attached to the house extends to 25.58 acres. The reduced covered area is 16,000 sq ft.

In April 1975 Coupland Castle was sold for a price in the region of £60,000 and in 1978 was again sold, for a price in the region of £130,000. There were extensive repairs and improvements between 1971 and 1978. The castle is located about 10 miles south of Carham.

Eccles House, Coldstream (respondents' comparable) was
D sold in April 1972 for £28,924. It is a substantial stone-built house built in 1893. It contains five reception rooms, four bedroom suites, five bedrooms and a lodge. The reduced covered area is 10,000 sq ft. The ground attached to the house extends to 25 acres of which about 14 acres are paddocks. The view to the south is of the Cheviot Hills.

The house was sold in March 1976 for £71,750, but extensive improvements had been carried out since 1972. The house was rewired and a new central-heating system installed. A stable and garage block had been built. A separate flat in the east wing was created.

Budle Hall, Bamburgh (appellants' comparable) was sold in
E April 1973 for £55,000. It is a stone-built house of late-18th-century Georgian architecture and is listed as being of special architectural and historic interest grade II. It contains four reception rooms and six bedrooms, two of which have dressing rooms. In the north wing there is a kitchen, bedroom and bathroom; in the south wing there is a kitchen, two reception rooms, two bedrooms. There is also a lodge. It has a reduced covered area of 10,000 sq ft. There is ground extending to 4.13 acres. It is located in an area of outstanding natural beauty and is 2 miles from Bamburgh. The distance from Newcastle is 50 miles. The house was in need of renovation.

Harleyburn, Melrose (appellants' comparable) was sold in
F November 1971 for £18,650 in lots. The main house together with 15 acres sold for £8,750. The other separate lots were (a) the stables and a four-roomed cottage, (b) land extending to 8 acres, and (c) a four-roomed lodge cottage. The house was built in 1875 and contains four reception rooms, seven bedrooms and a four-roomed staff flat. It has a reduced covered area of 11,300 sq ft. Oil central heating was installed and the house had been rewired in 1957. Planning permission had been obtained for use classes nos 12, 13, 14, 15 and 17 as contained in the Town and Country Planning (Use Classes) (Scotland) Order 1950. These classes cover hotel and institutional uses. The schedule of particulars showed on a plan tentative proposals for a road line which would have G substantially affected the house. Interested parties were advised to make their own inquiries as to the likelihood of this development taking place. In the event, another line was chosen which does not involve the acquisition of any land belonging to the house. The house and 15 acres were sold in August 1976 for £37,500.

Fowberry Tower, Wooler (respondents' comparable) is on the market at an asking price of £100,000. It has six reception rooms and nine bedrooms. There are two cottages with stabling and garage. The reduced covered area is 14,600 sq ft and the land extends to 33 acres.

Maxton, Newton St Boswells (respondents' comparable) was H sold in 1978 for £72,500. It has seven reception rooms and 10 bedrooms. There is a service flat, a stable cottage, a lodge and three garages. The reduced covered area is 11,000 sq ft and the land extends to 14 acres.

Lennel House, Coldstream (respondents' comparable) was sold in 1979 for £100,000. It has seven public rooms and 13 bedrooms. There are three flats, a lodge and outbuildings. The reduced covered area is 25,000 sq ft and the land extends to 7 acres.

Evidence for the appellants

J John M Clark MA JP FRICS gave evidence for the appellants. He qualified as a chartered surveyor in 1962. He is a partner in J M Clark & Partners, chartered surveyors and land agents, who have offices in Newcastle, Hexham, Carlisle, Haltwhistle and elsewhere. He is valuer to the Agricultural Mortgage Corporation. He lives at Featherstone Castle, Haltwhistle, which has about 60 rooms.

He had accompanied his father, Col John Clark, at an inspection made on August 29 1973. Col Clark, who was senior partner of J M Clark & Partners, was president of the Land Agents' Society in 1957 and president of the Royal Institution of Chartered Surveyors in 1969. Col Clark valued Carham Hall K at £15,000 as at August 2 1973, and his signed report of that time was submitted as an exhibit in the case. Col Clark died in 1977.

John Clark said that, in his opinion also, the value of Carham Hall as at August 2 1973 was £15,000.

He had valued the property as a private residence as he considered there was no demand for it for institutional purposes due to its location. He did not consider it practicable or economic to divide the property into flats or separate houses, especially for letting. The property was generally in good order at 1973 but the maintenance costs would be high mainly due to the complex roof structure. There was cracking in the plaster in some bedrooms and subsidence in the back yard and L he considered these "would make a potential purchaser nervous." From his own experience of living in a large house he estimated that heating would cost between £2,000 and £2,500 a year in 1973. The exceptional length of Carham Hall was a disadvantage in this respect compared with a more compact squarish house. If a purchaser bought the property he would have to try to reduce the unit and so the central heating would have to be altered. He considered that the house needed rewiring.

There was no view of the Tweed from the house and to the south one looked into trees because of the rising ground. This southern view, however, was in his opinion not unattractive. M

Most people wanting a house of this type in Northumberland would have inquired at his offices, but in 1973 they had no inquiries for this type of house. He agreed that the price of agricultural land and houses connected with agriculture increased dramatically between 1972 and 1973, but this increase did not apply to large houses which had no agricultural land attached to them. He had little experience in dealing with the smaller urban houses although he accepted that there was an upsurge in prices in that period. He considered, however, that the general increase in house prices did not apply to the type of house being dealt with in this appeal. In the late 1960s he said that it was difficult to sell a house like Carham at all, so that when Coupland Castle was sold in November 1971 for a price in the region of £20,000 it was " exciting news for agents " involved in that type of property.

There was little evidence to indicate what increase in price was likely for the large country house without agricultural land between 1971 and 1973. He considered the increase marginal because of the political and economic climate towards the autumn of 1973. People would not want to commit themselves to the substantial expense of running such large houses if there was a possibility of a recession in the country.

He had placed before the tribunal all the transactions of large houses which he knew about. He had considered and examined them to determine whether they supported his valuation.

(a) *Coupland Castle*

He considered that the location of the property was superior to Carham. It was on the edge of an area of outstanding natural beauty with a good view to the south. The castle was 2,000 sq ft smaller and easier to live in. The grounds were spacious with more than one access, which made it possible to convert the stables into flats or sell off one of the cottages without loss of privacy. He did not dispute that it was in poor condition in 1971 and indicated that the property would have fetched £24,000 if on the market in 1973.

(b) *Eccles House*

He said that this was not his comparable and had been put forward by the respondents. He drew attention, however, to the fact that it was about one half of the size of Carham and more adaptable with scope for division.

He estimated that the price of £28,924 included a sum of £7,000 for the land, which £7,000 was represented by 14 acres at £500 per acre. The net price for the house was therefore about £22,000. He did not consider that the sale price was much influenced by the fact that substantial improvements had to be carried out after the sale. In his view the improvements made the house easier to run.

(c) *Budle Hall*

Mr Clark said his firm had acted as agents during the sale and that there had been two serious offers for the property. One offer was from a farmer with large agricultural interests locally and the other from a person who wished to develop it as an hotel with caravans in the grounds. The farmer, in his view, would not want the opportunity to buy to go past and would not want an hotel and caravan development near his farmland. A high price was thus obtained because of the competition. Mr Clarke was aware that the hall needed modernising and that the lodge in particular was not in good condition. In his view the main structure was sound but he agreed that £30,000 may have been spent on it. The lack of central heating or a gas supply or mains drainage would not make much, if any, difference to the price. He said that no heating may be an advantage, ie an old heating installation in a large house may be a disadvantage because it has to be taken out. He said that, being a square house, it was relatively compact and so easier to maintain and heat, and it was, in his view, a house of manageable size.

Bamburgh is a coastal and holiday resort.

(d) *Harleyburn*

His firm had acted as selling agents for this house and had found it difficult to sell. It was three years on the market before it was sold in lots. He was unaware whether the tentative road line had dissuaded potential purchasers. He thought it had not.

On a general point Mr Clark said that purchasers of large and old houses have to adapt them and considerable work is usually needed. He described the buying and adapting of a large house as a " type of adventure." The condition of the house and its original state is therefore not usually a substantial factor in the purchase price.

Evidence for the respondents

John R Duffield FRICS, district valuer and valuation officer, Northumberland, since July 1973, gave evidence for the respondents. He has been 24 years in the surveying profession and prior to coming to Northumberland had been in the West Midlands. He said that he was experienced in dealing with country estates.

He first inspected Carham Hall on May 16 1977, although a valuer in his office had dealt with it earlier when it was referred to the valuation office in 1975. In Mr Duffield's opinion, the value at August 2 1973 was £80,000.

He said that it was not possible to make a mathematical comparison with other houses and that the valuation of this type of property was more a matter of " feel." This " feel " was derived to some extent from other negotiations with professional advisers.

Carham Hall was in his view a very attractive residence. It was occupied as a private residence up to the date of death. The structure was good, the woodwork of high quality and the house had central heating which would not have to be renewed. The plaster and subsidence cracks would not put anyone off buying the house. He agreed with Mr Clark that the roofs could involve high maintenance cost, but against that there was no external woodwork at the windows to be maintained.

He had not made any special additions to the value for other uses, such as institutional. He agreed that Carham could not be lotted although he thought it could be divided into flats. He considered that any sale prices before 1972 were of no help because in his view there was a dramatic increase from 1972 till the autumn of 1973. The prices of medium-sized houses doubled during that period and the prices of larger houses perhaps even more than that. His opinion about the increase was based on general experience of the exceptional conditions in that period. In his view prices dropped after 1973 and did not reach the 1973 level again until 1978 although he conceded that he was unable to produce any evidence to establish this assertion.

Eccles House, Coldstream, was the comparison he had put forward in support of his opinion. He did not agree with Mr Clark that an apportionment of £7,000 for the land was correct. He said that £300 per acre was more realistic in 1972 instead of Mr Clark's £500 per acre. The price of £71,750 paid in 1976 supported his opinion because he considered prices in 1976 were lower than in 1973. He conceded, however, that that was an opinion based on his experience and there was no comparable evidence to establish that contention. In his opinion the value of Eccles, in its 1976 condition and including the land, would have been at least £80,000 in 1973.

He submitted that the sale of Budle Hall supported his opinion of value of £80,000 for Carham Hall. He had talked to the present owner after the sale in 1973 and he had told him that the roof was renewed, the wiring and plumbing replaced and electrical off-peak heating installed. He estimated that it had cost £30,000 to put it in good condition, so that added to the £55,000 price was £85,000. He did not know about the competition for it at the time of the sale. He accepted that it was an attractive location with golf course, beaches and historic interest close to it, but he had made no specific adjustment for these factors.

So far as Coupland Castle was concerned he inspected it in 1979. It needed, he was informed by his staff, extensive repairs caused by dry rot and in his opinion would have fetched £100,000 in 1973 if no repairs were required. The price in the region of £20,000 in 1971, therefore, had to be adjusted for condition. The price in the region of £60,000 in 1975 corroborated his value because the condition had improved and the market had slipped back. He agreed that having 25 acres is an advantage over Carham. He also agreed that Coupland Castle was visually attractive and that the peel tower was a main feature, especially from an historic point of view.

Harleyburn was not in his opinion a helpful comparison. The road line would affect the market and he could not rely, in any event, on 1971 evidence.

Mr Duffield referred to Fowberry Tower, Maxton and Lennel House, which were on the market in 1978 and 1979. He said that the prices involved corroborated his value in 1973 in view of his contention that prices in 1978 were at the same level as in 1973.

Tribunal's decision on Carham Hall

So far as Col Clark's valuation is concerned, Philip Ridd, solicitor for the respondents, said I should give no weight to it as the opinion could not be cross-examined. Malcolm Spence QC, for the appellants, said that the valuation was lodged as the opinion, at the valuation date, of a person of distinction and that weight should be given to it. In arriving at my decision, I have relied mainly on Mr Clark's evidence.

The gap between the opposing views of £15,000 and £80,000 is dramatic and difficult to reconcile, especially between experienced valuers. It is clear from this wide difference of opinion that this large house is a difficult property to value in the changing market conditions of 1972 and 1973. The difference of opinion is due to (a) the effect on the price paid for a large house of the cost of repairs, adaptation and modernisation, and (b) the assumptions made by each valuer about the effect of the property boom in 1973 on large country houses with no agricultural land attached to them.

Mr Duffield said that the low prices listed were due to the condition of the houses, and to ascertain the proper price in good condition, it was necessary to add on to the price paid the cost of the necessary work. Mr Clark's view was that all large houses like Carham Hall need considerable money spent on them by purchasers and therefore the condition, provided the structure was good, did not materially affect the price.

Mr Clark and Mr Duffield were in agreement that the market for such large houses, without agricultural land, in Northumberland in the late 1960s was very limited and that it was only in 1971 that a market appeared with the sale of Coupland Castle. Mr Clark said that the market for this type of house was still restricted during 1972 and 1973, and he had no inquiries in his offices for such properties. It was not until 1975 when Coupland Castle was sold for £65,000 that a high price level was revealed. Mr Duffield said that the property boom in 1973, which particularly affected agricultural land and urban houses, would have affected larger houses such as Carham Hall.

I have inspected externally the four main comparables.

Coupland Castle I find attractively situated with an outstanding view. It is set in spacious grounds and the tower and keep give it character. Apart from the keep, it is a more compact house than Carham, having half the number of rooms. I have no detailed evidence about the condition in 1971 as Mr Duffield said he was informed by his staff that there was extensive dry rot. Whether the purchaser knew about this when he bought it or to what extent it affected the price, I do not know. We do know, however, that it was sold in November 1971 for a price in the region of £20,000. Mr Duffield says it was worth £100,000 in 1973 after repair. This is a dramatic increase in respect of a market less than two years old and in the absence of some detailed evidence about repair and sale prices at that level in 1973 I find that opinion difficult to accept.

Eccles I find to be a very attractive house and much more compact than Carham. It has a very good outlook to the south and stands in attractive gardens and grounds. The net price in April 1972 for the house, depending on the deduction for the land is in the region of £22,000 to £25,000. Mr Duffield said that substantial work was needed, but this may have been the choice of the purchaser to make the house more manageable. The rewiring and plumbing, the new central heating, the new stables and garage and the creation of a self-contained flat bring it up to a modern standard. In my view this house, even in its original condition, would sell for a higher price than Carham Hall.

I have looked at Budle Hall and despite the trees there is a clear view of the frontage from a small road on the west. Again, this is a more compact house than Carham, being substantially smaller in accommodation and size. The location is exceptional, overlooking Budle Bay. Bamburgh itself is an attractive holiday village with a few large private houses. I can accept that there would be competition for this house. I consider Budle Hall to be in a different market from Carham Hall and hesitate to draw the same conclusion as Mr Duffield, namely, that the price of £55,000 established that the market for large country houses with no agricultural land became very buoyant in 1973. It does indicate, however, that a substantial price could be obtained in 1973.

Harleyburn is not particularly helpful for comparable purposes, except it indicates that in November 1971 there was not a brisk demand for this type of property. The road line confused the issue although Mr Clark said that aspect did not arise in dealing with prospective purchasers. The subsequent sale in August 1976 for £37,500 indicates that a more buoyant market existed at that time and it links with the higher price for Coupland of £60,000 in 1975 and for Eccles of £71,750 in 1976.

Condition of mansion

Firstly, on the question of condition. As a result of my view, I accept Mr Clark's opinion that rewiring and alteration of the heating system would be necessary and that a considerable amount of money would be required to modernise the house and make it more manageable. I am not convinced that the addition of the cost of modernisation on a house of this type to the price paid necessarily equates with value. If the effect of the cost is to modernise and improve the house to a high standard then it does not seem correct to compare the eventual cost with the value of a house of lesser standard. Mr Duffield's main comparison was Eccles House. I do not find myself persuaded that his estimate of £80,000 for it, as modernised, in 1973 can be taken as a guide to the value of Carham Hall unmodernised as at the same date. After all, Eccles and the land sold for £71,750 in March 1976 with very extensive modernisation. Not all improvements, however, would be to every purchaser's requirements or taste and are not always fully reflected in enhanced market value according to the actual expenditure incurred. The condition, moreover, of Eccles after all this work had been carried out would make it a much better house than Carham. I have to value Carham, as at August 1973, in an unmodernised condition. Its very size (without prospect of easy subdivision) and the inferior setting would make the value considerably less.

Mr Duffield's value of £80,000 in 1973 for Eccles, in an improved condition, is anyway dependent on the correctness of his view that there was a dramatic increase between 1972 and 1973, followed by a later depression in prices.

Pattern of prices 1971-78

Secondly, Mr Duffield's view that the property market had a "boom period" in 1973, followed by a slump through 1976, to a recovery in 1978 to the 1973 level was not established by factual or comparable evidence. Mr Clark accepted that this was the pattern in the price of agricultural land and of houses with such land but he did not accept that this pattern applied to large country houses which had no agricultural land. His view was that the price of these large houses would have increased only marginally in the period 1972 to 1973.

Mr Clark was acting as an agent in the market during these years and in advising on value he would be conscious that he might have been given instructions to sell and have to obtain the price which he advised. His experience in the market in the late 1960s and the early 1970s would influence his opinion and tend to make him cautious. Apart from Budle Hall, which he knew about, there was no evidence in August 1973 of a sale price above £24,000 in the Northumberland area. He knew that Coupland Castle had sold in 1971 for a price in the region of £20,000 and that Harleyburn had sold, also in 1971, for £18,650 in lots with the mansion house and 15 acres of land at £8,750. He apparently did not know of Eccles selling in 1972 for £22,000 to £25,000 but that probably would have confirmed his view about the general level of values at that time.

Mr Duffield's opinion about the boom period in 1973 is supported to a certain extent by the sale of Budle Hall at £55,000. Although I have distinguished it from Carham Hall, it is a price level which, even allowing for the competition, is substantially higher than the 1971 and 1972 prices.

In order to help me to assess which opinion about trends is correct, I have considered the agreed sales information in 1975 and 1976. These show substantial increases from 1971 to 1972. It was argued on behalf of the appellants that, as the market was so limited in 1970 and 1971, it was unlikely that a dramatic increase would occur as soon as 1973. Equally it would seem unlikely that the substantial prices in 1975 and 1976 would have occurred dramatically in the period from 1973.

I did not find myself convinced by Mr Duffield's opinion, especially as there was no factual evidence to establish it, that prices of large country houses without agricultural land were the same in 1978 as they were in 1973. With inflation, this would mean a dramatic reduction in value in 1973 money terms.

I have come to the conclusion, on the evidence before me, that Mr Duffield has overestimated the value of Carham Hall, although I agree with him that Eccles House (as unimproved) is the most helpful of the comparables. From the evidence of the sale price for Eccles of £22,000-25,000 in April 1972 and taking into account the evidence of upward trends, I consider that the value of Eccles House in its unimproved condition in August 1973 would have been in the region of £40,000. For reasons already given, however, I consider Eccles is a more desirable and valuable property. In my opinion, after consideration of all the evidence, the value of £15,000 placed by Mr Clark on Carham is too conservative a figure and I find its value on August 2 1973 to be £30,000.

I therefore determine the principal value at August 2 1973 as follows:

Value of the property in accordance with section 7(5) of the Finance Act 1894, including timber	£30,000
Deduct commercial value of timber as agreed	£ 3,000
	£27,000

In the agreed statement of facts, there was information about the furnished let of the house and about 1978-79 sale or asking prices. This information was, in my view, too remote from the matters relevant to the appeal.

The 120.694 acres of land adjoining Carham Hall
Statement of facts

The following facts are agreed:

1. The 120.694 acres comprises:

River Walk and Wark Strip	8.251	acres
The Holme	57.991	„
Garden Strip	.957	„
Front Park	14.509	„
North Deer Park	22.680	„
South Deer Park	16.306	„
	120.694	„

2. On March 2 1971, 111.486 acres were let on an annual agricultural tenancy at a rent of £780 per annum. The tenancy agreement was produced. It contains the normal conditions. One special condition is that the tenant shall maintain the Front Park and The Holme field in permanent grass unless otherwise agreed with the landlord. Another is that the obligation to maintain the fence, hedge and gates along the main Cornhill to Kelso Road shall be the landlord's.

3. The River Walk and Wark Strip and the Garden Strip which contain 9.208 acres were woodlands and in hand. The land at the date of the valuation did not include any part of the bed of the River Tweed or any rights of fishing in that river.

4. The value of the land was returned for capital gains tax purposes at £15,300 and the liability for the year 1970-71 was assessed by the inspector of taxes, Newcastle 5th district, on the basis of this valuation.

5. The rent had been negotiated at £600 pa from November 11 1967 and again at £780 pa from November 11 1970.

6. The commercial valuation of the timber is agreed at £2,000 within the 120.694 acres.

7. The farms of Jardinefield, Royalty, Emerick and High Letham, which are located between Carham and Berwick-upon-Tweed and belong to Greenwich Hospital Estate, had the rents fixed at May 13 1971. The cumulo rent was £4,492 and the acreage 778. This represents a rate of £5.77 per acre. The previous rent fixed in May 1967 was £3,660, representing a rate of £4.70 per acre.

8. The rents of the farms on Carham Estate were all fixed at November 11 1970 as follows. The previous rents were fixed at November 1967. The farms are adjacent to Carham Hall but extend to the south.

Farm	*Acres*	*Rent at Nov 1970*		*Rent at Nov 1967*	
		Total	*Per acre*	*Total*	*Per acre*
Part Shidlaw and Home Farm	335.651	£2,400	£7.15	£1,900	£5.66
Part Shidlaw and Home Farm	395.835	£2,780	£7.02	£2,200	£5.55
Wark Common	473.65	£3,310	£6.98	£2,600	£5.49
Howburn	288.66	£1,890	£6.55	£1,460	£5.05
Sunilaws	301.33	£1,950	£6.47	£1,500	£4.98

9. The £780 rent for the 111.486 acres under appeal represents £6.99 per acre and the previous rent of £600 represents £5.30 per acre.

10. Norham East Mains and Norham West Mains farms were sold for £150,000 in April 1971 to the sitting tenant.

Norham East Mains extends to 309.359 acres of which 23.455 acres were woodland in hand. The let area was 285.904 acres. The rent was £2,080 (representing £7.27 per acre). It has a four-bedroomed farmhouse, two modernised cottages and a compact range of farmbuildings.

Norham West Mains extends to 605.905 acres of which 15.897 acres were woodland in hand. The let area was 590.008 acres. The rent was £3,100 (representing £5.25 per acre). It has a five-bedroomed farmhouse, four modernised cottages and a substantial range of farmbuildings.

Notices to review the farm rents from May 13 1971 were served on the tenant on May 5 1970.

11. Tillmouth Farm, described as an exceptional arable unit and extending to 789 acres, had the rent reviewed at November 12 1971 at £7,473 a year. This represents an average of £9.47 per acre. The lease runs from year to year on a full repairing tenant's obligation basis. The farmhouse has five bedrooms: there are four modern cottages and excellent farmbuildings of recent construction. One is a cattle shed 210 ft long and 170 ft wide, steel framed with stone and concrete walls and corrugated asbestos roof.

12. Recent lets and reviews on Carham Estate are:

Description	*Acreage*	*Present rent*		*Review*	*Remarks*
		Total	*Per acre*		
Wark Common	764.31	£21,783	£28.50	11.11.80	Sale by owner
	89.19	£2,540	£28.50	11.11.82	and leaseback
Shidlaw and	319.42	£5,385	£16.88	11.11.79	
Home Farm	396.61	£6,530	£16.46	11.11.79	
	38.43	£714	£18.58	11.11.79	North and South Deer Park (part of appeal land)
Sunilaws	301.37	£5,000	£16.61	11.11.79	
Nottylees	19.27	£220	£11.41	11.11.79	Bare land

A 13. All the agricultural land in England contained in these agreed facts is classified grade II.

Evidence for the appellants

Col Clark valued the land adjoining Carham Hall at March 2 1971 "in order to arrive at a figure at which it should be taken over by W J Straker-Smith." A signed copy of his valuation was produced at the hearing. He described the area as 113.659 acres of pasture land let at £780 per annum and he valued it at £15,300. As will be noted in the agreed facts this value was the basis of the capital gains liability as assessed by the inspector of taxes. Although the area is slightly larger B than the let area, the rent is that for the 111.486 acres.

John Clark walked the 120.694 acres on August 29 1973 with his father. His opinion of value was also £15,300 calculated as follows:

Rent	£	780 per annum
Less Landlord's maintenance costs of trimming hedge on the main road		30
Net rent		750
Years' purchase		20
(£135 per acre)		15,000
C 9.208 acres of woodland—prairie value		300
		£15,300

He considered that the rent fixed at November 1970 was a proper one. The land is a sandy loam and easy to work. Wind blow shelter was important for this type of soil. The north part may have been part of the river at one time and in his view it was not so good as the remainder and thus the lower value would affect the overall rate. The special condition that the area of 72 acres was to be maintained as permanent pasture did not materially affect the value.

D In support of his contention that the rent was a proper one, he referred to the rents settled for the rest of the Carham Estate. Sunilaws and Wark Common were marginally better than Carham because the land had more body in it. He also referred to the Greenwich Hospital Estate Farms; Jardinefield was a dairy farm and the best of the four; Royalty was a little run down; Emerick had 25 acres which were badly drained. The overall rate of £5.77 per acre supported the £6.99 for the appeal land. The general quality was the same as Carham.

The rent of Tillmouth was fixed in November 1971. He considered, however, that Tillmouth Farm with its modern E fixed equipment was much better than Carham. The rent included the modern cattle court which had been used for pigs. In his view the rental equivalent of the cattle court was about £1 per acre. The tenant had a 300-acre farm elsewhere in Northumberland and therefore the working of the two farms had considerable advantages; for example he could harvest earlier at Tillmouth.

Mr Clark did not accept that the rents being paid in 1978 in Carham Estate showed that the rents were low in 1970. The high rent for Wark Common was the result of a sale and leaseback arrangement and could not be relied on. A slightly higher rate for the North and South Deer Park was accounted F for by these being the better part of the 111.486 acres.

As I understood his analysis of the Norham sale it was as follows:

Sale price to sitting tenant	£150,000
As other offers ranged from £120,000 to £142,000 assume	£142,000 as market price
Deduct woodlands 40 acres at £250	£ 10,000
	£132,000
Assume rent at May 1971 would be 875 acres @ £7 per acre, say	£ 6,100
So years' purchase, say	22

The £150,000 for the 875 acres leased represents £171 per acre.

In his view the larger unit was more likely to attract an G investor and so a higher years' purchase than Carham could be expected. He also said that, in his opinion, the Norham farms were better than the Carham land.

His valuation of the woodland at Carham was £300 and in his view the woodland and the timber on it provided shelter for the land and in particular the arable fields of the North and South Deer Park. He produced photographs taken recently showing the effect on the crop of the removal of the trees from the east boundary on the fields. Mr Clark said that if the land and the woodlands were on the market, an additional price of only £300 would be paid. If the woodland was sold to another purchaser the loss of control by the owner of the H land and the possibility that the timber would be cut would cause a reduction in the rent. This in turn would reduce the price received for the land.

Evidence for the respondents

Mr Duffield's valuation is as follows:

Rent passing	£780 pa	
Less Cost of trimming hedge on the Kelso-Cornhill Road 42 ch @ 50p, say	20	
Net rent	£760	
YP for 3 years @ 3%	2.8286	£2,150 J
Reversion to rent of £10 per acre	£1,115 pa	
Less hedge trimming	20	
	£1,095	
YP in perpetuity @ 4% deferred for 3 years	22.2249	£24,336
	(£237 per acre)	£26,486
9.208 acres of amenity woodland @ £400 per acre		£ 3,683 K
		£30,169
	Total value, say	£30,000
Deduct Commercial value of timber		£ 2,000
Principal value and agricultural value of the agricultural property		£28,000

He said that the value of £15,300, used as a basis for the capital gains tax assessment, had not been referred to his office by the inspector of taxes. In his opinion the rent of £780 (about £7 per acre) was a low one when settled in 1970 and a purchaser would anticipate a higher figure, in L terms of the value of money in 1970, at the next rent review in November 1973. From his general experience in fixing agricultural rents he considered that a rent of £10 per acre, in 1970 terms, was appropriate. In support of that opinion he referred to the rent fixed for Tillmouth in November 1971. He conceded that he had not made a detailed inspection of the land, but from the schedule he had deduced that 74 acres was poorer land than the remainder. His analysis of the Tillmouth rent was as follows:

715 acres @ £10	£7,150
74 acres @ £4.36	323 M
	£7,473

As this rent was on a full-repairing basis he said that £1 per acre should be added to the rent per acre to bring it into line with Carham Estate. He agreed, however, that a bidder for Carham at the date of valuation, viz March 1971, would not know about the Tillmouth rent. The summary of the recent rents at Carham Estate was his production and he said that the figures supported his view that the rents in 1970 were low. He was not in a position, however, to establish by evidence that a sale and leaseback transaction produced a market rent.

The 4 per cent or 25 years' purchase was his opinion based on experience. He was unable to produce any comparables because there were few tenanted farm sales. His comment on the Norham sale was that if Mr Clark adopted £7 per acre it was necessary to deduct £1 per acre as a repairs allowance. A net rent of £6 per acre applied to 875 acres gives the sum of £5,250, which divided into the price of £150,000 represents 28 years' purchase.

Mr Duffield said that, as a check on his opinion, his value of £26,486 for the appeal land represented £237 per acre, which was realistic in relation to the national average level for tenanted farms in 1971.

In valuing the amenity woodlands he had used his experience to assess the value. His opinion of £3,683 was on the assumption that the owner of Carham Hall would be the most likely purchaser; not only would the woodland be important for the amenity of Carham Hall but it would give access to the river bank. In his view the woodland was not an asset to the arable land and might in fact cut off light from the field.

Decision

The main dispute between the valuers is whether the £7 per acre was a proper rent or a low rent in 1970.

I inspected on October 22 1979 all the farms referred to.

So far as the capital gains tax valuation is concerned, I can only have general regard to it. I have to determine the appeal on the evidence placed before me.

I have to consider what view a bidder would take about the prospects of a rent increase in deciding what price he would offer. He would know that he could not obtain an increase until November 1973. At that time, he would either agree a rent with the tenant or have the matter referred to arbitration if he wished to involve such procedure. The tenant was used to being at the same level as the rest of the tenanted farms on Carham Estate and was also the tenant of Shidlaw and Home Farm on the estate itself. If a purchaser decided he might go to arbitration no doubt he would take professional advice on the evidence which he would place before the arbitrator. The only evidence about rents being paid which was available at March 1971 was that provided by Mr Clark.

I do not consider that a purchaser would take the view from that evidence that the rent fixed in 1970 was too low and anticipate that in the rent review in 1973 evidence about rents would be available not only to establish that an adjustment would be made for inflation but also for an increase in terms of 1970 money values. Mr Duffield's opinion based on experience might have persuaded a purchaser, but in the absence of comparable evidence at the relevant date to support his view and the existence of comparable evidence to support Mr Clark, I favour Mr Clark's opinion on the rent level.

The deduction for the cost of trimming the hedge is a sum I thought could have been agreed at £25. It is a compromise between the £20 and the £30 and so I determine that the deduction be £25.

The dispute about the years' purchase is whether it should be 20 or 25 years' purchase. These were both opinions based on experience. The only sale placed before me is Norham.

The price paid by the sitting tenant was £150,000. I consider there is no justification for reducing that figure for that reason, as a sitting tenant at that price level is the most likely purchaser in the market. Mr Clark said that a deduction of £10,000 should be made for the woodlands. In his valuation of the appeal land he said that as the shelter belt was important to the farm it was unlikely to add much to the price of the farm. I think the woodlands would also be important as shelter belts to the Norham farms and although there may be commercial value I would hesitate to make such a substantial deduction from the price for them.

The rent was to be reviewed almost immediately and Mr Clark adopted, as I understood his analysis, a net rent of £7 per acre overall to compare with the appeal land. Mr Duffield criticised this £7 and said that a repair allowance of £1 should be deducted to arrive at a net rent. I think Mr Clark's £7 adopted for calculation purposes would only be subject, to compare like with like, to a deduction similar to the hedge trimming cost. My view is that with the superior fixed equipment a net rent of £7 is modest for Norham. If I assume, say, £145,000 as the price of the farm, making a deduction of about £5,000 for the woodland, and take a net rent of £6,100 this represents 23.7 years' purchase. Having regard to the opinion evidence as well as the analysis of the Norham sale and making allowances for the attractiveness of the larger unit, which I accept, I determine the years' purchase at 22.

Finally, there is the dispute about the amenity woodlands. I accept Mr Clark's view that the timber is a shelter belt to the fields and if it were felled, the loss of that shelter could affect the rent. Nevertheless, there could be competition for the ownership of the woodlands between the owner of Carham Hall and the purchaser of the farmland. In these circumstances, I do not consider that only a nominal price of £300 would be realised. The purchaser of the farmland would primarily have regard to the capitalisation of the reduction in rent whereas the owner of Carham Hall would want to preserve the amenity and gain control of the land along the river. The area of woodland on the east, which appears to extend to about 3½ acres, is about ¼-mile away from the house and is screened by other timber. That particular area was not likely to be of much amenity interest to the owner of Carham Hall—a view confirmed by the fact that the timber has now been felled—and therefore I would value it, as Mr Clark had done, at " prairie value."

It was a matter of agreement that the market in 1971 for Carham Hall would have been limited. I consider that the price paid for the amenity woodland would have had some relationship to the value of the house at that time. I have held that the value of the house in 1973 has been overestimated by Mr Duffield and I have the impression also that his opinion of the value of the amenity woodland is overestimated having regard to the probable value of Carham Hall in 1971. I therefore determine the amenity woodland value at £1,500 calculated as follows:

5.6 acres @ £250 per acre	£1,400
3.5 acres @ £30 per acre, say	100
	£1,500

My determination of the value of the 120.694 acres of land adjoining Carham Hall is thus:

Rent passing	£780			
Less hedge trimming cost	25			
Net rent	£755			
Years' purchase	22			
		£16,610		
9.208 acres of amenity woodland, say		£ 1,500		
Total value		£18,110	say	£18,000
Deduct commercial value of timber				£ 2,000
Principal value and agricultural value				£16,000

I heard submission as to costs. The respondents will pay to the appellants the costs of this appeal such costs if not agreed to be taxed by the Registrar of the Lands Tribunal on the High Court Scale.